Pressure:

$1.00 \text{ N/m}^2 = 1.4504 \times 10^{-4} \text{ lbf/in}^2 = 2.0886 \times 10^{-2} \text{ lbf/ft}^2$

$1.00 \text{ bar} = 10^5 \text{ N/m}^2 = 10^5 \text{ Pa}$

$1.00 \text{ lbf/in}^2 = 6.8947 \times 10^3 \text{ N/m}^2$

$1.00 \text{ lbf/in}^2 = 144.00 \text{ lbf/ft}^2$

Specific enthalpy:

$1.00 \text{ N} \cdot \text{m/kg} = 1.00 \text{ J/kg} = 1.00 \text{ m}^2/\text{s}^2$

$1.00 \text{ N} \cdot \text{m/kg} = 4.3069 \times 10^{-4} \text{ Btu/lbm} = 1.0781 \text{ ft lbf/slug}$

$1.00 \text{ Btu/lbm} = 2.3218 \times 10^3 \text{ N} \cdot \text{m/kg}$

Specific heat:

$1.00 \text{ N} \cdot \text{m/kg} \cdot \text{K} = 1.00 \text{ J/kg} \cdot \text{K} = 2.388 \times 10^{-4} \text{ Btu/lbm} \cdot {}^\circ\text{R}$
$\qquad = 5.979 \text{ ft} \cdot \text{lbf/slug} \cdot {}^\circ\text{R}$

$1.00 \text{ Btu/lbm} \cdot {}^\circ\text{R} = 32.174 \text{ Btu/slug} \cdot {}^\circ\text{R} = 4.1879 \times 10^3 \text{ N} \cdot \text{m/kg} \cdot \text{K}$
$\qquad = 4.1879 \times 10^3 \text{ J/kg} \cdot \text{K}$

Temperature:

The temperature of the ice point is 273.15 K (491.67°R).

$1.00 \text{ K} = 1.80{}^\circ\text{R}$

$\text{K} = {}^\circ\text{C} + 273.15$

${}^\circ\text{R} = {}^\circ\text{F} + 459.67$

$T{}^\circ\text{F} = 1.8 \, (T{}^\circ\text{C}) + 32$

$T{}^\circ\text{R} = 1.8 \, (T\text{K})$

Velocity:

$1.00 \text{ m/s} = 3.60 \text{ km/h}$

$1.00 \text{ km/h} = 0.2778 \text{ m/s} = 0.6214 \text{ mile/h} = 0.9113 \text{ ft/s}$

$1.00 \text{ ft/s} = 0.6818 \text{ mile/h} = 0.59209 \text{ knot}$

$1.00 \text{ mile/h} = 1.467 \text{ ft/s} = 1.609 \text{ km/h} = 0.4470 \text{ m/s}$

$1.00 \text{ knot} = 1.15155 \text{ mile/h}$

Viscosity:

$1.00 \text{ kg/m} \cdot \text{s} = 0.67197 \text{ lbm/ft} \cdot \text{s} = 2.0886 \times 10^{-2} \text{ lbf} \cdot \text{s/ft}^2$

$1.00 \text{ lbm/ft} \cdot \text{s} = 3.1081 \times 10^{-2} \text{ lbf} \cdot \text{s/ft}^2 = 1.4882 \text{ kg/m} \cdot \text{s}$

$1.00 \text{ lbf} \cdot \text{s/ft}^2 = 47.88 \text{ N} \cdot \text{s/m}^2 = 47.88 \text{ Pa} \cdot \text{s}$

$1.00 \text{ centipoise} = 0.001 \text{ Pa} \cdot \text{s} = 0.001 \text{ kg/m} \cdot \text{s} = 6.7197 \times 10^{-4} \text{ lbm/ft} \cdot \text{s}$

$1.00 \text{ stoke} = 1.00 \times 10^{-4} \text{ m}^2/\text{s}$ (kinematic viscosity)

Volume:

$1.00 \text{ liter} = 1000.0 \text{ cm}^3$

$1.00 \text{ barrel} = 5.6146 \text{ ft}^3$

$1.00 \text{ ft}^3 = 1728 \text{ in.}^3 = 0.03704 \text{ yd}^3 = 7.481 \text{ gal}$
$\qquad = 28.32 \text{ liters}$

$1.00 \text{ gal} = 3.785 \text{ liters} = 3.785 \times 10^{-3} \text{ m}^3$

$1.00 \text{ bushel} = 3.5239 \times 10^{-2} \text{ m}^3$

AERODYNAMICS FOR ENGINEERS

Photograph of a maneuvering F-18
(from the collection of J. Periaux)

Second Edition

AERODYNAMICS FOR ENGINEERS

JOHN J. BERTIN
Department of Aerospace Engineering
and Engineering Mechanics
The University of Texas at Austin

MICHAEL L. SMITH
Department of Aeronautics
United States Air Force Academy

PRENTICE HALL, Englewood Cliffs, New Jersey 07632

Library of Congress Cataloging-in-Publication Data

BERTIN, JOHN J.
   Aerodynamics for engineers, 2nd ed.

   Includes bibliography and index.
   1. Aerodynamics. I. Smith, Michael L. II. Title.
TL570.B42 1989   629.132'3—dc19   88-19044
   ISBN 0-13-018243-5

Editorial/production supervision and
   interior design: John Fleming
Cover design: Joel Mitnick Design, Inc.
Manufacturing buyer: Mary Noonan

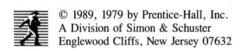
Printed in the United States of America
10 9 8 7 6 5

ISBN 0-13-018243-5

PRENTICE-HALL INTERNATIONAL (UK) LIMITED, *London*
PRENTICE-HALL OF AUSTRALIA PTY. LIMITED, *Sydney*
PRENTICE-HALL CANADA INC., *Toronto*
PRENTICE-HALL HISPANOAMERICANA, S.A., *Mexico*
PRENTICE-HALL OF INDIA PRIVATE LIMITED, *New Delhi*
PRENTICE-HALL OF JAPAN, INC., *Tokyo*
SIMON & SCHUSTER ASIA PTE. LTD., *Singapore*
EDITORA PRENTICE-HALL DO BRASIL, LTDA., *Rio de Janeiro*

The authors would like to dedicate this book to their teachers.

To my parents, Andrew and Yolanda,
and to my wife, Ruth E. Bertin.

To my wife Carol and my daughters
Summer and Elaine Smith.

CONTENTS

PREFACE

This text is designed for use in intermediate and advanced courses in aerodynamics for undergraduate and graduate students in mechanical engineering and aerospace engineering. All of the material presented in the first edition has been retained (including all of the homework problems from the first edition). The second edition contains roughly 50 percent more material than the first edition including an extended discussion and sample problems to clarify the basic concepts of fluid mechanics, complete chapters devoted to new topics (such as Computational Fluid Dynamics and Hypersonic Flows), and new topics and applications within existing chapters (such as Leading-Edge Extensions and Airframe Modification Procedures).

Although it is presumed that the reader has had basic courses in thermodynamics, physics and mathematics, including partial differential equations, the book contains introductory level discussions of basic fluid mechanics: Fluid Properties (Chapter 1), Fundamentals of Fluid Mechanics (Chapter 2), Dynamics of An Incompressible, Inviscid Flow Field (Chapter 3), and Incompressible, Viscous Boundary Layers (Chapter 4). For the reader who has had a course in fluid mechanics, these four chapters (together with Chapter 8) provide a comprehensive review and an introduction to the nomenclature and style of the present text. The nomenclature and parameters for airfoil and wing aerodynamics are presented in Chapter 5. Incompressible flows about airfoils and wings are discussed in Chapters 6 and 7, respectively. After reviewing the basic concepts of compressible flows in Chapter 8, these precepts are applied to compressible subsonic and transonic flows (Chapter 9), to supersonic flows around airfoils (Chapter 10) and to supersonic flows around wings (Chapter 11). The last three chapters provide an introduction to Computational Fluid Dynamics (Chapter 12), to Hypersonic Flows (Chapter 13), and to Aerodynamic Design Considerations (Chapter 14).

The general approach taken in this book is to present a background discussion of each new topic followed by a presentation of the theory. The assumptions (and, therefore, the restrictions) incorporated into the development of the theory are carefully noted. The applications of the theory are illustrated by working one or more example problems. Solutions are obtained using numerical techniques in order to apply the theory, for those flows where closed-form solutions are impractical or impossible. To illustrate the numerical techniques, the example problems are worked for relatively simple flow fields which can be adequately defined by a few control points and which can be worked using personal electronic calculators. The theoretical solutions and/or the computed results are compared with experimental data from the open literature to illustrate both the validity of the theoretical analysis and its limitations (or, equivalently, the range of conditions for which the theory is applicable).

Extensive discussions of the effects of viscosity, compressibility, shock/boundary-layer interactions, turbulence and other practical aspects of contemporary aerodynamic design are also presented. Problems at the end of each chapter are designed to complement the material presented within the chapter and to develop the student's understanding of the relative importance of various phenomena. The text places emphasis on practical problems and the fundamental principles of techniques through which solutions to these problems can be obtained. Because both SI units and English units are commonly used in the aerospace industry, both are used in this text. Conversion factors between SI units and English units are included.

The authors are indebted to their many friends and colleagues for their help in preparing the two editions of this text. We thank them for their suggestions and for copies of photographs and illustrations. The authors are indebted to L. C. Squire of Cambridge University, V. G. Szebehely of The University of Texas at Austin, F. A. Wierum of Rice University, T. J. Mueller of the University of Notre Dame, R. G. Bradley and C. Smith of General Dynamics, G. E. Erickson of Northrop, L. E. Ericsson of Lockheed Missiles and Space, L. Lemmerman and A. S. W. Thomas of Lockheed Georgia, J. Periaux of Avions Marcel Dassault, H. W. Carlson and M. L. Spearman of the Langley Research Center, D. Kanipe of the Johnson Space Center, and R. C. Maydew, S. McAlees, and W. H. Rutledge of the Sandia National Labs.

The authors are indebted to Lynn McKay, Jan Shrode, Tim Valdez, and Diane Williams for typing and graphics support during the preparation of the numerous drafts of the text. Finally, thank you, Margaret Baker.

John J. Bertin
The University of Texas at Austin

Michael L. Smith
United States Air Force Academy

1

FLUID PROPERTIES

A fundamental problem facing the aerodynamicist is to predict the aerodynamic forces and moments acting on a vehicle in flight. In order to predict these aerodynamic forces and moments with suitable accuracy, it is necessary to be able to describe the pattern of flow around the vehicle. The resultant flow pattern depends on the geometry of the vehicle, its orientation with respect to the undisturbed free stream, and the altitude and speed at which the vehicle is traveling. In analyzing the various flows that an aerodynamicist may encounter, assumptions about the fluid properties may be introduced. In some applications, the temperature variations are so small that they do not affect the velocity field. In addition, for those applications where the temperature variations have a negligible effect on the flow field, it is often assumed that the density is essentially constant. However, in analyzing high-speed flows, the density variations cannot be neglected. Since density is a function of pressure and temperature, it may be expressed in terms of these two parameters. In fact, for a gas in thermodynamic equilibrium, any thermodynamic property may be expressed as a function of two other thermodynamic properties. Thus, it is possible to formulate the governing equations using the enthalpy and the entropy as the flow properties instead of the pressure and the temperature.

CONCEPT OF A FLUID

From the point of view of fluid mechanics, matter can be in one of two states, either solid or fluid. The technical distinction between these two states lies in their response to an applied shear, or tangential, stress. A solid can resist a shear stress by a static deformation; a fluid cannot. A *fluid* is a substance that deforms continuously under the action of shearing forces. An important corollary of this definition is that there can be no shear stresses acting on fluid particles if there is no relative motion within the fluid; that is, such fluid particles are not deformed. Thus, if the fluid particles are at rest or if they are all moving at the same velocity, there are no shear stresses in the fluid. This zero shear stress condition is known as the *hydrostatic stress condition*.

A fluid can be either a liquid or a gas. A liquid is composed of relatively close packed molecules with strong cohesive forces. As a result, a given mass of liquid will occupy a definite volume of space. If a liquid is poured into a container, it assumes the shape of the container up to the volume it occupies and will form a free surface in a gravitational field if unconfined from above. The upper (or free) surface is planar and perpendicular to the direction of gravity. Gas molecules are widely spaced with relatively small cohesive forces. Therefore, if a gas is placed in a closed container, it will expand until it fills the entire volume of the container. A gas has no definite volume. Thus, if it is unconfined, it forms an atmosphere that is essentially hydrostatic.

FLUID AS A CONTINUUM

When developing equations to describe the motion of a system of fluid particles, one can either define the motion of each and every molecule or one can define the average behavior of the molecules within a given elemental volume. The size of the elemental volume is important, but only in relation to the number of fluid particles contained in the volume and to the physical dimensions of the flow field. Thus, the elemental volume should be large compared with the volume occupied by a single molecule so that it contains a large number of molecules at any instant of time. Furthermore, the number of molecules within the volume will remain essentially constant even though there is a continuous flux of molecules through the boundaries. If the elemental volume is too large, there could be a noticeable variation in the fluid properties determined statistically at various points in the volume.

In problems of interest to this text, our primary concern is not with the motion of individual molecules, but with the general behavior of the fluid. Thus, we are concerned with describing the fluid motion in spaces that are very large compared to molecular dimensions and that, therefore, contain a large number of molecules. The fluid in these problems may be considered to be a continuous material whose properties can be determined from a statistical average for the particles in the volume, that is, a macroscopic representation. The assumption of a continuous fluid is valid when the smallest volume of fluid that is of interest contains so many molecules that statistical averages are meaningful.

The number of molecules in a cubic meter of air at room temperature and at sea-level pressure is approximately 2.5×10^{25}. Thus, there are 2.5×10^{10} molecules in a cube 0.01 mm on a side. The mean free path at sea level is 6.6×10^{-8} m. There are sufficient molecules in this volume for the fluid to be considered a continuum, and the fluid properties can be determined from statistical averages. However, at an altitude of 130 km, there are only 1.6×10^{17} molecules in a cube 1 m on a side. The mean free path at this altitude is 10.2 m. Thus, at this altitude the fluid cannot be considered a continuum.

FLUID PROPERTIES

By employing the concept of a continuum, we can describe the gross behavior of the fluid motion using certain observable, macroscopic properties. Properties used to describe a general fluid motion include the temperature, the pressure, the density, the viscosity, and the speed of sound.

Temperature

We are all familiar with *temperature* in qualitative terms; that is, an object feels hot (or cold) to the touch. However, because of the difficulty in quantitatively defining the temperature, we define the equality of temperature. Two bodies have equality of temperature when no change in any observable property occurs when they are in thermal contact. Further, two bodies respectively equal in temperature to a third body must be equal in temperature to each other. It follows that an arbitrary scale of temperature can be defined in terms of a convenient property of a standard body.

Pressure

Because of the random motion due to their thermal energy, the individual molecules of our fluid would continuously strike a surface that is placed in the fluid. These collisions occur even though the surface is at rest relative to the fluid. By Newton's second law, a force is exerted on the surface equal to the time rate of change of the momentum of the rebounding molecules. *Pressure* is the magnitude of this force per unit area of surface. Since a fluid that is at rest cannot sustain tangential forces, the pressure on the surface must act in the direction perpendicular to that surface. Furthermore, the pressure acting at a point in a fluid at rest is the same in all directions.

   Standard atmospheric pressure is defined as the pressure that can support a column of mercury 760 mm in length when the density of the mercury is 13.5951 g/cm^3 and the acceleration due to gravity is the standard value. The standard atmospheric pressure is 1.01325×10^5 N/m^2. In English units, the standard atmospheric pressure is 14.696 lbf/in^2 or 2116.22 lbf/ft^2.

   In many aerodynamic applications, we are interested in the difference between the local pressure and the atmospheric pressure. Many pressure gages indicate the difference between the absolute pressure and the atmospheric pressure existing at the gage. This difference, which is referred to as *gage pressure*, is illustrated in Fig. 1.1.

Density

The *density* of a fluid at a point in space is the mass of the fluid contained in an incremental volume surrounding the point. As is the case when evaluating the other fluid properties, the incremental volume must be large compared to molecular dimen-

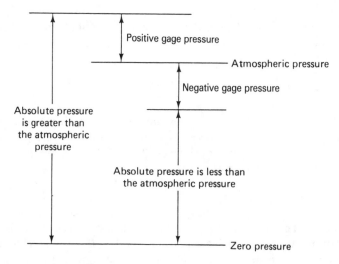

Figure 1.1   Terms used in pressure measurements.

sions yet very small relative to the dimensions of the vehicle whose flow field we seek to analyze. Thus, provided that the fluid may be assumed to be a continuum, the density at a point is defined as

$$\rho = \lim_{\delta(\text{vol}) \to 0} \frac{\delta(\text{mass})}{\delta(\text{vol})} \tag{1.1}$$

The dimensions of density are (mass)/(length)³.

In general, the density of a gas is a function of the composition of the gas, its temperature, and its pressure. The relation

$$\rho(\text{composition}, T, p) \tag{1.2}$$

is known as an *equation of state*. For a thermally perfect gas, the equation of state is

$$\rho = \frac{p}{RT} \tag{1.3}$$

The gas constant R has a particular value for each substance. The gas constant for air has the value 287.05 N·m/kg·K in SI units and 53.34 ft·lbf/lbm·°R or 1716.16 ft²/s²·°R in English units.

Example 1.1

Calculate the density of air when the pressure is 1.01325×10^5 N/m² and the temperature is 288.15 K. Since air at this pressure and temperature behaves as a perfect gas, we can use equation (1.3).

Solution

$$\rho = \frac{1.01325 \times 10^5 \text{ N/m}^2}{(287.05 \text{ N·m/kg·K})(288.15 \text{ K})}$$

$$= 1.2250 \text{ kg/m}^3$$

Example 1.2

Calculate the density of air when the pressure is 2116.22 lbf/ft² and the temperature is 518.67°R. Since air at this pressure and temperature behaves as a perfect gas, we can use equation (1.3). Note that throughout the remainder of this book, air will be assumed to behave as a perfect gas unless specifically stated otherwise.

Solution

$$\rho = \frac{2116.22 \, \dfrac{\text{lbf}}{\text{ft}^2}}{\left(53.34 \, \dfrac{\text{ft} \cdot \text{lbf}}{\text{lbm} \cdot °\text{R}}\right)(518.67°\text{R})} = 0.07649 \, \frac{\text{lbm}}{\text{ft}^3}$$

Alternatively,

$$\rho = \frac{2116.22 \, \dfrac{\text{lbf}}{\text{ft}^2}}{\left(1716.16 \, \dfrac{\text{ft}^2}{\text{s}^2 \cdot °\text{R}}\right)(518.67°\text{R})} = 0.002377 \, \frac{\text{lbf} \cdot \text{s}^2}{\text{ft}^4}$$

The unit lbf·s²/ft⁴ is often written as slugs/ft³, where slugs are alternative units of density in the English system.

For vehicles that are flying at approximately 100 m/s, or less, the density of the air flowing past the vehicle is assumed constant when obtaining a solution for the flow field. Rigorous application of equation (1.3) would require that the pressure and the temperature remain constant (or change proportionally) in order for the density to remain constant throughout the flow field. We know that the pressure around the vehicle is not constant, since the aerodynamic forces and moments in which we are interested are the result of pressure variations associated with the flow pattern. However, the assumption of constant density for velocities below 100 m/s is a valid approximation because the pressure changes that occur from one point to another in the flow field are small relative to the absolute value of the pressure.

Viscosity

In all real fluids, a shearing deformation is accompanied by a shearing stress. The fluids of interest in this text are *Newtonian* in nature; that is, the shearing stress is proportional to the rate of shearing deformation. The constant of proportionality is called the *coefficient of viscosity*, μ. Thus,

$$\text{shear stress} = \mu \times \text{transverse gradient of velocity} \tag{1.4}$$

There are many problems of interest to us in which the effects of viscosity can be neglected. In such problems, the magnitude of the coefficient of viscosity of the fluid and of the velocity gradients in the flow field are such that their product is negligible relative to the inertia and to the pressure forces acting on the fluid particles. We shall use the term *inviscid flow* in these cases to emphasize the fact that it is the character of both the flow field and the fluid which allows us to neglect viscous effects. No real fluid has a zero coefficient of viscosity.

The viscosity of a fluid relates to the transport of momentum in the direction of the velocity gradient (but opposite in sense). Therefore, viscosity is a transport property. In general, the coefficient of viscosity is a function of the composition of the gas, its temperature, and its pressure. For temperatures below 3000 K, the viscosity of air is independent of pressure. In this temperature range, we could use Sutherland's equation to calculate the coefficient of viscosity:

$$\mu = 1.458 \times 10^{-6} \frac{T^{1.5}}{T + 110.4} \qquad (1.5a)$$

where T is the temperature (in K) and the units for μ are kg/s·m.

Example 1.3

Calculate the viscosity of air when the temperature is 288.15 K.

Solution

$$\mu = 1.458 \times 10^{-6} \frac{(288.15)^{1.5}}{288.15 + 110.4}$$

$$= 1.7894 \times 10^{-5} \text{ kg/s·m}$$

For temperatures below 5400°R, the viscosity of air is independent of pressure. In this temperature range, Sutherland's equation for the viscosity of air in English units is

$$\mu = 2.27 \times 10^{-8} \frac{T^{1.5}}{T + 198.6} \qquad (1.5b)$$

where T is the temperature in °R and the units for μ are lbf·s/ft².

Example 1.4

Calculate the viscosity of air when the temperature is 59.0°F.

Solution   First, convert the temperature to the absolute scale for English units, °R, 59.0°F + 459.67 = 518.67°R.

$$\mu = 2.27 \times 10^{-8} \frac{(518.67)^{1.5}}{518.67 + 198.6}$$

$$= 3.7383 \times 10^{-7} \frac{\text{lbf·s}}{\text{ft}^2}$$

Equations used to calculate the coefficient of viscosity depend on the model used to describe the intermolecular forces of the gas molecules, so that it is necessary to define the potential energy of the interaction of the colliding molecules. Svehla (Ref. 1.1) noted that the potential for the Sutherland model is described physically as a rigid, impenetrable sphere, surrounded by an inverse-power attractive force. This model is qualitatively correct in that the molecules attract one another when they are far apart and exert strong repulsive forces upon one another when they are close together.

Chapman and Cowling (Ref. 1.2) note that equations (1.5a) and (1.5b) closely represent the variation of μ with temperature over a "fairly" wide range of temperatures. They caution, however, that the success of Sutherland's equation in representing the variation of μ with temperature for several gases does not establish the validity of Sutherland's molecular model for those gases. "In general it is not adequate to represent the core of a molecule as a rigid sphere, or to take molecular attractions into account to a first order only. The greater rapidity of the experimental increase of μ with T, as compared with that for non-attracting rigid spheres, has to be explained as due partly to the 'softness' of the repulsive field at small distances, and partly to attractive forces which have more than a first-order effect. The chief value of Sutherland's formula seems to be as a simple interpolation formula over restricted ranges of temperature."

The Lennard-Jones model for the potential energy of an interaction, which takes into account both the softness of the molecules and their mutual attraction at large distances, has been used by Svehla (Ref. 1.1) to calculate the viscosity and the thermal conductivity of gases at high temperatures. The coefficients of viscosity for air as tabulated by Svehla are compared with the values calculated using equation (1.5a) in Table 1.1. These comments are made to emphasize the fact that even the basic fluid properties may involve approximate models that have a limited range of applicability.

Kinematic Viscosity

The aerodynamicist may encounter many applications where the ratio μ/ρ has been replaced by a single parameter. Because this ratio appears frequently, it has been given a special name, the kinematic viscosity. The symbol used to represent the kinematic viscosity is v:

$$v = \frac{\mu}{\rho} \tag{1.6}$$

In this ratio, the force units (or, equivalently, the mass units) cancel. Thus, v has the dimensions of L^2/T (e.g., square meters per second or square feet per second).

Example 1.5

Using the results of Examples 1.2 and 1.4, calculate the kinematic viscosity of air when the temperature is 518.67°R and the pressure is 2116.22 lbf/ft².

Solution   From Example 1.2, $\rho = 0.07649$ lbm/ft³ $= 0.002377$ lbf·s²/ft⁴; while, from Example 1.4, $\mu = 3.7383 \times 10^{-7}$ lbf·s/ft². Thus,

$$v = \frac{\mu}{\rho} = \frac{3.7383 \times 10^{-7} \dfrac{\text{lbf·s}}{\text{ft}^2}}{0.002377 \dfrac{\text{lbf·s}^2}{\text{ft}^4}} = 1.573 \times 10^{-4} \frac{\text{ft}^2}{\text{s}}$$

If we use the alternative units for the density, we must employ the factor g_c which is equal to 32.174 ft·lbm/lbf·s² to arrive at the appropriate units.

$$\nu = \frac{\mu}{\rho} = \frac{3.7383 \times 10^{-7} \dfrac{\text{lbf·s}}{\text{ft}^2}}{0.07649 \dfrac{\text{lbm}}{\text{ft}^3}} \left(32.174 \ \frac{\text{ft·lbm}}{\text{lbf·s}^2}\right)$$

$$= 1.573 \times 10^{-4} \ \text{ft}^2/\text{s}$$

Speed of Sound

The speed at which a disturbance of infinitesimal proportions propagates through a fluid that is at rest is known as the *speed of sound*, which is designated in this book as a. The speed of sound is established by the properties of the fluid. For a perfect gas $a = \sqrt{\gamma RT}$, where γ is the ratio of specific heats (see Chapter 8) and R is the

TABLE 1.1 COMPARISON OF THE COEFFICIENT OF VISCOSITY FOR AIR AS TABULATED BY SVEHLA (REF. 1.1) AND AS CALCULATED USING SUTHERLAND'S EQUATION [EQUATION (1.5a)]

T (K)	$\mu \times 10^5$ (kg/m·s)*	$\mu \times 10^5$ (kg/m·s)†
200	1.360	1.329
400	2.272	2.285
600	2.992	3.016
800	3.614	3.624
1000	4.171	4.152
1200	4.695	4.625
1400	5.197	5.057
1600	5.670	5.456
1800	6.121	5.828
2000	6.553	6.179
2200	6.970	6.512
2400	7.373	6.829
2600	7.765	7.132
2800	8.145	7.422
3000	8.516	7.702
3200	8.878	7.973
3400	9.232	8.234
3600	9.579	8.488
3800	9.918	8.734
4000	10.252	8.974
4200	10.580	9.207
4400	10.902	9.435
4600	11.219	9.657
4800	11.531	9.874
5000	11.838	10.087

* From Ref. 1.1.
† Calculated using equation (1.5a).

gas constant. For the range of temperature over which air behaves as a perfect gas, $\gamma = 1.4$ and the speed of sound is given by

$$a = 20.047 \sqrt{T} \tag{1.7a}$$

where T is the temperature (in K) and the unit for the speed of sound is m/s. In English units

$$a = 49.02 \sqrt{T} \tag{1.7b}$$

where T is the temperature (in °R) and the unit for the speed of sound is ft/s.

TABLE 1.2 U.S. STANDARD ATMOSPHERE, 1962
a. Metric Units

Altitude (km)	Pressure (mm Hg)	Temperature (K)	Density (kg/m³)	Viscosity (kg/s·m) × 10⁵	Speed of sound (m/s)
0	760.000	288.150	1.2250	1.7894	340.294
1	674.127	281.651	1.1117	1.7579	336.435
2	596.309	275.154	1.0066	1.7260	332.532
3	525.952	268.659	0.9092	1.6938	328.583
4	462.491	262.166	0.8194	1.6612	324.589
5	405.395	255.676	0.7364	1.6282	320.545
6	354.161	249.187	0.6601	1.5949	316.452
7	308.315	242.700	0.5900	1.5612	312.306
8	267.409	236.215	0.5258	1.5271	308.105
9	231.024	229.733	0.4671	1.4926	303.848
10	198.765	223.252	0.4135	1.4577	299.532
11	170.263	216.774	0.3648	1.4223	295.154
12	145.508	216.650	0.3119	1.4216	295.069
13	124.357	216.650	0.2666	1.4216	295.069
14	106.286	216.650	0.2279	1.4216	295.069
15	90.846	216.650	0.1948	1.4216	295.069
16	77.653	216.650	0.1665	1.4216	295.069
17	66.378	216.650	0.1423	1.4216	295.069
18	56.744	216.650	0.1216	1.4216	295.069
19	48.510	216.650	0.1040	1.4216	295.069
20	41.473	216.650	0.0889	1.4216	295.069
21	35.470	217.581	0.0757	1.4267	295.703
22	30.359	218.574	0.0645	1.4322	296.377
23	26.004	219.567	0.0550	1.4376	297.049
24	22.290	220.560	0.0469	1.4430	297.720
25	19.121	221.552	0.0401	1.4484	298.389
26	16.414	222.544	0.0343	1.4538	299.056
27	14.101	223.536	0.0293	1.4592	299.722
28	12.123	224.527	0.0251	1.4646	300.386
29	10.429	225.518	0.0215	1.4699	301.048
30	8.978	226.509	0.0184	1.4753	301.709

b. English Units[a]

Altitude $(10^3$ ft)	Pressure, p/p_0	Temperature (°R)	Density, ρ/ρ_0	Viscosity, μ/μ_0	Speed of sound (ft/s)
0	1.00000	518.67	1.0000	1.00000	1116.45
5	0.83209	500.84	0.8617	0.97307	1097.10
10	0.68783	483.03	0.7386	0.94570	1077.40
15	0.56459	465.22	0.6295	0.91786	1057.36
20	0.45991	447.42	0.5332	0.88953	1036.93
25	0.37158	429.62	0.4486	0.86070	1016.10
30	0.29754	411.84	0.3747	0.83135	994.85
35	0.23596	394.06	0.3106	0.80144	973.14
40	0.18577	389.97	0.2471	0.79447	968.08
45	0.14623	389.97	0.1945	0.79447	968.08
50	0.11512	389.97	0.1531	0.79447	968.08
55	0.09063	389.97	0.1206	0.79447	968.08
60	0.07137	389.97	0.0949	0.79447	968.08
65	0.05620	389.97	0.0747	0.79447	968.08
70	0.04429	392.25	0.0586	0.79835	970.90
75	0.03496	394.97	0.0459	0.80298	974.26
80	0.02765	397.69	0.0361	0.80760	977.62
85	0.02190	400.42	0.0284	0.81219	980.95
90	0.01738	403.14	0.0224	0.81678	984.28
95	0.01381	405.85	0.0177	0.82134	987.59
100	0.01100	408.57	0.0140	0.82589	990.90

[a] The corresponding sea-level values are: $p_0 = 2116.22$ lbf/ft^2, $\rho_0 = 0.076474$ lbm/ft$^3 = 0.002376$ slug/ft^3, and $\mu_0 = 1.2024 \times 10^{-5}$ lbm/ft·s $= 3.740 \times 10^{-7}$ lbf·s/ft^2.

PRESSURE VARIATION IN A STATIC FLUID MEDIUM

In order to compute the forces and moments or the heat-transfer rates acting on a vehicle or to determine the flight path (i.e., the trajectory) of the vehicle, the engineer will often develop an analytic model of the atmosphere instead of using a table look-up procedure such as Table 1.2. To do this, let us develop the equations describing the pressure variation in a static fluid medium. If fluid particles, when viewed as a continuum, are either all at rest or all moving with the same velocity, the fluid is said to be a *static medium*. Thus, the term *static fluid properties* may be applied to situations in which the elements of the fluid are moving, provided that there is no relative motion between finite elements. Since there is no relative motion between adjacent layers of the fluid, there are no shear forces. Thus, with no relative motion between fluid elements, the viscosity of the fluid is of no concern. For these inviscid flows, the only forces acting on the surface of a fluid element are pressure forces.

Consider the small fluid element whose center is defined by the coordinates x, y, and z, as shown in Fig. 1.2. A first-order Taylor's series expansion is used to evaluate the pressure at each face. Thus, the pressure at the back face of the element is $p - (\partial p/\partial x)(\Delta x/2)$; that at the front face is $p + (\partial p/\partial x)(\Delta x/2)$. If the fluid is not accelerating,

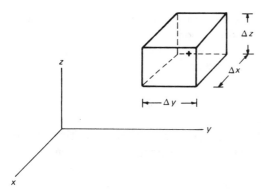

Figure 1.2 Derivation of equations (1.8) through (1.10).

the element must be in equilibrium. For equilibrium, the sum of the forces in any direction must be zero. Thus,

$$-\left(p + \frac{\partial p}{\partial x}\frac{\Delta x}{2}\right)\Delta y\,\Delta z + \left(p - \frac{\partial p}{\partial x}\frac{\Delta x}{2}\right)\Delta y\,\Delta z = 0 \qquad (1.8a)$$

$$-\left(p + \frac{\partial p}{\partial y}\frac{\Delta y}{2}\right)\Delta x\,\Delta z + \left(p - \frac{\partial p}{\partial y}\frac{\Delta y}{2}\right)\Delta x\,\Delta z = 0 \qquad (1.8b)$$

$$-\left(p + \frac{\partial p}{\partial z}\frac{\Delta z}{2}\right)\Delta x\,\Delta y + \left(p - \frac{\partial p}{\partial z}\frac{\Delta z}{2}\right)\Delta x\,\Delta y - \rho g\,\Delta x\,\Delta y\,\Delta z = 0 \qquad (1.8c)$$

Note that the coordinate system has been chosen such that gravity acts in the negative z direction. Combining terms and dividing by $\Delta x\,\Delta y\,\Delta z$ gives us

$$\frac{\partial p}{\partial x} = 0 \qquad (1.9a)$$

$$\frac{\partial p}{\partial y} = 0 \qquad (1.9b)$$

$$\frac{\partial p}{\partial z} = -\rho g \qquad (1.9c)$$

The three equations can be written as one using vector notation:

$$\nabla p = \rho\vec{f} = -\rho g\hat{k} \qquad (1.10)$$

where, $\vec{f}$ represents the body force per unit mass. For the cases of interest to this book, the body force is gravity.

These equations illustrate two important principles for a nonaccelerating, hydrostatic, or shear-free, flow: (1) There is no pressure variation in the horizontal direction that is, the pressure is constant in a plane perpendicular to the direction of gravity, and (2) the vertical pressure variation is proportional to gravity, density, and change in depth. Furthermore, as the element shrinks to zero volume (i.e., as $\Delta z \to 0$), it can be seen that the pressure is the same on all faces. That is, pressure at a point in a static fluid is independent of orientation.

Since the pressure varies only with z, that is, it is not a function of x or y, an ordinary derivative may be used and equation (1.9c) may be written

$$\frac{dp}{dz} = -\rho g \qquad (1.11)$$

Let us assume that the air behaves as a perfect gas. Thus, the expression for density given by equation (1.3) can be substituted into equation (1.11) to give

$$\frac{dp}{dz} = -\rho g = -\frac{pg}{RT} \qquad (1.12)$$

If the temperature is assumed to be constant, separating the variables and integrating between two points yields

$$\int \frac{dp}{p} = \ln \frac{p_2}{p_1} = -\frac{g}{RT} \int dz = -\frac{g}{RT}(z_2 - z_1)$$

where the integration reflects the fact that the temperature has been assumed constant. Rearranging yields

$$p_2 = p_1 \exp \left[\frac{g(z_1 - z_2)}{RT} \right] \qquad (1.13)$$

The pressure variation described by equation (1.13) is a reasonable approximation of that in the atmosphere near the earth's surface.

An improved correlation for pressure variation in the earth's atmosphere can be obtained if one accounts for the temperature variation with altitude. The earth's mean atmospheric temperature decreases almost linearly with z up to an altitude of nearly 11,000 m. That is,

$$T = T_0 - Bz \qquad (1.14)$$

where T_0 is the sea-level temperature (absolute) and B is the lapse rate, both of which vary from day to day. The following standard values will be assumed to apply from 0 to 11,000 m:

$$T_0 = 288.15 \text{ K} \qquad \text{and} \qquad B = 0.0065 \text{ K/m}$$

Substituting equation (1.14) into the relation

$$\int \frac{dp}{p} = -\int \frac{g \, dz}{RT}$$

and integrating, we obtain

$$p = p_0 \left(1 - \frac{Bz}{T_0} \right)^{g/RB} \qquad (1.15)$$

The exponent g/RB, which is dimensionless, is equal to 5.26 for air.

Example 1.6

Using equations (1.14) and (1.15), calculate the temperature and pressure of air at an altitude of 10 km. Compare the tabulated values with those presented in Table 1.2.

Solution The ambient temperature at 10,000 m is

$$T = T_0 - Bz = 288.15 - 0.0065(10^4) = 223.15 \text{ K}$$

The tabulated value from Table 1.2 is 223.252 K. The calculated value for the ambient pressure is

$$p = p_0 \left(1 - \frac{Bz}{T_0}\right)^{g/RB}$$

$$= 1.01325 \times 10^5 \left[1 - \frac{0.0065(10^4)}{288.15}\right]^{5.26}$$

$$= 1.01325 \times 10^5 (0.26063) = 2.641 \times 10^4 \text{ N/m}^2 = 198.08 \text{ mm Hg}$$

The comparable value in Table 1.2 is 198.765 mm Hg.

Example 1.7

Develop equations for the pressure and for the density as a function of altitude from 0 to 65,000 ft. The analytical model of the atmosphere should make use of the hydrostatic equations [i.e., equation (1.11)], for which the density is eliminated through the use of the equation of state for a thermally perfect gas. Assume that the temperature of air from 0 to 36,100 ft is given by

$$T = 518.67 - 0.003565 z$$

and that the temperature from 36,100 to 65,000 ft is constant at 389.97°R.

Solution From 0 to 36,100 ft, the temperature varies linearly as described in general by equation (1.14). Specifically,

$$T = 518.67 - 0.003565 z$$

Thus, $T_0 = 518.67°R$ and $B = 0.003565°R/ft$. Using English unit terms in equation (1.15) gives us

$$p = p_0 \left(1 - \frac{Bz}{T_0}\right)^{g/RB}$$

$$= 2116.22(1.0 - 6.873 \times 10^{-6}z)^{5.26}$$

For a thermally perfect gas, the density is

$$\rho = \frac{p}{RT} = \frac{2116.22(1.0 - 6.873 \times 10^{-6}z)^{5.26}}{53.34(518.67 - 0.003565 z)}$$

Dividing by ρ_0, the value of the density at standard sea-level conditions,

$$\rho_0 = \frac{p_0}{RT_0} = \frac{2116.22}{(53.34)(518.67)}$$

one obtains the nondimensionalized density:

$$\frac{\rho}{\rho_0} = (1.0 - 6.873 \times 10^{-6}z)^{4.26}$$

Since the temperature is constant from 36,100 to 65,000 ft, equation (1.13) can be used to express the pressure, with the values at 36,100 ft serving as the reference values p_1 and z_1:

$$p_{36,100} = 2116.22(1.0 - 6.873 \times 10^{-6}z)^{5.26} = 472.19 \ \text{lbf/ft}^2$$

Thus,

$$p = 472.19 \exp\left[\frac{g(36,100 - z)}{RT}\right]$$

In English units,

$$\frac{g}{RT} = \frac{32.174 \ \dfrac{\text{ft}}{\text{s}^2}}{\left(53.34 \ \dfrac{\text{ft·lbf}}{\text{lbm·°R}}\right)(389.97°\text{R})}$$

However, to have the correct units, multiply by $(1/g_c)$, so that

$$\frac{g}{RT} = \frac{32.174 \ \dfrac{\text{ft}}{\text{s}^2}}{\left(53.34 \ \dfrac{\text{ft·lbf}}{\text{lbm·°R}}\right)(389.97°\text{R})\left(32.174 \ \dfrac{\text{ft·lbm}}{\text{lbf·s}^2}\right)} = 4.8075 \times 10^{-5}/\text{ft}$$

Thus,

$$\frac{p}{p_0} = 0.2231 \exp(1.7355 - 4.8075 \times 10^{-5}z)$$

The nondimensionalized density is

$$\frac{\rho}{\rho_0} = \frac{p}{p_0} \frac{T_0}{T}$$

Since $T = 389.97°\text{R} = 0.7519T_0$,

$$\frac{\rho}{\rho_0} = 0.2967 \exp(1.7355 - 4.8075 \times 10^{-5}z)$$

SUMMARY

Specific values and equations for fluid properties (e.g., viscosity, density, and speed of sound) have been presented in this chapter. The reader should note that it may be necessary to use alternative relations for calculating fluid properties. For instance, for the relatively high temperatures associated with hypersonic flight, it may be necessary to account for real-gas effects (e.g., dissociation). Numerous references present the thermodynamic properties and transport properties of gases at high temperatures and pressures (e.g., Refs. 1.4 through 1.6).

PROBLEMS

1.1. Using the values for the pressure and for the temperature given in Table 1.2, calculate the density [equation (1.3)] and the viscosity [equation (1.5a)] at 20 km. Compare the calculated values with those given in Table 1.2. What is the kinematic viscosity at this altitude [equation (1.6)]?

1.2. Using the values for the pressure and for the temperature given in Table 1.2, calculate the density [equation (1.3)] and the viscosity [equation (1.5b)] at 35,000 ft. Compare the calculated values with those given in Table 1.2. What is the kinematic viscosity at this altitude [equation (1.6)]?

1.3. The conditions in the reservoir (or stagnation chamber) of Tunnel B at the Arnold Engineering Development Center (AEDC) are that the pressure (p_{t1}) is 5.723×10^6 N/m² and the temperature (T_t) is 750 K. Using the perfect-gas relations, what are the density and the viscosity in the reservoir?

1.4. The air in Tunnel B is expanded through a convergent/divergent nozzle to the test section, where the Mach number is 8, the free-stream static pressure is 586 N/m², and the free-stream temperature is 54.3 K. Using the perfect-gas relations, what are the corresponding values for the test-section density, viscosity, and velocity? Note that $U_\infty = M_\infty a_\infty$.

1.5. The conditions in the reservoir (or stagnation chamber) of Aerothermal Tunnel C at the Arnold Engineering Development Center (AEDC) are that the pressure (p_{t1}) is 24.5 psia and the temperature (T_t) is 1660°R. Using the perfect-gas relations, what are the density and the viscosity in the reservoir?

1.6. The air in Tunnel C accelerates through a convergent/divergent nozzle until the Mach number is 4 in the test section. The corresponding values for the free-stream pressure and the free-stream temperature in the test section are 23.25 lbf/ft² abs. and $-65°F$, respectively. What are the corresponding values for the free-stream density, viscosity, and velocity in the test section? Note that $M_\infty = U_\infty/a_\infty$. Using the values for the static pressure given in Table 1.2, what is the pressure altitude simulated in the wind tunnel by this test condition?

1.7. The pilot announces that you are flying at a velocity of 470 knots at an altitude of 35,000 ft. What is the velocity of the airplane in km/h? In ft/s?

1.8. Using equations (1.14) and (1.15), calculate the temperature and pressure of the atmosphere at 7000 m. Compare the tabulated values with those presented in Table 1.2.

1.9. Using an approach similar to that used in Example 1.7, develop metric-unit expressions for the pressure, the temperature, and the density of the atmosphere from 11,000 to 20,000 m. The temperature is constant and equal to 216.650 K over this range of altitude.

1.10. Using the expressions developed in Problem 1.9, what are the pressure, the density, the viscosity, and the speed of sound for the ambient atmospheric air at 18 km? Compare these values with the corresponding values tabulated in Table 1.2.

1.11. Using the expressions developed in Example 1.7, calculate the pressure, the temperature, and the density for the ambient atmospheric air at 10,000, 30,000, and 65,000 ft. Compare these values with the corresponding values presented in Table 1.2.

1.12. If the water in a lake is everywhere at rest, what is the pressure as a function of the distance from the surface? The air above the surface of the water is at standard atmospheric conditions. How far down must one go before the pressure is 1 atm greater than the pressure at the surface? Use equation (1.11).

1.13. A U-tube manometer is used to measure the pressure at the stagnation point of a model in a wind tunnel. One side of the manometer goes to an orifice at the stagnation point; the other side is open to the atmosphere (Fig. P1.13). If there is a difference of 3.0 cm in the mercury levels in the two tubes, what is the stagnation pressure?

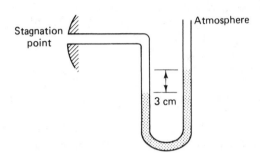

Figure P1.13

REFERENCES

1.1. Svehla, R. A., "Estimated Viscosities and Thermal Conductivities of Gases at High Temperatures," *Technical Report R-132*, NASA, 1962.

1.2. Chapman, S., and T. G. Cowling, *The Mathematical Theory of Non-uniform Gases*, Cambridge University Press, Cambridge, 1960.

1.3. *U.S. Standard Atmosphere, 1962*, Government Printing Office, Washington, D.C., Dec. 1962.

1.4. Moeckel, W. E., and K. C. Weston, "Composition and Thermodynamic Properties of Air in Chemical Equilibrium," *TN 4265*, NACA, Aug. 1958.

1.5. Hansen, C. F., "Approximations for the Thermodynamic and Transport Properties of High-Temperature Air," *Technical Report R-50*, NACA, Nov. 1957.

1.6. Yos, J. M., "Transport Properties of Nitrogen, Hydrogen, Oxygen, and Air to 30,000 K," *Technical Memorandum RAD-TM-63-7*, AVCO Corporation, Mar. 22, 1963.

2

FUNDAMENTALS
OF FLUID MECHANICS

As noted in Chapter 1, to accurately predict the aerodynamic forces and moments that act on a vehicle in flight, it is necessary to be able to describe the pattern of flow around the configuration. The resultant flow pattern depends on the geometry of the vehicle, its orientation with respect to the undisturbed free stream, and the altitude and speed at which the vehicle is traveling. The fundamental physical laws used to solve for the fluid motion in a general problem are:

1. Conservation of mass (or continuity equation)
2. Conservation of linear momentum (or Newton's second law of motion)
3. Conservation of energy (or the first law of thermodynamics)

Because the flow patterns are often very complex, it may be necessary to use experimental investigations as well as theoretical analysis to describe the resultant flow. The theoretical descriptions may utilize simplifying approximations in order to obtain any solution at all. The validity of the simplifying approximations for a particular application should be verified experimentally. Thus, it is important that we understand the fundamental laws that govern the fluid motion so we can relate the theoretical solutions obtained using approximate flow models with the experimental results, which usually involve scale models.

INTRODUCTION TO FLUID DYNAMICS

To calculate the aerodynamic forces acting on an airplane, it is necessary to solve the equations governing the flow field about the vehicle. The flow-field solution can be formulated from the point of view of an observer on the ground or from the point of view of the pilot. Provided that the two observers apply the appropriate boundary conditions to the governing equations, both observers will obtain the same values for the aerodynamic forces acting on the airplane.

To an observer on the ground, the airplane is flying into a mass of air substantially at rest (assuming there is no wind). The neighboring air particles are accelerated and decelerated by the airplane and the reaction of the particles to the acceleration results in a force on the airplane. The motion of a typical air particle is shown in Fig. 2.1. The particle, which is initially at rest well ahead of the airplane, is accelerated by the passing airplane. The description of the flow field in the ground-observer-fixed-coordinate system must represent the time-dependent motion (i.e., a nonsteady flow).

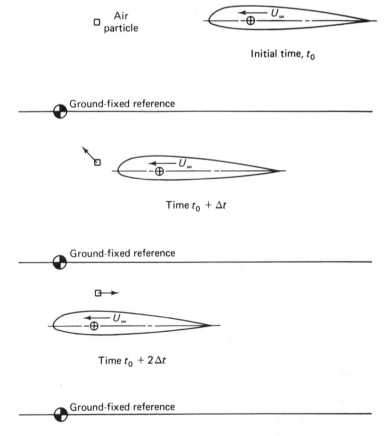

Figure 2.1 (Nonsteady) airflow around a wing in the ground-fixed coordinate system.

As viewed by the pilot, the air is flowing past the airplane and moves in response to the geometry of the vehicle. If the airplane is flying at constant altitude and constant velocity, the terms of the flow-field equations that contain derivatives with respect to time are zero in the vehicle-fixed-coordinate system. Thus, as shown in Fig. 2.2, the velocity and the flow properties of the air particles that pass through a specific location relative to the vehicle are independent of time. The flow field is steady relative to a set of axes fixed to the vehicle (or pilot). Therefore, the equations are usually easier to solve in the vehicle (or pilot)-fixed-coordinate system than in the ground-observer-fixed-coordinate system. Because of the resulting simplification of the mathematics, many problems in aerodynamics are formulated as the flow of a stream of fluid past a body at rest. Note that the subsequent locations of the air particle which passed through our control volume at time t_0 are included for comparison with Fig. 2.1.

In this text we shall use the vehicle (or pilot)-fixed-coordinate system. Thus, instead of describing the fluid motion around a vehicle flying through the air, we will examine air flowing around a fixed vehicle. At points far from the vehicle (i.e., the undisturbed free stream), the fluid particles are moving toward the vehicle with the

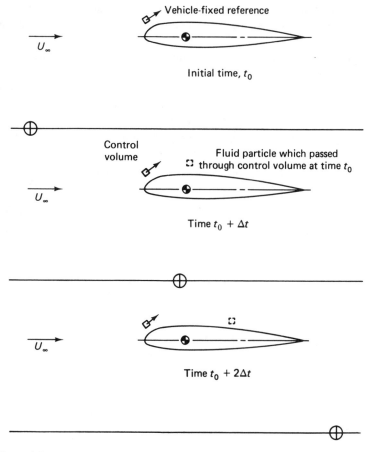

Figure 2.2   (Steady) airflow around a wing in a vehicle-fixed-coordinate system.

velocity U_∞ (see Fig.2.2), which is in reality the speed of the vehicle (see Fig. 2.1). The subscript ∞ will be used to denote the undisturbed (or free-stream) flow conditions (i.e., those conditions far from the vehicle). Since all the fluid particles in the free stream are moving with the same velocity, there is no relative motion between them, and hence there are no shearing stresses in the free-stream flow. When there is no relative motion between the fluid particles, the fluid is termed a *static medium*. The values of the static fluid properties (e.g., pressure and temperature) are the same for either coordinate system.

THE CONTINUITY EQUATION

Let us apply the principle of conservation of mass to a small volume of space through which the fluid can move freely. For convenience, we shall use a cartesian coordinate system (x, y, z). Furthermore, in the interest of simplicity, we shall treat a two-dimensional flow, that is, one in which there is no flow along the z axis. Flow patterns are the same for any xy plane. As indicated in the sketch of Fig. 2.3, the component of the fluid velocity in the x direction will be designated by u, and that in the y direction by v. The net outflow of mass through the surface surrounding the volume must be equal to the decrease of mass within the volume. The mass-flow rate through a surface bounding the element is equal to the product of the density, the velocity component normal to the surface, and the area of that surface. Flow out of the volume is considered positive. A first-order Taylor's series expansion is used to evaluate the flow properties at the faces of the element, since the properties are a function of position. Referring to Fig. 2.3, the net outflow of mass per unit time per unit depth (into the paper) is

$$\left[\rho u + \frac{\partial(\rho u)}{\partial x}\frac{\Delta x}{2}\right]\Delta y + \left[\rho v + \frac{\partial(\rho v)}{\partial y}\frac{\Delta y}{2}\right]\Delta x$$

$$-\left[\rho u - \frac{\partial(\rho u)}{\partial x}\frac{\Delta x}{2}\right]\Delta y - \left[\rho v - \frac{\partial(\rho v)}{\partial y}\frac{\Delta y}{2}\right]\Delta x$$

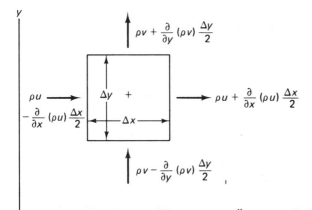

Figure 2.3 Velocities and densities for the mass flow balance through a fixed volume element in two dimensions.

which must equal the rate at which the mass contained within the element decreases:

$$-\frac{\partial \rho}{\partial t} \Delta x \, \Delta y$$

Equating the two expressions, combining terms, and dividing by $\Delta x \, \Delta y$, we obtain

$$\frac{\partial \rho}{\partial t} + \frac{\partial}{\partial x}(\rho u) + \frac{\partial}{\partial y}(\rho v) = 0$$

If the approach were extended to include flow in the z direction, we would obtain the general differential form of the continuity equation:

$$\frac{\partial \rho}{\partial t} + \frac{\partial}{\partial x}(\rho u) + \frac{\partial}{\partial y}(\rho v) + \frac{\partial}{\partial z}(\rho w) = 0 \qquad (2.1)$$

In vector form, the equation is

$$\frac{\partial \rho}{\partial t} + \nabla \cdot (\rho \vec{V}) = 0 \qquad (2.2)$$

As has been discussed, the pressure variations that occur in relatively low speed flows are sufficiently small, so the density is essentially constant. For these incompressible flows, the continuity equation becomes

$$\frac{\partial u}{\partial x} + \frac{\partial v}{\partial y} + \frac{\partial w}{\partial z} = 0 \qquad (2.3)$$

In vector form, this equation is

$$\nabla \cdot \vec{V} = 0 \qquad (2.4)$$

Using boundary conditions, such as the requirement that there is no flow through a solid surface (i.e., the normal component of the velocity is zero at a solid surface), we can solve equation (2.4) for the velocity field. In so doing, we obtain a detailed picture of the velocity as a function of position.

Example 2.1

Consider the case where a steady, incompressible, uniform flow whose velocity is U_∞ (i.e., a free-stream flow) approaches a flat plate. In the viscous region near the surface, the streamwise component of velocity is given by

$$u = U_\infty \left(\frac{y}{\delta}\right)^{1/7}$$

where δ, the boundary-layer thickness at a given station, is a function of x. Is a horizontal line parallel to the plate and a distance Δ from the plate (where Δ is equal to δ at the downstream station) a streamline? (See Fig. 2.4.)

Solution By continuity for this steady, incompressible flow,

$$\frac{\partial u}{\partial x} + \frac{\partial v}{\partial y} = 0$$

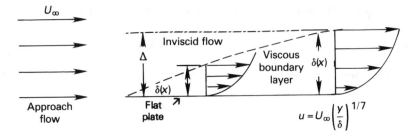

Figure 2.4   Flow diagram for Example 2.1.

since $u = U_\infty (y/\delta)^{1/7}$ and $\delta(x)$,

$$\frac{\partial v}{\partial y} = -\frac{\partial u}{\partial x} = \frac{U_\infty}{7} \frac{y^{1/7}}{\delta^{8/7}} \frac{d\delta}{dx}$$

Integrating with respect to y yields

$$v = \frac{U_\infty}{8} \frac{y^{8/7}}{\delta^{8/7}} \frac{d\delta}{dx} + C$$

where C, the constant of integration, can be set equal to zero since $v = 0$ when $y = 0$ (i.e., there is no flow through the wall). Thus, when $y = \Delta$,

$$v_e = \frac{U_\infty}{8} \left(\frac{\Delta}{\delta}\right)^{8/7} \frac{d\delta}{dx}$$

Since v is not equal to zero, there is flow across the horizontal line which is Δ above the surface, and this line is not a streamline.

___

If the details of the flow are not of concern, the mass conservation principle can be applied directly to the entire region. Integrating equation (2.2) over a fixed finite volume in our fluid space (see Fig. 2.5) yields

$$\iiint_{vol} \frac{\partial \rho}{\partial t} \, d(\text{vol}) + \iiint_{vol} \nabla \cdot (\rho \vec{V}) \, d(\text{vol}) = 0$$

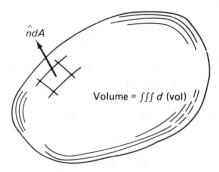

Figure 2.5   Nomenclature for the integral form of the continuity equation.

The second volume integral can be transformed into a surface integral using Gauss' theorem, which is

$$\iiint\limits_{\text{vol}} \nabla \cdot (\rho \vec{V}) \, d(\text{vol}) = \oiint\limits_{A} \hat{n} \cdot \rho \vec{V} \, dA$$

where $\hat{n} \, dA$ is a vector normal to the surface dA which is positive when pointing outward from the enclosed volume and which is equal in magnitude to the surface area. The circle through the integral sign for the area indicates that the integration is to be performed over the entire surface bounding the volume. The resultant equation is the general integral expression for the conservation of mass:

$$\frac{\partial}{\partial t} \iiint\limits_{\text{vol}} \rho \, d(\text{vol}) + \oiint\limits_{A} \rho \vec{V} \cdot \hat{n} \, dA = 0 \tag{2.5}$$

In words, the time rate of change of the mass within the volume plus the net efflux (outflow) of mass through the surface bounding the volume must be zero.

The volumetric flux Q is the flow rate through a particular surface and is equal to $\iint \vec{V} \cdot \hat{n} \, dA$.

For a sample problem using the integral form of the continuity equation, see Example 2.3.

CONSERVATION OF LINEAR MOMENTUM

The equation for the conservation of linear momentum is obtained by applying Newton's second law: the net force acting on a fluid particle is equal to the time rate of change of the linear momentum of the fluid particle. As the fluid element moves in space, its shape and volume may change, but its mass is conserved. Thus, using a coordinate system that is neither accelerating nor rotating, called an *inertial coordinate system*, we may write

$$\vec{F} = m \frac{d\vec{V}}{dt} \tag{2.6}$$

The velocity V of a fluid particle is, in general, an explicit function of time t as well as of its position x, y, z. Furthermore, the position coordinates x, y, z of the fluid particle are themselves a function of time. Since the time differentiation of equation (2.6) follows a given particle in its motion, the derivative is frequently termed the *particle*, *total*, or *substantial derivative* of $\vec{V}$. Since $\vec{V}(x, y, z, t)$ and $x(t)$, $y(t)$, and $z(t)$,

$$\frac{d\vec{V}}{dt} = \frac{\partial \vec{V}}{\partial x} \frac{dx}{dt} + \frac{\partial \vec{V}}{\partial y} \frac{dy}{dt} + \frac{\partial \vec{V}}{\partial z} \frac{dz}{dt} + \frac{\partial \vec{V}}{\partial t} \tag{2.7}$$

However,

$$\frac{dx}{dt} = u \qquad \frac{dy}{dt} = v \qquad \frac{dz}{dt} = w$$

Therefore, the acceleration of a fluid particle is

$$\frac{d\vec{V}}{dt} = \frac{\partial \vec{V}}{\partial t} + u \frac{\partial \vec{V}}{\partial x} + v \frac{\partial \vec{V}}{\partial y} + w \frac{\partial \vec{V}}{\partial z} \tag{2.8}$$

or

$$\frac{d\vec{V}}{dt} = \frac{\partial \vec{V}}{\partial t} + (\vec{V} \cdot \nabla)\vec{V} \tag{2.9}$$

total   local   convective

Thus, the substantial derivative is the sum of the local, time-dependent changes that occur at a point in the flow field and of the changes that occur because the fluid particle moves around in space. Problems where the local, time-dependent changes are zero,

$$\frac{\partial \vec{V}}{\partial t} = 0$$

are known as *steady-state flows*. Note that even when this is true, $d\vec{V}/dt$ is not necessarily zero, since the fluid particle may accelerate as it moves to different points in space.

The principal forces with which we are concerned are those which act directly on the mass of the fluid element, the *body forces*, and those which act on its surface, the *pressure forces* and *shear forces*. The stress system acting on an element of the surface is illustrated in Fig. 2.6. The stress components τ acting on the small cube are assigned subscripts. The first subscript indicates the direction of the normal to the surface on which the stress acts and the second indicates the direction in which the stress acts. Thus, τ_{xy} denotes a stress acting in the y direction on the surface whose normal points in the x direction. Similarly, τ_{xx} denotes a normal stress acting on that surface. The stresses are described in terms of a right-hand coordinate system in which the outwardly directed surface normal indicates the positive direction.

The properties of most fluids have no preferred direction in space; that is, fluids are isotropic. Furthermore, the stresses at a point do not explicitly depend either on the position coordinate or on the velocity of the fluid. As a result,

$$\tau_{xy} = \tau_{yx} \qquad \tau_{yz} = \tau_{zy} \qquad \tau_{zx} = \tau_{xz} \tag{2.10}$$

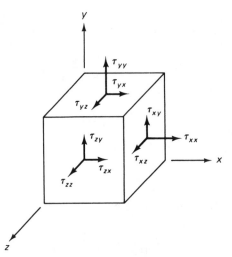

Figure 2.6   Nomenclature for the normal stresses and the shear stresses acting on a fluid element.

In general, the various stresses change from point to point. Thus, they produce net forces on the fluid particle, which cause it to accelerate. The forces acting on each surface are obtained by taking into account the variations of stress with position by using the center of the element as a reference point. To simplify the illustration of the force balance on the fluid particle we shall again consider a two-dimensional flow, as indicated in Fig. 2.7. The resultant force in the x direction (for a unit depth in the z direction) is

$$\rho f_x\, \Delta x\, \Delta y + \frac{\partial}{\partial x}(\tau_{xx})\, \Delta x \Delta y + \frac{\partial}{\partial y}(\tau_{yx})\, \Delta y \Delta x$$

where f_x is the body force per unit mass in the x direction. The most common body force for the flow fields of interest to this text is that of gravity.

Including flow in the z direction, the resultant force in the x direction would be

$$F_x = \rho f_x\, \Delta x\, \Delta y\, \Delta z + \frac{\partial}{\partial x}(\tau_{xx})\, \Delta x\, \Delta y\, \Delta z + \frac{\partial}{\partial y}(\tau_{yx})\, \Delta y\, \Delta x\, \Delta z$$

$$+ \frac{\partial}{\partial z}(\tau_{zx})\, \Delta z\, \Delta y\, \Delta x$$

which, by equation (2.6), is equal to

$$ma_x = \rho\, \Delta x\, \Delta y\, \Delta z\, \frac{du}{dt} = \rho\, \Delta x\, \Delta y\, \Delta z \left[\frac{\partial u}{\partial t} + (\vec{V} \cdot \nabla)u \right]$$

Equating the two and dividing by the volume of the fluid particle $\Delta x\, \Delta y\, \Delta z$ yields

$$\rho\, \frac{du}{dt} = \rho f_x + \frac{\partial}{\partial x}\, \tau_{xx} + \frac{\partial}{\partial y}\, \tau_{yx} + \frac{\partial}{\partial z}\, \tau_{zx} \qquad (2.11a)$$

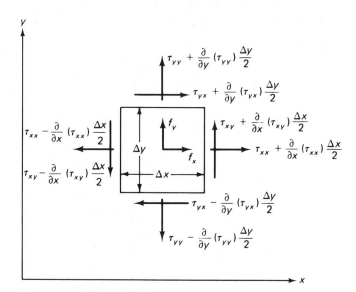

Figure 2.7 Stresses acting on a two-dimensional element of fluid.

Similarly, we obtain, the equation of motion for the y direction,

$$\rho \frac{dv}{dt} = \rho f_y + \frac{\partial}{\partial x} \tau_{xy} + \frac{\partial}{\partial y} \tau_{yy} + \frac{\partial}{\partial z} \tau_{zy} \qquad (2.11b)$$

and for the z direction,

$$\rho \frac{dw}{dt} = \rho f_z + \frac{\partial}{\partial x} \tau_{xz} + \frac{\partial}{\partial y} \tau_{yz} + \frac{\partial}{\partial z} \partial_{zz} \qquad (2.11c)$$

Next, we need to relate the stresses to the motion of the fluid. For a fluid at rest or for an inviscid fluid motion, there is no shearing stress, and the normal stress is in the nature of a pressure. For fluid particles, the stress is related to the rate of strain by a physical law based on the following assumptions:

1. Stress components may be expressed as a linear function of the components of the rate of strain. The friction law for one-dimensional flow of a Newtonian fluid is a special case of this linear stress/rate-of-strain relation: $\tau = \mu(\partial u/\partial y)$. The viscosity μ is more precisely called the first viscosity coefficient. In a more rigorous development, one should include the second viscosity coefficient (λ), which would appear in the normal stress terms. However, λ disappears completely when the flow is incompressible and is equal to zero for a perfect, monatomic gas. The second viscosity coefficient is of significance in a few specialized problems, such as the analysis of the shock wave structure where extremely large changes in pressure and in temperature take place over very short distances.
2. The relations between the stress components and the rate-of-strain components must be invariant to a coordinate transformation consisting of either a rotation or a mirror reflection of axes, since a physical law cannot depend upon the choice of the coordinate system.
3. When all velocity gradients are zero (i.e., the shear stress vanishes), the stress components must reduce to the hydrostatic pressure, p.

For a fluid that satisfies these criteria:

$$\tau_{xx} = -p - \frac{2}{3} \mu \vec{\nabla} \cdot \vec{V} + 2\mu \frac{\partial u}{\partial x}$$

$$\tau_{yy} = -p - \frac{2}{3} \mu \vec{\nabla} \cdot \vec{V} + 2\mu \frac{\partial v}{\partial y}$$

$$\tau_{zz} = -p - \frac{2}{3} \mu \vec{\nabla} \cdot \vec{V} + 2\mu \frac{\partial w}{\partial z}$$

$$\tau_{xy} = \tau_{yx} = \mu \left(\frac{\partial u}{\partial y} + \frac{\partial v}{\partial x} \right)$$

$$\tau_{xz} = \tau_{zx} = \mu \left(\frac{\partial u}{\partial z} + \frac{\partial w}{\partial x} \right)$$

$$\tau_{yz} = \tau_{zy} = \mu \left(\frac{\partial v}{\partial z} + \frac{\partial w}{\partial y} \right)$$

With the appropriate expressions for the surface stresses substituted into equation (2.11), we obtain

$$\rho\frac{\partial u}{\partial t} + \rho(\vec{V}\cdot\nabla)u = \rho f_x - \frac{\partial p}{\partial x} + \frac{\partial}{\partial x}\left(2\mu\frac{\partial u}{\partial x} - \frac{2}{3}\mu\nabla\cdot\vec{V}\right)$$
$$+ \frac{\partial}{\partial y}\left[\mu\left(\frac{\partial u}{\partial y} + \frac{\partial v}{\partial x}\right)\right] + \frac{\partial}{\partial z}\left[\mu\left(\frac{\partial w}{\partial x} + \frac{\partial u}{\partial z}\right)\right] \tag{2.12a}$$

$$\rho\frac{\partial v}{\partial t} + \rho(\vec{V}\cdot\nabla)v = \rho f_y + \frac{\partial}{\partial x}\left[\mu\left(\frac{\partial u}{\partial y} + \frac{\partial v}{\partial x}\right)\right]$$
$$- \frac{\partial p}{\partial y} + \frac{\partial}{\partial y}\left(2\mu\frac{\partial v}{\partial y} - \frac{2}{3}\mu\nabla\cdot\vec{V}\right) \tag{2.12b}$$
$$+ \frac{\partial}{\partial z}\left[\mu\left(\frac{\partial w}{\partial y} + \frac{\partial v}{\partial z}\right)\right]$$

$$\rho\frac{\partial w}{\partial t} + \rho(\vec{V}\cdot\nabla)w = \rho f_z + \frac{\partial}{\partial x}\left[\mu\left(\frac{\partial w}{\partial x} + \frac{\partial u}{\partial z}\right)\right]$$
$$+ \frac{\partial}{\partial y}\left[\mu\left(\frac{\partial v}{\partial z} + \frac{\partial w}{\partial y}\right)\right] - \frac{\partial p}{\partial z} \tag{2.12c}$$
$$+ \frac{\partial}{\partial z}\left(2\mu\frac{\partial w}{\partial z} - \frac{2}{3}\mu\nabla\cdot\vec{V}\right)$$

These general, differential equations for the conservation of linear momentum are known as the *Navier-Stokes equations*. Note that the viscosity μ is considered to be dependent on the spatial coordinates. This is done since, for a compressible flow, the changes in velocity and pressure, together with the heat due to friction, bring about considerable temperature variations. The temperature dependence of viscosity in the general case should, therefore, be incorporated into the governing equations.

For a general application, the unknown parameters that appear in the Navier-Stokes equations are the three velocity components (u, v, and w), the pressure (p), the density (ρ), and the viscosity (μ). As we discussed in Chapter 1, for a fluid of known composition that is in equilibrium, the density and viscosity are unique functions of pressure and temperature. Thus, there are five primary (or primitive) variables for a general flow problem: the three velocity components, the pressure, and the temperature. However, at present we have only four equations: the continuity equation, equation (2.2), and the three components of the momentum equation, equations (2.12a) through (2.12c). To solve for a general flow involving all five variables, we would need to introduce the energy equation.

Since equations (2.12a) through (2.12c) are the general differential equations for the conservation of linear momentum, the equations for a static medium can be obtained by neglecting the terms relating to the acceleration of the fluid particles and to the viscous forces. Neglecting these terms in equations (2.12a) through (2.12c) and assuming the body force is gravity and that it acts in the z direction, the reader would obtain equations (1.9a) through (1.9c).

The integral form of the momentum equation can be obtained by returning to Newton's law. The sum of the forces acting on a system of fluid particles is equal to the rate of change of momentum of the fluid particles. Thus, the sum of the body forces and of the surface forces equals the time rate of change of momentum within the volume plus the net efflux of momentum through the surface bounding the volume. In vector form,

$$\vec{F}_{body} + \vec{F}_{surface} = \frac{\partial}{\partial t} \iiint_{vol} \rho \vec{V} \, d(vol) + \oiint_A \vec{V}(\rho \vec{V} \cdot \hat{n} \, dA) \tag{2.13}$$

APPLICATIONS OF CONSTANT-PROPERTY FLOWS

For many flows, temperature variations are sufficiently small that the density and viscosity may be assumed constant throughout the flow field. Such flows will be termed *constant-property* flows in this text. The terms *low-speed* and/or *incompressible* flows will also be used in the description of these flows. A gas flow is considered incompressible if the Mach number is less than 0.3 to 0.5, depending upon the application. For these flows, there are only four unknowns: the three velocity components (u, v, and w) and the pressure (p). Thus, we have a system of four independent equations that can be solved for the four unknowns; that is, the energy equation is not needed to obtain the velocities and pressure of a constant-property flow.

Let us consider two constant-property flows, one for which the solution will be obtained using differential equations and one for which the integral equations are used.

Example 2.2

Consider a steady, low-speed flow of a viscous fluid in an infinitely long, two-dimensional channel of height h (Fig. 2.8). This is known as *Poiseuille flow*. Since the flow is low-speed, we will assume that the viscosity and the density are constant. Because the channel is infinitely long, the velocity components do not change in the x direction. In this text, such a flow is termed *fully developed flow*. Let us assume that the body forces are negligible. We are to determine the velocity profile and the shear-stress distribution.

Solution   For a two-dimensional flow, $w \equiv 0$ and all the derivatives with respect to z are zero. The continuity equation for this steady-state, constant-property flow yields

$$\frac{\partial u}{\partial x} + \frac{\partial v}{\partial y} = 0$$

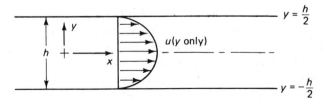

Figure 2.8   Flow diagram for Example 2.2.

Since the velocity components do not change in the x direction,

$$\frac{\partial v}{\partial y} = 0$$

Further, since $v = 0$ at both walls (i.e., there is no flow through the walls or, equivalently, the walls are streamlines) and v does not depend on x or z, $v \equiv 0$ everywhere. Thus, the flow is everywhere parallel to the x axis.

At this point, we know that $v = 0$ everywhere, $w = 0$ everywhere, and u is a function of y only, since it does not depend on x, z, or t. We can also neglect the body forces. Thus, all the terms in equations (2.12b) and (2.12c) are zero and need not be considered further. Expanding the acceleration term of equation (2.12a), we obtain

$$\rho \frac{\partial u}{\partial t} + \rho u \frac{\partial u}{\partial x} + \rho v \frac{\partial u}{\partial y} + \rho w \frac{\partial u}{\partial z} = \rho f_x - \frac{\partial p}{\partial x} + \frac{\partial}{\partial x} \left(2\mu \frac{\partial u}{\partial x} - \frac{2}{3} \mu \, \nabla \cdot \vec{V} \right)$$
$$+ \frac{\partial}{\partial y} \left[\mu \left(\frac{\partial u}{\partial y} + \frac{\partial v}{\partial x} \right) \right] + \frac{\partial}{\partial z} \left[\mu \left(\frac{\partial u}{\partial z} + \frac{\partial w}{\partial x} \right) \right]$$

(2.14)

Because we are considering low-speed flow of a simple fluid, μ is constant throughout the flow field. Thus, we can rewrite the viscous terms of this equation as follows:

$$\frac{\partial}{\partial x} \left(2\mu \frac{\partial u}{\partial x} - \frac{2}{3} \mu \, \nabla \cdot \vec{V} \right) + \frac{\partial}{\partial y} \left[\mu \left(\frac{\partial u}{\partial y} + \frac{\partial v}{\partial x} \right) \right] + \frac{\partial}{\partial z} \left[\mu \left(\frac{\partial u}{\partial z} + \frac{\partial w}{\partial x} \right) \right]$$
$$= \mu \frac{\partial^2 u}{\partial x^2} + \mu \frac{\partial}{\partial x} \left(\frac{\partial u}{\partial x} \right) - \frac{2}{3} \mu \frac{\partial}{\partial x} (\nabla \cdot \vec{V}) + \mu \frac{\partial^2 u}{\partial y^2}$$
$$+ \mu \frac{\partial}{\partial x} \left(\frac{\partial v}{\partial y} \right) + \mu \frac{\partial^2 u}{\partial z^2} + \mu \frac{\partial}{\partial x} \left(\frac{\partial w}{\partial z} \right)$$
$$= \mu \left(\frac{\partial^2 u}{\partial x^2} + \frac{\partial^2 u}{\partial y^2} + \frac{\partial^2 u}{\partial z^2} \right) + \mu \frac{\partial}{\partial x} \left(\frac{\partial u}{\partial x} + \frac{\partial v}{\partial y} + \frac{\partial w}{\partial z} \right) - \frac{2}{3} \mu \frac{\partial}{\partial x} (\nabla \cdot \vec{V})$$

Noting that

$$\frac{\partial u}{\partial x} + \frac{\partial v}{\partial y} + \frac{\partial w}{\partial z} = 0$$

and that $\nabla \cdot \vec{V} = 0$ (since these are two ways of writing the continuity equation for a constant density flow), we can write equation (2.14) as

$$\rho \frac{\partial u}{\partial t} + \rho u \frac{\partial u}{\partial x} + \rho v \frac{\partial u}{\partial y} + \rho w \frac{\partial u}{\partial z} = \rho f_x - \frac{\partial p}{\partial x} + \mu \left(\frac{\partial^2 u}{\partial x^2} + \frac{\partial^2 u}{\partial y^2} + \frac{\partial^2 u}{\partial z^2} \right) \qquad (2.15)$$

However,

$$\rho \frac{\partial u}{\partial t} = 0 \qquad\qquad\qquad \text{because the flow is steady}$$

$$\rho u \frac{\partial u}{\partial x} = 0 \qquad\qquad\qquad \text{because } u = u \ (y \text{ only})$$

$$\rho v \frac{\partial u}{\partial y} = 0 \qquad\qquad\qquad \text{because } v = 0$$

$$\rho w \frac{\partial u}{\partial z} = \mu \frac{\partial^2 u}{\partial x^2} = \mu \frac{\partial^2 u}{\partial z^2} = 0 \qquad \text{because } u = u \, (y \text{ only})$$

$$\rho f_x = 0 \qquad\qquad\qquad\qquad \text{because body forces are negligible}$$

Thus,

$$0 = -\frac{\partial p}{\partial x} + \mu \frac{\partial^2 u}{\partial y^2}$$

$$0 = -\frac{\partial p}{\partial y}$$

$$0 = -\frac{\partial p}{\partial z}$$

These equations require that the pressure is a function of x only. Recall that u is a function of y only. These two statements can be true only if

$$\mu \frac{d^2 u}{dy^2} = \frac{dp}{dx} = \text{constant}$$

Integrating twice gives

$$u = \frac{1}{2\mu} \frac{dp}{dx} y^2 + C_1 y + C_2$$

To evaluate the constants of integration, we apply the viscous-flow boundary condition that the fluid particles at a solid surface move with the same speed as the surface (i.e., do not slip relative to the surface). Thus,

$$\text{At } y = -\frac{h}{2} \qquad u = 0$$

$$\text{At } y = +\frac{h}{2} \qquad u = 0$$

When we do this, we find that

$$C_1 = 0$$

$$C_2 = -\frac{1}{2\mu} \frac{dp}{dx} \frac{h^2}{4}$$

so that

$$u = +\frac{1}{2\mu} \frac{dp}{dx} \left(y^2 - \frac{h^2}{4} \right)$$

The velocity profile is parabolic, with the maximum velocity at the center of the channel. The shear stress at the two walls is

$$\tau = \mu \frac{du}{dy} \bigg|_{y = \pm h/2} = +\frac{h}{2} \frac{dp}{dx}$$

The pressure must decrease in the x direction (i.e., dp/dx must be negative) to have a velocity in the direction shown. The negative, or favorable, pressure gradient results because of the viscous forces.

Examination of the integral momentum equation (2.13) verifies that a change in pressure must occur to balance the shear forces. Let us verify this by using equation (2.13) on the control volume shown in Fig. 2.9.

Since the flow is fully developed, the positive momentum efflux at station 2 is balanced by the negative momentum influx at station 1:

$$\oiint \vec{V}(\rho \vec{V} \cdot \hat{n} \, dA) = 0$$

Thus, for this steady flow with negligible body forces,

$$\vec{F}_{\text{surface}} = 0$$

or

$$p_1 h - p_2 h + 2\tau \, \Delta x = 0$$

where the factor of 2 accounts for the existence of shear forces at the upper and lower walls. Finally, as shown in the approach using the differential equation,

$$\tau = +\frac{p_2 - p_1}{\Delta x}\frac{h}{2} = +\frac{dp}{dx}\frac{h}{2}$$

Note, as discussed above, that there are subtle implications regarding the signs in these terms, since for the velocity profile shown, the shear acts to retard the fluid motion and the pressure must decrease in the x direction.

Example 2.3

A steady, uniform ($\vec{V} = U_\infty \hat{\imath}$), low-speed flow approaches a very thin, flat-plate "airfoil" whose length is c. Because of viscosity, the flow near the plate slows, such that velocity measurements at the trailing edge of the plate indicate that the x-component of the velocity (above the plate) varies as

$$u = U_\infty \left(\frac{y}{\delta}\right)^{1/7}$$

Below the plate, the velocity is a mirror image of this profile. The pressure was found to be uniform over the entire control surface. Neglecting the body forces, what is the drag coefficient for this flow if $\delta = 0.01c$? The drag coefficient is the drag per unit span (unit depth into the page) divided by the free-stream dynamic pressure times the reference area per unit span (which is the chord length c):

$$C_d = \frac{d}{\frac{1}{2}\rho_\infty U_\infty^2 c}$$

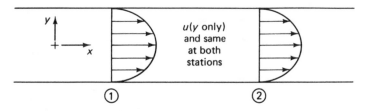

Figure 2.9 Another flow diagram for Example 2.2.

Solution   Let us apply the integral form of the momentum equation [i.e., equation (2.13)] to this flow. Noting that the momentum equation is a vector equation, let us consider the x component only, since that is the direction in which the drag force acts. Furthermore, since the flow is planar symmetric, only the control volume above the plate will be considered (i.e., from $y = 0$ to $y = \delta$). Since the pressure is uniform over the control surface and since the body forces are negligible, the only force acting on the fluid in the control volume is the retarding force of the plate on the fluid, which is $-d/2$.

Because the flow is steady, the first term on the right-hand side of equation (2.13) is zero. Thus,

$$ -\frac{d}{2} = \oiint (\rho \vec{V} \cdot \hat{n}\, dA)V_x $$

Noting that $V_x \equiv u$ and using the values of which are shown in Fig. 2.10, we obtain

$$ -\frac{d}{2} = \int_0^\delta [\rho(U_\infty \hat{i}) \cdot (-\hat{i}\, dy)]U_\infty + \int_0^c [\rho(U_\infty \hat{i} + v_e \hat{j}) \cdot (\hat{j}\, dx)]U_\infty $$

$$ \underbrace{\qquad\qquad}_{(1)} \qquad\qquad \underbrace{\qquad\qquad}_{(2)} $$

$$ + \int_0^\delta \left\{ \rho \left[U_\infty \left(\frac{y}{\delta}\right)^{1/7} \hat{i} + v\hat{j} \right] \cdot (\hat{i}\, dy) \right\} U_\infty \left(\frac{y}{\delta}\right)^{1/7} $$

$$ \underbrace{\qquad\qquad\qquad}_{(3)} $$

A comment about each of the terms describing the momentum efflux from the control volume: (1) Since $\hat{n}\, dA$ is a vector normal to the surface dA which is positive when pointing outward from the enclosed volume, $\hat{n}$ is in the $-x$ direction and dA per unit span is equal to dy. (2) Because viscosity slows the air particles near the plate, line 2 is not a streamline, and the velocity vector along this surface is

$$ \vec{V} = U_\infty \hat{i} + v_e \hat{j} $$

The volumetric efflux (per unit depth) across this surface is $\int_0^c v_e\, dx$, which will be represented by the symbol Q_2. (3) Because of viscosity, the velocity vector has a y component, which is a function of y. However, this y component of velocity does not transport fluid across the area $\hat{i}\, dy$. (4) There is no flow across this surface of the control volume since it is at a solid wall. Furthermore, because the flow is low speed, the density will be assumed constant, and the momentum equation becomes

$$ -\frac{d}{2} = -\rho U_\infty^2\, \delta + \rho U_\infty Q_2 + \frac{7}{9}\rho U_\infty^2 \delta $$

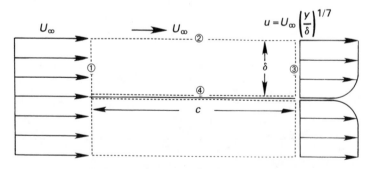

Figure 2.10 Flow diagram for Example 2.3.

To obtain an expression for Q_2, let us use the integral form of the continuity equation, equation (2.5).

$$\oint \rho \vec{V} \cdot \hat{n} \, dA = \int_0^{\delta} [\rho(U_\infty \hat{i}) \cdot (-\hat{i} \, dy)]$$

$$\xleftarrow{\hspace{1cm}} (1) \xrightarrow{\hspace{1cm}}$$

$$+ \int_0^c [\rho(U_\infty \hat{i} + v_e \hat{j}) \cdot (\hat{j} \, dx)] + \int_0^{\delta} \left\{ \rho \left[U_\infty \left(\frac{y}{\delta} \right)^{1/7} \hat{i} + v\hat{j} \right] \cdot (\hat{i} \, dy) \right\} = 0$$

$$\xleftarrow{\hspace{1cm}} (2) \xrightarrow{\hspace{1cm}} \qquad \xleftarrow{\hspace{2cm}} (3) \xrightarrow{\hspace{2cm}}$$

$$- \rho U_\infty \delta + \rho \int_0^c v_e \, dx + \rho U_\infty (\tfrac{7}{8}\delta) = 0 \qquad \int_0^c v_e \, dx = Q_2 = \tfrac{1}{8} U_\infty \delta$$

Substituting this expression into the momentum equation yields

$$-\frac{d}{2} = -\rho U_\infty^2 \delta + \frac{1}{8} \rho U_\infty^2 \delta + \frac{7}{9} \rho U_\infty^2 \delta$$

$$d = \left(2 - \frac{1}{4} - \frac{14}{9} \right) \rho U_\infty^2 \delta = \frac{7}{36} \rho U_\infty^2 \delta$$

$$C_d = \frac{d}{\frac{1}{2} \rho_\infty U_\infty^2 c} = \frac{\frac{7}{36} \rho U_\infty^2 (0.01c)}{\frac{1}{2} \rho_\infty U_\infty^2 c} = 0.00389$$

REYNOLDS NUMBER AND MACH NUMBER AS SIMILARITY PARAMETERS

Because of the difficulty of obtaining theoretical solutions of the flow field around a vehicle, numerous experimental programs have been conducted to measure directly the parameters that define the flow field. Some of the objectives of such test programs are:

1. To obtain information necessary to develop a flow model that could be used in theoretical solutions.
2. To investigate the effect of various geometric parameters on the flow field (such as determining the best location for the engines on a supersonic transport).
3. To verify theoretical predictions of the aerodynamic characteristics for a particular configuration.
4. To measure directly the aerodynamic characteristics of a complete vehicle.

Usually, either scale models of the complete vehicle or large-scale simulations of elements of the vehicle (such as the wing section) have been used in these wind-tunnel programs. Furthermore, in many test programs, the free-stream conditions (such as the velocity, the static pressure, etc.) for the wind-tunnel tests were not equal to the values for the flight condition that was to be simulated.

It is important, then, to determine under what conditions the experimental results obtained for one flow are applicable to another flow which is confined by boundaries that are geometrically similar (but of different size). To do this, consider the *x*-momentum equation as applied to the two flows of Fig. 2.11. For simplicity, let us limit ourselves to constant-property flows. Since the body-force term is usually negligible in aerodynamic problems, equation (2.15) can be written

$$\rho \frac{\partial u}{\partial t} + \rho u \frac{\partial u}{\partial x} + \rho v \frac{\partial u}{\partial y} + \rho w \frac{\partial u}{\partial z} = -\frac{\partial p}{\partial x} + \mu \frac{\partial^2 u}{\partial x^2} + \mu \frac{\partial^2 u}{\partial y^2} + \mu \frac{\partial^2 u}{\partial z^2} \qquad (2.16)$$

Let us divide each of the thermodynamic properties by the value of that property at a point far from the vehicle (i.e., the free-stream value of the property) for each of the two flows. Thus, for the first flow,

$$p_1^* = \frac{p}{P_{\infty,1}} \qquad \rho_1^* = \frac{\rho}{\rho_{\infty,1}} \qquad \mu_1^* = \frac{\mu}{\mu_{\infty,1}}$$

and for the second flow,

$$p_2^* = \frac{p}{P_{\infty,2}} \qquad \rho_2^* = \frac{\rho}{\rho_{\infty,2}} \qquad \mu_2^* = \frac{\mu}{\mu_{\infty,2}}$$

Note that the free-stream values for all three nondimensionalized (*) thermodynamic properties are unity for both cases. Similarly, let us divide the velocity components by the free-stream velocity. Thus, for the first flow,

$$u_1^* = \frac{u}{U_{\infty,1}} \qquad v_1^* = \frac{v}{U_{\infty,1}} \qquad w_1^* = \frac{w}{U_{\infty,1}}$$

and for the second flow,

$$u_2^* = \frac{u}{U_{\infty,2}} \qquad v_2^* = \frac{v}{U_{\infty,2}} \qquad w_2^* = \frac{w}{U_{\infty,2}}$$

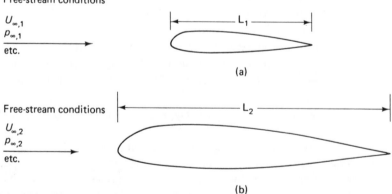

Free-stream conditions

$U_{\infty,1}$
$p_{\infty,1}$
etc.

$\overleftarrow{} L_1 \overrightarrow{}$

(a)

Free-stream conditions

$U_{\infty,2}$
$p_{\infty,2}$
etc.

$\overleftarrow{} L_2 \overrightarrow{}$

(b)

Figure 2.11 Flow around geometrically similar (but different size) configurations: (a) first flow; (b) second flow.

With the velocity components thus nondimensionalized, the free-stream boundary conditions are the same for both flows: that is, at points far from the vehicle

$$u_1^* = u_2^* = 1 \quad \text{and} \quad v_1^* = v_2^* = w_1^* = w_2^* = 0$$

A characteristic dimension L is used to nondimensionalize the independent variables.

$$x_1^* = \frac{x}{L_1} \qquad y_1^* = \frac{y}{L_1} \qquad z_1^* = \frac{z}{L_1} \qquad t_1^* = \frac{tU_{\infty,1}}{L_1}$$

and

$$x_2^* = \frac{x}{L_2} \qquad y_2^* = \frac{y}{L_2} \qquad z_2^* = \frac{z}{L_2} \qquad t_2^* = \frac{tU_{\infty,2}}{L_2}$$

In terms of these dimensionless parameters, the x-momentum equation (2.16) becomes

$$\rho_1^* \frac{\partial u_1^*}{\partial t_1^*} + \rho_1^* u_1^* \frac{\partial u_1^*}{\partial x_1^*} + \rho_1^* v_1^* \frac{\partial u_1^*}{\partial y_1^*} + \rho_1^* w_1^* \frac{\partial u_1^*}{\partial z_1^*}$$

$$= -\left(\frac{p_{\infty,1}}{\rho_{\infty,1}U_{\infty,1}^2}\right)\frac{\partial p_1^*}{\partial x_1^*} + \left(\frac{\mu_{\infty,1}}{\rho_{\infty,1}U_{\infty,1}L_1}\right)\left(\mu_1^* \frac{\partial^2 u_1^*}{\partial x_1^{*2}} + \mu_1^* \frac{\partial^2 u_1^*}{\partial y_1^{*2}} + \mu_1^* \frac{\partial^2 u_1^*}{\partial z_1^{*2}}\right) \tag{2.17a}$$

for the first flow. For the second flow,

$$\rho_2^* \frac{\partial u_2^*}{\partial t_2^*} + \rho_2^* u_2^* \frac{\partial u_2^*}{\partial x_2^*} + \rho_2^* v_2^* \frac{\partial u_2^*}{\partial y_2^*} + \rho_2^* w_2^* \frac{\partial u_2^*}{\partial z_2^*}$$

$$= -\left(\frac{p_{\infty,2}}{\rho_{\infty,2}U_{\infty,2}^2}\right)\frac{\partial p_2^*}{\partial x_2^*} + \left(\frac{\mu_{\infty,2}}{\rho_{\infty,2}U_{\infty,2}L_2}\right)\left(\mu_2^* \frac{\partial^2 u_2^*}{\partial x_2^{*2}} + \mu_2^* \frac{\partial^2 u_2^*}{\partial y_2^{*2}} + \mu_2^* \frac{\partial^2 u_2^*}{\partial z_2^{*2}}\right) \tag{2.17b}$$

Both the dependent variables and the independent variables have been nondimensionalized, as indicated by the * quantities. The dimensionless *boundary-condition values* for the dependent variables are the same for the two flows around geometrically similar configurations. As a consequence, the solutions of the two problems in terms of the dimensionless variables will be identical provided that the differential equations are identical. The differential equations will be identical if the parameters in the parentheses have the same values for both problems. In this case, the flows are said to be dynamically similar as well as geometrically similar.

Let us examine the first similarity parameter from equation (2.17),

$$\frac{p_\infty}{\rho_\infty U_\infty^2} \tag{2.18}$$

Recall that for a perfect gas, the equation of state is

$$p_\infty = \rho_\infty R T_\infty$$

and the free-stream speed of sound is given by

$$a_\infty = \sqrt{\gamma R T_\infty}$$

Substituting these relations into equation (2.18) yields

$$\frac{p_\infty}{\rho_\infty U_\infty^2} = \frac{RT_\infty}{U_\infty^2} = \frac{a_\infty^2}{\gamma U_\infty^2} = \frac{1}{\gamma M_\infty^2} \tag{2.19}$$

since $M_\infty = U_\infty/a_\infty$. Thus, the first dimensionless similarity parameter can be interpreted in terms of the free-stream Mach number.

The inverse of the second similarity parameter is written

$$\mathrm{Re}_{\infty,L} = \frac{\rho_\infty U_\infty L}{\mu_\infty} \tag{2.20}$$

which is the *Reynolds number*, a measure of the ratio of inertia forces to viscous forces.

As has been discussed, the free-stream values of the fluid properties, such as the static pressure and the static temperature, are a function of altitude. Thus, once the velocity, the altitude, and the characteristic dimension of the vehicle are defined, the free-stream Mach number and the free-stream Reynolds number can be calculated as a function of velocity and altitude. This has been done using the values presented in Table 1–2. The free-stream Reynolds number is defined by equation (2.20) with the characteristic length L (e.g., the chord of the wing or the diameter of the missile) chosen to be 1.0 m for the correlations of Fig. 2.12. The correlations represent altitudes up to 30 km (9.84×10^4 ft) and velocities up to 2500 km/h (1554 mi/h or 1350 knots). Note that 1 knot $\equiv$ 1 nautical mile per hour.

Example 2.4

An airplane is flying at a Mach number of 2 at an altitude of 40,000 ft. If the characteristic length for the aircraft is 14 ft, what is the velocity in mi/h, and what is the Reynolds number for this flight condition?

Solution   The density, the viscosity, and the speed of sound of the free-stream flow at 40,000 ft can be found in Table 1.2.

$$\mu_\infty = 0.79447\mu_o = 2.9713 \times 10^{-7} \frac{\mathrm{lbf} \cdot \mathrm{s}}{\mathrm{ft}^2}$$

$$\rho_\infty = 0.2471\rho_o = 5.8711 \times 10^{-4} \frac{\mathrm{slug}}{\mathrm{ft}^3}$$

$$a_\infty = 968.08 \text{ ft/s}$$

Since the Mach number is 2.0,

$$U_\infty = M_\infty a_\infty = 2.0(968.08) = 1936.16 \frac{\mathrm{ft}}{\mathrm{s}} \left(\frac{3600\frac{\mathrm{s}}{\mathrm{h}}}{5280 \frac{\mathrm{ft}}{\mathrm{mi}}} \right) = 1320.11 \frac{\mathrm{mi}}{\mathrm{h}}$$

The corresponding Reynolds number is

$$\mathrm{Re}_{\infty,L} = \frac{\rho_\infty U_\infty L}{\mu_\infty} = \frac{\left(5.8711 \times 10^{-4} \frac{\mathrm{lbf} \cdot \mathrm{s}^2}{\mathrm{ft}^4} \right) \left(1936.16 \frac{\mathrm{ft}}{\mathrm{s}} \right) (14 \text{ ft})}{2.9713 \times 10^{-7} \frac{\mathrm{lbs} \cdot \mathrm{s}}{\mathrm{ft}^2}}$$

$$= 5.3560 \times 10^7$$

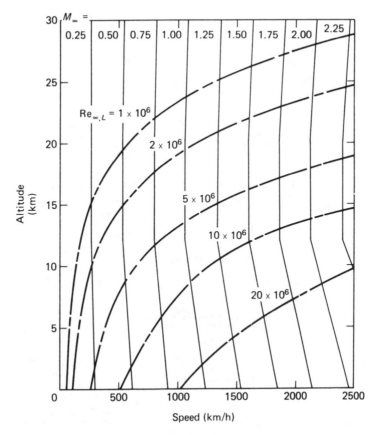

Figure 2.12 Reynolds number/Mach number correlations as a function of velocity and altitude for U.S. Standard Atmosphere.

CONCEPT OF THE BOUNDARY LAYER

For many high-Reynolds-number flows (such as those of interest to the aerodynamicist), the flow field may be divided into two regions: (1) a viscous *boundary layer* adjacent to the surface of the vehicle and (2) the essentially inviscid flow outside the boundary layer. The velocity of the fluid particles increases from a value of zero (in a vehicle-fixed-coordinate system) at the wall to the value that corresponds to the external "frictionless" flow outside the boundary layer, as shown in Fig. 2.13. Because of the resultant velocity gradients, the shear forces are relatively large in the boundary layer. Outside the boundary layer, the velocity gradients become so small that the shear stresses acting on a fluid element are negligible. Thus, the effect of the viscous terms may be ignored in the solution for the flow field external to the boundary layer. To generate a solution for the inviscid portion of the flow field, we require that the velocity of the fluid particles at the surface be parallel to the surface (but not necessarily of zero magnitude). This represents the physical requirement that there is no flow through

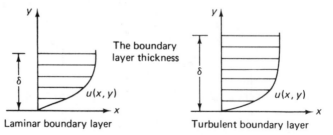

. Relatively thin layer with limited
  mass transfer

. Relatively low velocity gradient near
  the wall

. Relatively low skin friction

. Thicker layer with considerable
  mass transport

. Higher velocities near the surface

. Higher skin friction

The boundary
layer thickness

y

δ

$u(x, y)$

x

Laminar boundary layer

y

δ

$u(x, y)$

x

Turbulent boundary layer

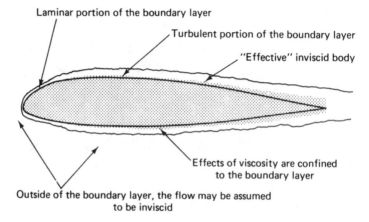

Laminar portion of the boundary layer

Turbulent portion of the boundary layer

"Effective" inviscid body

Effects of viscosity are confined
to the boundary layer

Outside of the boundary layer, the flow may be assumed
to be inviscid

Figure 2.13   Viscous boundary layer on an airfoil.

a solid surface. The analyst may approximate the effects of the boundary layer on the inviscid solution by defining the geometry of the surface to be that of the actual surface plus a displacement due to the presence of the boundary layer. The "effective" inviscid body (the actual configuration plus the displacement thickness) is represented by the shaded area of Fig. 2.13. The solution of the boundary-layer equations and the subsequent determination of a corresponding displacement thickness are dependent on the velocity at the edge of the boundary layer (which is, in effect, the velocity at the surface from the inviscid solution). The process of determining the interaction of the solutions provided by the inviscid-flow equations with those for the boundary-layer equations requires a thorough understanding of the problem (e.g., refer to Ref. 2.1).

For many problems involving flow past streamlined shapes such as airfoils and wings (at low angles of attack), the presence of the boundary layer causes the actual pressure distribution to be only negligibly different from the inviscid pressure distribution. Let us consider this statement further by studying the x and y components of equations (2.12) for a two-dimensional incompressible flow. The resultant equations, which define the flow in the boundary layer shown in Fig. 2.13, are

$$\rho \frac{\partial u}{\partial t} + \rho u \frac{\partial u}{\partial x} + \rho v \frac{\partial u}{\partial y} = - \frac{\partial p}{\partial x} + \mu \frac{\partial^2 u}{\partial x^2} + \mu \frac{\partial^2 u}{\partial y^2}$$

and

$$\rho \frac{\partial v}{\partial t} + \rho u \frac{\partial v}{\partial x} + \rho v \frac{\partial v}{\partial y} = - \frac{\partial p}{\partial y} + \mu \frac{\partial^2 v}{\partial x^2} + \mu \frac{\partial^2 v}{\partial y^2}$$

where the x coordinate is measured parallel to the airfoil surface and the y coordinate is measured perpendicular to it. Solving for the pressure gradients gives us

$$- \frac{\partial p}{\partial x} = \left(\rho \frac{\partial}{\partial t} + \rho u \frac{\partial}{\partial x} + \rho v \frac{\partial}{\partial y} - \mu \frac{\partial^2}{\partial x^2} - \mu \frac{\partial^2}{\partial y^2} \right) u$$

$$- \frac{\partial p}{\partial y} = \left(\rho \frac{\partial}{\partial t} + \rho u \frac{\partial}{\partial x} + \rho v \frac{\partial}{\partial y} - \mu \frac{\partial^2}{\partial x^2} - \mu \frac{\partial^2}{\partial y^2} \right) v$$

Near a solid surface, the normal component of velocity is usually much less than the streamwise component of velocity (i.e., $v < u$). Thus,

$$\frac{\partial p}{\partial y} < \frac{\partial p}{\partial x}$$

The pressure variation across the boundary layer is usually negligible. Therefore, the pressure distribution around the airfoil is essentially that of the inviscid flow (accounting for the displacement effect of the boundary layer).

The assumption that the pressure variation across the boundary layer is negligible breaks down, for turbulent boundary layers at very high Mach numbers. Bushnell et al. (Ref. 2.2) cite data for which the wall pressure is significantly greater than the edge value for turbulent boundary layers where the edge Mach number is approximately 20. The characteristics distinguishing laminar and turbulent boundary layers are discussed in Chapter 4.

When the combined action of an adverse pressure gradient and the viscous forces causes the boundary layer to separate from the vehicle surface (which may occur for blunt bodies or for streamlined shapes at high angles of attack), the flow field is very sensitive to the Reynolds number. The Reynolds number, therefore, also serves as an indicator of how much of the flow can be accurately described by the inviscid-flow equations. For detailed discussions of the viscous portion of the flow field, the reader is referred to Chapter 4 and to Refs. 2.3 and 2.4.

THE CONSERVATION OF ENERGY

There are many flows that involve sufficient temperature variations so that convective heat transfer is important, but for which the constant-property assumption is reasonable. An example is flow in a heat exchanger. For such flows, the temperature field is obtained by solving the energy equation after the velocity field has been determined by solving the continuity equation and the momentum equation.

We must also include the energy equation in the solution algorithm for compressible flows. Compressible flows are those in which the pressure and temperature variations are sufficiently large that we must account for changes in the other fluid properties (e.g., density and viscosity). Recall the discussion relating to equations (2.12a) through (2.12c).

In the remainder of this chapter, we derive the energy equation and discuss its application to various flows.

FIRST LAW OF THERMODYNAMICS

Consider a system of fluid particles. Everything outside the group of particles is called the *surroundings* of the system. The *first law of thermodynamics* results from the fundamental experiments of James Joule. Joule found that, for a cyclic process, that is, one in which the initial state and the final state of the fluid are identical,

$$\oint \delta q - \oint \delta w = 0 \qquad (2.21)$$

Thus, Joule has shown that the heat transferred from the surroundings to the system less the work done by the system on its surroundings during a cyclic process is zero. In equation (2.21), we have adopted the convention that heat transfer to the system is positive and that work done by the system is positive. The use of lowercase symbols to represent the parameters means that we are considering the magnitude of the parameter per unit mass of the fluid. We use the symbols δq and δw to designate that the incremental heat transfer to the system and the work done by the system are not exact differentials but depend on the process used in going from state 1 to state 2. Equation (2.21) is true for any and all cyclic processes. Thus, if we apply it to a process that takes place between any two states (1 and 2), then

$$\delta q - \delta w = de = e_2 - e_1 \qquad (2.22)$$

Note that de is an exact differential and the energy is, therefore, a property of the fluid. The energy is usually divided into three components: (1) kinetic energy, (2) potential energy, and (3) all other energy. The internal energy of the fluid is part of the third component. Since we are normally only concerned with changes in energy rather than its absolute value, an arbitrary zero energy (or datum) state can be assigned.

In terms of the three energy components, equation (2.22) becomes

$$\delta q - \delta w = dke + dpe + du_e \qquad (2.23)$$

Note that u_e is the symbol used for specific internal energy (i.e., the internal energy per unit mass).

Work

In mechanics, work is defined as the effect that is produced by a system on its surroundings when the system moves the surroundings in the direction of the force exerted by the system on its surroundings. The magnitude of the effect is measured by the product of the displacement times the component of the force in the direction of the motion. Thermodynamics deals with phenomena considerably more complex than covered by this definition from mechanics. Thus, we may say that work is done by a system on its surroundings if we can postulate a process in which the system passes through the same series of states as in the original process, but in which the sole effect on the surroundings is the raising of a weight.

In an inviscid flow, the only forces acting on a fluid system (providing we neglect gravity) are the pressure forces. Consider a small element of the surface dA of a fluid system, as shown in Fig. 2.14. The force acting on dA due to the fluid in the system is $p\,dA$. If this force displaces the surface a differential distance ds in the direction of the force, the work done is $p\,dA\,ds$. Differential displacements are assumed so that the process is reversible; that is, there are no dissipative factors such as friction and/ or heat transfer. But the product of dA times ds is just $d(\text{vol})$, the change in volume of the system. Thus, the work per unit mass is

$$\delta w = + p\,dv \tag{2.24a}$$

where v is the volume per unit mass (or specific volume). It is, therefore, the reciprocal of the density. The reader should not confuse this use of the symbol v with the y component of velocity. Equivalently,

$$w = + \int_1^2 p\,dv \tag{2.24b}$$

where the work done by the system on its surroundings in going from state 1 to state 2 (a finite process), as given by equation (2.24b), is positive when dv represents an increase in volume.

DERIVATION OF THE ENERGY EQUATION

Having discussed the first law and its implications, we are now ready to derive the differential form of the energy equation for a viscous, heat-conducting compressible flow. Consider the fluid particles shown in Fig. 2.15. Differentiating equation (2.23) with respect to time, we can describe the energy balance on the particle as it moves along in the flow:

$$\rho\dot{q} - \rho\dot{w} = \rho\frac{d}{dt}(e) = \rho\frac{d}{dt}(ke) + \rho\frac{d}{dt}(pe) + \rho\frac{d}{dt}(u_e) \tag{2.25}$$

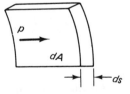

Figure 2.14 Incremental work done by the pressure force.

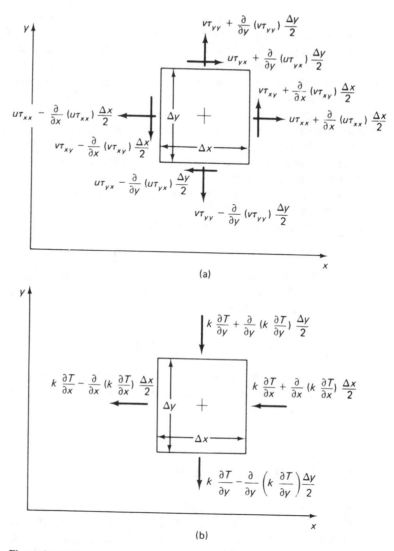

Figure 2.15 Heat-transfer and flow-work terms for the energy equation for a two-dimensional fluid element: (a) work done by stresses acting on a two-dimensional element; (b) heat transfer to a two-dimensional element.

where the overdot notation denotes differentiation with respect to time. Recall that the substantial (or total) derivative is

$$\frac{d}{dt} = \frac{\partial}{\partial t} + \vec{V} \cdot \nabla$$

and therefore represents the local, time-dependent changes, as well as those due to convection through space.

To simplify the illustration of the energy balance on the fluid particle, we shall again consider a two-dimensional flow, as shown in Fig. 2.15. The rate at which work is done by the system on its surrounding is equal to the negative of the product of the forces acting on a boundary surface times the flow velocity (i.e., the displacement per unit time) at that surface. The work done by the body forces is not included in this term. It is accounted for in the potential energy term, see equation (2.27). Thus, using the nomenclature of Fig. 2.15a, we can evaluate the rate at which work is done by the system (per unit depth)

$$-\dot{W} = \frac{\partial}{\partial x}(u\tau_{xx}) \, \Delta x \, \Delta y + \frac{\partial}{\partial x}(v\tau_{xy}) \, \Delta x \, \Delta y + \frac{\partial}{\partial y}(u\tau_{yx}) \, \Delta y \, \Delta x + \frac{\partial}{\partial y}(v\tau_{yy}) \, \Delta y \, \Delta x$$

Using the constitutive relations for τ_{xx}, τ_{xy}, τ_{yx}, and τ_{yy} given earlier in this chapter and dividing by $\Delta x \, \Delta y$, we obtain

$$-\rho \dot{w} = 2\mu \left[\left(\frac{\partial u}{\partial x}\right)^2 + \left(\frac{\partial v}{\partial y}\right)^2 \right] - p \left(\frac{\partial u}{\partial x} + \frac{\partial v}{\partial y}\right) - \frac{2}{3}\mu \, (\nabla \cdot \vec{V})^2$$

$$+ \mu \left[\left(\frac{\partial u}{\partial y} + \frac{\partial v}{\partial x}\right)^2 \right] + u\frac{\partial \tau_{xx}}{\partial x} + v\frac{\partial \tau_{yy}}{\partial y} + v\frac{\partial \tau_{xy}}{\partial x} + u\frac{\partial \tau_{yx}}{\partial y}$$

(2.26a)

From the component momentum equations (2.11),

$$u\frac{\partial \tau_{xx}}{\partial x} + u\frac{\partial \tau_{yx}}{\partial y} = u\rho\frac{du}{dt} - u\rho f_x$$

(2.26b)

$$v\frac{\partial \tau_{xy}}{\partial x} + v\frac{\partial \tau_{yy}}{\partial y} = v\rho\frac{dv}{dt} - v\rho f_y$$

(2.26c)

From Fourier's law of heat conduction,

$$\dot{Q} = -k\hat{n}A \cdot \nabla T$$

we can evaluate the rate at which heat is added to the system (per unit depth). Note that the symbol T will be used to denote temperature, the symbol t, time, and Q, the total heat flux rate. Referring to Fig. 2.15b and noting that, if the temperature is increasing in the outward direction, heat is added to the particle (which is positive by our convention),

$$\dot{Q} = + \frac{\partial}{\partial x}\left(k\frac{\partial T}{\partial x}\right) \Delta x \, \Delta y + \frac{\partial}{\partial y}\left(k\frac{\partial T}{\partial y}\right) \Delta y \, \Delta x$$

Therefore,

$$\rho \dot{q} = \frac{\partial}{\partial x}\left(k\frac{\partial T}{\partial x}\right) + \frac{\partial}{\partial y}\left(k\frac{\partial T}{\partial y}\right)$$

(2.26d)

Substituting equations (2.26) into equation (2.25), we obtain

$$\frac{\partial}{\partial x}\left(k\frac{\partial T}{\partial x}\right) + \frac{\partial}{\partial y}\left(k\frac{\partial T}{\partial y}\right) + 2\mu \left[\left(\frac{\partial u}{\partial x}\right)^2 + \left(\frac{\partial v}{\partial y}\right)^2 \right] - p\nabla \cdot \vec{V}$$

$$- \frac{2}{3}\mu \, (\nabla \cdot \vec{V})^2 + \mu \left[\left(\frac{\partial u}{\partial y} + \frac{\partial v}{\partial x}\right)^2 \right] + \rho u\frac{du}{dt} - \rho u f_x + \rho v\frac{dv}{dt} - \rho v f_x \qquad (2.27)$$

$$= \rho \frac{d[(u^2 + v^2)/2]}{dt} + \rho \frac{d(pe)}{dt} + \rho \frac{d(u_e)}{dt}$$

From the continuity equation,

$$\nabla \cdot \vec{V} = -\frac{1}{\rho}\frac{d\rho}{dt}$$

and by definition

$$\rho\frac{d(p/\rho)}{dt} = \frac{dp}{dt} - \frac{p}{\rho}\frac{d\rho}{dt}$$

Thus,

$$-p\nabla \cdot \vec{V} = -\rho\frac{d(p/\rho)}{dt} + \frac{dp}{dt} \tag{2.28a}$$

For a conservative force field,

$$\rho\frac{d(pe)}{dt} = \rho\vec{V} \cdot \nabla F = -\rho u f_x - \rho v f_y \tag{2.28b}$$

Substituting equations (2.28) into equation (2.27), we obtain

$$\frac{\partial}{\partial x}\left(k\frac{\partial T}{\partial x}\right) + \frac{\partial}{\partial y}\left(k\frac{\partial T}{\partial y}\right) + 2\mu\left[\left(\frac{\partial u}{\partial x}\right)^2 + \left(\frac{\partial v}{\partial y}\right)^2\right] - \frac{2}{3}\mu\,(\nabla \cdot \vec{V})^2$$

$$+ \mu\left[\left(\frac{\partial u}{\partial y} + \frac{\partial v}{\partial x}\right)^2\right] - \rho\frac{d(p/\rho)}{dt} + \frac{dp}{dt} + \rho u\frac{du}{dt} + \rho v\frac{dv}{dt} - \rho u f_x \tag{2.29}$$

$$- \rho v f_y = \rho u\frac{du}{dt} + \rho v\frac{dv}{dt} - \rho u f_x - \rho v f_y + \rho\frac{d(u_e)}{dt}$$

Since the terms u_e and p/ρ appear as a sum in many flow applications, it is convenient to introduce a symbol for this sum. Let us introduce the definition that

$$h = u_e + \frac{p}{\rho} \tag{2.30}$$

where h is called the specific enthalpy. Using equation (2.30) and combining terms, we can write equation (2.29) as

$$\frac{\partial}{\partial x}\left(k\frac{\partial T}{\partial x}\right) + \frac{\partial}{\partial y}\left(k\frac{\partial T}{\partial y}\right) + 2\mu\left[\left(\frac{\partial u}{\partial x}\right)^2 + \left(\frac{\partial v}{\partial y}\right)^2\right]$$

$$- \frac{2}{3}\mu(\nabla \cdot \vec{V})^2 + \mu\left[\left(\frac{\partial u}{\partial y} + \frac{\partial v}{\partial x}\right)^2\right] = \rho\frac{dh}{dt} - \frac{dp}{dt} \tag{2.31}$$

This is the energy equation for a general, compressible flow in two dimensions. The process can be extended to a three-dimensional flow field to yield

$$\rho\frac{dh}{dt} - \frac{dp}{dt} = \nabla \cdot (k\,\nabla T) + \phi \tag{2.32a}$$

where

$$\phi = -\frac{2}{3}\mu(\nabla \cdot \vec{V})^2 + 2\mu\left[\left(\frac{\partial u}{\partial x}\right)^2 + \left(\frac{\partial v}{\partial y}\right)^2 + \left(\frac{\partial w}{\partial z}\right)^2\right]$$
$$+ \mu\left[\left(\frac{\partial u}{\partial y} + \frac{\partial v}{\partial x}\right)^2 + \left(\frac{\partial v}{\partial z} + \frac{\partial w}{\partial y}\right)^2 + \left(\frac{\partial w}{\partial x} + \frac{\partial u}{\partial z}\right)^2\right]$$

(2.32b)

Equation (2.32b) defines the dissipation function ϕ.

Integral Form of the Energy Equation

The integral form of the energy equation is

$$\dot{Q} - \dot{W} = \iiint \frac{\partial}{\partial t}(\rho e)\, d(\text{vol}) + \oiint e\rho\vec{V} \cdot \hat{n}\, dA$$

(2.33)

That is, the net heat added to the system less the work done by the system is equal to the time rate of change of energy within the control volume plus the net efflux of energy across the system boundary. Note that the heat added to the system is positive. So, too, is the work done by the system. Conversely, heat transferred from the system or work done on the system is negative by this convention.

Energy of the System

As noted earlier, the energy of the system can take a variety of forms. They are usually grouped as follows:

1. *Kinetic energy (ke)*: energy associated with the directed motion of the mass
2. *Potential energy (pe)*: energy associated with the position of the mass in the external field
3. *Internal energy (u_e)*: molecular and atomic energy associated with the internal fields of the mass

Thus, the energy of the system may be written as

$$e = ke + pe + u_e$$

(2.34a)

Let us further examine the terms that comprise the energy of the system. The kinetic energy per unit mass is given by

$$ke = \tfrac{1}{2} V^2$$

(2.34b)

Note that the change in kinetic energy during a process clearly depends only on the initial velocity and final velocity of the system of fluid particles. Assuming that the external force field is that of gravity, the potential energy per unit mass is given by

$$pe = gz$$

(2.34c)

Note that the change in the potential energy depends only on the initial and final elevations. Furthermore, the change in internal energy is a function of the values at the endpoints only.

Substituting equations (2.34) into equation (2.33), we obtain

$$\dot{Q} - \dot{W} = \frac{\partial}{\partial t} \iiint \rho \left(\frac{V^2}{2} + gz + u_e \right) d(\text{vol})$$
$$+ \oiint \rho \left(\frac{V^2}{2} + gz + u_e \right) \vec{V} \cdot \hat{n}\, dA \tag{2.35}$$

It should be noted that, whereas the changes in the energy components are a function of the states, the amount of heat transferred and the amount of work done during a process are path dependent. That is, the changes depend not only on the initial and final states but on the process that takes place between these states.

Let us consider further the work term, $\dot{W}$. For convenience, the work is divided into flow work ($\dot{W}_f$) and shaft work ($\dot{W}_s$).

Flow Work

Flow work is the work done by the pressure forces on the surroundings as the fluid moves through space. Consider flow through the streamtube shown in Fig. 2.16. The pressure p_2 acts over the differential area $\hat{n}_2\, dA_2$ at the right end (i.e., the downstream end) of the control volume. Recall that the pressure is a compressive force acting on the system of particles. Thus, the force acting on the right end surface is $-p_2\hat{n}_2\, dA_2$. In moving the surrounding fluid through the distance $\vec{V}_2\, \Delta t$ for the velocity shown in Fig. 2.16, the system does work on the surroundings (which is positive by our sign convention). Thus, the work done is the dot product of the force times the distance:

$$p_2\hat{n}_2\, dA_2 \cdot \vec{V}_2\, \Delta t$$
$$\dot{W}_{f,2} = p_2\vec{V}_2 \cdot \hat{n}_2\, dA_2 = \frac{p_2}{\rho_2} \rho_2\vec{V}_2 \cdot \hat{n}_2\, dA_2 \tag{2.36a}$$

The positive sign is consistent with the assumed directions of Fig. 2.16; that is, the velocity and the area vectors are in the same direction, and with the convention that work done by the system on the surroundings is positive.

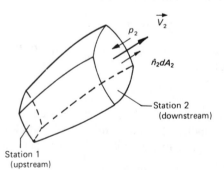

Station 1
(upstream)

Figure 2.16 Streamtube for derivation of equation (2.36).

In a similar manner, it can be shown that the flow work done on the surrounding fluid at the upstream end (station 1) is

$$\dot{W}_{f,1} = -\frac{p_1}{\rho^1} \rho_1 \vec{V}_1 \cdot \hat{n}_1 \, dA_1 \tag{2.36b}$$

The negative sign results because the pressure force is compressive (acts on the system of particles), and the assumed velocity represents movement of the fluid particles in that direction. Thus, since work is done by the surroundings on the system at the upstream end, it is negative by our sign convention.

Shaft Work

Shaft work is defined as any other work done by the system other than the flow work. This usually enters or leaves the system through the action of a shaft (from which the term originates), which either takes energy out of or puts energy into the system. Since a turbine extracts energy from the system, the system does work on the surroundings and $\dot{W}_s$ is positive. In the case where the shaft is that of a pump, the surroundings are doing work on the system and $\dot{W}_s$ is negative.

Thus, the energy equation can be written

$$\dot{Q} - \dot{W}_s = \frac{\partial}{\partial t} \iiint \rho \left(\frac{V^2}{2} + gz + u_e \right) d(\text{vol})$$
$$+ \oiint \rho \left(\frac{V^2}{2} + gz + u_e + \frac{p}{\rho} \right) \vec{V} \cdot \hat{n} \, dA \tag{2.37}$$

Note that the flow work, as represented by equation (2.36) has been incorporated into the second integral of the right-hand side of equation (2.37).

For a steady, adiabatic flow ($\dot{Q} = 0$) with no shaft work ($\dot{W}_s = 0$), equation (2.37) can be written

$$\oiint \rho \left(\frac{V^2}{2} + gz + h \right) \vec{V} \cdot \hat{n} \, dA = 0 \tag{2.38}$$

where the definition for the enthalpy has been used.

Example 2.5

Consider the steady, inviscid, one-dimensional flow of water in the curved pipe shown in Fig. 2.17. If water drains to the atmosphere at station 2 at the rate of 0.001π m³/s, what is the static pressure at station 1? There is no shaft work or heat transfer, and there are no perceptible changes in the internal energy.

Solution   We will apply equation (2.37) to the control volume that encloses the fluid in the pipes between stations 1 and 2. Applying the conditions and assumptions in the problem statement,

$$\text{No heat transfer: } \dot{Q} = 0$$

$$\text{No shaft work: } \dot{W}_s = 0$$

$$\text{Steady flow: } \frac{\partial}{\partial t} = 0$$

Thus,

$$\oint \left(\frac{V^2}{2} + gz + u_e + \frac{p}{\rho} \right) \rho \vec{V} \cdot \hat{n} \, dA = 0$$

Since the properties for the inviscid, one-dimensional flow are uniform over the plane of each station and since the velocities are perpendicular to the cross-sectional area, the integral can readily be evaluated. It is

$$\left(\frac{V_1^2}{2} + gz_1 + u_{e1} + \frac{p_1}{\rho_1} \right) \rho_1 V_1 A_1 = \left(\frac{V_2^2}{2} + gz_2 + u_{e2} + \frac{p_2}{\rho_2} \right) \rho_2 V_2 A_2$$

By continuity, $\rho_1 V_1 A_1 = \rho_2 V_2 A_2$. Since water is incompressible, $\rho_1 = \rho_2$. Furthermore, we are told that there are no perceptible changes in the internal energy (i.e., $u_{e1} = u_{e2}$). Thus,

$$\frac{V_1^2}{2} + gz_1 + \frac{p_1}{\rho} = \frac{V_2^2}{2} + gz_2 + \frac{p_2}{\rho}$$

Note that the resultant form of the energy equation for this flow is Bernoulli's equation.

$$V_2 = \frac{Q}{A_2} = \frac{0.001\pi}{[\pi(0.02)^2]/4} = 10 \text{ m/s}$$

$$V_1 = \left(\frac{D_2}{D_1} \right)^2 V_2 = 0.16(10) = 1.6 \text{ m/s}$$

Thus,

$$\frac{(1.6)^2}{2} + 9.8066(0.3) + \frac{p_1}{1000} = \frac{10^2}{2} + \frac{p_{atm}}{1000}$$

$$p_1 = p_{atm} + (50 - 1.28 - 2.94)1000$$
$$= 4.58 \times 10^4 \text{ N/m}^2, \text{ gage}$$

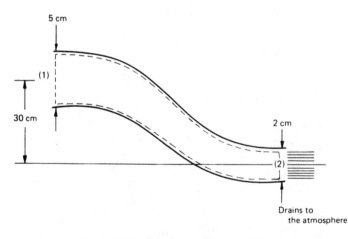

Figure 2.17 Pipe flow for Example 2.5.

SUMMARY

Both the differential and integral forms of the equation of motion for a compressible, viscous flow have been developed in this chapter. Presented were: the continuity equation, equation (2.2) or (2.5); the momentum equation, equation (2.12) or (2.13); and the energy equation, equation (2.32) or (2.33). The dependent variables, or unknowns, in these equations include pressure, temperature, velocity, density, viscosity, thermal conductivity, internal energy, and enthalpy. For many applications, simplifying assumptions can be introduced to eliminate one or more of the dependent variables. For example, a common assumption is that the variation in the fluid properties as the fluid moves through the flow field are very small when the Mach number is less than 0.4. Thus, assuming that the density, the viscosity, and the thermal conductivity are constant, we can eliminate them as variables and still obtain solutions of suitable accuracy. Examples of constant-property flows will be worked out in subsequent chapters.

PROBLEMS

2.1. Derive the continuity equation in cylindrical coordinates, starting with the general vector form

$$\frac{\partial \rho}{\partial t} + \nabla \cdot (\rho \vec{V}) = 0$$

where

$$\nabla = \hat{e}_r \frac{\partial}{\partial r} + \frac{\hat{e}_\theta}{r} \frac{\partial}{\partial \theta} + \hat{e}_z \frac{\partial}{\partial z}$$

in cylindrical coordinates.

$$\frac{1}{r} \frac{\partial (\rho r v_r)}{\partial r} + \frac{1}{r} \frac{\partial (\rho v_\theta)}{\partial \theta} + \frac{\partial (\rho v_z)}{\partial z} = 0$$

2.2. Which of the following flows are physically possible, that is, satisfy the continuity equation? Substitute the expressions for density and for the velocity field into the continuity equation to substantiate your answer.

(a) Water, which has a density of 1.0 g/cm³, is flowing radially outward from a source in a plane such that $\vec{V} = (K/2\pi r)\hat{e}_r$. Note that $v_\theta = v_z = 0$. Note also that, in cylindrical coordinates,

$$\nabla = \hat{e}_r \frac{\partial}{\partial r} + \frac{\hat{e}_\theta}{r} \frac{\partial}{\partial \theta} + \hat{e}_z \frac{\partial}{\partial z}$$

(b) A gas is flowing at relatively low speeds (so that its density may be assumed constant) where

$$u = -\frac{2xyz}{(x^2 + y^2)^2} U_\infty L$$

$$v = \frac{(x^2 - y^2)z}{(x^2 + y^2)^2} U_\infty L$$

$$w = \frac{y}{x^2 + y^2} U_\infty L$$

Here U_∞ and L are a reference velocity and a reference length, respectively.

2.3. For the two-dimensional flow of incompressible air near the surface of a flat plate, the streamwise (or x) component of the velocity may be approximated by the relation

$$u = a_1 \frac{y}{\sqrt{x}} - a_2 \frac{y^3}{x^{1.5}}$$

Using the continuity equation, what is the velocity component v in the y direction? Evaluate the constant of integration by noting that $v = 0$ at $y = 0$.

2.4. Consider a one-dimensional steady flow along a streamtube. Differentiate the resultant integral continuity equation to show that

$$\frac{d\rho}{\rho} + \frac{dA}{A} + \frac{dV}{V} = 0$$

For a low-speed, constant-density flow, what is the relation between the change in area and the change in velocity?

2.5. Water flows through a circular pipe, as shown in Fig. P2.5, at a constant volumetric flow rate of 0.5 m³/s. Assuming that the velocities at stations 1, 2, and 3 are uniform across the cross section (i.e., that the flow is one-dimensional), use the integral form of the continuity equation to calculate the velocities, U_1, U_2, and U_3. The corresponding diameters are $d_1 = 0.4$ m, $d_2 = 0.2$ m, and $d_3 = 0.6$ m.

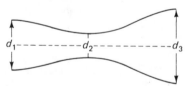

Figure P2.5

2.6. A long pipe (with a reducer section) is attached to a large tank, as shown in Fig. P2.6. The diameter of the tank is 5.0 m; the diameter of the pipe is 20 cm at station 1 and

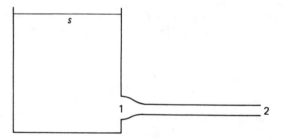

Figure P2.6

10 cm at station 2. The effects of viscosity are such that the velocity (u) may be considered constant across the cross section at the surface (s) and at station 1, but varies with the radius at station 2 such that

$$u = U_0 \left(1 - \frac{r^2}{R_2^2} \right)$$

where U_0 is the velocity at the centerline, R_2 the radius of the pipe at station 2, and r the radial coordinate. If the density is 0.85 g/cm³ and the mass flow rate is 10 kg/s, what are the velocities at s and 1, and what is the value of U_0?

A note for Problems 2.7 through 2.10 and 2.18 through 2.22. The drag force acting on an airfoil can be calculated by determining the change in the momentum of the fluid as it flows past the airfoil. As part of this exercise, one measures the velocity distribution well upstream of the airfoil and well downstream of the airfoil. The integral equations of motion can be applied to either a rectangular control volume or a control volume bounded by streamlines, as discussed in Problems 2.7 through 2.10 and 2.18 through 2.22.

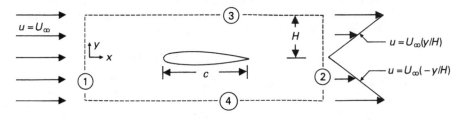

Figure P2.7

2.7. Velocity profiles are measured at the upstream end (surface 1) and at the downstream end (surface 2) of a rectangular control volume, as shown in Fig. P2.7. If the flow is incompressible, two-dimensional, and steady, what is the total volumetric flow rate $\left(\int \vec{V} \cdot \hat{n} \, dA \right)$ across the horizontal surfaces (surfaces 3 and 4)?

2.8. Velocity profiles are measured at the upstream end (surface 1) and at the downstream end (surface 2) of the control volume shown in Fig. P2.8. The flow is incompressible, two-dimensional, and steady. If surfaces 3 and 4 are streamlines, what is the vertical dimension of the upstream station (H_u)?

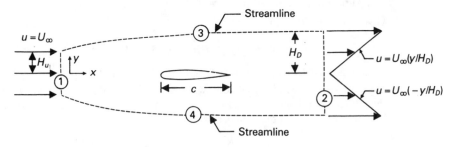

Figure P2.8

2.9. Velocity profiles are measured at the upstream end (surface 1) and at the downstream end (surface 2) of a rectangular control volume, as shown in Fig. P2.9. If the flow is incompressible, two-dimensional, and steady, what is the total volumetric flow rate $\left(\int \vec{V} \cdot \hat{n} \, dA \right)$ across the horizontal surfaces (surfaces 3 and 4)?

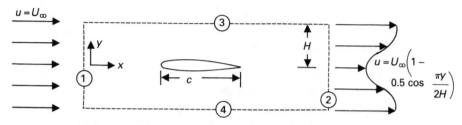

Figure P2.9

2.10. Velocity profiles are measured at the upstream end (surface 1) and at the downstream end (surface 2) of the control volume shown in Fig. P2.10. The flow is incompressible, two-dimensional, and steady. If surfaces 3 and 4 are streamlines, what is the vertical dimension of the upstream station (H_u)?

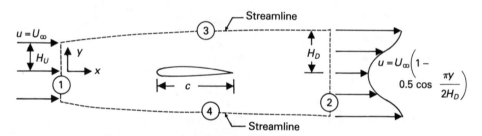

Figure P2.10

2.11. Given the velocity field

$$\vec{V} = - \frac{2xyz}{(x^2 + y^2)^2} U_\infty L \hat{i} + \frac{(x^2 - y^2)z}{(x^2 + y^2)^2} U_\infty L \hat{j} + \frac{y}{x^2 + y^2} U_\infty L \hat{k}$$

in a compressible flow where $\rho = \rho_0 xt$. Using equation (2.8), what is the total acceleration of a particle at $(1, 1, 1)$ at time $t = 10$?

2.12. Given the velocity field

$$\vec{V} = (6 + 2xy + t^2)\hat{i} - (xy^2 + 10t)\hat{j} + 25\hat{k}$$

what is the acceleration of a particle at $(3, 0, 2)$ at time $t = 1$?

2.13. Consider steady two-dimensional flow about a cylinder of radius R (Fig. P2.13). Using cylindrical coordinates, we can express the velocity field for steady, inviscid, incompressible flow around the cylinder as

$$\vec{V}(r,\theta) = U_\infty \left(1 - \frac{R^2}{r^2} \right) \cos \theta \, \hat{e}_r - U_\infty \left(1 + \frac{R^2}{r^2} \right) \sin \theta \, \hat{e}_\theta$$

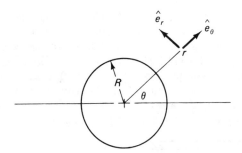

Figure P2.13

where U_∞ is the velocity of the undisturbed stream (and is, therefore, a constant). Derive the expression for the acceleration of a fluid particle at the surface of the cylinder (i.e., at points where $r = R$). Use equation (2.8) and the definition that

$$\nabla = \hat{e}_r \frac{\partial}{\partial r} + \frac{\hat{e}_\theta}{r} \frac{\partial}{\partial \theta} + \hat{e}_z \frac{\partial}{\partial z}$$

and

$$\vec{V} = v_r \hat{e}_r + v_\theta \hat{e}_\theta + v_z \hat{e}_z$$

2.14. Consider the one-dimensional motion of a fluid particle moving on the centerline of the converging channel, as shown in Fig. P2.14. The vertical dimension of the channel (and, thus, the area per unit depth) varies as

$$2y = 2h - h \sin\left(\frac{\pi}{2} \frac{x}{L}\right)$$

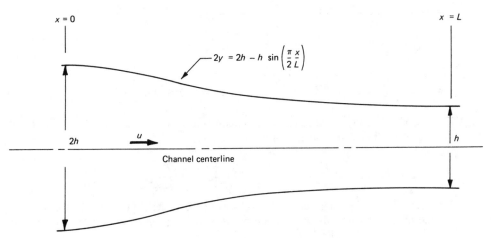

Figure P2.14

Assume that the flow is steady and incompressible. Use the integral form of the continuity equation to describe the velocity along the channel centerline. Also determine the corresponding axial acceleration. If u at $x = 0$ is 2 m/s, h is 1 m, and $L = 1$ m, calculate the acceleration when $x = 0$ and when $x = 0.5L$.

2.15. Consider steady, low-speed flow of a viscous fluid in an infinitely long, two-dimensional channel of height h (i.e., the flow is fully developed; Fig. P2.15). Since this is a low-speed flow, we will assume that the viscosity and the density are constant. Assume the body forces to be negligible. The upper plate (which is at $y = h$) moves in the x direction at the speed U_0, while the lower plate (which is at $y = 0$) is stationary.
  (a) Develop expressions for u, v, and w (which satisfy the boundary conditions) as functions of U_0, h, μ, dp/dx, and y.
  (b) Write the expression for dp/dx in terms of μ, U_0, and h, if $u = 0$ at $y = h/2$.

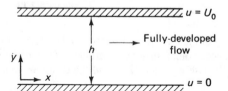

Figure P2.15

2.16. Consider steady, laminar, incompressible flow between two parallel plates, as shown in Fig. P2.16. The upper plate moves at velocity U_0 to the right and the lower plate is stationary. The pressure gradient is zero. The lower half of the region between the plates (i.e., $0 \le y \le h/2$) is filled with fluid with density ρ_1 and viscosity μ_1, and the upper half ($h/2 \le y \le h$) is filled with fluid of density ρ_2 and viscosity μ_2.
  (a) State the condition that the shear stress must satisfy for $0 < y < h$.
  (b) State the conditions that must be satisfied by the fluid velocity at the walls and at the interface of the two fluids.
  (c) Obtain the velocity profile in each of the two regions and sketch the result for $\mu_1 > \mu_2$.
  (d) Calculate the shear stress at the lower wall.

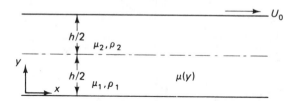

Figure P2.16

2.17. Consider the fully developed flow in a circular pipe, as shown in Fig. P2.17. The velocity u is a function of the radial coordinate only:

$$u = U_{\text{C.L.}} \left(1 - \frac{r^2}{R^2} \right)$$

where $U_{\text{C.L.}}$ is the magnitude of the velocity at the centerline (or axis) of the pipe. Use the integral form of the momentum equation [i.e., equation (2.13)] to show how the

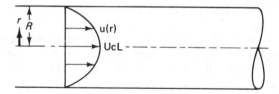

Figure P2.17

pressure drop per unit length dp/dx changes if the radius of the pipe were to be doubled while the mass flux through the pipe is held constant at the value $\dot{m}$. Neglect the weight of the fluid in the control volume and assume that the fluid properties are constant.

2.18. Velocity profiles are measured at the upstream end (surface 1) and at the downstream end (surface 2) of a rectangular control volume, as shown in Fig. P2.18. If the flow is incompressible, two-dimensional, and steady, what is the drag coefficient for the airfoil? The vertical dimension H is $0.025c$ and

$$C_d = \frac{d}{\frac{1}{2}\rho_\infty U_\infty^2 c}$$

The pressure is p_∞ (a constant) over the entire surface of the control volume. (This problem is an extension of Problem 2.7.)

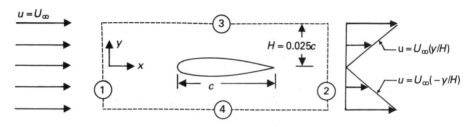

Figure P2.18

2.19. Velocity profiles are measured at the upstream end (surface 1) and at the downstream end (surface 2) of the control volume shown in Fig. P2.19. Surfaces 3 and 4 are streamlines. If the flow is incompressible, two-dimensional, and steady, what is the drag coefficient for the airfoil? The vertical dimension H_D is $0.025c$. The pressure is p_∞ (a constant) over the entire surface of the control volume. (This problem is an extension of Problem 2.8.)

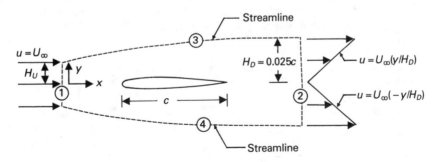

Figure P2.19

2.20. Velocity profiles are measured at the upstream end (surface 1) and at the downstream end (surface 2) of a rectangular control volume, as shown in Fig. P2.20. If the flow is incompressible, two-dimensional, and steady, what is the drag coefficient? The vertical dimension $H = 0.025c$. The pressure is p_∞ (a constant) over the entire control volume. (This problem is a variation of Problem 2.18.)

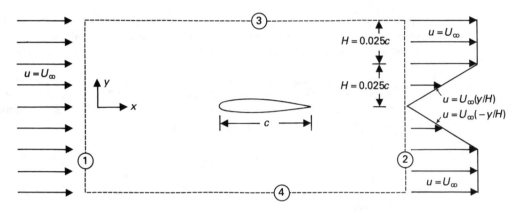

Figure P2.20

At the upstream end (surface 1), $\vec{V} = U_\infty \hat{i}$. At the downstream end (surface 2):

$$0 \leq y \leq H \qquad \vec{V} = \frac{U_\infty y}{H}\hat{i} + v\hat{j}$$

$$H \leq y \leq 2H \qquad \vec{V} = U_\infty\hat{i} + v_0\hat{j}$$

$$-H \leq y \leq 0 \qquad \vec{V} = -\frac{U_\infty y}{H}\hat{i} - v\hat{j}$$

$$-2H \leq y \leq -H \qquad \vec{V} = U_\infty\hat{i} - v_0\hat{j}$$

where $v(x, y)$ and $v_0(x)$ are y components of the velocity which are not measured.

2.21. Velocity profiles are measured at the upstream end (surface 1) and at the downstream end (surface 2) of a rectangular control volume, as shown in Fig. P2.21. If the flow is incompressible, two-dimensional, and steady, what is the drag coefficient for the airfoil? The vertical dimension H is $0.025c$. The pressure is p_∞ (a constant) over the entire surface of the control volume. (This is an extension of Problem 2.9.)

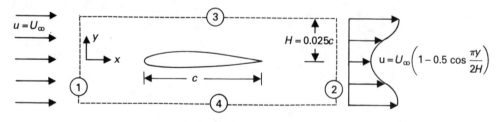

Figure P2.21

2.22. Velocity profiles are measured at the upstream end (surface 1) and at the downstream end (surface 2) of the control volume shown in Fig. P2.22. Surfaces 3 and 4 are streamlines. If the flow is incompressible, two-dimensional, and steady, what is the drag coefficient for the airfoil? The vertical dimension $H_D = 0.025c$. The pressure is p_∞ (a

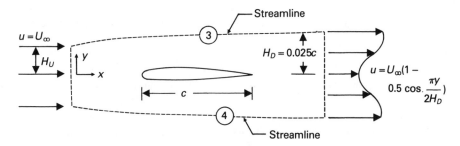

Figure P2.22

constant) over the entire surface of the control volume. (This problem is an extension of Problem 2.10.)

2.23. What are the free-stream Reynolds number [as given by equation 2.20)] and the freestream Mach number [as given by equation (2.19)] for the following flows?
   (a) A golf ball whose characteristic length (i.e., its diameter) is 1.7 in moves through the standard sea-level atmosphere at 200 ft/s.
   (b) A hypersonic transport flies at a Mach number of 6.0 at an altitude of 30 km. The characteristic length of the transport is 32.8 m.

2.24. **(a)** An airplane has a characteristic chord length of 10.4 m. What is the free-stream Reynolds number for the Mach 3 flight at an altitude of 20 km?
   (b) What is the characteristic free-stream Reynolds number of an airplane flying 160 mi/h in a standard sea-level environment? The characteristic chord length is 4.0 ft.

2.25. To illustrate the point that the two integrals in equation (2.21) are path-dependent, consider a system consisting of air contained in a piston/cylinder arrangement. The system of air particles is made to undergo two cyclic processes. Note that all properties (p, T, ρ, etc.) return to their original value (i.e., undergo a net change of zero), since the processes are cyclic.
   (a) Assume that both cycles are reversible and determine (1) $\oint \delta q$ and (2) $\oint \delta w$ for each cycle.
   (b) Describe what occurs physically with the piston/cylinder/air configuration during each leg of each cycle.
   (c) Using the answers to part (a), what is the value of ($\oint \delta q - \oint \delta w$) for each cycle?
   (d) Is the first law satisfied for this system of air particles?

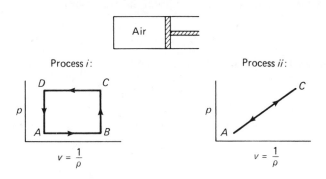

Figure P2.25

2.26. In Problem 2.25, the entropy change in going from A to C directly (i.e., following process ii) is

$$s_C - s_A$$

Going via B (i.e., following process i) the entropy change is

$$s_C - s_A = (s_C - s_B) + (s_B - s_A)$$

(a) Is the net entropy change ($s_C - s_A$) the same for both paths?

(b) Processes AC and ABC were specified to be reversible. What is $s_C - s_A$ if the processes are irreversible? Does $s_C - s_A$ depend on the path if the process is irreversible?

2.27. What assumptions were made in deriving equation (2.32)?

2.28. Show that for an adiabatic, inviscid flow, equation (2.32a) can be written as $ds/dt = 0$.

2.29. Consider the wing-leading edge of a Cessna 172 flying at 130 mi/h through the standard atmosphere at 10,000 ft. Using the integral form of the energy equation for a steady, one-dimensional, adiabatic flow,

$$H_t = h_\infty + \tfrac{1}{2} U_\infty^2$$

For a perfect gas,

$$h_\infty = c_p T_\infty \quad \text{and} \quad H_t = c_p T_t$$

where T_∞ is the free-stream static temperature and T_t is the total (or stagnation point) temperature. If $c_p = 0.2404$ Btu/lbm · °R, compare the total temperature with the free-stream static temperature for this flow. Is convective heating likely to be a problem for this aircraft?

2.30. Consider the wing-leading edge of an SR-71 flying at Mach 3 at 80,000 ft. Using the integral form of the energy equation for a steady, one-dimensional, adiabatic flow,

$$H_t = h_\infty + \tfrac{1}{2} U_\infty^2$$

For a perfect gas

$$h_\infty = c_p T_\infty \quad \text{and} \quad H_t = c_p T_t$$

where T_∞ is the free-stream static temperature and T_t is the total (or stagnation point) temperature. If $c_p = 0.2404$ Btu/lbm · °R, compare the total temperature with the free-stream static temperature for this flow. Is convective heating likely to be a problem for this aircraft?

2.31. Start with the integral form of the energy equation for a one-dimensional, steady, adiabatic flow:

$$H_t = h + \tfrac{1}{2} U^2$$

and the equation for the entropy change for a perfect gas:

$$s - s_t = c_p \ln \frac{T}{T_t} - R \ln \frac{p}{p_t}$$

and develop the expression relating the local pressure to the stagnation pressure:

$$\frac{p}{p_{t1}} = \left(1 + \frac{\gamma - 1}{2} M^2 \right)^{-\gamma/(\gamma - 1)}$$

Carefully note the assumptions made at each step of the derivation. Under what conditions is this valid?

REFERENCES

2.1. Brune, G. W., P. W. Rubbert, and T. C. Nark, Jr., "A New Approach to Inviscid Flow/Boundary Layer Matching," *AIAA Paper 74–601*, presented at the 7th Fluid and Plasma Dynamics Conference, Palo Alto, Calif., June 1974.

2.2. Bushnell, D. M., A. M. Cary, Jr., and J. E. Harris, "Calculation Methods for Compressible Turbulent Boundary Layers—1976," *SP-422*, NASA, 1977.

2.3. Schlichting, H., *Boundary Layer Theory*, McGraw-Hill Book Company, New York, 1968.

2.4. White, F. M., *Viscous Fluid Flow*, McGraw-Hill Book Company, New York, 1974.

3

DYNAMICS OF AN INCOMPRESSIBLE, INVISCID FLOW FIELD

The engineer must be able to determine the flow field about the vehicle of interest with suitable accuracy and speed. Not only should the solution be of reasonable accuracy, but speed of execution is important if the designer is to be able to review a number of different configurations and/or flow conditions. Because of the rapid development of computational hardware and software, approximate numerical solutions of the Navier–Stokes equations have been generated for three-dimensional flows. However, even with the most advanced computers available at this writing, such solutions may require 1,000,000 grid points (see Chapter 12) and more than 50 hours of computational time.

Flow fields for the leeside of highly swept, sharp-leading-edge delta wings in a supersonic flow were numerically simulated (Ref. 3.1) using Navier–Stokes solvers and using Euler solvers (which neglect the viscous terms). For these flows, "the Euler code was seen to be adequate only in predicting primary flow structures (leading edge vortex and cross flow shock), whereas the Navier–Stokes code was capable of predicting secondary flow structures (i.e., secondary vortex)." However, for many applications, flow fields of suitable accuracy can be obtained by solving the Euler equations. For example, numerical solutions for the lift coefficient and for the pitching moment coefficients were generated for a high-speed research model over a range of angle of attack from $-6°$ to $+6°$ (Ref. 3.2). The computed coefficients were in good agreement with test data.

Thus, for many applications, solutions of the inviscid region can provide important design information. Furthermore, once the inviscid flow field has been defined, it can be used as boundary conditions for a thin, viscous boundary layer adjacent to the surface. For the majority of this text, the analysis of the flow field will make use of a two-region flow model (one region in which the viscous forces are negligible and one in which viscous forces cannot be neglected).

INVISCID FLOWS

As noted in Chapter 2, the shearing stresses may be expressed as the product of the viscosity μ times the shearing stress velocity gradient. There are no real fluids for which the viscosity is zero. However, there are many situations where the product of the viscosity times the shearing velocity gradient is sufficiently small that the shear stress terms may be neglected when compared to the other terms in the governing equations. Let us use the term *inviscid flow* to describe the flow in those regions of the flow field where the viscous shear stresses are negligibly small. By using the term *inviscid flow* instead of *inviscid fluid*, we emphasize that we are assuming that the combined product of viscosity and the shearing velocity gradients has a small effect on the flow field and not that the fluid viscosity is zero. In fact, once the solution for the inviscid flow field is obtained, the engineer may want to solve the boundary-layer equations and calculate the skin friction drag on the configuration.

In regions of the flow field where the viscous shear stresses are negligibly small (i.e., in regions where the flow is *inviscid*), equation (2.12) becomes

$$\rho \frac{du}{dt} = \rho f_x - \frac{\partial p}{\partial x} \tag{3.1a}$$

$$\rho \frac{dv}{dt} = \rho f_y - \frac{\partial p}{\partial y} \tag{3.1b}$$

$$\rho \frac{dw}{dt} = \rho f_z - \frac{\partial p}{\partial z} \tag{3.1c}$$

In vector form, the equation is

$$\frac{\overrightarrow{dV}}{dt} = \frac{\overrightarrow{\partial V}}{\partial t} + (\vec{V} \cdot \nabla)\vec{V} = \vec{f} - \frac{1}{\rho} \nabla p \tag{3.2}$$

No assumption has been made about density, so these equations apply to a compressible flow as well as to an incompressible one. These equations, derived in 1755 by Euler, are called the *Euler equations*.

In this chapter we develop fundamental concepts for describing the flow around configurations in a low-speed stream. Let us assume that the viscous boundary layer is thin and therefore has a negligible influence on the inviscid flow field. (The effect of violating this assumption will be discussed when we compare theoretical results with data.) We will seek the solution for the inviscid portion of the flow field (i.e., the flow outside the boundary layer). The momentum equation is Euler's equation.

BERNOULLI'S EQUATION

As has been discussed, the density is essentially constant in the flow field around a vehicle at relatively low speeds. Further, let us consider only body forces that are conservative (such as is the case for gravity),

$$\vec{f} = -\nabla F \tag{3.3}$$

and flows that are steady (or steady state):

$$\frac{\partial \vec{V}}{\partial t} = 0$$

using the vector identity that

$$(\vec{V} \cdot \nabla)\vec{V} = \nabla\left(\frac{U^2}{2}\right) - \vec{V} \times (\nabla \times \vec{V})$$

equation (3.2) becomes (for these assumptions)

$$\nabla\left(\frac{U^2}{2}\right) + \nabla F + \frac{1}{\rho}\nabla p - \vec{V} \times (\nabla \times \vec{V}) = 0 \tag{3.4}$$

In these equations U is the scalar magnitude of the velocity $\vec{V}$. Consideration of the subsequent applications leads us to use U (rather than V).

Let us calculate the change in the magnitude of each of these terms along an arbitrary path whose length and direction are defined by the vector $\vec{dr}$. To do this, we take the dot product of each term in equation (3.4) and the vector $\vec{dr}$. The result is

$$d\left(\frac{U^2}{2}\right) + dF + \frac{dp}{\rho} - \vec{V} \times (\nabla \times \vec{V}) \cdot \vec{dr} = 0 \tag{3.5}$$

Note that, since $\vec{V} \times (\nabla \times \vec{V})$ is a vector perpendicular to $\vec{V}$, the last term is zero (1) for any displacement $\vec{dr}$ if the flow is irrotational or (2) for a displacement along a streamline if the flow is rotational. Thus, for a flow that is:

1. Inviscid,
2. Incompressible,
3. Steady, and
4. Irrotational (or, if the flow is rotational, we consider only displacements along a streamline), and for which
5. The body forces are conservative,

the first integral of Euler's equation is

$$\int d\left(\frac{U^2}{2}\right) + \int dF + \int \frac{dp}{\rho} = \text{constant} \tag{3.6}$$

Since each term involves an exact differential,

$$\frac{U^2}{2} + F + \frac{p}{\rho} = \text{constant} \tag{3.7}$$

The force potential most often encountered is that due to gravity. Let us take the z axis to be positive when pointing upward and normal to the surface of the earth. The force per unit mass due to gravity is directed downward and is of magnitude g. Therefore, referring to equation (3.3),

$$\vec{f} = -\frac{\partial F}{\partial z}\hat{k} = -g\hat{k}$$

so

$$F = gz \tag{3.8}$$

The momentum equation becomes

$$\frac{U^2}{2} + gz + \frac{p}{\rho} = \text{constant} \tag{3.9}$$

Equation (3.9) is known as *Bernoulli's equation*.

Because the density has been assumed constant, it is not necessary to include the energy equation in the procedure to solve for the velocity and pressure fields. In fact, equation (3.9), which is a form of the momentum equation, can be derived from equation (2.37), which is the integral form of the energy equation. As indicated in Example 2.5, there is an interrelation between the total energy of the flow and the flow work. Note that, in deriving equation (3.9), we have assumed that dissipative mechanisms do not significantly affect the flow. As a corollary, Bernoulli's equation is valid only for flows where there is no mechanism for dissipation, such as viscosity. In thermodynamics, the flow process would be called *reversible*.

Note that, if the acceleration is zero throughout the entire flow field, the pressure variation in a static fluid as given by equation (3.9) is identical to that given by equation (1.10). This is as it should be, since the five conditions required for Bernoulli's equation are valid for the static fluid.

For aerodynamic problems, the changes in potential energy are negligible. Neglecting the change in potential energy, equation (3.9) may be written

$$p + \tfrac{1}{2}\rho U^2 = \text{constant} \tag{3.10}$$

This equation establishes a direct relation between the pressure and the velocity. Thus, if either parameter is known, the other can be uniquely determined provided that the flow does not violate the assumptions listed above. The equation can be used to relate the flow at various points around the vehicle, e.g., (1) a point far from the vehicle (i.e., the free stream), (2) a point where the velocity relative to the vehicle is zero (i.e., a stagnation point), and (3) a general point just outside the boundary layer. The nomenclature for these points is illustrated in Fig. 3.1.

$$p_\infty + \tfrac{1}{2}\rho_\infty U_\infty^2 = p_t = p_3 + \tfrac{1}{2}\rho_\infty U_3^2 \tag{3.11}$$

The *stagnation* (or total) *pressure*, which is the constant of equation (3.10), is the sum of the free-stream static pressure (p_∞) and the free-stream *dynamic pressure* ($\tfrac{1}{2}\rho_\infty U_\infty^2$, which is designated by the symbol q_∞).

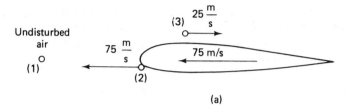

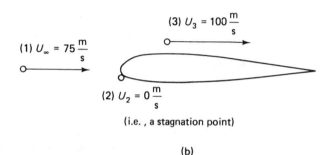

Figure 3.1   Velocity field around an airfoil: (a) ground-fixed coordinate system; (b) vehicle-fixed coordinate system.

Example 3.1

The airfoil of Fig. 3.1a moves through the air at 75 m/s at an altitude of 2 km. The fluid at point 3 moves downstream at 25 m/s relative to the ground-fixed-coordinate system. What are the values of the static pressure at points (1), (2), and (3)?

Solution   To solve this problem, let us superimpose a velocity of 75 m/s to the right so that the airfoil is at rest in the transformed coordinate system. In this vehicle-fixed-coordinate system, the fluid "moves" past the airfoil, as shown in Fig. 3.1b. The velocity at point 3 is 100 m/s relative to the stationary airfoil. The resultant flow is steady. p_∞ is found directly in Table 1.2.

$$\text{Point 1: } p_\infty = \frac{596.309 \text{ mm Hg}}{760.000 \text{ mm Hg/atm}} \left(1.01325 \times 10^5 \frac{\text{N/m}^2}{\text{atm}} \right)$$
$$= 79{,}501 \text{ N/m}^2$$

$$\text{Point 2: } p_t = p_\infty + \tfrac{1}{2}\rho_\infty U_\infty^2$$
$$= 79{,}501 \text{ N/m}^2 + \tfrac{1}{2}(1.0066 \text{ kg/m}^3)(75 \text{ m/s})^2$$
$$= 82{,}332 \text{ N/m}^2$$

$$\text{Point 3: } p_3 + \tfrac{1}{2}\rho_\infty U_3^2 = p_\infty + \tfrac{1}{2}\rho_\infty U_\infty^2$$
$$p_3 = 82{,}332 \text{ N/m}^2 - \tfrac{1}{2}(1.0066 \text{ kg/m}^3)(100 \text{ m/s})^2$$
$$= 77{,}299 \text{ N/m}^2$$

USE OF BERNOULLI'S EQUATION
TO DETERMINE AIRSPEED

Equation (3.11) indicates that a pitot-static probe (see Fig. 3.2) can be used to obtain a measure of the vehicle's airspeed. The pitot head has no internal flow velocity, and the pressure in the pitot tube is equal to the total pressure of the airstream (p_t). The purpose of the static ports is to sense the true static pressure of the free stream (p_∞). When the aircraft is operated through a large angle of attack range, the surface pressure may vary markedly, and, as a result, the pressure sensed at the static port may be significantly different from the free-stream static pressure. The total-pressure and the static-pressure lines can be attached to a differential pressure gage in order to determine the airspeed using the value of the free-stream density for the altitude at which the vehicle is flying:

$$U_\infty = \sqrt{\frac{2(p_t - p_\infty)}{\rho_\infty}}$$

As indicated in Fig. 3.2, the measurements of the local static pressure are often made using an orifice flush-mounted at the vehicle's surface. Although the orifice opening is located on the surface beneath the viscous boundary layer, the static pressure measurement is used to calculate the velocity at the (outside) edge of the boundary layer (i.e., the velocity of the inviscid stream). Nevertheless, the use of Bernoulli's equation, which is valid only for an inviscid flow is appropriate. It is appropriate because (as discussed in Chapter 2) the analysis of the y-momentum equation reveals that the pressure is essentially constant across a thin boundary layer. As a result, the value of the static pressure measured at the wall is essentially equal to the value of the static pressure in the inviscid stream (immediately outside the boundary layer).

There can be many conditions of flight where the airspeed indicator may not reflect the actual velocity of the vehicle relative to the air. The definitions for various terms associated with airspeed are given below.

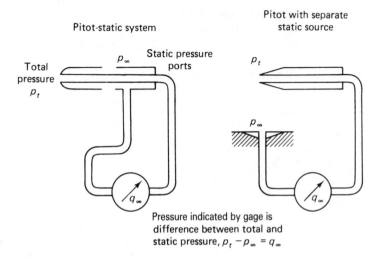

Figure 3.2 Pitot-static probes that can be used to "measure" air speed.

1. *Indicated airspeed (IAS)*. Indicated airspeed is equal to the pitot static airspeed indicator reading as installed in the airplane without correction for airspeed indicator system errors but including the sea-level standard adiabatic compressible flow correction. (The latter correction is included in the calibration of the airspeed instrument dials.)

2. *Calibrated airspeed (CAS)*. CAS is the result of correcting IAS for errors of the instrument and errors due to position or location of the installation. The instrument error may be small by design of the equipment and is usually negligible in equipment that is properly maintained and cared for. The position error of the installation must be small in the range of airspeed involving critical performance conditions. Position errors are most usually confined to the static source in that the actual static pressure sensed at the static port may be different from the free airstream static pressure.

3. *Equivalent airspeed (EAS)*. Equivalent airspeed is equal to the airspeed indicator reading corrected for position error, instrument error, and for adiabatic compressible flow for the particular altitude. The equivalent airspeed (EAS) is the flight speed in the standard sea-level air mass that would produce the same free stream dynamic pressure as the actual flight condition.

4. *True airspeed (TAS)*. The true airspeed results when the EAS is corrected for density altitude. Since the airspeed indicator is calibrated for the dynamic pressures corresponding to air speeds at standard sea-level conditions, variations in air density must be accounted for. To relate EAS and TAS requires consideration that the EAS coupled with standard sea-level density produces the same dynamic pressure as the TAS coupled with the actual air density of the flight condition. From this reasoning, it can be shown that

$$TAS = EAS \sqrt{\frac{\rho_{sl}}{\rho}}$$

where TAS = true airspeed
     EAS = equivalent airspeed
       ρ = actual air density
     ρ_{sl} = standard sea-level air density

TABLE 3.1 DYNAMIC PRESSURE AND EAS AS A FUNCTION OF ALTITUDE AND TAS

	Altitude					
	Sea level $(\rho = 1.0000\rho_{sl})$		10,000 m $(\rho = 0.3376\rho_{sl})$		20,000 m $(\rho = 0.0726\rho_{sl})$	
TAS (km/h)	q_∞ (N/m²)	EAS (km/h)	q_∞ (N/m²)	EAS (km/h)	q_∞ (N/m²)	EAS (km/h)
200	1.89×10^3	200	6.38×10^2	116.2	1.37×10^2	53.9
400	7.56×10^3	400	2.55×10^3	232.4	5.49×10^2	107.8
600	1.70×10^4	600	5.74×10^3	348.6	1.23×10^3	161.6
800	3.02×10^4	800	1.02×10^4	464.8	2.20×10^3	215.5
1000	4.73×10^4	1000	1.59×10^4	581.0	3.43×10^3	269.4

The result shows that the EAS is a function of TAS and density altitude. Table 3.1 presents the EAS and the dynamic pressure as a function of TAS and altitude. The free-stream properties are those of the U.S. standard atmosphere (Ref. 3.3).

THE PRESSURE COEFFICIENT

The engineer often uses experimental data or theoretical solutions for one flow condition to gain insight into the flow field which exists at another flow condition. Wind-tunnel data, where scale models are exposed to flow conditions that simulate the design flight environment, are used to gain insight to describe the full-scale flow field at other flow conditions. Therefore, it is most desirable to present (experimental or theoretical) correlations in terms of dimensionless coefficients which depend only upon the configuration geometry and upon the angle of attack. One such dimensionless coefficient is the *pressure coefficient*:

$$C_p = \frac{p - p_\infty}{\frac{1}{2}\rho_\infty U_\infty^2} = \frac{p - p_\infty}{q_\infty} \tag{3.12}$$

The choice of parameters that are used to nondimensionalize the local static pressure can be seen by referring to Bernoulli's equation, equations (3.10) and (3.11). Rearranging gives us

$$C_p = \frac{p - p_\infty}{\frac{1}{2}\rho_\infty U_\infty^2} = 1 - \frac{U^2}{U_\infty^2} \tag{3.13}$$

Thus, at the stagnation point, where the local velocity is zero, $C_p = C_{p,t} = 1.0$ for an incompressible flow. Note that the stagnation-point value is independent of the free-stream flow conditions or the configuration geometry.

Example 3.2

Consider flow in a subsonic wind tunnel with an open test section; that is, the region where the model is located is open to the room in which the tunnel is located. Thus, as shown in Fig. 3.3, the air accelerates from a reservoir (or stagnation chamber) where the velocity is essentially zero, through a converging nozzle, exhausting into the room (i.e., the test section) in a uniform, parallel stream of velocity U_∞. Using a barometer located on the wall, we know that the barometric pressure in the room is 29.5 in Hg.

The model is a cylinder of infinite span (i.e., the dimension normal to the paper is infinite). There are two pressure orifices flush with the wind-tunnel walls and two orifices flush with the model surface, as shown in Fig. 3.3. The pressure sensed at orifice 3, which is at the stagnation point of the cylindrical model, is $+2.0$ in of water, gage. Furthermore, we know that the pressure coefficient for point 4 is -1.2.

What is the pressure sensed by orifice 1 in the stagnation chamber? What is the pressure sensed by orifice 2, located in the exit plane of the wind-tunnel nozzle? What is the free-stream velocity, U_∞? What is the static pressure at point 4? What is the velocity of an air particle just outside the boundary layer at point 4?

Solution As discussed in Chapter 1, the standard atmospheric pressure is defined as that pressure which can support a column of mercury 760 mm high. It is also equal to 2116.22 lbf/ft². Thus, an equivalency statement may be written:

$$760 \text{ mm Hg} = 29.92 \text{ in Hg} = 2116.22 \text{ lbf/ft}^2$$

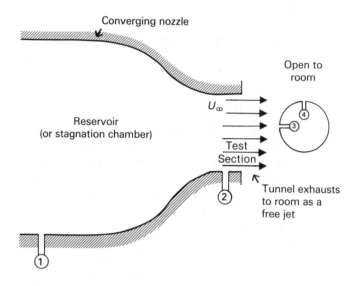

Figure 3.3   Open test-section wind-tunnel used in Example 3.2.

Since the barometric pressure is 29.5 in Hg, the pressure in the room is

$$p_{\text{room}} = 29.5 \text{ in Hg} \left(\frac{2116.22 \text{ lbf/ft}^2/\text{atm}}{29.92 \text{ in Hg/atm}} \right) = 2086.51 \text{ lbf/ft}^2$$

Furthermore, since the nozzle exhausts into the room at subsonic speeds and since the streamlines are essentially straight (so that there is no pressure variation normal to the streamlines), the pressure in the room is equal to the free-stream static pressure for the test section (p_∞) and is equal to the static pressure in the exit plane of the nozzle (p_2).

$$p_{\text{room}} = p_\infty = p_2 = 2086.51 \text{ lbf/ft}^2$$

Because $p_\infty = 2086.51 \text{ lbf/ft}^2$, we will assume that the free-stream density is reduced proportionally:

$$\rho_\infty = (2086.51 \text{ lbf/ft}^2) \left(\frac{0.002376 \text{ slug/ft}^3}{2116.22 \text{ lbf/ft}^2} \right) = 0.00234 \frac{\text{slug}}{\text{ft}^3}$$

Since the pressure measurement sensed by orifice 3 is given as a gage pressure, it is the difference between the stagnation pressure and the free-stream pressure, that is,

$$p_3 - p_\infty = 2 \text{ in H}_2\text{O, gage} \left(\frac{2116.22 \text{ lbf/ft}^2/\text{atm}}{407.481 \text{ in H}_2\text{O/atm}} \right)$$

$$= 10.387 \text{ lbf/ft}^2, \text{ gage}$$

where 407.481 in H_2O is the column of water equivalent to 760 mm Hg if the density of water is 1.937 slugs/ft^3. But since $p_3 = p_t$, we can rearrange Bernoulli's equation:

$$p_3 - p_\infty = p_t - p_\infty = \tfrac{1}{2}\rho_\infty U_\infty^2$$

Equating these expressions and solving for U_∞,

$$U_\infty = \sqrt{\frac{2(10.387 \text{ lbf/ft}^2)}{0.00234 \text{ lbf} \cdot \text{s}^2/\text{ft}^4}} = 94.22 \text{ ft/s}$$

Since $\frac{1}{2}\rho_\infty U_\infty^2$ is the dynamic pressure, we can rearrange the definition for the pressure coefficient to find the static pressure at point 4:

$$p_4 = p_\infty + C_{p_4}q_\infty = 2086.51 + (-1.2)(10.387)$$
$$= 2074.05 \ \text{lbf/ft}^2$$

Since we seek the velocity of the air particles just outside the boundary layer above orifice 4, Bernoulli's equation is applicable, and we can use equation (3.13):

$$\frac{U_4^2}{U_\infty^2} = 1 - C_{p_4} = 2.2$$

Thus, $U_4 = 139.75$ ft/s.

CIRCULATION

The *circulation* is defined as the line integral of the velocity around any closed curve. Referring to the closed curve C of Fig. 3.4, the circulation is given by

$$-\Gamma = \oint_C \vec{V} \cdot \vec{dr} \tag{3.14}$$

where $\vec{V} \cdot \vec{dr}$ is the scalar product of the velocity vector and the differential vector length along the path of integration. As indicated by the circle through the integral sign, the integration is carried out for the complete closed path. The path of the integration is counterclockwise, so the area enclosed by the curve C is always on the left. A negative sign is used in equation (3.14) for convenience in the subsequent application to lifting-surface aerodynamics.

Consider the circulation around a small, square element in the xy plane, as shown in Fig. 3.5a. Integrating the velocity components along each of the sides and proceeding counterclockwise (i.e., keeping the area on the left of the path),

$$-\Delta\Gamma = u \ \Delta x + \left(v + \frac{\partial v}{\partial x} \Delta x\right) \Delta y - \left(u + \frac{\partial u}{\partial y} \Delta y\right) \Delta x - v \ \Delta y$$

Simplifying yields

$$-\Delta\Gamma = \left(\frac{\partial v}{\partial x} - \frac{\partial u}{\partial y}\right) \Delta x \ \Delta y$$

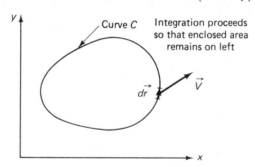

Figure 3.4 Concept of circulation.

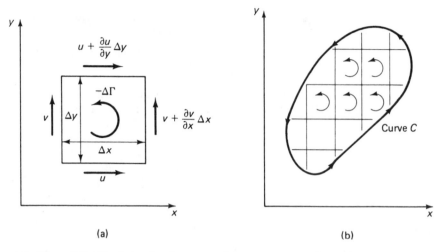

Figure 3.5 Circulation for elementary closed curves: (a) rectangular element; (b) general curve C.

This procedure can be extended to calculate the circulation around a general curve C in the xy plane, such as that of Fig. 3.5b. The result for this general curve in the xy plane is

$$-\Gamma = \oint_C (u\,dx + v\,dy) = \iint_A \left(\frac{\partial v}{\partial x} - \frac{\partial u}{\partial y}\right) dx\,dy \qquad (3.15)$$

Equation (3.15) represents *Green's lemma* for the transformation from a line integral to a surface integral in two-dimensional space. The transformation from a line integral to a surface integral in three-dimensional space is governed by *Stokes's theorem*:

$$\oint_C \vec{V} \cdot \vec{dr} = \iint_A (\nabla \times \vec{V}) \cdot \hat{n}\,dA \qquad (3.16)$$

where $\hat{n}\,dA$ is a vector normal to the surface, positive when pointing outward from the enclosed volume, and equal in magnitude to the surface area (see Fig. 3.6). Note that equation (3.15) is a planar simplification of the more general equation, equation (3.16). In words, the integral of the normal component of the curl of the velocity vector over any surface A is equal to the line integral of the tangential component of

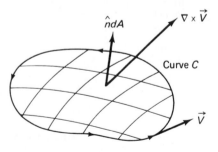

Figure 3.6 Nomenclature for Stokes's theorem.

the velocity around the curve C which bounds A. Stokes's theorem is valid when A represents a simply connected region in which $\vec{V}$ is continuously differentiable. Thus, equation (3.16) is not valid if the area A contains regions where the velocity is infinite.

IRROTATIONAL FLOW

By means of Stokes's theorem it is apparent that, if the curl of $\vec{V}$ (i.e., $\nabla \times \vec{V}$) is zero at all points in the region bounded by C, then the line integral of $\vec{V} \cdot \vec{dr}$ around the closed path is zero. If

$$\nabla \times \vec{V} \equiv 0 \tag{3.17}$$

and the flow contains no singularities, the flow is said to be *irrotational*. Stokes's theorem leads us to the conclusion that

$$\oint_C \vec{V} \cdot \vec{dr} = -\Gamma = 0$$

For this irrotational velocity field, the line integral

$$\int \vec{V} \cdot \vec{dr}$$

is independent of path. A necessary and sufficient condition that

$$\int \vec{V} \cdot \vec{dr}$$

be independent of path is that the curl of $\vec{V}$ is everywhere zero. Thus, its value depends only on its limits. However, a line integral can be independent of the path of integration only if the integrand is an exact differential. Therefore,

$$\vec{V} \cdot \vec{dr} = d\phi \tag{3.18}$$

where $d\phi$ is an exact differential. Expanding equation (3.18) in cartesian coordinates,

$$u\,dx + v\,dy + w\,dz = \frac{\partial\phi}{\partial x}\,dx + \frac{\partial\phi}{\partial y}\,dy + \frac{\partial\phi}{\partial z}\,dz$$

it is apparent that

$$\vec{V} = \nabla\phi \tag{3.19}$$

Thus, a *velocity potential* $\phi(x, y, z)$ exists for this flow such that the partial derivative of ϕ in any direction is the velocity component in that direction. That equation (3.19) is a valid representation of the velocity field for an irrotational flow can be seen by noting that

$$\nabla \times \nabla\phi \equiv 0 \tag{3.20}$$

That is, the curl of any gradient is necessarily zero. Thus, an irrotational flow is termed a *potential flow*.

KELVIN'S THEOREM

Having defined the necessary and sufficient condition for the existence of a flow that has no circulation, let us examine a theorem first demonstrated by Lord Kelvin. In an inviscid, homogeneous flow with conservative body forces, the circulation around a closed fluid line remains constant with respect to time. A homogeneous (or barotropic) fluid is one in which the density depends only on the pressure.

The time derivative of the circulation along a closed fluid line (i.e., a fluid line that is composed of the same fluid particles) is

$$-\frac{d\Gamma}{dt} = \frac{d}{dt}\left(\oint_c \vec{V} \cdot \vec{dr}\right) = \oint_c \frac{\vec{dV}}{dt} \cdot \vec{dr} + \oint_c \vec{V} \cdot \frac{d}{dt}(\vec{dr}) \tag{3.21}$$

Again, the negative sign is used for convenience, as discussed in equation (3.14). Euler's equation, equation (3.2), which is the momentum equation for an inviscid flow yields

$$\frac{\vec{dV}}{dt} = \vec{f} - \frac{1}{\rho}\nabla p$$

Using the constraint that the body forces are conservative (as is true for gravity, the body force of most interest to this text), we have

$$\vec{f} = -\nabla F$$

and

$$\frac{\vec{dV}}{dt} = -\nabla F - \frac{1}{\rho}\nabla p \tag{3.22}$$

where F is the body-force potential. Since we are following a particular fluid particle, the order of time and space differentiation does not matter.

$$\frac{d}{dt}(\vec{dr}) = d\left(\frac{\vec{dr}}{dt}\right) = \vec{dV} \tag{3.23}$$

Substituting equations (3.22) and (3.23) into equation (3.21) yields

$$\frac{d}{dt}\oint_c \vec{V} \cdot \vec{dr} = -\oint_c dF - \oint_c \frac{dp}{\rho} + \oint_c \vec{V} \cdot \vec{dV} \tag{3.24}$$

Since the density is a function of the pressure only, all the terms on the right-hand side involve exact differentials. The integral of an exact differential around a closed curve is zero. Thus,

$$\frac{d}{dt}\left(\oint_c \vec{V} \cdot \vec{dr}\right) = 0 \tag{3.25}$$

Or, as given in the statement of Kelvin's theorem, the circulation remains constant along the closed fluid line for the conservative flow.

Implication of Kelvin's Theorem

If the fluid starts from rest, or if the fluid in some region is uniform and parallel, the rotation in this region is zero. Kelvin's theorem leads to the important conclusion that the entire flow remains irrotational in the absence of viscous forces and of discontinuities provided that the fluid is homogeneous and the body forces can be described by a potential function.

In many flow problems (those of interest to this text), the undisturbed, free-stream flow is a uniform parallel flow in which there are no shear stresses. Kelvin's theorem implies that, although the fluid particles in the subsequent flow patterns may follow curved paths, the flow remains irrotational except in those regions where the dissipative viscous forces are an important factor.

INCOMPRESSIBLE, IRROTATIONAL FLOW

Kelvin's theorem states that for an inviscid flow having a conservative force field, the circulation must be constant around a path that moves so as always to touch the same particles and which contains no singularities. Thus, since the free-stream flow is irrotational, a homogeneous flow around the vehicle will remain irrotational provided that viscous effects are not important. For an irrotational flow, the velocity may be expressed in terms of a potential function;

$$\vec{V} = \nabla \phi \qquad (3.19)$$

For relatively low speed flows (i.e., incompressible flows), the continuity equation is

$$\nabla \cdot \vec{V} = 0 \qquad (2.4)$$

Combining equations (2.4) and (3.19), one finds that for an incompressible, irrotational flow,

$$\nabla^2 \phi = 0 \qquad (3.26)$$

Thus, the governing equation, which is known as *Laplace's equation*, is a linear, second-order partial differential equation.

Boundary Conditions

Within certain constraints on geometric slope continuity, a bounded, simply-connected velocity field is uniquely determined by the distribution on the flow boundaries either of the normal component of the total velocity $\nabla \phi \cdot \hat{n}$ or of the total potential ϕ. These boundary value problems are respectively designated Neumann or Dirichlet problems. For applications in this book, the Neumann formulation will be used since most practical cases involve prescribed normal velocity boundary conditions. Specifically, the flow tangency requirement associated with inviscid flow past a solid body is expressed mathematically as:

$$\nabla \phi \cdot \hat{n} = 0$$

STREAM FUNCTION IN A TWO-DIMENSIONAL, INCOMPRESSIBLE FLOW

Just as the condition of irrotationality is the necessary and sufficient condition for the existence of a velocity potential, so the equation of continuity for an incompressible, two-dimensional flow is the necessary and sufficient condition for the existence of a stream function. The flow need be two-dimensional only in the sense that it requires only two spatial coordinates to describe the motion. Therefore, stream functions exist both for plane flow and for axially symmetric flow. The reader might note, although it is not relevant to this chapter, that stream functions exist for compressible, two-dimensional flows, if they are steady.

Examining the continuity equation for an incompressible, two-dimensional flow in cartesian coordinates:

$$\nabla \cdot \vec{V} = \frac{\partial u}{\partial x} + \frac{\partial v}{\partial y} = 0$$

It is obvious that the equation is satisfied by a stream function ψ, for which the velocity components can be calculated as

$$u = \frac{\partial \psi}{\partial y} \tag{3.27a}$$

$$v = -\frac{\partial \psi}{\partial x} \tag{3.27b}$$

A corollary to this is that the existence of a stream function is a necessary condition for a physically possible flow (i.e., one that satisfies the continuity equation).

Since ψ is a point function,

$$d\psi = \frac{\partial \psi}{\partial x}\, dx + \frac{\partial \psi}{\partial y}\, dy$$

so that

$$d\psi = -v\, dx + u\, dy \tag{3.28}$$

Since a streamline is a curve whose tangent at every point coincides with the direction of the velocity vector, the definition of a streamline in a two-dimensional flow is

$$\frac{dx}{u} = \frac{dy}{v}$$

Rearranging, it can be seen that

$$u\, dy - v\, dx = 0 \tag{3.29}$$

along a streamline. Equating equations (3.28) and (3.29), we find that

$$d\psi = 0$$

along a streamline. Thus, the change in ψ is zero along a streamline or, equivalently, ψ is a constant along a streamline. A corollary statement is that lines of constant ψ are streamlines of the flow. It follows, then, that the volumetric flow rate (per unit depth) between any two points in the flow is the difference between the values of the stream function at the two points of interest.

Referring to Fig. 3.7, it is clear that the product $v(-dx)$ represents the volumetric flow rate per unit depth across AO and the product $u\,dy$ represents the volumetric flow rate per unit depth across OB. By continuity, the fluid crossing lines AO and OB must cross the curve AB. Therefore, $d\psi$ is a measure of the volumetric flow rate per unit depth across AB. A line can be passed through A for which $\psi = \psi_A$ (a constant), while a line can be passed through B for which $\psi = \psi_B = \psi_A + d\psi$ (a different constant). As noted previously, the difference $d\psi$ is the volumetric flow rate (per unit depth) between the two streamlines. The fact that the flow is always tangent to a streamline and has no component of velocity normal to it has an important consequence. Any streamline in an inviscid flow can be replaced by a solid boundary of the same shape without affecting the remainder of the flow pattern.

If the flow is also irrotational,

$$\nabla \times \vec{V} = 0$$

Then writing the velocity components in terms of the stream function, as defined in equation (3.27), we obtain

$$\nabla^2 \psi = 0 \qquad\qquad (3.30)$$

Thus, for an irrotational, two-dimensional incompressible flow, the stream function is also governed by Laplace's equation.

RELATION BETWEEN STREAMLINES AND EQUIPOTENTIAL LINES

If a flow is incompressible, irrotational, and two-dimensional, the velocity field may be calculated using either a potential function or a stream function. Using the potential function, the velocity components in cartesian coordinates are

$$u = \frac{\partial \phi}{\partial x} \qquad v = \frac{\partial \phi}{\partial y}$$

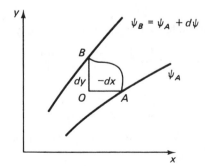

Figure 3.7 The significance of the stream function.

For a potential function,

$$d\phi = \frac{\partial \phi}{\partial x} dx + \frac{\partial \phi}{\partial y} dy = u\, dx + v\, dy$$

Therefore, for lines of constant potential ($d\phi = 0$),

$$\left(\frac{dy}{dx}\right)_{\phi=C} = -\frac{u}{v} \tag{3.31}$$

Since a streamline is everywhere tangent to the local velocity, the slope of a streamline, which is a line of constant ψ, is

$$\left(\frac{dy}{dx}\right)_{\psi=C} = \frac{v}{u} \tag{3.32}$$

Comparing equations (3.31) and (3.32) yields

$$\left(\frac{dy}{dx}\right)_{\phi=C} = -\frac{1}{(dy/dx)_{\psi=C}} \tag{3.33}$$

The slope of an equipotential line is the negative reciprocal of the slope of a streamline. Therefore, streamlines (ψ = constant) are perpendicular to equipotential lines (ϕ = constant), except at stagnation points, where the components vanish simultaneously.

Example 3.3

Consider the incompressible, irrotational, two-dimensional flow, where the stream function is

$$\psi = 2xy$$

(a) What is the velocity at $x = 1$, $y = 1$? At $x = 2$, $y = \frac{1}{2}$? (Note that both points are on the same streamline.)
(b) Sketch the streamline pattern and discuss the significance of the spacing between the streamlines.
(c) What is the velocity potential for this flow?
(d) Sketch the lines of constant potential. How do the lines of equipotential relate to the streamlines?

Solution

(a) The stream function can be used to calculate the velocity components:

$$u = \frac{\partial \psi}{\partial y} = 2x \qquad v = -\frac{\partial \psi}{\partial x} = -2y$$

Therefore,

$$\vec{V} = 2x\hat{\imath} - 2y\hat{\jmath}$$

At $x = 1$, $y = 1$, $\vec{V} = 2\hat{\imath} - 2\hat{\jmath}$, and the magnitude of the velocity is

$$U = 2.8284$$

At $x = 2$, $y = \frac{1}{2}$, $\vec{V} = 4\hat{\imath} - \hat{\jmath}$, and the magnitude of the velocity is

$$U = 4.1231$$

(b) A sketch of the streamline pattern is presented in Fig. 3.8. Results are presented only for the first quadrant (x positive, y positive). Mirror-image patterns would exist in other quadrants. Note that the $x = 0$ and the $y = 0$ axes represent the $\psi = 0$ "streamline."

Since the flow is incompressible and steady, the integral form of the continuity equation (2.5) indicates that the product of the velocity times the distance between the streamlines is a constant. That is, since $\rho = $ constant,

$$\oiint \vec{V} \cdot \hat{n}\, dA = 0$$

Therefore, the distance between the streamlines decreases as the magnitude of the velocity increases.

(c) Since $u = \partial\phi/\partial x$ and $v = \partial\phi/\partial y$,

$$\phi = \int u\, dx + g(y) = \int 2x\, dx + g(y)$$

Also,

$$\phi = \int v\, dy + f(x) = -\int 2y\, dy + f(x)$$

The potential function which would satisfy both of these equations is

$$\phi = x^2 - y^2 + C$$

where C is an arbitrary constant.

(d) The equipotential lines are included in Fig. 3.8, where C, the arbitrary constant, has been set equal to zero. The lines of equipotential are perpendicular to the streamlines.

SUPERPOSITION OF FLOWS

Since equation (3.26) for the potential function and equation (3.30) for the stream function are linear, functions that individually satisfy these (Laplace's) equations may be added together to describe a desired, complex flow. The boundary conditions are that the resultant velocity is equal to the free-stream value at points far from the surface and that the component of the velocity normal to the surface is zero (i.e., the surface is a streamline). There are numerous two-dimensional and axisymmetric solutions available through "inverse" methods. These inverse methods do not begin with a prescribed boundary surface and directly solve for the potential flow, but instead assume a set of known singularities in the presence of an onset flow. The total potential function (or stream function) for the singularities and the onset flow are then used to determine the streamlines, any one of which may be considered to be a "boundary surface." If the resultant boundary surface corresponds to that of the desired configuration, the

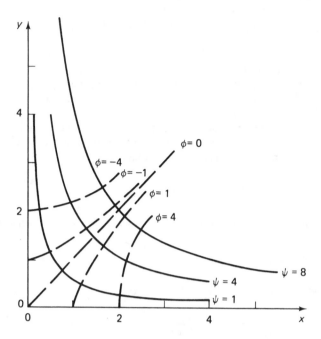

Figure 3.8 Equipotential lines and streamlines for Example 3.3.

desired solution has been obtained. The singularities most often used in such approaches, which were suggested by Rankine in 1871, include a source, a sink, a doublet, and a vortex.

For a constant-density potential flow, the velocity field can be determined using only the continuity equation and the condition of irrotationality. Thus, the equation of motion is not used, and the velocity may be determined independently of the pressure. Once the velocity field has been determined, Bernoulli's equation can be used to calculate the corresponding pressure field. It is important to note that the pressures of the component flows cannot be superimposed (or added together), since they are nonlinear functions of the velocity. Referring to equation (3.10), the reader can see that the pressure is a quadratic function of the velocity.

ELEMENTARY FLOWS

Uniform Flow

The simplest flow is a *uniform stream* moving in a fixed direction at a constant speed. Thus, the streamlines are straight and parallel to each other everywhere in the flow field (see Fig. 3.9). Using a cylindrical coordinate system, the potential function for a uniform flow moving parallel to the x axis is

$$\phi = U_\infty r \cos \theta \tag{3.34}$$

where U_∞ is the velocity of the fluid particles. Using a cartesian coordinate system, the potential function for the uniform stream of Fig. 3.9 is

$$\phi = U_\infty x \tag{3.35}$$

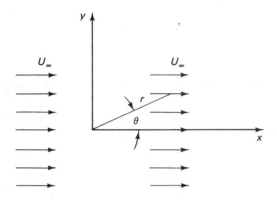

Figure 3.9 Streamlines for a uniform flow parallel to the x axis.

Source or Sink

A *source* is defined as a point from which fluid issues and flows radially outward (see Fig. 3.10) such that the continuity equation is satisfied everywhere but at the singularity that exists at the source's center. The potential function for the two-dimensional (planar) source centered at the origin is

$$\phi = \frac{K}{2\pi} \ln r \tag{3.36}$$

where r is the radial coordinate from the center of the source and K is the source strength. The resultant velocity field in cylindrical coordinates is

$$\vec{V} = \nabla\phi = \hat{e}_r \frac{\partial \phi}{\partial r} + \frac{\hat{e}_\theta}{r} \frac{\partial \phi}{\partial \theta} \tag{3.37}$$

Since

$$\vec{V} = \hat{e}_r v_r + \hat{e}_\theta v_\theta$$

$$v_r = \frac{\partial \phi}{\partial r} = \frac{K}{2\pi r}$$

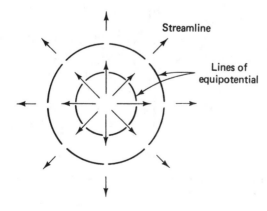

Streamline

Lines of equipotential

Figure 3.10 Equipotential lines and streamlines for flow from a two-dimensional source.

and

$$v_\theta = \frac{1}{r}\frac{\partial \phi}{\partial \theta} = 0$$

Note that the resultant velocity has only a radial component and that this component varies inversely with the radial distance from the source.

A sink is a negative source. That is, fluid flows into a sink along radial streamlines. Thus, for a sink of strength K centered at the origin,

$$\phi = -\frac{K}{2\pi}\ln r \qquad (3.38)$$

Example 3.4

Show that the mass-flow rate passing through a circle of radius r is proportional to K, the strength of the two-dimensional source, and is independent of the radius.

Solution

$$\dot{m} = \iint \rho \vec{V} \cdot \hat{n}\, dA$$

$$= \int_0^{2\pi} \rho\left(\frac{K}{2\pi r}\right) r\, d\theta$$

$$= K\rho$$

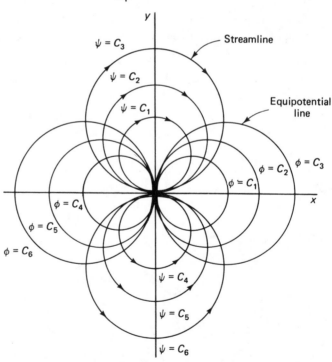

Figure 3.11 Equipotential lines and streamlines for a doublet (flow proceeds out from the origin in the negative x direction).

Doublet

A *doublet* is defined to be the singularity resulting when a source and a sink of equal strength are made to approach each other (see Fig. 3.11) such that the product of their strengths (K) and their distance apart (a) remains constant at a preselected finite value in the limit as the distance between them approaches zero. The line along which the approach is made is called the *axis of the doublet* and is considered to have a positive direction when oriented from sink to source. The potential for a doublet for which the flow proceeds out from the origin in the negative x direction is

$$\phi = \frac{B}{r} \cos \theta \qquad (3.39)$$

where B is a constant.

Potential Vortex

A *potential vortex* is defined as a singularity about which fluid flows with concentric streamlines (see Fig. 3.12). The potential for a vortex centered at the origin is

$$\phi = -\frac{\Gamma \theta}{2\pi} \qquad (3.40)$$

where Γ is the strength of the vortex. We have used a minus sign to represent a vortex with clockwise circulation. Differentiating the potential function, one finds the velocity distribution about an isolated vortex to be

$$v_r = \frac{\partial \phi}{\partial r} = 0$$

$$v_\theta = \frac{1}{r} \frac{\partial \phi}{\partial \theta} = -\frac{\Gamma}{2\pi r}$$

Thus, there is no radial velocity component and the circumferential component varies with the reciprocal of the radial distance from the vortex.

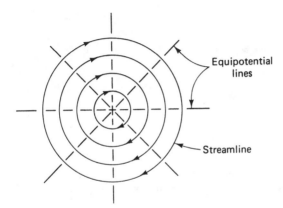

Equipotential lines

Streamline

Figure 3.12 Equipotential lines and streamlines for a potential vortex.

The curl of the velocity vector for the potential vortex can be found using the definition for the curl of $\vec{V}$ in cylindrical coordinates

$$\nabla \times \vec{V} = \frac{1}{r} \begin{vmatrix} \hat{e}_r & r\hat{e}_\theta & \hat{e}_z \\ \dfrac{\partial}{\partial r} & \dfrac{\partial}{\partial \theta} & \dfrac{\partial}{\partial z} \\ v_r & rv_\theta & v_z \end{vmatrix}$$

we find that

$$\nabla \times \vec{V} = 0$$

Thus, although the flow is irrotational, we must remember that the velocity is infinite at the origin (i.e., when $r = 0$).

Let us calculate the circulation around a closed curve C_1 which encloses the origin. We shall choose a circle of radius r_1, as shown in Fig. 3.13a. Using equation (3.14), the circulation is

$$-\Gamma_{C_1} = \oint_{C_1} \vec{V} \cdot \vec{dr} = \int_0^{2\pi} \left(-\frac{\Gamma}{2\pi r_1} \hat{e}_\theta\right) \cdot r_1 \, d\theta \, \hat{e}_\theta$$

$$= \int_0^{2\pi} (-)\frac{\Gamma}{2\pi} \, d\theta = -\Gamma$$

Recall that Stokes's theorem [equation (3.16)] is not valid if the region contains points where the velocity is infinite.

$$-\Gamma_{C_1} = \oint_{C_1} \vec{V} \cdot \vec{dr} = \int_0^{2\pi} \left(-\frac{\Gamma}{2\pi r_1} \hat{e}_\theta\right) \cdot r_1 \, d\theta \, \hat{e}_\theta$$

$$= \int_0^{2\pi} (-)\frac{\Gamma}{2\pi} \, d\theta = -\Gamma$$

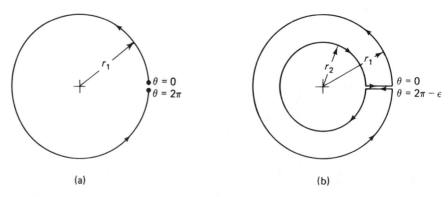

(a)   (b)

Figure 3.13 Paths for the calculation of the circulation for a potential vortex: (a) closed curve C_1, which encloses origin; (b) closed curve C_2, which does not enclose the origin.

However, if we calculate the circulation around a closed curve C_2 which does not enclose the origin, such as that shown in Fig. 3.13b, we find that

$$-\Gamma_{C_2} = \oint_{C_2} \vec{V} \cdot \vec{dr} = \int_0^{2\pi-\epsilon} (-)\frac{\Gamma}{2\pi r_1} r_1 \, d\theta + \int_{2\pi-\epsilon}^0 (-)\frac{\Gamma}{2\pi r_2} r_2 \, d\theta$$

or

$$-\Gamma_{C_2} = 0$$

Thus, the circulation around a closed curve not containing the origin is zero.

The reader may be familiar with the rotation of a two-dimensional, solid body about its axis, such as the rotation of a record on a turntable. For solid-body rotation,

$$v_r = 0$$

$$v_\theta = r\omega$$

where ω is the angular velocity. Substituting these velocity components into the definition

$$\nabla \times \vec{V} = \frac{1}{r}\left[\frac{\partial(rv_\theta)}{\partial r} - \frac{\partial v_r}{\partial \theta}\right]\hat{e}_z$$

we find that

$$\nabla \times \vec{V} = 2\omega\hat{e}_z$$

We see that the velocity field which describes two-dimensional solid-body rotation is not irrotational and, therefore, cannot be defined using a potential function.

Vortex lines (or filaments) will have an important role in the study of the flow around wings. Therefore, let us summarize the vortex theorems of Helmholtz. If there exists a potential for all forces acting on an inviscid fluid, the following statements are true:

1. The circulation around a given vortex line (i.e., the strength of the vortex filament) is constant along its length.

2. A vortex filament cannot end in a fluid. It must form a closed path, end at a boundary, or go to infinity. Examples of these three kinds of behavior are a smoke ring, a vortex bound to a two-dimensional airfoil that spans from one wall to the other in a wind tunnel (see Chapter 6), and the downstream ends of the horseshoe vortices representing the loading on a three-dimensional wing (see Chapter 7).

3. No fluid particle can have rotation, if it did not originally rotate. Or, equivalently, in the absence of rotational external forces, a fluid that is initially irrotational remains irrotational. In general, we can conclude that vortices are preserved as time passes. Only through the action of viscosity (or some other dissipative mechanism) can they decay or disappear.

Table 3.2 summarizes the potential functions and the stream functions for the elementary flows discussed above.

TABLE 3.2　STREAM FUNCTIONS AND POTENTIAL
FUNCTIONS FOR ELEMENTARY FLOWS

Flow	ψ	ϕ
Uniform flow	$U_\infty r \sin \theta$	$U_\infty r \cos \theta$
Source	$\dfrac{K\theta}{2\pi}$	$\dfrac{K}{2\pi} \ln r$
Doublet	$-\dfrac{B}{r} \sin \theta$	$\dfrac{B}{r} \cos \theta$
Vortex (with clockwise circulation)	$\dfrac{\Gamma}{2\pi} \ln r$	$-\dfrac{\Gamma\theta}{2\pi}$
90° corner flow	Axy	$\frac{1}{2}A(x^2 - y^2)$
Solid-body rotation	$\frac{1}{2}\omega r^2$	Does not exist

ADDING ELEMENTARY FLOWS TO DESCRIBE FLOW AROUND A CYLINDER

Velocity Field

Consider the case where a uniform flow is superimposed on a doublet whose axis of development is parallel to the direction of the uniform flow and is so oriented that the direction of the efflux opposes the uniform flow (see Fig. 3.14). Substituting the potential function for a uniform flow [equation (3.34)] and that for the doublet [equation (3.39)] into the expression for the velocity field [equation (3.37)], one finds that

$$v_\theta = \frac{1}{r}\frac{\partial\phi}{\partial\theta} = -U_\infty \sin\theta - \frac{B}{r^2}\sin\theta$$

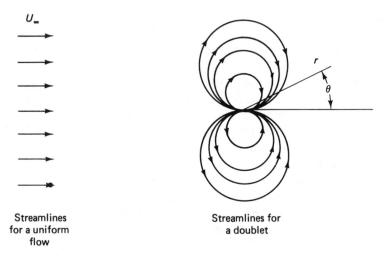

Streamlines for a uniform flow

Streamlines for a doublet

Figure 3.14　Streamlines for the two elementary flows which, when superimposed, describe the flow around a cylinder.

and

$$v_r = \frac{\partial \phi}{\partial r} = U_\infty \cos \theta - \frac{B}{r^2} \cos \theta$$

Note that $v_r = 0$ at every point where $r = \sqrt{B/U_\infty}$, which is a constant. Since the velocity is always tangent to a streamline, the fact that velocity component (v_r) perpendicular to a circle of $r = R = \sqrt{B/U_\infty}$ is zero means that the circle may be considered as a streamline of the flow field. Replacing B by $R^2 U_\infty$ allows us to write the velocity components as

$$v_\theta = -U_\infty \sin \theta \left(1 + \frac{R^2}{r^2} \right) \tag{3.41a}$$

$$v_r = U_\infty \cos \theta \left(1 - \frac{R^2}{r^2} \right) \tag{3.41b}$$

The velocity field not only satisfies the surface boundary condition that the inviscid flow is tangent to a solid wall, but the velocity at points far from the cylinder is equal to the undisturbed free-stream velocity U_∞. Streamlines for the resultant inviscid flow field are illustrated in Fig. 3.15. The resultant two-dimensional, irrotational (inviscid), incompressible flow is that around a cylinder of radius R whose axis is perpendicular to the free-stream direction.

The velocity at the surface of the cylinder is equal to

$$v_\theta = -2U_\infty \sin \theta \tag{3.42}$$

and, of course, as noted earlier, $v_r = 0$. Since the solution is for the inviscid model of the flow field, it is not inconsistent that the fluid particles next to the surface move relative to the surface (i.e., violate the no-slip requirement). When $\theta = 0$ or π (points B and A, respectively, of Fig. 3.15) the fluid is at rest with respect to the cylinder (i.e., $v_r = v_\theta = 0$). These points are, therefore, termed *stagnation points*.

Pressure Distribution

Because the velocity at the surface of the cylinder is a function of θ, the local static pressure will also be a function of θ. Once the pressure distribution has been defined, it can be used to determine the forces and the moments acting on the configuration. Using Bernoulli's equation [equation (3.10)], we obtain the expression for the dimensional, local-static pressure distribution:

$$p = p_\infty + \tfrac{1}{2}\rho_\infty U_\infty^2 - 2\rho_\infty U_\infty^2 \sin^2 \theta \tag{3.43}$$

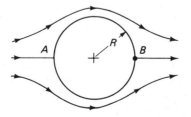

Figure 3.15 Two-dimensional, inviscid flow around a cylinder with zero circulation.

Expressing the pressure in terms of the dimensionless pressure coefficient, we have

$$C_p = 1 - 4 \sin^2 \theta \qquad (3.44)$$

The pressure coefficients, thus calculated, are presented in Fig. 3.16 as a function of θ. Recall that, in the nomenclature of this chapter, θ = 180° corresponds to the plane of symmetry for the windward surface or forebody (i.e., that facing the free stream). Starting with the undisturbed free-stream flow and following the streamline that wets the surface, the flow is decelerated from the free-stream velocity to zero velocity at the (windward) stagnation point in the plane of symmetry. The flow then accelerates reaching a maximum velocity, equal in magnitude to twice the free-stream velocity. From these maxima (which occur at θ = 90° and at 270°), the flow tangent to the leeward surface decelerates to a stagnation point at the surface in the leeward plane of symmetry.

However, even though the viscosity of air is relatively small, the actual flow field is radically different from the inviscid solution described in the previous paragraphs. When the air particles in the boundary layer, which have already been slowed by the action of viscosity, encounter the relatively large adverse pressure gradient associated with the deceleration of the leeward flow for this blunt configuration, boundary-layer separation occurs. The photograph of the smoke patterns for flow around a cylinder presented in Fig. 3.17 clearly illustrates the flow separation. Note that separation results when the fluid particles in the boundary layer (already slowed by viscosity) encounter an adverse pressure gradient that they cannot overcome. However, not all boundary layers separate when they encounter an adverse pressure gradient. There is a relation between the characteristics of the boundary layer and the magnitude of the adverse pressure gradient required to produce separation. Thus, a turbulent boundary layer, which has relatively fast moving particles near the wall, would remain attached longer than a laminar boundary layer, which has slower-moving particles near the wall for

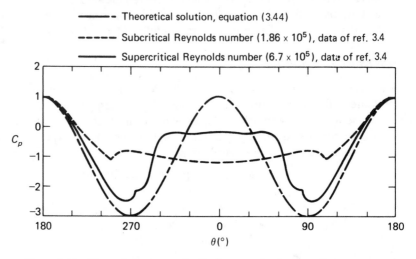

Figure 3.16 Theoretical pressure distribution around a circular cylinder, compared with data for a subcritical Reynolds number and that for a supercritical Reynolds number. [From *Boundary Layer Theory* by H. Schlichting (1968), used with permission of McGraw-Hill Book Company.]

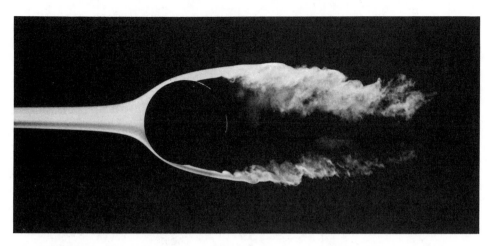

Figure 3.17 Smoke visualization of flow pattern for subcritical flow around a cylinder. (Photograph by F. N. M. Brown, courtesy of T. J. Mueller of University of Notre Dame.)

the same value of the edge velocity. (Boundary layers are discussed in more detail in Chapter 4.) Therefore, the separation location, the size of the wake, and the surface pressure in the wake region depend on the character of the forebody boundary layer.

The experimentally determined separation locations for a circular cylinder as reported by Achenbach in Ref. 3.5 are presented as a function of Reynolds number in Fig. 3.18. As discussed in Chapter 2, the Reynolds number is a dimensionless parameter (in this case, $\mathrm{Re}_d = \rho_\infty U_\infty d/\mu_\infty$) that relates to the viscous characteristics of the flow. At subcritical Reynolds numbers (i.e., less than approximately 3×10^5), the boundary layer on the windward surface (or forebody) is laminar and separation occurs for $\theta \sim 100°$, that is, $80°$ from the windward stagnation point. Note the occurrence of separation so alters the flow that separation actually occurs on the windward surface, where the inviscid solution, as given by equation (3.44) and presented in Fig. 3.16, indicates that there still should be a favorable pressure gradient (i.e., one for which the static pressure decreases in the streamwise direction). If the pressure were actually decreasing in the streamwise direction, separation would not occur. Thus, the occurrence of separation alters the pressure distribution on the forebody (windward surface) of the cylinder. Above the critical Reynolds number, the forebody boundary layer is turbulent. Due to the higher levels of energy for the fluid particles near the surface in a turbulent boundary layer, the flow is able to run longer against the pressure gradient. In the critical region, Achenbach observed an intermediate ''separation bubble'' with final separation not occurring until $\theta = 40°$ (i.e., $140°$ from the stagnation point). For $\mathrm{Re}_d > 1.5 \times 10^6$, the separation bubble no longer occurs, indicating that the supercritical state of flow has been reached. For supercritical Reynolds numbers, separation occurs in the range $60° < \theta < 70°$. (The reader should note that the critical Reynolds number is sensitive both to the turbulence level in the free stream and to surface roughness.)

Experimental pressure distributions are presented in Fig. 3.16 for the cases where the forebody boundary layer is laminar (a subcritical Reynolds number) and where the forebody boundary layer is turbulent (a supercritical Reynolds number). The subcritical pressure-coefficient distribution is essentially unchanged over a wide range of Reynolds

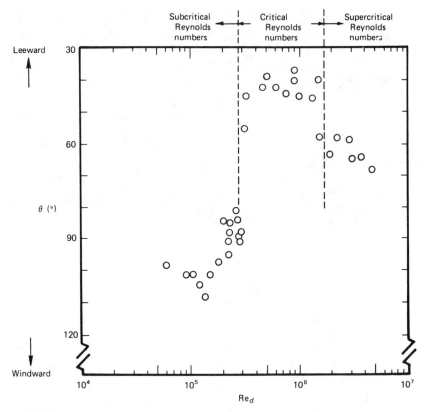

Figure 3.18   Location of the separation points on a circular cylinder as a function of the Reynolds number. (Data from Achenbach, *Journal of Fluid Mechanics*, Ref. 3.5.)

numbers below the critical Reynolds number. Similarly, the supercritical pressure-coefficient distribution is independent of Reynolds number over a wide range of Reynolds numbers above the critical Reynolds number. For the flow upstream of the separation location, the boundary layer is thin, and the pressure-coefficient distribution is essentially independent of the character of the boundary layer for the cylinder. However, because the character of the attached boundary layer affects the separation location, it affects the pressure in the separated region. If the attached boundary layer is turbulent, separation is delayed and the pressure in the separated region is higher and closer to the inviscid level.

Lift and Drag

The motion of the air particles around the cylinder produces forces that may be viewed as a normal (or pressure) component and a tangential (or shear) component. It is conventional to resolve the resultant force on the cylinder into a component perpendicular to the free-stream velocity direction (called the *lift*) and a component parallel to the free-stream velocity direction (called the *drag*). The nomenclature is illustrated in Fig. 3.19.

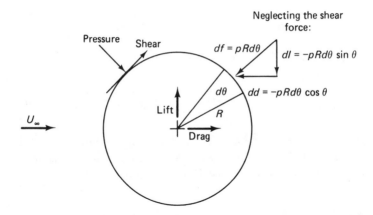

Figure 3.19   Forces acting on a cylinder whose axis is perpendicular to the free-stream flow.

Since the expressions for the velocity distribution [equation (3.42)] and for the pressure distribution [equation (3.43) or (3.44)] were obtained for an inviscid flow, we shall consider only the contribution of the pressure to the lift and to the drag. As shown in Fig. 3.19, the lift per unit span of the cylinder is

$$l = -\int_0^{2\pi} p \sin \theta \, R \, d\theta \tag{3.45}$$

Using equation (3.43) to define the static pressure as a function of θ, one finds that

$$l = 0 \tag{3.46}$$

It is not surprising that there is zero lift per unit span of the cylinder, since the pressure distribution is symmetric about the x axis.

Instead of using equation (3.43), which is the expression for the static pressure, the aerodynamicist might be more likely to use equation (3.44), which is the expression for the dimensionless pressure coefficient. To do this, note that the net force in any direction due to a constant pressure acting on a closed surface is zero. As a result,

$$\int_0^{2\pi} p_\infty \sin \theta \, R \, d\theta = 0 \tag{3.47}$$

Adding equations (3.45) and (3.47) yields

$$l = -\int_0^{2\pi} (p - p_\infty) \sin \theta \, R \, d\theta$$

Dividing both sides of this equation by the product $q_\infty 2R$, which is (dynamic pressure) times (area per unit span in the x plane), yields

$$\frac{l}{q_\infty 2R} = -0.5 \int_0^{2\pi} C_p \sin \theta \, d\theta \tag{3.48}$$

Both sides of equation (3.48) are dimensionless. The expression of the left-hand side is known as the *section lift coefficient* for a cylinder:

$$C_l = \frac{l}{q_\infty 2R}$$
(3.49)

Using equation (3.44) to define C_p as a function of θ,

$$C_l = -0.5 \int_0^{2\pi} (1 - 4 \sin^2 \theta) \sin \theta \, d\theta = 0$$

which, of course, is the same result as was obtained by integrating the pressure directly.

Referring to Fig. 3.19 and following a similar procedure, we can calculate the drag per unit span of the cylinder for the inviscid flow. Thus, the drag per unit span is

$$d = -\int_0^{2\pi} p \cos \theta \, R \, d\theta$$
(3.50)

Substituting equation (3.43) for the local pressure,

$$d = -\int_0^{2\pi} (p_\infty + \tfrac{1}{2}\rho_\infty U_\infty^2 - 2\rho_\infty U_\infty^2 \sin^2 \theta) \cos \theta \, R \, d\theta$$

we find that

$$d = 0$$
(3.51)

A drag of zero is an obvious contradiction to the reader's experience (and is known as *d'Alembert's paradox*). Note that the actual pressure in the separated, wake region near the leeward plane of symmetry (in the vicinity of $\theta = 0$ in Fig. 3.16) is much less than the theoretical value. It is the resultant difference between the high pressure acting near the windward plane of symmetry (in the vicinity of $\theta = 180°$, i.e., the stagnation point) and the relatively low pressures acting near the leeward plane of symmetry which produces the large drag component.

A drag force that represents the streamwise component of the pressure force integrated over the entire configuration is termed *pressure* (or *form*) *drag*. The drag force that is obtained by integrating the streamwise component of the shear force over the vehicle is termed *skin friction drag*. Note that in the case of real flow past a cylinder, the skin friction drag is small. However, significant form drag results because of the action of viscosity, which causes the boundary layer to separate and therefore radically alters the pressure field. The pressure near the leeward plane of symmetry is higher (and closer to the inviscid values) when the forebody boundary layer is turbulent. Thus, the difference between the pressure acting on the forward surface and that acting on the leeward surface is less in the turbulent case. As a result, the form drag for a turbulent boundary layer is markedly less than the corresponding value for a laminar (forebody) boundary layer.

The *drag coefficient* per unit span for a cylinder is

$$C_d = \frac{d}{q_\infty 2R}$$
(3.52)

Experimental drag coefficients for a circular cylinder in a low-speed stream (Ref. 3.4) are presented as a function of Reynolds number in Fig. 3.20. For Reynolds numbers below 300,000, the drag coefficient is essentially constant (approximately 1.2), independent of Reynolds number. Recall that when we were discussing the experimental values of C_p presented in Fig. 3.16, it was noted that the subcritical pressure-coefficient distribution is essentially unchanged over a wide range of Reynolds number. Since both the drag coefficient and the pressure coefficient are essentially independent of Reynolds number below the critical Reynolds number, the pressure (or form) drag is the dominant component for blunt bodies. Thus,

$$C_d \approx -0.5 \int_0^{2\pi} C_p \cos \theta \, d\theta$$

Above the critical Reynolds number (when the forebody boundary layer is turbulent), the drag coefficient is significantly lower. Reviewing the supercritical pressure distribution, we recall that the pressure in the separated region is closer to the inviscid level. In a situation where the Reynolds number is subcritical, it may be desirable to induce boundary-layer transition by roughening the surface. Examples of such transition-promoting roughness elements are the dimples on a golf ball or the vortex generators on aerodynamic surfaces. The dimples on a golf ball are intended to reduce drag; the vortex generators are designed to delay separation.

LIFT AND DRAG COEFFICIENTS AS DIMENSIONLESS FLOW-FIELD PARAMETERS

The formulas for the drag coefficient [equation (3.52)] and for the lift coefficient for a cylinder [equation (3.49)] have the same elements. Thus, we can define a *force coefficient* as

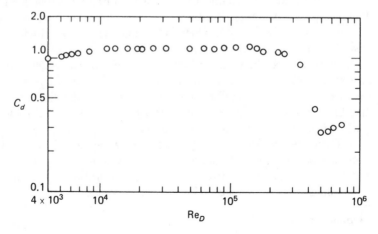

Figure 3.20   Drag coefficient for a circular cylinder as a function of the Reynolds number. [From *Boundary Layer Theory* by H. Schlichting (1968), used with permission of McGraw-Hill Book Company.]

$$C_F = \frac{\text{force}}{\underset{\substack{\text{dynamic}\\\text{pressure}}}{(\frac{1}{2}\rho_\infty U_\infty^2)} \, (\text{reference area})} \tag{3.53}$$

Note that, for a configuration of infinite span, a force per unit span would be divided by the reference area per unit span. Ideally, the force coefficient would be a function of configuration geometry and of attitude only. However, the effects of viscosity and compressibility cause variations in the force coefficients. These effects can be correlated in terms of parameters, such as the Reynolds number and Mach number. Such variations are evident in the drag coefficient measurements presented in this chapter.

From equation (3.53), it is clear that an aerodynamic force is proportional to the square of the free-stream velocity, to the free-stream density, to the size of the object, and to the force coefficient. An indication of the effect of configuration geometry on the total drag and on the drag coefficient is given in Figs. 3.21 and 3.22, which are taken from Ref. 3.6. The actual drag for several incompressible, flow condition/configuration geometry combinations are presented in Fig. 3.21. Compare the results for configurations (a), (b), and (c), which are configurations having the same dimension and exposed to the same Reynolds number stream. Streamlining produces dramatic reductions in the pressure (or form) drag with only a slight increase in skin friction drag at this Reynolds number. Thus, streamlining reduces the drag coefficient.

Note that the diameter of the small cylinder is one-tenth that of the other configurations. Therefore, in the same free-stream flow as configuration (b), the small cylinder operates at a Reynolds number of 10^4. Because the size is reduced, the drag forces for (d) are an order of magnitude less than for (b). However, over this range of Reynolds number, the drag coefficients are essentially equal (see Fig. 3.20). As shown in Fig. 3.21, the total drag of the small cylinder is equal to that of the much thicker streamlined shape. The student can readily imagine how much additional drag was produced by the interplane wire bracing of the World War I biplanes.

When the Reynolds number is increased to 10^7 (corresponding to the supercritical flow of Fig. 3.20), the pressure drag is very large. However, the drag coefficient for this condition (e) is only 0.6, which is less than the drag coefficient for the subcritical flow (b) even though the pressure drag is significantly greater. Note that since the cylinder diameter is the same for both (b) and (e), the two order of magnitude increase in Reynolds number is accomplished by increasing the free-stream density and the free-stream velocity. Therefore, the denominator of equation (3.52) increases more than the numerator. As a result, even though the dimensional force is increased, the nondimensionalized force coefficient is decreased.

There are a variety of sources of aerodynamic data. However, Refs. 3.7 and 3.8 offer the reader unique and entertaining collections of data. In these volumes the reader will find aerodynamic coefficients for flags, World War II airplanes, and vintage automobiles as well as more classical configurations.

Example 3.5

You are to design a quonset hut to serve as temporary housing near the seashore. The quonset hut may be considered to be a closed (no leaks) semicylinder, whose radius is 5 m, mounted on tie-down blocks, as shown in Fig. 3.23. Neglect viscous effects and

(a) Flat plate broadside to the flow (height = d), $Re_d = 10^5$

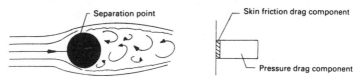

(b) Large cylinder with subcritical flow (diameter = d), $Re_d = 10^5$

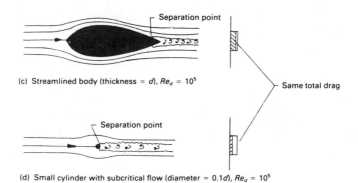

(c) Streamlined body (thickness = d), $Re_d = 10^5$

Same total drag

(d) Small cylinder with subcritical flow (diameter = $0.1d$), $Re_d = 10^5$

(e) Large cylinder with supercritical flow (diameter = d), $Re_d = 10^7$

Figure 3.21 Comparison of the drag components for various shapes and flows. (From Ref. 3.6.)

assume that the flow field over the top of the hut is identical to the flow over the cylinder for $0 \leq \theta \leq \pi$. When calculating the flow over the upper surface of the hut, neglect the presence of the air space under the hut. The air under the hut is at rest and the pressure is equal to the stagnation pressure, p_t.

What is the net lift force acting on the quonset hut? The wind speed is 50 m/s and the static free-stream properties are those for standard sea-level conditions.

Solution Since we are to assume that the flow over the upper surface of the quonset hut is identical to an inviscid flow over a cylinder, the velocity is given by equation (3.42):

$$U_u = v_\theta = -2U_\infty \sin \theta$$

and the pressure is given by equation (3.43):

$$p_u = p_\infty + \tfrac{1}{2}\rho_\infty U_\infty^2 - 2\rho_\infty U_\infty^2 \sin^2\theta$$

The pressure on the lower surface (under the hut) is

$$p_l = p_t = p_\infty + \tfrac{1}{2}\rho_\infty U_\infty^2$$

Equation (3.45) and Fig. 3.19 can be used to calculate the lifting contribution of the upper surface but not the lower surface, since it is a "flat plate" rather than a circular arc. The lift per unit depth of the quonset hut is

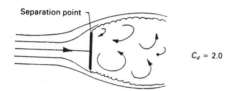

(a) Flat plate broadside to the flow (height = d), $Re_d = 10^5$

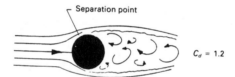

(b) Large cylinder with subcritical flow (diameter = d), $Re_d = 10^5$

(c) Streamlined body (thickness = d), $Re_d = 10^5$

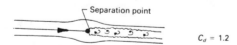

(d) Small cylinder with subcritical flow (diameter = $0.1d$), $Re_d = 10^5$

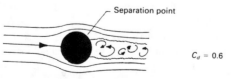

(e) Large cylinder with supercritical flow (diameter = d), $Re_d = 10^7$

Figure 3.22   Comparison of section drag coefficients for various shapes and flows. (From Ref. 3.6.)

$$l = -\int_0^\pi p_u \sin\theta\, R\, d\theta + p_l(2R)$$

$$= -R\int_0^\pi (p_\infty \sin\theta + \tfrac{1}{2}\rho_\infty U_\infty^2 \sin\theta - 2\rho_\infty U_\infty^2 \sin^3\theta)\, d\theta + p_\infty 2R + \tfrac{1}{2}\rho_\infty U_\infty^2 2R$$

$$= + p_\infty R \cos\theta \Big|_0^\pi + \tfrac{1}{2}\rho_\infty U_\infty^2 R \cos\theta \Big|_0^\pi + 2\rho_\infty U_\infty^2 R(-\cos\theta + \tfrac{1}{3}\cos^3\theta)\Big|_0^\pi$$

$$+ p_\infty 2R + \tfrac{1}{2}\rho_\infty U_\infty^2 2R$$

$$= \tfrac{8}{3}\rho_\infty U_\infty^2 R$$

Note that the lift coefficient is

$$C_l = \frac{l}{\tfrac{1}{2}\rho_\infty U_\infty^2 (2R)} = \frac{\tfrac{8}{3}\rho_\infty U_\infty^2 R}{\rho_\infty U_\infty^2 R} = \frac{8}{3}$$

which is independent of Mach number and Reynolds number (a desirable result) for this "inviscid," incompressible flow.

The actual lift is

$$l = \frac{8}{3}\left(1.225\,\frac{\text{kg}}{\text{m}^3}\right)\left(50\,\frac{\text{m}}{\text{s}}\right)^2 (5\text{ m}) = 40{,}833\text{ N/m}$$

By symmetry, the drag is zero, which reflects the fact that we neglected the effects of viscosity.

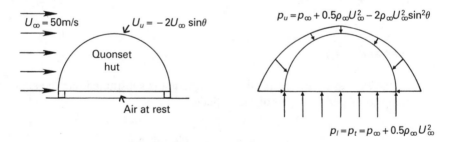

Figure 3.23 Inviscid flow model for quonset hut of Example 3.5.

FLOW AROUND A CYLINDER WITH CIRCULATION

Velocity Field

Let us consider the flow field that results if a vortex with clockwise circulation is superimposed on the doublet/uniform-flow combination discussed above. The resultant potential function is

$$\phi = U_\infty r \cos\theta + \frac{B}{r}\cos\theta - \frac{\Gamma\theta}{2\pi} \tag{3.54}$$

Thus,

$$v_r = \frac{\partial \phi}{\partial r} = U_\infty \cos \theta - \frac{B \cos \theta}{r^2} \tag{3.55a}$$

and

$$v_\theta = \frac{1}{r}\frac{\partial \phi}{\partial \theta} = \frac{1}{r}\left(-U_\infty r \sin \theta - \frac{B \sin \theta}{r} - \frac{\Gamma}{2\pi}\right) \tag{3.55b}$$

Again, $v_r = 0$ at every point where $r = \sqrt{B/U_\infty}$, which is a constant and will be designated as R. Since the velocity is always tangent to a streamline, the fact that the velocity component (v_r) perpendicular to a circle of radius R is zero means that the circle may be considered as a streamline of the flow field. Thus, the resultant potential function also represents flow around a cylinder. For this flow, however, the streamline pattern away from the surface is not symmetric about the horizontal plane. The velocity at the surface of the cylinder is equal to

$$v_\theta = -U = -2U_\infty \sin \theta - \frac{\Gamma}{2\pi R} \tag{3.56}$$

The resultant irrotational flow about the cylinder is uniquely determined once the magnitude of the circulation around the body is specified. Using the definition for the pressure coefficient [equation (3.13)], we obtain

$$C_p = 1 - \frac{U^2}{U_\infty^2} = 1 - \frac{1}{U_\infty^2}\left[4U_\infty^2 \sin^2 \theta + \frac{2\Gamma U_\infty \sin \theta}{\pi R} + \left(\frac{\Gamma}{2\pi R}\right)^2\right] \tag{3.57}$$

Lift and Drag

If the expression for the pressure distribution is substituted into the expression for the drag force per unit span of the cylinder,

$$d = -\int_0^{2\pi} p(\cos \theta)R \, d\theta = 0$$

The prediction of zero drag may be generalized to apply to any general, two-dimensional body in an irrotational, steady, incompressible flow. In any real two-dimensional flow, a drag force does exist. For incompressible flow, drag is due to viscous effects, which produce the shear force at the surface and which may also produce significant changes in the pressure field (causing form drag).

Integrating the pressure distribution to determine the lift force per unit span for the cylinder, one obtains

$$l = -\int_0^{2\pi} p(\sin \theta)R \, d\theta = \rho_\infty U_\infty \Gamma \tag{3.58}$$

Thus, the lift per unit span is directly related to the circulation about the cylinder. This result, which is known as the *Kutta-Joukouski theorem*, applies to the potential flow about closed cylinders of arbitrary cross section. To see this, consider the circulating flow field around the closed configuration to be represented by the superposition of a uniform flow and a unique set of sources, sinks, and vortices within the body. For a closed body, continuity requires that the sum of the source strengths be equal to the sum of the sink strengths. When one considers the flow field from a point far from the surface of the body, the distance between the sources and sinks becomes negligible and the flow field appears to be that generated by a single doublet with circulation equal to the sum of the vortex strengths within the body. Thus, in the limit, the forces acting are independent of the shape of the body and

$$l = \rho_\infty U_\infty \Gamma$$

The locations of the stagnation points (see Fig. 3.24) also depend on the circulation. To locate the stagnation points, we need to find where

$$v_r = v_\theta = 0$$

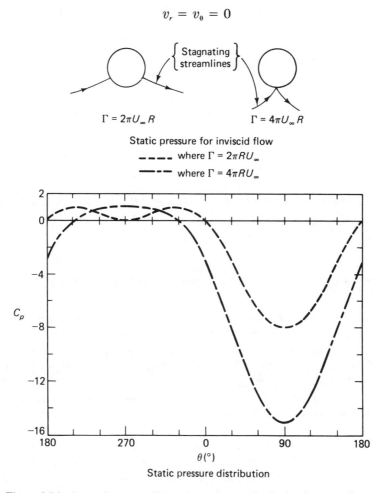

Figure 3.24 Stagnating streamlines and the pressure distribution for a two-dimensional circulating flow around a cylinder. (a) $\Gamma = 2\pi U_\infty R$; (b) $\Gamma = 4\pi U_\infty R$; (c) static pressure distribution.

Since $v_r = 0$ at every point on the cylinder, the stagnation points occur when $v_\theta = 0$. Therefore,

$$-2U_\infty \sin \theta - \frac{\Gamma}{4\pi R} = 0$$

or

$$\theta = \sin^{-1} \left(-\frac{\Gamma}{4\pi R U_\infty} \right) \tag{3.59}$$

If $\Gamma < 4\pi R U_\infty$, there are two stagnation points on the surface of the cylinder. They are symmetrically located about the y axis and both are below the x axis (see Fig. 3.24). If $\Gamma = 4\pi U_\infty R$, only one stagnation point exists on the cylinder and it exists at $\theta = 270°$. For this magnitude of the circulation, the lift per unit span is

$$l = \rho_\infty U_\infty \Gamma = \rho_\infty U_\infty^2 R 4\pi \tag{3.60}$$

The lift coefficient per unit span of the cylinder is

$$C_l = \frac{l}{q_\infty 2R}$$

Thus,

$$C_l = \frac{\rho_\infty U_\infty^2 R 4\pi}{\frac{1}{2}\rho_\infty U_\infty^2 2R} = 4\pi \tag{3.61}$$

The value 4π represents the maximum lift coefficient that can be generated for a circulating flow around a cylinder unless the circulation is so strong that no stagnation point exists on the body.

SOURCE DENSITY DISTRIBUTION ON THE BODY SURFACE

Thus far, we have studied fundamental fluid phenomena, such as the Kutta-Joukowski theorem, using the inverse method. Flow fields for other configurations, such as axisymmetric shapes in a uniform stream parallel to the axis of symmetry, can be represented by a source distribution along the axis of symmetry. An "exact" solution for the flow around an arbitrary configuration can be approached using a direct method in a variety of ways, all of which must finally become numerical and make use of a computing machine. The reader is referred to Ref. 3.9 for an extensive review of the problem.

Consider a two-dimensional configuration in a uniform stream, such as shown in Fig. 3.25. The coordinate system used in this section (i.e., x in the chordwise direction and y in the spanwise direction), will be used in subsequent chapters on wing and airfoil aerodynamics. The configuration is represented by a finite number (M) of linear segments, or panels. The effect of the jth panel on the flow field is characterized by a distributed source whose strength is uniform over the surface of

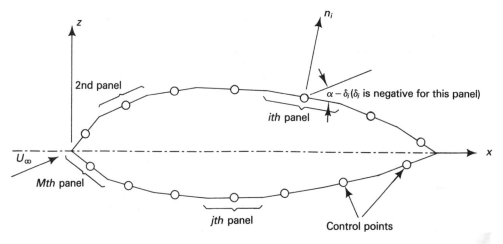

Figure 3.25   Source density distribution of the body surface.

the panel. Referring to equation (3.36), a source distribution on the jth panel causes an induced velocity whose potential at a point (x, z) is given by

$$\phi(x, z) = \int \frac{k_j \, ds_j}{2\pi} \ln r \qquad (3.62)$$

where k_j is defined as the volume of fluid discharged per unit area of the panel and the integration is carried out over the length of the panel ds_j. Note also that

$$r = \sqrt{(x - x_j)^2 + (z - z_j)^2} \qquad (3.63)$$

Since the flow is two-dimensional, all calculations are for a unit length along the y axis, or span.

Each of the M panels can be represented by similar sources. To determine the strengths of the various sources k_j, we need to satisfy the physical requirement that the surface must be a streamline. Thus, we require that the sum of the source-induced velocities and the free-stream velocity is zero in the direction normal to the surface of the panel at the surface of each of the M panels. The points at which the requirement that the resultant flow is tangent to the surface will be numerically satisfied are called the *control points*. The control points are chosen to be the midpoints of the panels, as shown in Fig. 3.25.

At the control point of the ith panel, the velocity potential for the flow resulting from the superposition of the M source panels and the free-stream flow is

$$\phi(x_i, z_i) = U_\infty x_i \cos \alpha + U_\infty z_i \sin \alpha + \sum_{j=1}^{M} \frac{k_j}{2\pi} \int \ln r_{ij} \, ds_j \qquad (3.64)$$

where r_{ij} is the distance from the control point of the ith panel to a point on the jth panel.

$$r_{ij} = \sqrt{(x_i - x_j)^2 + (z_i - z_j)^2} \qquad (3.65)$$

Note that the source strength k_j has been taken out of the integral, since it is constant over the jth panel. Each term in the summation represents the contribution of the jth panel (integrated over the length of the panel) to the potential at the control point of the ith panel.

The boundary conditions require that the resultant velocity normal to the surface be zero at each of the control points. Thus,

$$\frac{\partial}{\partial n_i} \phi_i(x_i, z_i) = 0 \qquad (3.66)$$

must be satisfied at each and every control point. Care is required in evaluating the spatial derivatives of equation (3.64), because the derivatives become singular when the contribution of the ith panel is evaluated. Referring to equation (3.65), we have

$$r_{ij} = 0$$

where $j = i$. A rigorous development of the limiting process is given by Kellogg (Ref. 3.10). Although the details will not be repeated here, the resultant differentiation indicated in equation (3.66) yields

$$\frac{k_i}{2} + \sum_{\substack{j=1 \\ (j \ne i)}}^{M} \frac{k_j}{2\pi} \int \frac{\partial}{\partial n_i} (\ln r_{ij}) \, ds_j = -U_\infty \sin (\alpha - \delta_i) \qquad (3.67)$$

where δ_i is the slope of the ith panel relative to the x axis. Note that the summation is carried out for all values of j except $j = i$. The two terms of the left side of equation (3.67) have a simple interpretation. The first term is the contribution of the source density of the ith panel to the outward normal velocity at the point (x_i, z_i), that is, the control point of the ith panel. The second term represents the contribution of the remainder of the boundary surface to the outward normal velocity at the control point of the ith panel.

Evaluating the terms of equation (3.67) for a particular ith control point yields a linear equation in terms of the unknown source strengths k_j (for $j = 1$ to M, including $j = i$). Evaluating the equation for all values of i (i.e., for each of the M control points) yields a set of M simultaneous equations which can be solved for the source strengths. Once the panel source strengths have been determined, the velocity can be determined at any point in the flow field using equations (3.64) and (3.65). With the velocity known, Bernoulli's equation can be used to calculate the pressure field.

Example 3.6

Let us apply the surface source density distribution to describe the flow around a cylinder in a uniform stream, where the free-stream velocity is U_∞. The radius of the cylinder is unity. The cylinder is represented by eight, equal-length linear segments, as shown in Fig. 3.26. The panels are arranged such that panel 1 is perpendicular to the undisturbed stream.

Solution Let us calculate the contribution of the source distribution on panel 2 to the normal velocity at the control point of panel 3. A detailed sketch of the two panels involved in this sample calculation is presented in Fig. 3.27. Referring to equation (3.67), we are to evaluate the integral:

$$\int \frac{\partial}{\partial n_i} (\ln r_{ij}) \, ds_j$$

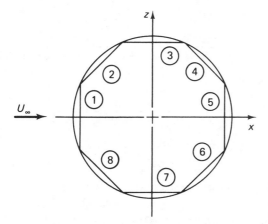

Figure 3.26 Representation of flow around a cylinder of unit radius by eight surface source panels.

where $i = 3$ and $j = 2$. We will call this integral I_{32}. Note that

$$\frac{\partial}{\partial n_3} (\ln r_{32}) = \frac{1}{r_{32}} \frac{\partial r_{32}}{\partial n_3}$$

(3.68)

$$= \frac{(x_3 - x_2) \dfrac{\partial x_3}{\partial n_3} + (z_3 - z_2) \dfrac{\partial z_3}{\partial n_3}}{(x_3 - x_2)^2 + (z_3 - z_2)^2}$$

where $x_3 = 0.00$ and $z_3 = 0.92388$ are the coordinates of the control points of panel 3. Note also that

$$\frac{\partial x_3}{\partial n_3} = 0.00, \qquad \frac{\partial z_3}{\partial n_3} = 1.00$$

Furthermore, for the source line represented by panel 2:

$$x_2 = -0.92388 + 0.70711 s_2$$

$$z_2 = +0.38268 + 0.70711 s_2$$

and the length of the panel is

$$l_2 = 0.76537$$

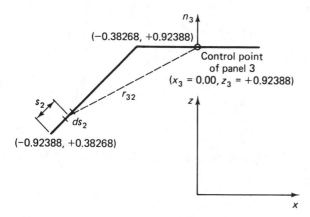

Figure 3.27 Detailed sketch for calculation of the contribution of the source distribution on panel 2 to the normal velocity at the control point of panel 3.

Combining these expressions, we obtain

$$I_{32} = \int_0^{0.76537} \frac{(0.92388 - 0.38268 - 0.70711s_2)\, ds_2}{(0.92388 - 0.70711s_2)^2 + (0.92388 - 0.38268 - 0.70711s_2)^2}$$

This equation can be rewritten

$$I_{32} = 0.54120 \int_0^{0.76537} \frac{ds_2}{1.14645 - 2.07195s_2 + 1.00002s_2^2}$$

$$- 0.70711 \int_0^{0.76537} \frac{s_2\, ds_2}{1.14645 - 2.07195s_2 + 1.00002s_2^2}$$

Using the integral tables to evaluate these expressions, we obtain

$$I_{32} = -0.70711 \left[\tfrac{1}{2} \ln (s_2^2 - 2.07195s_2 + 1.14645) \right]_{s_2=0}^{s_2=0.76537}$$

$$- 0.70711 \left[\tan^{-1} \left(\frac{2s_2 - 2.07195}{\sqrt{0.29291}} \right) \right]_{s_2=0}^{s_2=0.76537}$$

Thus,

$$I_{32} = 0.3528$$

In a similar manner, we could calculate the contributions of source panels 1, 4, 5, 6, 7, and 8 to the normal velocity at the control point of panel 3. Substituting the values of these integrals into equation (3.67), we obtain a linear equation of the form

$$I_{31}k_1 + I_{32}k_2 + \pi k_3 + I_{34}k_4 + I_{35}k_5 + I_{36}k_6 + I_{37}k_7 + I_{38}k_8 = 0.00 \qquad (3.69)$$

The right-hand side is zero since $\alpha = 0$ and $\delta_3 = 0$.

Repeating the process for all eight control points, we would obtain a set of eight linear equations involving the eight unknown source strengths. Solving the system of equations, we would find that

$$k_1 = 2\pi U_\infty(+0.3765)$$

$$k_2 = 2\pi U_\infty(+0.2662)$$

$$k_3 = 0.00$$

$$k_4 = 2\pi U_\infty(-0.2662)$$

$$k_5 = 2\pi U_\infty(-0.3765)$$

$$k_6 = 2\pi U_\infty(-0.2662)$$

$$k_7 = 0.00$$

$$k_8 = 2\pi U_\infty(+0.2662)$$

Note there is a symmetrical pattern in the source distribution, as should be expected. Also,

$$\sum k_i = 0 \qquad (3.70)$$

as must be true since the sum of the strengths of the sources and sinks (negative sources) must be zero if we are to have a closed configuration.

INCOMPRESSIBLE, AXISYMMETRIC FLOW

The irrotational flows discussed thus far have been planar, two-dimensional. That is, the flow field that exists in the plane of the paper will exist in any and every plane parallel to the plane of the paper. Thus, although the sketches of the flow field defined by equations (3.41) through (3.61) depict the flow around a circle of radius R, in reality they represent the flow around a cylinder whose axis is perpendicular to the plane of the paper. For these flows, $w \equiv 0$ and $\partial/\partial z \equiv 0$.

Let us now consider another type of "two-dimensional" flow: an axisymmetric flow. The coordinate system is illustrated in Fig. 3.28. There are no circumferential variations in an axisymmetric flow; that is,

$$v_\theta \equiv 0 \quad \text{and} \quad \frac{\partial}{\partial \theta} \equiv 0$$

Thus, the incompressible, continuity equation becomes

$$\frac{\partial v_r}{\partial r} + \frac{v_r}{r} + \frac{\partial v_z}{\partial z} = 0$$

Noting that r and z are the independent coordinates (i.e., variables), we can rewrite this expression as

$$\frac{\partial}{\partial r}(rv_r) + \frac{\partial}{\partial z}(rv_z) = 0 \tag{3.71}$$

As has been discussed, a stream function will exist for an incompressible, two-dimensional flow. The flow need be two-dimensional only in the sense that it requires only two spatial coordinates to describe the motion. The stream function that identically satisfies equation (3.71) is

$$\frac{\partial \psi}{\partial z} = rv_r \quad \text{and} \quad \frac{\partial \psi}{\partial r} = -rv_z$$

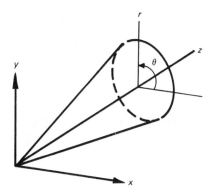

Figure 3.28 Coordinate system for an axisymmetric flow.

Thus, in the coordinate system of Fig. 3.28,

$$v_r = \frac{1}{r} \frac{\partial \psi}{\partial z} \quad \text{and} \quad v_z = -\frac{1}{r} \frac{\partial \psi}{\partial r} \tag{3.72}$$

Flow Around a Sphere

To describe a steady, inviscid, incompressible flow around a sphere, we will add the axisymmetric potential functions for a uniform flow and for a doublet. We will first introduce the necessary relations in spherical coordinates. For a spherical coordinate system:

$$v_r = \frac{\partial \phi}{\partial r} \qquad v_\omega = \frac{1}{r} \frac{\partial \phi}{\partial \omega} \qquad v_\theta = \frac{1}{r \sin \omega} \frac{\partial \phi}{\partial \theta} \tag{3.73}$$

for an irrotational flow where $\vec{V} = \nabla \phi$. In equation (3.73), ϕ represents the potential function, and r, θ, and ω represent the independent coordinates. By symmetry,

$$v_\theta = 0 \quad \text{and} \quad \frac{\partial}{\partial \theta} = 0$$

The velocity potential for an axisymmetric doublet is

$$\phi = +\frac{B}{4\pi r^2} \cos \omega$$

where the doublet is so oriented that the source is placed upstream. The potential function for a uniform flow is

$$\phi = U_\infty r \cos \omega$$

Thus, the sum of the potential functions is

$$\phi = U_\infty r \cos \omega + \frac{B}{4\pi r^2} \cos \omega \tag{3.74}$$

The velocity components for this potential function are

$$v_r = \frac{\partial \phi}{\partial r} = U_\infty \cos \omega - \frac{B}{2\pi r^3} \cos \omega \tag{3.75a}$$

and

$$v_\omega = \frac{1}{r} \frac{\partial \phi}{\partial \omega} = -U_\infty \sin \omega - \frac{B}{4\pi r^3} \sin \omega \tag{3.75b}$$

As we did when modeling the inviscid flow around a cylinder, we note that

$$v_r = 0$$

when

$$r^3 = \frac{B}{2\pi U_\infty} = \text{constant} = R^3$$

Thus, if $B = 2\pi U_\infty R^3$, we can use the potential function described by equation (3.74) to describe steady, inviscid, incompressible flow around a sphere of radius R. For this flow,

$$v_r = U_\infty \left(1 - \frac{R^3}{r^3}\right) \cos \omega \qquad (3.76a)$$

and

$$v_\omega = -U_\infty \left(1 + \frac{R^3}{2r^3}\right) \sin \omega \qquad (3.76b)$$

On the surface of the sphere (i.e., for $r = R$), the resultant velocity is given by

$$U = |\bar{V}| = v_\omega = -\tfrac{3}{2} U_\infty \sin \omega \qquad (3.77)$$

The static pressure acting at any point on the sphere can be calculated using equation (3.77) to represent the local velocity in Bernoulli's equation:

$$p = p_\infty + \tfrac{1}{2}\rho_\infty U_\infty^2 - \tfrac{1}{2}\rho_\infty U_\infty^2 \left(\tfrac{9}{4}\sin^2 \omega\right) \qquad (3.78)$$

Rearranging the terms, we obtain the expression for the pressure coefficient for steady, inviscid, incompressible flow around a sphere:

$$C_p = 1 - \tfrac{9}{4}\sin^2 \omega \qquad (3.79)$$

Compare this expression with equation (3.44) for flow around a cylinder of infinite span whose axis is perpendicular to the free-stream flow:

$$C_p = 1 - 4\sin^2 \theta$$

Note that both θ and ω represent the angular coordinate relative to the axis, one for the two-dimensional flow, the other for axisymmetric flow. Thus, although the configurations have the same cross section in the plane of the paper (a circle) and both are described in terms of two coordinates, the flows are significantly different.

The drag coefficients for a sphere, as reported in Ref. 3.4, are presented as a function of the Reynolds number in Fig. 3.29. The drag coefficient for a sphere is defined as

$$C_D = \frac{\text{drag}}{q_\infty \left(\pi d^2/4\right)} \qquad (3.80)$$

The Reynolds number dependence of the drag coefficient for a sphere is similar to that for a cylinder. Again, a significant reduction in drag occurs as the critical Reynolds number is exceeded and the windward boundary layer becomes turbulent.

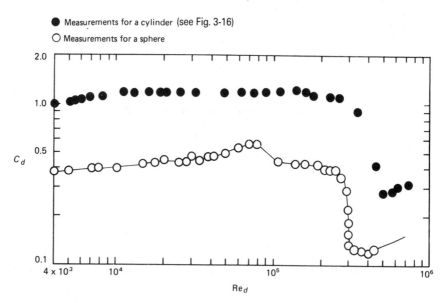

Figure 3.29　Drag coefficient for a sphere as a function of the Reynolds number. (From Schlichting, Boundary-Layer Theory (1968) with permission of McGraw-Hill.)

SUMMARY

In most flow fields of interest to the aerodynamicist, there are regions where the product of the viscosity times the shearing velocity gradient is sufficiently small that we may neglect the shear stress terms in our analysis. The momentum equation for these inviscid flows is known as Euler's equation. From Kelvin's theorem, we know that a flow remains irrotational in the absence of viscous forces and discontinuities provided that the fluid is homogeneous and the body forces are conservative. Potential functions can be used to describe the velocity field for such flows. If we assume further that the flow is incompressible (i.e., low speed), we can linearly add potential functions to obtain the velocity field for complex configurations and use Bernoulli's equation to determine the corresponding pressure distribution. The inviscid flow field solutions, thus obtained, form the outer (edge) boundary conditions for the thin viscous (boundary) layer adjacent to the wall. The characteristics of the boundary layer and techniques for analyzing it are described in the next chapter.

PROBLEMS

3.1. What conditions must be satisfied before we can use Bernoulli's equation to relate the flow characteristics between two points in the flow field?

3.2. Water fills the circular tank (which is 20.0 ft in diameter) shown in Fig. P3.2. Water flows out of a hole which is 1.0 in. in diameter and which is located in the side of the tank, 15.0 ft from the top and 15.0 ft from the bottom. Consider the water to be inviscid. $\rho_{H_2O} = 1.940$ slug/ft³.

(a) Calculate the static pressure and the velocity at points 1, 2, and 3. For these calculations you can assume that the fluid velocities are negligible at points more than 10.0 ft from the opening.

(b) Having calculated U_3 in part (a), what is the velocity U_1? Was the assumption of part (a) valid?

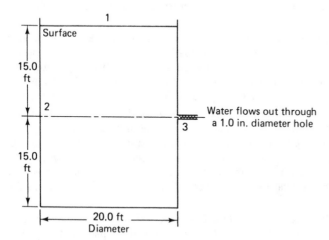

Figure P3.2

3.3. Consider a low-speed, steady flow around the thin airfoil shown in Fig. P3.3. We know the velocity and altitude at which the vehicle is flying. Thus, we know p_∞ (i.e., p_1) and U_∞. We have obtained experimental values of the local static pressure at points 2 through 6. At which of these points can we use Bernoulli's equation to determine the local velocity? If we cannot, why not?

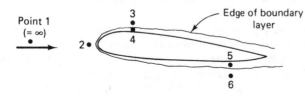

Figure P3.3

Point 2: at the stagnation point of the airfoil
Point 3: at a point in the inviscid region just outside the laminar boundary layer
Point 4: at a point in the laminar boundary layer
Point 5: at a point in the turbulent boundary layer
Point 6: at a point in the inviscid region just outside the turbulent boundary layer

3.4. Assume that the airfoil of problem 3.3 is moving at 300 km/h at an altitude of 3 km. The experimentally determined pressure coefficients are:

Point	2	3	4	5	6
C_p	1.00	−3.00	−3.00	+0.16	+0.16

    (a) What is the Mach number and the Reynolds number for this configuration? Assume that the characteristic dimension for the airfoil is 1.5 m.

    (b) Calculate the local pressure in N/m² and in lbf/in² at all five points. What is the percentage change in the pressure relative to the free-stream value? That is, what is $(p_{local} - p_\infty)/p_\infty$? Was it reasonable to assume that the pressure changes are sufficiently small that the density is approximately constant?

    (c) Why are the pressures at points 3 and 4 equal and at points 5 and 6 equal?

    (d) At those points where Bernoulli's equation can be used validly, calculate the local velocity.

3.5. A pitot-static probe is used to determine the airspeed of an airplane that is flying at an altitude of 6000 m. If the stagnation pressure is 4.8540×10^4 N/m², what is the airspeed? What is the pressure recorded by a gage that measures the difference between the stagnation pressure and the static pressure (such as that shown in Fig. 3.2)? How fast would an airplane have to fly at sea level to produce the same reading on this gage?

3.6. A high-rise office building located in Albuquerque, New Mexico, at an altitude of 1 mi is exposed to a wind of 40 mi/h. What is the static pressure in the airstream away from the influence of the building? What is the maximum pressure acting on the building? Pressure measurements indicate that a value of $C_p = -5$ occurs near the corner of the wall parallel to the wind direction. If the internal pressure is equal to the free-stream static pressure, what is the total force on a pane of glass 3 ft × 8 ft located in this region?

3.7. You are working as a flight-test engineer at the Dryden Research Center. During the low-speed phase of the test program for the X-30, you know that the plane is flying at an altitude of 8 km. The pressure at gage 1 is 1550 N/m², gage; the pressure at gage 2 is -3875 N/m², gage.

    (a) If gage 1 is known to be at the stagnation point, what is the velocity of the airplane? What is its Mach number?

    (b) What is the free-stream dynamic pressure for this test condition?

    (c) What is the velocity of the air at the edge of the boundary layer at the second point relative to the airplane? What is the velocity relative to the ground? What is C_p for this gage?

3.8. Air flows through a converging pipe section, as shown in Fig. P3.8. Since the centerline of the duct is horizontal, the change in potential energy is zero. The pitot probe at the upstream station provides a measure of the total pressure (or stagnation pressure). The downstream end of the U-tube provides a measure of the static pressure at the second station. Assuming the density of air to be 0.00238 slug/ft³ and neglecting the effects of

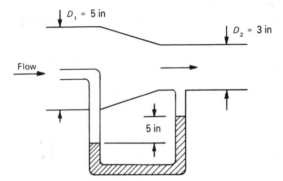

Figure P3.8

viscosity, compute the volumetric flow rate in ft³/s. The fluid in the manometer is unity weight oil (ρ_{oil} = 1.9404 slug/ft³).

3.9. An in-draft wind tunnel (Fig. P3.9) takes air from the quiescent atmosphere (outside the tunnel) and accelerates it in the converging section, so that the velocity of the air at a point in the test section but far from the model is 50 m/s. What is the static pressure at this point? What is the pressure at the stagnation point on a model in the test section? Use Table 1.2 to obtain the properties of the ambient air.

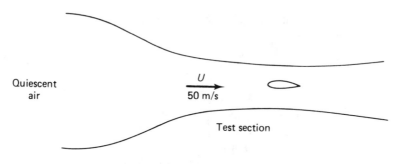

Quiescent
air

$\underrightarrow{U}$
50 m/s

Test section

Figure P3.9

3.10. What is the circulation around a circle of constant radius R_1 for the velocity field given as

$$\vec{V} = \frac{\Gamma}{2\pi r} \hat{e}_\theta$$

3.11. The velocity field for the fully developed viscous flow discussed in Example 2.2 is

$$u = \frac{1}{2\mu} \frac{dp}{dx} \left(y^2 - \frac{h^2}{4} \right)$$

$$v = 0$$

$$w = 0$$

Is the flow rotational or irrotational? Why?

3.12. Find the integral along the path $\vec{r}$ between the points (0, 0) and (1, 2) of the component of $\vec{V}$ in the direction of $\vec{r}$ for the following three cases:
(a) $\vec{r}$ a straight line.
(b) $\vec{r}$ a parabola with vertex at the origin and opening to the right.
(c) $\vec{r}$ a portion of the x axis and a straight line perpendicular to it.

The components of $\vec{V}$ are given by the expressions

$$u = x^2 + y^2$$

$$v = 2xy^2$$

3.13. Consider the velocity field given in Problem 3.12:

$$\vec{V} = (x^2 + y^2)\hat{i} + 2xy^2\hat{j}$$

Is the flow rotational or irrotational? Calculate the circulation around the right triangle shown in Fig. P3.13.

$$\oint \vec{V} \cdot \vec{dr} = ?$$

What is the integral of the component of the curl $\vec{V}$ over the surface of the triangle? That is,

$$\iint (\nabla \times \vec{V}) \cdot \hat{n} \, \vec{dA} = ?$$

Are the results consistent with Stokes's theorem?

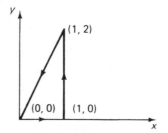

Figure P3.13

3.14. The absolute value of the velocity and the equation of the potential function lines in a two-dimensional velocity field (Fig. P3.14) are given by the expressions

$$|V| = \sqrt{4x^2 + 4y^2}$$
$$\phi = x^2 - y^2 + C$$

Evaluate both the left-hand side and the right-hand side of equation (3.16) to demonstrate the validity of Stokes's theorem for this irrotational flow.

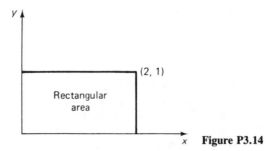

Figure P3.14

3.15. Consider the incompressible, irrotational two-dimensional flow where the potential function is

$$\phi = K \ln \sqrt{x^2 + y^2}$$

where K is an arbitrary constant.

(a) What is the velocity field for this flow? Verify that the flow is irrotational. What is the magnitude and direction of the velocity at (2, 0), at $(\sqrt{2}, \sqrt{2})$, and at (0, 2)?

(b) What is the stream function for this flow? Sketch the streamline pattern.

(c) Sketch the lines of constant potential. How do the lines of equipotential relate to the streamlines?

3.16. The stream function of a two-dimensional, incompressible flow is given by

$$\psi = \frac{\Gamma}{2\pi} \ln r$$

(a) Graph the streamlines.
(b) What is the velocity field represented by this stream function? Does the resultant velocity field satisfy the continuity equation?
(c) Find the circulation about a path enclosing the origin. For the path of integration, use a circle of radius 3 with a center at the origin. How does the circulation depend on the radius?

3.17. The absolute value of the velocity and the equation of the streamlines in a velocity field are given by

$$|\vec{V}| = \sqrt{2y^2 + x^2 + 2xy}$$

$$\psi = y^2 + 2xy = \text{constant}$$

Find u and v.

3.18. The absolute value of the velocity and the equation of the streamlines in a two-dimensional velocity field (Fig. P3.18) are given by the expressions

$$|\vec{V}| = \sqrt{5y^2 + x^2 + 4xy}$$

$$\psi = xy + y^2 = C$$

Find the integral over the surface shown of the normal component of curl $\vec{V}$ by two methods.

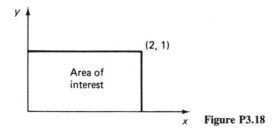

Figure P3.18

3.19. Given an incompressible, steady flow, where the velocity is

$$\vec{V} = (x^2y - xy^2)\hat{i} + \left(\frac{y^3}{3} - xy^2\right)\hat{j}$$

(a) Does the velocity field satisfy the continuity equation? Does a stream function exist? If a stream function exists, what is it?
(b) Does a potential function exist? If a potential function exists, what is it?
(c) For the region shown in Fig. P3.19, evaluate

$$\iint (\nabla \times \vec{V}) \cdot \hat{n} \, dA = ?$$

and

$$\oint \vec{V} \cdot \vec{dr} = ?$$

to demonstrate that Stokes's theorem is valid.

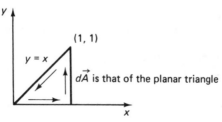

Circulation around the triangle   **Figure P3.19**

3.20. Consider the superposition of a uniform flow and a source of strength K. If the distance from the source to the stagnation point is R, calculate the strength of the source in terms of U_∞ and R.

(a) Determine the equation of the streamline that passes through the stagnation point. Let this streamline represent the surface of the configuration of interest.

(b) Noting that

$$v_r = \frac{1}{r} \frac{\partial \psi}{\partial \theta} \qquad v_\theta = -\frac{\partial \psi}{\partial r}$$

complete the following table for the surface of the configuration.

θ	$\dfrac{r}{R}$	$\dfrac{U}{U_\infty}$	C_p
30°			
45°			
90°			
135°			
150°			
180°			

3.21. A two-dimensional free vortex is located near an infinite plane at a distance h above the plane (Fig. P3.21). The pressure at infinity is p_∞ and the velocity at infinity is U_∞ parallel to the plane. Find the total force (per unit depth normal to the paper) on the plane if the pressure on the underside of the plane is p_∞. The strength of the vortex is Γ. The fluid is incompressible and perfect. To what expression does the force simplify if h becomes very large?

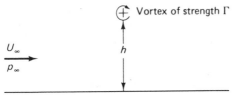

p_∞ **Figure P3.21**

3.22. A perfect, incompressible irrotational fluid is flowing past a wall with a sink of strength K per unit length at the origin (Fig. P3.22). At infinity the flow is parallel and of uniform velocity U_∞. Determine the location of the stagnation point x_0 in terms of U_∞ and K. Find the pressure distribution along the wall as a function of x. Taking the free-stream static pressure at infinity to be p_∞, express the pressure coefficient as a function of x/x_0. Sketch the resulting pressure distribution.

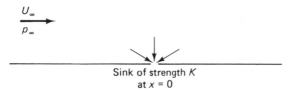

Sink of strength K
at $x = 0$ **Figure P3.22**

3.23. What is the stream function that represents the potential flow about a cylinder whose radius is 1 m and which is located in an air stream where the free-stream velocity is 50 m/s? What is the change in pressure from the free-stream value to the value at the top of the cylinder (i.e., $\theta = 90°$)? What is the change in pressure from the free-stream value to that at the stagnation point (i.e., $\theta = 180°$)? Assume that the free-stream conditions are those of the standard atmosphere at sea level.

3.24. A cylindrical tube with three radially drilled orifices, as shown in Fig. P3.24, can be used as a flow-direction indicator. Whenever the pressure on the two side holes is equal, the pressure at the center hole is the stagnation pressure. The instrument is called a *direction-finding pitot tube*, or a *cylindrical yaw probe*.

   (a) If the orifices of a direction-finding pitot tube were to be used to measure the free-stream static pressure, where would they have to be located if we use our solution for flow around a cylinder?
   (b) For a direction-finding pitot tube with orifices located as calculated in part (a), what is the sensitivity? Let the sensitivity be defined as the pressure change per unit angular change (i.e., $\partial p/\partial \theta$).

Figure P3.24

3.25. Consider the flow around the quonset hut shown in Fig. P3.25 to be represented by superimposing a uniform flow and a doublet. Assume steady, incompressible, potential flow. The ground plane is represented by the plane of symmetry and the hut by the upper half of the cylinder. The free-stream velocity is 175 km/h; the radius R_0 of the hut is 6 m. The door is not well sealed and the static pressure inside the hut is equal to that on the outer surface of the hut, where the door is located.

(a) If the door to the hut is located at ground level (i.e., at the stagnation point), what is the net lift acting on the hut? What is the lift coefficient?

(b) Where should the door be located (i.e., at what angle θ_0 relative to the ground) so that the net force on the hut will vanish?

For both parts of the problem, the opening is very small compared to the radius R_0. Thus, the pressure on the door is essentially constant and equal to the value of the angle θ_0 at which the door is located. Assume that the wall is negligibly thin.

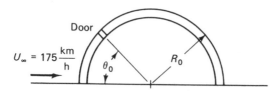

Figure P3.25

3.26. Consider an incompressible flow around a semicylinder as shown in Fig. P3.26. Assume that velocity distribution for the windward surface of the cylinder is given by the inviscid solution

$$\vec{V} = -2U_\infty \sin\theta\, \hat{e}_\theta$$

Calculate the lift and drag coefficients if the base pressure (i.e., the pressure on the flat, or leeward, surface) is equal to the pressure at the separation point, p_{corner}.

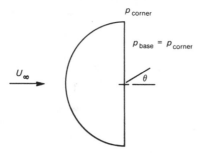

Figure P3.26

3.27. A semicylindrical tube, as shown in Fig. P3.26, is submerged in a stream of air where $\rho_\infty = 1.22$ kg/m³ and $U_\infty = 75$ m/s. The radius is 0.3 m. What are the lift and drag forces acting on the tube using the equations developed in Problem 3.26?

3.28. You are to design quonset huts for a military base in the mideast. The design wind speed is 100 ft/s. The static free-stream properties are those for standard sea-level conditions. The quonset hut may be considered to be a closed (no leaks) semicylinder, whose radius is 15 ft, mounted on tie-down blocks, as shown in Fig. P3.28. The flow is such that the velocity distribution and, thus, the pressure distribution over the top of the hut (the semicircle of the sketch) is represented by the potential function

$$\phi = U_\infty r \cos \theta + \frac{B}{r} \cos \theta$$

When calculating the flow over the hut, neglect the presence of the air space under the hut. The air under the hut is at rest and the pressure is equal to the stagnation pressure, $p_t \, (= p_\infty + \frac{1}{2}\rho_\infty U_\infty^2)$.

(a) What is the value of B for the 15-ft-radius (R) quonset hut?

(b) What is the net lift force acting on the quonset hut?

(c) What is the net drag force acting on the quonset hut?

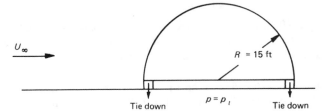

U_∞ $R = 15$ ft

Tie down $p = p_t$ Tie down   **Figure P3.28**

3.29. Using equation (3.56) to define the surface velocity distribution for inviscid flow around a cylinder with circulation, derive the expression for the local static pressure as a function of θ. Substitute the pressure distribution into the expression for the lift to verify that equation (3.58) gives the lift force per unit span. Using the definition that

$$C_l = \frac{1}{q_\infty 2R}$$

what is the section lift coefficient?

3.30. Combining equations (3.45) and (3.49), it has been shown that the section lift coefficient for inviscid flow around a cylinder is

$$C_l = -\frac{1}{2}\int_0^{2\pi} C_p \sin \theta \, d\theta \qquad (3.48)$$

Using equation (3.57) to define the pressure coefficient distribution for inviscid flow with circulation, calculate the section lift coefficient for this flow.

3.31. There were early attempts in the development of the airplane to use rotating cylinders as airfoils. Consider such a cylinder having a diameter of 1 m and a length of 10 m. If this cylinder is rotated at 100 rpm while the plane moves at a speed of 100 km/h through the air at 2-km standard atmosphere, estimate the maximum lift that could be developed disregarding end effects.

3.32. Using the procedures illustrated in Example 3.6, calculate the contribution of the source distribution on panel 3 to the normal velocity at the control point of panel 4. The configuration geometry is illustrated in Fig. 3.26.

3.33. Consider the pressure distribution shown in Fig. P3.33 for the windward and leeward surfaces of a thick disk whose axis is parallel to the free-stream flow. What is the corresponding drag coefficient?

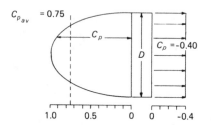

$C_{p_{av}} = 0.75$

C_p

$C_p = -0.40$

D

1.0   0.5   0   0   -0.4   **Figure P3.33**

3.34. Consider an incompressible flow around a hemisphere, as shown in Fig. P3.26. Assume that the velocity distribution for the windward surface of the cylinder is given by the inviscid solution

$$V = -\tfrac{3}{2} U_\infty \sin \omega \qquad (3.77)$$

Calculate the lift and drag coefficients if the base pressure (i.e., the pressure on the flat, or leeward surface) is equal to the pressure at the separation point, p_{corner}. How does the drag coefficient for a hemisphere compare with that for a hemicylinder (i.e., Problem 3.26)?

3.35. A hemisphere, as shown in Fig. P3.26, is submerged in an airstream where $\rho_\infty = 0.002376$ slug/ft^3 and $U_\infty = 200$ ft/s. The radius is 1.0 ft. What are the lift and drag forces on the hemisphere using the equations developed in Problem 3.34?

3.36. Consider air flowing past a hemisphere resting on a flat surface, as shown in Fig. P3.36. Neglecting the effects of viscosity, if the internal pressure is p_i, find an expression for the pressure force on the hemisphere. At what angular location should a hole be cut in the surface of the hemisphere so that the net pressure force will be zero?

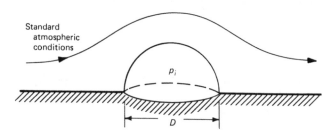

Standard
atmospheric
conditions

p_i

D

Figure P3.36

3.37. A major league pitcher is accused of hiding sandpaper in his back pocket in order to scuff up an otherwise "smooth" baseball. Why would he do this? To estimate the Reynolds number of the baseball, assume its speed to be 90 mi/h and its diameter, 2.75 in.

REFERENCES

3.1. McMillin, S. N., J. L. Thomas, and E. M. Murman, "Euler and Navier-Stokes Solutions for the Leeside Flow over Delta Wings at Supersonic Speeds," *AIAA Paper 87–2270*, presented at the AIAA 5th Applied Aerodynamics Conference, Monterey, Calif., Aug. 1987.

3.2. Verhoff, A., and P. J. O'Neil, "Accurate, Efficient Prediction Method for Supersonic/ Hypersonic Inviscid Flow," *AIAA Paper 87-1165*, presented at the AIAA 19th Fluid Dynamics, Plasma Dynamics, and Lasers Conference, Honolulu, Hawaii, June 1987.

3.3. *U.S. Standard Atmosphere, 1962*, Government Printing Office, Washington, D.C., Dec. 1962.

3.4. Schlichting, H., *Boundary Layer Theory*, McGraw-Hill Book Company, New York, 1968.

3.5. Achenbach, E., "Distribution of Local Pressure and Skin Friction Around a Circular Cylinder in Cross-Flow up to Re $= 5 \times 10^6$," *Journal of Fluid Mechanics*, Vol. 34, Part 4, 1968, pp. 625–639.

3.6. Talay, T. A., "Introduction to the Aerodynamics of Flight," *SP-367*, NASA, 1975.

3.7. Hoerner, S. F., *Fluid Dynamic Drag*, 1958, published by the author.

3.8. Hoerner, S. F., and H. V. Borst, *Fluid Dynamic Lift*, 1975, published by the author.

3.9. Hess, J. L., and A. M. O. Smith, *Calculation of Potential Flow About Arbitrary Bodies*, Progress in Aeronautical Sciences, Vol. 8, Pergamon Press, New York, 1966.

3.10. Kellogg, O. D., *Foundations of Potential Theory*, Dover Publications, New York, 1953.

4

INCOMPRESSIBLE, VISCOUS BOUNDARY LAYERS

The equation for the conservation of linear momentum was developed in Chapter 2 by applying Newton's law, which states that the net force acting on a fluid particle is equal to the time rate of change of the linear momentum of the fluid particle. The principal forces considered were those that act directly on the mass of the fluid element (i.e., the body forces) and those that act on its surface (i.e., the pressure forces and shear forces). The resultant equations are known as the Navier-Stokes equations. Even today, there are no general solutions for the complete Navier-Stokes equations. Nevertheless, reasonable approximations can be introduced to describe the motion of a viscous fluid if the viscosity is either very large or very small. The latter case is of special interest to us, since two important fluids, water and air, have very small viscosities. Not only is the viscosity of these fluids very small, but the velocity for many of the practical applications relevant to this text is such that the Reynolds number is very large.

Even in the limiting case where the Reynolds number is large, it is not permissible simply to omit the viscous terms completely, because the solution of the simplified equation could not be made to satisfy the complete boundary conditions. However, for many high Reynolds number flows, the flow field may be divided into two regions: (1) a viscous boundary layer adjacent to the surface of the vehicle and (2) the essentially inviscid flow outside the boundary layer. The velocity of the fluid particles increases from a value of zero (in a vehicle-fixed coordinate system) at the wall to the value that corresponds to the external "frictionless" flow outside the boundary layer.

Outside the boundary layer, the transverse velocity gradients become so small that the shear stresses acting on a fluid element are negligibly small. Thus, the effect of the viscous terms may be ignored when solving for the flow field external to the boundary layer. The solution for the inviscid portion of the flow field must satisfy the boundary conditions, that the velocity of the fluid particles far from the body be equal to the free-stream value and that the velocity of the fluid particles adjacent to the body be parallel to the "surface" (but not necessarily of zero magnitude). This latter condition represents the physical requirement that there is no flow through a solid surface. When using the two-region flow model to solve for the flow field, the first step is to calculate the inviscid flow field subject to the boundary condition that the flow be parallel to the actual surface of the body. The second step is to calculate the resultant boundary layer and its displacement thickness (which will be defined in this chapter). Then the inviscid flow field is recalculated using the effective surface as the boundary condition. The geometry of the effective surface is determined by adding the displacement thickness to the surface coordinate of the actual configuration. As discussed in Ref. 4.1, the iterative procedure required to converge to a solution requires an understanding of each region of the flow field and their interactions. In Chapter 3 we generated solutions for the inviscid flow field for a variety of configurations. In this chapter we examine the viscous region in detail, assuming that the inviscid flow field is known.

EQUATIONS GOVERNING THE BOUNDARY LAYER FOR A STEADY, TWO-DIMENSIONAL, INCOMPRESSIBLE FLOW

In this chapter we discuss techniques by which we can obtain engineering solutions when the boundary layer is either laminar or turbulent. Thus, for the purposes of this text, we shall assume that we know whether the boundary layer is laminar or turbulent. The transition process is quite complex and depends on many parameters (e.g., surface roughness, surface temperature, pressure gradient, and Mach number). A brief summary of the factors affecting transition is presented later in this chapter. For a more detailed discussion of the parameters that affect transition, the reader is referred to Refs. 4.2 and 4.3.

To simplify the development of the solution techniques, we will consider the flow to be steady, two-dimensional, and constant property (or, equivalently, incompressible for a gas flow). By restricting ourselves to such flows, we can concentrate on the development of the solution techniques themselves. As shown in Fig. 4.1, the coordinate system is fixed to the surface of the body. The x coordinate is measured in the streamwise direction along the surface of the configuration. The stagnation point (or the leading edge if the configuration is a sharp object) is at $x = 0$. The y coordinate is perpendicular to the surface. This coordinate system is used throughout this chapter.

Referring to equation (2.4), the differential form of the continuity equation for this flow is

$$\frac{\partial u}{\partial x} + \frac{\partial v}{\partial y} = 0 \qquad (4.1)$$

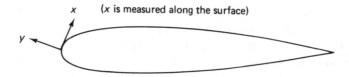

Figure 4.1   Coordinate system for the boundary-layer equations.

Referring to equation (2.16) and neglecting the body forces, the x component of momentum is

$$\rho u \frac{\partial u}{\partial x} + \rho v \frac{\partial u}{\partial y} = -\frac{\partial p}{\partial x} + \mu \frac{\partial^2 u}{\partial x^2} + \mu \frac{\partial^2 u}{\partial y^2} \tag{4.2}$$

Similarly, the y component of momentum is

$$\rho u \frac{\partial v}{\partial x} + \rho v \frac{\partial v}{\partial y} = -\frac{\partial p}{\partial y} + \mu \frac{\partial^2 v}{\partial x^2} + \mu \frac{\partial^2 v}{\partial y^2} \tag{4.3}$$

Note that, if the boundary layer is thin and the streamlines are not highly curved, then $u \gg v$. Thus, if we compare each term in equation (4.3) with the corresponding term in equation (4.2), we conclude that

$$\rho u \frac{\partial u}{\partial x} > \rho u \frac{\partial v}{\partial x} \qquad \rho v \frac{\partial u}{\partial y} > \rho v \frac{\partial v}{\partial y} \qquad \mu \frac{\partial^2 u}{\partial x^2} > \mu \frac{\partial^2 v}{\partial x^2} \qquad \mu \frac{\partial^2 u}{\partial y^2} > \mu \frac{\partial^2 v}{\partial y^2}$$

Thus, as discussed in Chapter 2, we conclude that

$$\frac{\partial p}{\partial x} > \frac{\partial p}{\partial y}$$

The essential information supplied by the y component of the momentum equation is that the pressure variation in the y direction may be neglected for most boundary layer flows. This is true whether the boundary layer is laminar, transitional, or turbulent. It is not true in wake flows, that is, separated regions in the lee side of blunt bodies such as those behind cylinders, which were discussed in Chapter 3. The assumption that the pressure variation across a thin boundary layer is negligible only breaks down for turbulent boundary layers at very high Mach numbers. The common assumption for thin boundary layers also may be written as

$$\frac{\partial p}{\partial y} \approx 0 \tag{4.4}$$

Thus, the local static pressure is a function of x only and is determined from the solution of the inviscid portion of the flow field. As a result, Euler's equation for a steady flow with negligible body forces, which relates the streamwise pressure gradient to the velocity gradient for the inviscid flow, can be used to evaluate the pressure gradient in the viscous region, also.

$$-\frac{\partial p}{\partial x} = -\frac{dp_e}{dx} = \rho_e u_e \frac{du_e}{dx} \tag{4.5}$$

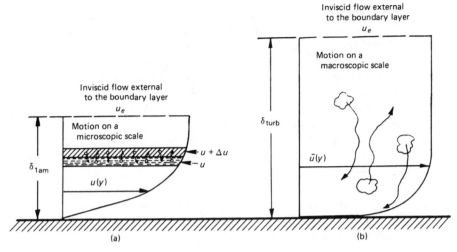

Figure 4.2 Momentum-transport models: (a) laminar boundary layer; (b) turbulent boundary layer.

Substituting equation (4.5) into equation (4.2), and noting that $\mu\ (\partial^2 u/\partial x^2) < \mu\ (\partial^2 u/\partial y^2)$, we obtain

$$\rho u \frac{\partial u}{\partial x} + \rho v \frac{\partial u}{\partial y} = \rho_e u_e \frac{du_e}{dx} + \mu \frac{\partial^2 u}{\partial y^2} \tag{4.6}$$

Let us examine equations (4.1) and (4.6). The assumption that the flow is constant property (or incompressible) implies that fluid properties, such as density ρ and viscosity μ, are constants. For low-speed flows of gases, the changes in pressure and temperature through the flow field are sufficiently small that ρ and μ are essentially constant. By limiting ourselves to incompressible flows, it is not necessary to include the energy equation in the formulation of our solution. For compressible (or high-speed) flows, the temperature changes in the flow field are sufficiently large that the temperature dependence of the viscosity and density must be included. As a result, the analysis of a compressible boundary layer involves the simultaneous solution of the continuity equation, the x momentum equation, and the energy equation. For a detailed treatment of compressible boundary-layers, the reader is referred to other sources (e.g., Refs. 4.2 and 4.4).

When the boundary layer is laminar, the transverse exchange of momentum (i.e., the momentum transfer in a direction perpendicular to the principal flow direction) takes place on a microscopic scale. As a result of the molecular movement, slower moving fluid particles from the lower layer (or lamina) of fluid move upward, slowing the particles in the upper layer. Conversely, when the faster moving fluid particles from the upper layer migrate downward, they tend to accelerate the fluid particles in that layer. This molecular interchange of momentum for a laminar flow is depicted in Fig. 4.2a. Thus, the shear stress at a point in a Newtonian fluid is that given by constitutive relations preceding equations (2.12).

For a turbulent boundary layer, there is a macroscopic transport of fluid particles, as shown in Fig. 4.2b. Thus, in addition to the laminar shear stress described in the preceding paragraph, there is an effective, *turbulent shear stress* that is due to the transverse transport of momentum and that is very large. Because slower moving fluid particles near the wall are transported well upward, the turbulent boundary layer is relatively thick. Because faster-moving fluid particles (which are normally located near the edge of the boundary layer) are transported toward the wall, they produce relatively high velocities for the fluid particles near the surface. Thus, the shear stress at the wall for a turbulent boundary layer is larger than that for a laminar boundary layer. Because the macroscopic transport of fluid introduces large localized variations in the flow at any instant, the values of the fluid properties and the velocity components are (in general) the sum of the ''average'' value and a fluctuating component.

We could introduce the fluctuating characteristics of turbulent flow at this point and treat both laminar and turbulent boundary layers in a unified fashion. For a laminar boundary layer the fluctuating components of the flow would be zero. However, to simplify the discussion, we will first discuss laminar flows and their analysis and then turbulent boundary layers and their analysis.

BOUNDARY CONDITIONS

Let us now consider the boundary conditions that we must apply in order to obtain the desired solutions. Since we are considering that portion of the flow field where the viscous forces are important, the condition of no slip on the solid boundaries must be satisfied. That is, at $y = 0$:

$$u(x, 0) = 0 \qquad (4.7a)$$

At a solid wall, the normal component of velocity must be zero. Thus,

$$v(x, 0) = 0 \qquad (4.7b)$$

Porous walls through which fluid can flow are treated in the problems at the end of the chapter. Furthermore, as we reach the edge of the boundary layer (i.e., at $y = \delta$, where δ is the local thickness of the boundary layer), the streamwise component of the velocity equals that given by the inviscid solution. In equation form,

$$u(x, \delta) = u_e(x) \qquad (4.8)$$

Note that throughout this chapter, the subscript e will be used to denote parameters evaluated at the edge of the boundary layer (i.e., those for the inviscid solution).

INCOMPRESSIBLE, LAMINAR BOUNDARY LAYER

In this section we analyze the boundary layer in the region from the stagnation point (or from the leading edge of a sharp object) to the onset of transition (i.e., that ''point'' at which the boundary layer becomes turbulent). The reader should note that, in reality, the boundary layer does not go from a laminar state to a turbulent state at a point but that the transition process takes place over a distance. The length of the transition zone may be as long as the laminar region. Typical velocity profiles for the laminar boundary layer are presented in Fig. 4.3. The streamwise (or x) component of velocity is presented as a function of distance from the wall (the y coordinate). Instead of presenting the dimensional parameter u, which is a function both of x and y, let us seek a dimensionless velocity parameter that perhaps can be written as a function of a single variable. Note that, at each station, the velocity varies from zero at $y = 0$ (i.e., at the wall) to u_e at $y = \delta$ (i.e., at the edge of the boundary layer). The local velocity at the edge of the boundary layer u_e is a function of x only. Thus, a logical dimensionless velocity parameter is u/u_e.

Instead of using the dimensional y coordinate, we will use a dimensionless coordinate η, which is proportional to y/δ for these incompressible, laminar boundary layers. The boundary-layer thickness δ at any x station depends not only on the magnitude of x but on the kinematic viscosity, the local velocity at the edge of the boundary layer, and the velocity variation from the origin to the point of interest. Thus, we will introduce the coordinate transformation for η:

$$\eta = \frac{u_e y}{\sqrt{2\nu s}} \tag{4.9a}$$

where ν is the kinematic viscosity, as defined in Chapter 1, and where

$$s = \int u_e\, dx \tag{4.9b}$$

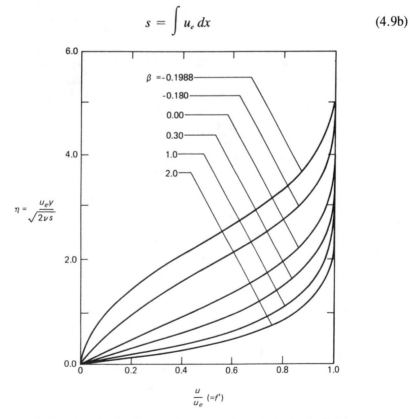

Figure 4.3 Solutions for the dimensionless streamwise velocity for the Falkner-Skan, laminar, similarity flows.

Note that for flow past a flat plate, where u_e is a constant (independent of x),

$$\eta = y \sqrt{\frac{u_e}{2\nu x}} \tag{4.10}$$

Those readers who are familiar with the transformations used in more complete treatments of boundary layers will recognize that this definition for η is consistent with that commonly used to transform the incompressible laminar boundary layer on a flat plate (Ref. 4.3). The flat-plate solution is the classical Blasius solution. This transformation is also consistent with more general forms used in the analysis of a compressible laminar flow (Ref. 4.4). By using this definition of s as the transformed x coordinate, we can account for the effect of the variation in u_e on the streamwise growth of the boundary layer.

Note that we have two equations [equations (4.1) and (4.6)] with two unknowns, the velocity components: u and v. Since the flow is two-dimensional and the density is constant, the necessary and sufficient conditions for the existence of a stream function are satisfied. (*Note*: Because of viscosity, the boundary-layer flow cannot be considered as irrotational. Therefore, potential functions cannot be used to describe the flow in the boundary layer.) We shall define the stream function such that

$$u = \left(\frac{\partial \psi}{\partial y}\right)_x \text{ and } v = -\left(\frac{\partial \psi}{\partial x}\right)_y$$

By introducing the stream function, the continuity equation (4.1) is automatically satisfied. Thus, we need to solve only one equation, the x component of the momentum equation, in terms of only one unknown, the stream function.

Let us now transform our equations from the x, y coordinate system to the s, η coordinate system. To do this, note that

$$\left(\frac{\partial}{\partial y}\right)_x = \left(\frac{\partial \eta}{\partial y}\right)_x \left(\frac{\partial}{\partial \eta}\right)_s = \frac{u_e}{\sqrt{2\nu s}} \left(\frac{\partial}{\partial \eta}\right)_s \tag{4.11a}$$

$$\left(\frac{\partial}{\partial x}\right)_y = \left(\frac{\partial s}{\partial x}\right)_y \left[\left(\frac{\partial \eta}{\partial s}\right)_y \left(\frac{\partial}{\partial \eta}\right)_s + \left(\frac{\partial}{\partial s}\right)_\eta\right] \tag{4.11b}$$

Thus, the streamwise component of velocity may be written in terms of the stream function as

$$u = \left(\frac{\partial \psi}{\partial y}\right)_x = \frac{u_e}{\sqrt{2\nu s}} \left(\frac{\partial \psi}{\partial \eta}\right)_s \tag{4.12a}$$

Let us introduce a transformed stream function f, which we define so that

$$u = u_e \left(\frac{\partial f}{\partial \eta}\right)_s \tag{4.12b}$$

Comparing equations (4.12a) and (4.12b), we see that

$$f = \frac{1}{\sqrt{2\nu s}} \psi \tag{4.13}$$

Similarly, we can develop an expression for the transverse component of velocity:

$$v = -\left(\frac{\partial \psi}{\partial x}\right)_y$$

$$= -u_e \sqrt{2vs} \left[\left(\frac{\partial \eta}{\partial s}\right)_y \left(\frac{\partial f}{\partial \eta}\right)_s + \left(\frac{\partial f}{\partial s}\right)_\eta + \left(\frac{f}{2s}\right) \right]$$

(4.14)

In equations (4.12b) and (4.14), we have written the two velocity components, which were the unknowns in the original formulation of the problem, in terms of the transformed stream function. We can rewrite equation (4.6) using the differentials of the variables in the s, η coordinate system. For example,

$$\frac{\partial^2 u}{\partial y^2} = \frac{\partial}{\partial y} \left[\frac{\partial \eta}{\partial y} \frac{\partial(u_e f')}{\partial \eta} \right] = \left(\frac{\partial \eta}{\partial y}\right)^2 u_e \frac{\partial^2 f'}{\partial \eta^2}$$

$$= \frac{u_e^2}{2vs} u_e f'''$$

where the prime ($'$) denotes differentiation with respect to η. Using these substitutions, the momentum equation becomes

$$ff'' + f''' + [1 - (f')^2] \frac{2s}{u_e} \frac{du_e}{ds} = 2s \left[f' \left(\frac{\partial f'}{\partial s}\right)_y - f'' \left(\frac{\partial f}{\partial s}\right)_y \right] \quad (4.15)$$

As discussed earlier, by using a stream function we automatically satisfy the continuity equation. Thus, we have reduced the formulation to one equation with one unknown.

For many problems, the parameter $(2s/u_e)(du_e/ds)$, which is represented by the symbol β, is assumed to be constant. The assumption that β is a constant implies that the s derivatives of f and f' are zero. As a result, the transformed stream function and its derivatives are functions of η only, and equation (4.15) becomes the ordinary differential equation:

$$ff'' + f''' + [1 - (f')^2]\beta = 0 \quad (4.16)$$

Because the dimensionless velocity function f' is a function of η only, the velocity profiles at one s station are the same as those at another. Thus, the solutions are called *similar solutions*. Note that the Reynolds number does not appear as a parameter when the momentum equation is written in the transformed coordinates. It will appear when our solutions are transformed back into the x, y coordinate system. There are no analytical solutions to this third-order equation, which is known as the *Falkner–Skan equation*. Nevertheless, there are a variety of well-documented numerical techniques available to solve it.

Let us examine the three boundary conditions necessary to solve the equation. Substituting the definition that

$$f' = \frac{u}{u_e}$$

into the boundary conditions given by equations (4.7) and (4.8),

$$f'(s, 0) = 0 \quad (4.17a)$$

and

$$f'(s, \eta_e) = 1.0 \tag{4.17b}$$

where η_e is given by

$$\eta_e = \frac{u_e \delta}{\sqrt{2\nu s}}$$

Using equations (4.14) and (4.7), the boundary condition that the transverse velocity be zero at the wall becomes

$$f(s, 0) = 0 \tag{4.17c}$$

Since f is the transformed stream function, this third boundary condition states that the stream function is constant along the wall (i.e., the surface is a streamline). This is consistent with the requirement that $v(x, 0) = 0$, which results because the component of velocity normal to a streamline is zero.

Numerical Solutions for the Falkner–Skan Problem

Numerical solutions of equation (4.16) that satisfy the boundary conditions represented by equation (4.17) have been generated for $-0.1988 \leq \beta \leq +2.0$. The resultant velocity profiles are presented in Fig. 4.3 and Table 4.1. Since

$$\beta = \frac{2s}{u_e} \frac{du_e}{ds}$$

these solutions represent a variety of inviscid flow fields and, therefore, represent the flow around different configurations. Note that when $\beta = 0$, $u_e = $ constant, and the solution is that for flow past a flat plate (known as the Blasius solution). Negative values of β correspond to cases where the inviscid flow is decelerating, which corresponds to an adverse pressure gradient [i.e., $(dp/dx) > 0$]. The positive values of β correspond to an accelerating inviscid flow, which results from a favorable pressure gradient [i.e., $(dp/dx) < 0$].

As noted in the discussion of flow around cylinder in Chapter 3, when the air particles in the boundary layer encounter a relatively large adverse pressure gradient, boundary-layer separation may occur. Separation results because the fluid particles in the viscous layer have been slowed to the point that they cannot overcome the adverse pressure gradient. The effect of an adverse pressure gradient is evident in the velocity profiles presented in Fig. 4.3. When $\beta = -0.1988$, not only is the streamwise velocity zero at the wall, but the velocity gradient $\partial u/\partial y$ is also zero at the wall. If the adverse pressure gradient were any larger, the laminar boundary layer would separate from the surface, and flow reversal would occur.

For the accelerating flows (i.e., positive β), the velocity increases rapidly with distance from the wall. Thus, $\partial u/\partial y$ at the wall is relatively large. Referring to equation (1.4), one would expect that the shear force at the wall would be relatively large. To calculate the shear force at the wall,

$$\tau = \left(\mu \frac{\partial u}{\partial y}\right)_{y=0} \tag{4.18}$$

TABLE 4.1 NUMERICAL VALUES OF THE DIMENSIONLESS STREAMWISE
VELOCITY $f'(\eta)$ FOR THE FALKNER-SKAN, LAMINAR, SIMILARITY FLOWS

η	β					
	-0.1988	-0.180	0.000	0.300	1.000	2.000
0.0	0.0000	0.0000	0.0000	0.0000	0.0000	0.0000
0.1	0.0010	0.0138	0.0470	0.0760	0.1183	0.1588
0.2	0.0040	0.0293	0.0939	0.1489	0.2266	0.2979
0.3	0.0089	0.0467	0.1408	0.2188	0.3252	0.4185
0.4	0.0159	0.0658	0.1876	0.2857	0.4145	0.5219
0.5	0.0248	0.0867	0.2342	0.3494	0.4946	0.6096
0.6	0.0358	0.1094	0.2806	0.4099	0.5663	0.6834
0.7	0.0487	0.1337	0.3265	0.4671	0.6299	0.7450
0.8	0.0636	0.1597	0.3720	0.5211	0.6859	0.7959
0.9	0.0804	0.1874	0.4167	0.5717	0.7351	0.8377
1.0	0.0991	0.2165	0.4606	0.6189	0.7779	0.8717
1.2	0.1423	0.2790	0.5452	0.7032	0.8467	0.9214
1.4	0.1927	0.3462	0.6244	0.7742	0.8968	0.9531
1.6	0.2498	0.4169	0.6967	0.8325	0.9323	0.9727
1.8	0.3127	0.4895	0.7611	0.8791	0.9568	0.9845
2.0	0.3802	0.5620	0.8167	0.9151	0.9732	0.9915
2.2	0.4510	0.6327	0.8633	0.9421	0.9839	0.9955
2.4	0.5231	0.6994	0.9011	0.9617	0.9906	
2.6	0.5946	0.7605	0.9306	0.9755	0.9946	
2.8	0.6635	0.8145	0.9529	0.9848		
3.0	0.7277	0.8606	0.9691	0.9909		
3.2	0.7858	0.8985	0.9804	0.9947		
3.4	0.8363	0.9285	0.9880			
3.6	0.8788	0.9514	0.9929			
3.8	0.9131	0.9681	0.9959			
4.0	0.9398	0.9798				
4.2	0.9597	0.9876				
4.4	0.9740	0.9927				
4.6	0.9838	0.9959				
4.8	0.9903					
5.0	0.9944					

let us introduce the transformation presented in equation (4.11a). Thus, the shear is

$$\tau = \frac{\mu u_e^2}{\sqrt{2vs}} f''(0) \tag{4.19}$$

Because of its use in equation (4.19), we will call f'' the *transformed shear function*.
Theoretical values of $f''(0)$ are presented in Fig. 4.4 and in Table 4.2. Note that $f''(0)$
is a unique function of β for these incompressible, laminar boundary layers. The
value does not depend on the stream conditions, such as velocity or Reynolds number.
When $\beta = 0$, $f''(0) = 0.4696$. Thus, for the laminar boundary layer on a flat
plate,

$$\tau = 0.332 \sqrt{\frac{\rho \mu u_e^3}{x}} \tag{4.20}$$

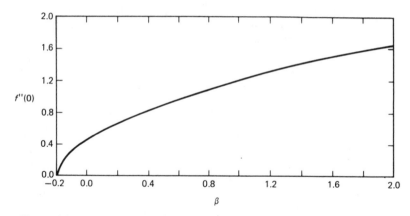

Figure 4.4 Transformed shear function at the wall for laminar boundary layers as a function of β.

For flow past a flat plate, the velocity at the edge of the boundary layer (u_e) is equal to the free-stream value (U_∞). We can express the shear in terms of the dimensionless skin-friction coefficient:

$$C_f = \frac{\tau}{\frac{1}{2}\rho_\infty U_\infty^2} = \frac{0.664}{\sqrt{Re_x}} \tag{4.21}$$

where

$$Re_x = \frac{\rho u_e x}{\mu} = \frac{\rho U_\infty x}{\mu} \tag{4.22}$$

Mentally substituting the values of $f''(0)$ presented in Fig. 4.4, we see that the shear is zero when $\beta = -0.1988$. Thus, this value of β corresponds to the onset of separation. Conversely, when the inviscid flow is accelerating, the shear is greater than that for a zero pressure gradient flow.

The transformed stream function (f), the dimensionless streamwise velocity (f'), and the shear function (f'') are presented as a function of η for a laminar boundary layer on a flat plate in Table 4.3. Note that as η increases (i.e., as y increases) the shear goes to zero and the function f' tends asymptotically to 1.0. If we define the boundary-layer thickness as that distance from the wall for which $u = 0.99u_e$, we see that

$$\eta_e \approx 3.5$$

TABLE 4.2 THEORETICAL VALUES OF THE TRANSFORMED SHEAR FUNCTION AT THE WALL FOR LAMINAR BOUNDARY LAYERS AS A FUNCTION OF β

β	−0.1988	−0.180	0.000	0.300	1.000	2.000
$f''(0)$	0.0000	0.1286	0.4696	0.7748	1.2326	1.6872

TABLE 4.3 SOLUTION FOR THE LAMI-
NAR BOUNDARY LAYER ON A FLAT
PLATE

η	f	f'	f''
0.0	0.0000	0.0000	0.4696
0.1	0.0023	0.0470	0.4696
0.2	0.0094	0.0939	0.4693
0.3	0.0211	0.1408	0.4686
0.4	0.0375	0.1876	0.4673
0.5	0.0586	0.2342	0.4650
0.6	0.0844	0.2806	0.4617
0.7	0.1147	0.3265	0.4572
0.8	0.1497	0.3720	0.4512
0.9	0.1891	0.4167	0.4436
1.0	0.2330	0.4606	0.4344
1.2	0.3336	0.5452	0.4106
1.4	0.4507	0.6244	0.3797
1.6	0.5829	0.6967	0.3425
1.8	0.7288	0.7610	0.3005
2.0	0.8868	0.8167	0.2557
2.2	1.0549	0.8633	0.2106
2.4	1.2315	0.9010	0.1676
2.6	1.4148	0.9306	0.1286
2.8	1.6032	0.9529	0.0951
3.0	1.7955	0.9691	0.0677
3.2	1.9905	0.9804	0.0464
3.4	2.1874	0.9880	0.0305
3.5	2.2863	0.9907	0.0244
4.0	2.7838	0.9978	0.0069
4.5	3.2832	0.9994	0.0015

independent of the specific flow properties of the free stream. Converting this to a physical distance, the corresponding boundary-layer thickness (δ) is

$$\delta = y_e = \eta_e \sqrt{\frac{2vx}{u_e}}$$

or

$$\frac{\delta}{x} = \frac{5.0}{\sqrt{Re_x}} \qquad (4.23)$$

Thus, the thickness of a laminar boundary layer is proportional to $\sqrt{x}$ and is inversely proportional to the square root of the Reynolds number.

Although the transverse component of velocity at the wall is zero, it is not zero at the edge of the boundary layer. Referring to equation (4.14), we can see that

$$\frac{v_e}{u_e} = \frac{1}{\sqrt{2}} \sqrt{\frac{v}{u_e x}} \left[\eta_e(f')_e - f_e \right] \tag{4.24}$$

Using the values given in Table 4.3,

$$\frac{v_e}{u_e} = \frac{0.84}{\sqrt{Re_x}} \tag{4.25}$$

This means that at the outer edge there is an outward flow, which is due to the fact that the increasing boundary-layer thickness causes the fluid to be displaced from the wall as it flows along it. There is no boundary-layer separation for flow past a flat plate, since the streamwise pressure gradient is zero.

Since the streamwise component of the velocity in the boundary layer asymptotically approaches the local free-stream value, the magnitude of δ is very sensitive to the ratio of u/u_e, which is chosen as the criterion for the edge of the boundary layer [e.g., 0.99 was the value used to develop equation (4.23)]. A more significant measure of the boundary layer is the displacement thickness δ^*, which is the distance by which the external streamlines are shifted due to the presence of the boundary layer. Referring to Fig. 4.5,

$$\rho_e u_e \delta^* = \int_0^\delta \rho(u_e - u)\, dy$$

Thus, for any incompressible boundary layer,

$$\delta^* = \int_0^\delta \left(1 - \frac{u}{u_e} \right) dy \tag{4.26}$$

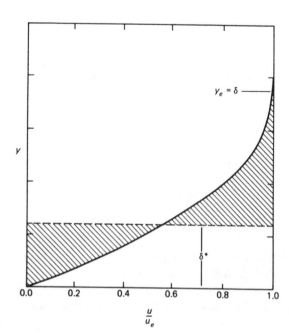

$y_e = \delta$

δ^*

0.0   0.2   0.4   0.6   0.8   1.0

$\dfrac{u}{u_e}$

Figure 4.5 Velocity profile for a laminar boundary layer on a flat plate illustrating the boundary-layer thickness δ and the displacement thickness δ^*.

Note that, since the integrand is zero for any point beyond δ, the upper limit for the integration does not matter providing it is equal to (or greater than) δ. Substituting the transformation of equation (4.10) for the laminar boundary layer on a flat plate,

$$\delta^* = \sqrt{\frac{2vx}{u_e}} \int_0^\infty (1 - f') \, d\eta$$

Using the values presented in Table 4.3, we obtain

$$\frac{\delta^*}{x} = \frac{\sqrt{2}(\eta_e - f_e)}{\sqrt{Re_x}} = \frac{1.72}{\sqrt{Re_x}} \tag{4.27}$$

Thus, for a flat plate at zero incidence in a uniform stream, the displacement thickness δ^* is on the order of one-third the boundary-layer thickness δ.

Example 4.1

A rectangular plate, whose streamwise dimension (or chord c) is 0.2 m and whose width (or span b) is 1.8 m, is mounted in a wind tunnel. The free-stream velocity is 40 m/s. The density of the air is 1.2250 kg/m³, and the absolute viscosity is 1.7894×10^{-5} kg/m · s. Graph the velocity profiles at $x = 0.0$ m, $x = 0.05$ m, $x = 0.10$ m, and $x = 0.20$ m. Calculate the chordwise distribution of the skin friction coefficient and the displacement thickness. What is the drag coefficient for the plate?

Solution Since the span (or width) of the plate is 9.0 times the chord (or streamwise dimension), let us assume that the flow is two-dimensional (i.e., it is independent of the spanwise coordinate). The maximum value of the local Reynolds number, which occurs when $x = c$, is

$$Re_c = \frac{(1.225 \text{ kg/m}^3)(40 \text{ m/s})(0.2 \text{ m})}{(1.7894 \times 10^{-5} \text{ kg/m} \cdot \text{s})} = 5.477 \times 10^5$$

This Reynolds number is close enough to the transition criteria for a flat plate that we will assume that the boundary layer is laminar for its entire length. Thus, we will use the relations developed in this section to calculate the required parameters.

Noting that

$$y = \sqrt{\frac{2vx}{u_e}} \, \eta = 8.546 \times 10^{-4} \sqrt{x} \, \eta$$

we can use the results presented in Table 4.3 to calculate the velocity profiles. The resultant profiles are presented in Fig. 4.6. At the leading edge of the flat plate (i.e., at $x = 0$), the velocity is constant (independent of y). The profiles at the other stations illustrate the growth of the boundary layer with distance from the leading edge. Note that the scale of the y coordinate is greatly expanded relative to that for the x coordinate. Even though the streamwise velocity at the edge of the boundary layer (u_e) is the same at all stations, the velocity within the boundary layer is a function of x and y. However, if the dimensionless velocity (u/u_e) is presented as a function of η, the profile is the same at all stations. Specifically, the profile is that for $\beta = 0.0$ in Fig. 4.3. Since the dimensionless profiles are similar at all x stations, the solutions are termed *similarity solutions*.

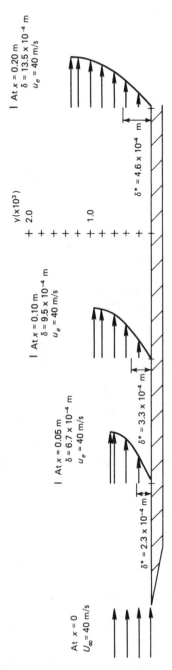

Figure 4.6 Velocity profile for the flat-plate laminar boundary layer $Re_c = 5.477 \times 10^5$.

At $x = 0$
$U_\infty = 40$ m/s

I At $x = 0.05$ m
$\delta = 6.7 \times 10^{-4}$ m
$u_e = 40$ m/s

$\delta^* = 2.3 \times 10^{-4}$ m

I At $x = 0.10$ m
$\delta = 9.5 \times 10^{-4}$ m
$u_e = 40$ m/s

$\delta^* = 3.3 \times 10^{-4}$ m

I At $x = 0.20$ m
$\delta = 13.5 \times 10^{-4}$ m
$u_e = 40$ m/s

$\delta^* = 4.6 \times 10^{-4}$ m

$y (\times 10^3)$
2.0

1.0

The displacement thickness in meters is

$$\delta^* = \frac{1.72x}{\sqrt{Re_x}} = 1.0394 \times 10^{-3} \sqrt{x}$$

The chordwise (or streamwise) distribution of the displacement thickness is presented in Fig. 4.5. These calculations verify the validity of the common assumption that the boundary layer is thin. Therefore, the inviscid solution obtained neglecting the boundary layer altogether and that obtained for the effective geometry (the actual surface plus the displacement thickness) are essentially the same.

The skin-friction coefficient is

$$C_f = \frac{0.664}{\sqrt{Re_x}} = \frac{4.013 \times 10^{-4}}{\sqrt{x}}$$

Let us now calculate the drag coefficient for the plate. Obviously, the pressure contributes nothing to the drag. Therefore, the drag force acting on the flat plate is due only to skin friction. Using general notation, we see that

$$D = 2b \int_0^c \tau \, dx \qquad (4.28)$$

We need integrate only in the x direction, since by assuming the flow to be two-dimensional, we have assumed that there is no spanwise variation in the flow. In equation (4.28), the integral, which represents the drag per unit width (or span) of the plate, is multiplied by b (the span) and by 2 (since friction acts on both the top and bottom surfaces of the plate). Substituting the expression for the laminar shear forces, given in equation (4.20),

$$D = 0.664b \sqrt{\rho\mu u_e^3} \int_0^c \frac{dx}{\sqrt{x}}$$
$$= 1.328b \sqrt{c\rho\mu u_e^3} \qquad (4.29)$$

Since the edge velocity (u_e) is equal to the free stream velocity (U_∞), the drag coefficient for the plate is, therefore,

$$C_D = \frac{D}{q_\infty cb} = \frac{2.656}{\sqrt{Re_c}} \qquad (4.30)$$

For the present problem, $C_D = 3.589 \times 10^{-3}$.

Example 4.2

The streamwise velocity component for a laminar boundary layer is often assumed to be roughly approximated by the linear relation

$$u = \frac{y}{\delta} u_e$$

where $\delta = 1.25 \times 10^{-2} \sqrt{x}$. Assume that we are trying to approximate the flow of air at standard sea-level conditions past a flat plate where $u_e = 2.337$ m/s. Calculate the streamwise distribution of the displacement thickness (δ^*), the velocity at the edge of the boundary layer (v_e), and the skin-friction coefficient (C_f). Compare the values obtained assuming a linear velocity profile with the more exact solutions presented in this chapter.

Solution   As given in Table 1.2, the standard atmospheric conditions at sea level include

$$\rho_\infty = 1.2250 \text{ kg/m}^3 \quad \text{and} \quad \mu_\infty = 1.7894 \times 10^{-5} \text{ kg/s} \cdot \text{m}$$

Thus, for constant-property flow past a flat plate,

$$\text{Re}_x = \frac{\rho_\infty u_e x}{\mu_\infty} = 1.60 \times 10^5 x$$

Using the definition for the displacement thickness of an incompressible boundary layer, equation (4.26),

$$\delta^* = \int_0^\delta \left(1 - \frac{u}{u_e}\right) dy = \delta \int_0^1 \left(1 - \frac{u}{u_e}\right) d\left(\frac{y}{\delta}\right)$$

Notice that, since we have u/u_e in terms of y/δ, we have changed our independent variable from y to y/δ. We must also change the upper limit on our integral from δ to 1. Thus, since

$$\frac{u}{u_e} = \frac{y}{\delta} \quad \text{and} \quad \delta = 1.25 \times 10^{-2} \sqrt{x}$$

then

$$\delta^* = 1.25 \times 10^{-2} \sqrt{x} \int_0^1 \left(1 - \frac{y}{\delta}\right) d\left(\frac{y}{\delta}\right) = 0.625 \times 10^{-2} \sqrt{x}$$

for the linear profile.

Using the equation for the more exact formulation [equation (4.27)], and noting that $\text{Re}_x = 1.60 \times 10^5 x$, we find that

$$\delta^* = 0.430 \times 10^{-2} \sqrt{x}$$

Using the continuity equation, we would find that the linear approximation gives a value for v_e of

$$v_e = \frac{3.125 \times 10^{-3}}{\sqrt{x}} u_e$$

Using the more exact formulation of equation (4.25) yields

$$v_e = \frac{0.84}{\sqrt{\text{Re}_x}} u_e = \frac{2.10 \times 10^{-3}}{\sqrt{x}} u_e$$

Finally, we find that the skin friction for the linear velocity approximation is given by

$$\tau = \mu \left(\frac{\partial u}{\partial y}\right)_{y=0} = \frac{\mu u_e}{\delta}$$

Thus, the skin friction coefficient is

$$C_f = \frac{\tau}{\frac{1}{2}\rho_\infty u_e^2} = \frac{2\mu_\infty}{\rho_\infty u_e \delta} = \frac{2}{1.60 \times 10^5 (1.25 \times 10^{-2} \sqrt{x})}$$

$$= \frac{1.00 \times 10^{-3}}{\sqrt{x}}$$

For the more exact formulation,

$$C_f = \frac{0.664}{\sqrt{Re_x}} = \frac{1.66 \times 10^{-3}}{\sqrt{x}}$$

Summarizing these calculations provides the following comparison:

	Linear approximation	More exact solution
δ^*	$0.625 \times 10^{-2} \sqrt{x}$	$0.430 \times 10^{-2} \sqrt{x}$
v_e	$(3.125 \times 10^{-3}\, u_e)/\sqrt{x}$	$(2.10 \times 10^{-3}\, u_e)/\sqrt{x}$
C_f	$(1.00 \times 10^{-3})/\sqrt{x}$	$(1.66 \times 10^{-3})/\sqrt{x}$

Comparing the velocity profiles, which are presented in Fig. 4.7, the reader should be able to use physical reasoning to determine that these relationships are intuitively correct. That is, if one uses a linear profile, the shear would be less than that for the exact solution, where δ^* and v_e would be greater for the linear profile.

In this example, we assumed that the boundary-layer thickness δ was $1.25 \times 10^{-2} \sqrt{x}$, which is the value obtained using the more exact formulation, i.e., equation (4.23). However, if we had used the integral approach to determine the value of δ for a linear profile, we would have obtained

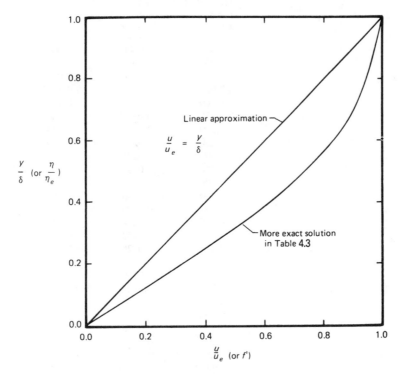

Figure 4.7 Comparison of velocity profiles for a laminar boundary layer on a flat plate.

$$\delta = \frac{3.464x}{\sqrt{Re_x}} = 0.866 \times 10^{-2} \sqrt{x}$$

Although this is considerably less than the assumed (more correct) value, the values of the other parameters (e.g., δ^* and C_f) would be in closer agreement with those given by the more exact solution.

Although the linear profile for the streamwise velocity component is a convenient approximation to use when demonstrating points about the continuity equation or about Kelvin's theorem, it clearly does not provide reasonable values for engineering parameters, such as δ^* and C_f. A more realistic approximation for the streamwise velocity component in a laminar boundary layer would be

$$\frac{u}{u_e} = \frac{3}{2}\left(\frac{y}{\delta}\right) - \frac{1}{2}\left(\frac{y}{\delta}\right)^3 \tag{4.31}$$

Example 4.3

Calculate the velocity gradient parameter β, which appears in the Falkner–Skan form of the momentum equation [equation (4.16)] for the NACA 65-006 airfoil. The coordinates of this airfoil section, which are given in Table 4.4, are given in terms of the coordinate

TABLE 4.4 PRESSURE DISTRIBUTION FOR THE NACA 65–006

$\bar{x}\ (= \bar{x}/c)$	$\bar{z}\ (= \bar{z}/c)$	C_p
0.000	0.0000	1.000
0.005	0.0048	−0.044
0.025	0.0096	−0.081
0.050	0.0131	−0.100
0.100	0.0182	−0.120
0.150	0.0220	−0.134
0.200	0.0248	−0.143
0.250	0.0270	−0.149
0.300	0.0285	−0.155
0.350	0.0295	−0.159
0.400	0.0300	−0.163
0.450	0.0298	−0.166
0.500	0.0290	−0.165
0.550	0.0274	−0.145
0.600	0.0252	−0.124
0.650	0.0225	−0.100
0.700	0.0194	−0.073
0.750	0.0159	−0.044
0.800	0.0123	−0.013
0.850	0.0087	+0.019
0.900	0.0051	+0.056
0.950	0.0020	+0.098
1.000	0.0000	+0.142

Source: I. H. Abbott and A. E. von Doenhoff, *Theory of Wing Sections*. New York: Dover Publications, 1949 (Ref. 4.5).

$\tilde{z}$

Note $\tilde{z}$ scale is five times the $\tilde{x}$ scale

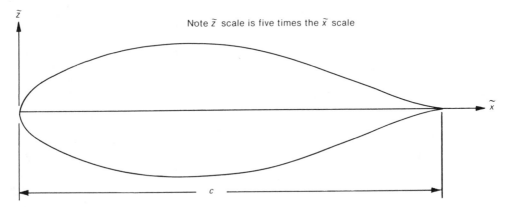

$\tilde{x}$

c

Figure 4.8   Cross section for a symmetric NACA 65-006 airfoil of Example 4.3.

system used in Fig. 4.8. Note that the maximum thickness is located relatively far aft
in order to maintain a favorable pressure gradient, which tends to delay transition. The
β distribution is required as an input to obtain the local similarity solutions for a laminar
boundary layer.

Solution   Using the definition for β gives us

$$\beta = \frac{2s}{u_e}\frac{du_e}{ds} = \frac{2\int u_e\,dx}{u_e}\frac{du_e}{dx}\frac{dx}{ds}$$

But

$$\frac{dx}{ds} = \frac{1}{u_e}$$

$$\frac{u_e}{U_\infty} = (1 - C_p)^{0.5}$$

Therefore, at any chordwise location for a thin airfoil

$$\beta = -\frac{\int_0^{\bar{x}} (1 - C_p)^{0.5}\,d\bar{x}}{(1 - C_p)^{1.5}}\frac{dC_p}{d\bar{x}}$$

where $\bar{x} = \tilde{x}/c$.

The resultant β distribution is presented in Fig. 4.9. Note that a favorable pressure
gradient acts over the first half of the airfoil. For $\bar{x} \geq 0.6$, the negative values of β exceed
that required for separation of a similar laminar boundary layer. Because of the large
streamwise variations in β, the nonsimilar character of the boundary layer should be
taken into account when establishing a separation criteria. Nevertheless, these calculations
indicate that, if the boundary layer were laminar along its entire length, it would separate,
even for this airfoil at zero angle of attack. Boundary-layer separation would result in
significant changes in the flow field. However, the experimental measurements of the
pressure distribution indicate that the actual flow field corresponds closely to the inviscid

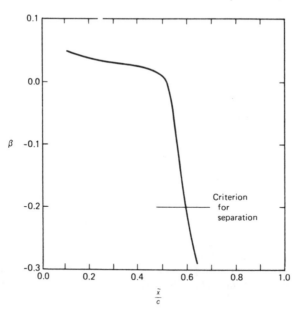

Figure 4.9 Distribution for a NACA 65-006 airfoil (assuming that the boundary layer does not separate).

flow field. Thus, boundary-layer separation apparently does not occur at zero angle of attack. The reason that separation does not occur is as follows. At the relatively high Reynolds numbers associated with airplane flight, the boundary layer is turbulent over a considerable portion of the airfoil. As discussed previously, a turbulent boundary layer can overcome an adverse pressure gradient longer, and separation is not as likely to occur.

The fact that a turbulent boundary layer can flow without separation into regions of much steeper adverse pressure gradients than can a laminar boundary layer is illustrated in Fig. 4.10. Incompressible boundary-layer solutions were generated for symmetrical Joukowski airfoils at zero angle of attack. The edge velocity, and therefore the corresponding inviscid pressure distributions, are shown in Fig. 4.10. At the conditions indicated, boundary-layer separation will occur for any Joukowski airfoil that is thicker than 4.6% if the flow is entirely laminar. However, if the boundary layer is turbulent, separation will not occur until a thickness of about 31% has been exceeded. The boundary layer effectively thickens the airfoil, especially near the trailing edge, since δ^* increases with distance. This thickening alleviates the adverse pressure gradients, which in turn permits somewhat thicker sections before separation occurs. To ensure that boundary-layer transition occurs and, thus, delay or avoid separation altogether, one might use vortex generators or other forms of surface roughness, such as shown in Fig. 4.11.

BOUNDARY-LAYER TRANSITION

As the boundary layer develops in the streamwise direction, it is subjected to numerous disturbances. The disturbances may be due to surface roughness, temperature irregularities, background noise, and so on. For some flows, these disturbances are damped

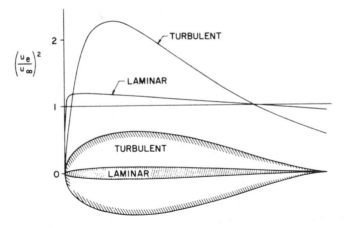

Figure 4.10 Thickest symmetrical Joukowski airfoils capable of supporting fully attached laminar and turbulent flows. The angle of attack is 0°, and the Mach number is 0. For turbulent flow, transition is assumed to occur at the velocity peak. The turbulent case is calculated for $\mathrm{Re}_c = 10^7$. Results for laminar flow are independent of Reynolds number. Maximum thickness for laminar flow is about 4.6%, for turbulent flow, 31%. If displacement-thickness effects on pressure distribution were included, the turbulent airfoil would increase to about 33%. The change in the laminar case would be negligible. (From Ref. 4.6.)

Figure 4.11 Row of vortex generators on the wing of NASA's Shuttle Trainer Aircraft (STA). (Courtesy of NASA.)

and the flow remains laminar. For other flows, the disturbances amplify and the boundary layer becomes turbulent. The onset of transition from a laminar boundary layer to a turbulent layer (if it occurs at all) depends on many parameters, such as the following:

1. Pressure gradient
2. Surface roughness
3. Compressibility effects (usually related to the Mach number)
4. Surface temperature
5. Suction or blowing at the surface
6. Free-stream turbulence

Obviously, no single criterion for the onset of transition can be applied to a wide variety of flow conditions. However, as a rule of thumb, adverse pressure gradients, surface roughness, blowing at the surface, and free-stream turbulence promote transition, that is, cause it to occur early. Conversely, favorable pressure gradients, increased Mach numbers, suction at the surface, and surface cooling delay transition. Although the parameters used and the correlation formula for the onset of transition depend on the details of the application, transition criteria incorporate a Reynolds number. For incompressible flow past a flat plate, a typical transition criterion is

$$\text{Re}_{x,tr} = 500,000 \tag{4.32}$$

Thus, the location for the onset of boundary-layer transition would occur at

$$x_{tr} = \frac{\text{Re}_{x,tr}}{\rho u_e / \mu} \tag{4.33}$$

Once the critical Reynolds number is exceeded, the flat-plate boundary layer would contain regions with the following characteristics as it transitioned from the laminar state to a fully turbulent flow:

1. Stable, laminar flow near the leading edge
2. Unstable flow containing two-dimensional Tollmien–Schlichting (T-S) waves
3. A region where three-dimensional unstable waves and hairpin eddies develop
4. A region where vortex breakdown produces locally high shear
5. Fluctuating, three-dimensional flow due to cascading vortex breakdown
6. A region where turbulent spots form
7. Fully turbulent flow

A sketch of the idealized transition process is presented in Fig. 4.12.

Stability theory predicts and experiment verifies that the initial instability is in the form of two-dimensional T-S waves that travel in the mean flow direction. Even though the mean flow is two-dimensional, three-dimensional unstable waves and hairpin eddies soon develop as the T-S waves begin to show spanwise variations. The experimental verification of the transition process is illustrated in the photograph of Fig. 4.13. A vibrating ribbon perturbs the low-speed flow upstream of the left margin of

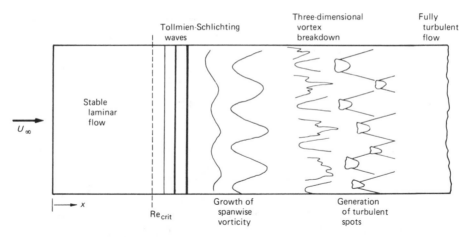

Figure 4.12 Idealized sketch of the transition process on a flat plate. (Based on the sketch from *Viscous Fluid Flow* by F. M. White. Copyright ©1974 by McGraw-Hill Book Company.)

the photograph. Smoke accumulation in the small recirculation regions associated with the T-S waves can be seen at the left edge of the photograph. The sudden appearance of three dimensionality is associated with the nonlinear growth region of the laminar instability. In the advanced stages of the transition process, intense local fluctuations

Figure 4.13 Flow visualization of the transition process on a flat plate. (Photograph supplied by A. S. W. Thomas, Lockheed Aeronautical Systems Company, Georgia Division.)

occur at various times and locations in the viscous layer. From these local intensities, true turbulence bursts forth and grows into a turbulent spot. Downstream of the region where the spots first form, the flow becomes fully turbulent.

Transition-promoting phenomena, such as an adverse pressure gradient and finite surface roughness, may short circuit the transition process, eliminating one or more of the five transitional regions described previously.

INCOMPRESSIBLE, TURBULENT BOUNDARY LAYER

Let us now consider flows where transition has occurred and the boundary layer is fully turbulent. A turbulent flow is one in which irregular fluctuations (mixing or eddying motions) are superimposed on the mean flow. Thus, the velocity at any point in a turbulent boundary layer is a function of time. The fluctuations occur in the direction of the mean flow and at right angles to it, and they affect macroscopic lumps of fluid. Thus, whereas momentum transport occurs on a microscopic (or molecular) scale in a laminar boundary layer, it occurs on a macroscopic scale in a turbulent boundary layer. It should be noted that, although the velocity fluctuations may be only several percent of the local streamwise values, they have a decisive effect on the overall motion. The size of these macroscopic lumps determines the scale of turbulence.

The effects caused by the fluctuations are as if the viscosity were increased by a factor of 10 or more. As a result, the shear forces at the wall and the skin-friction component of the drag are much larger when the boundary layer is turbulent. However, since a turbulent boundary layer can negotiate an adverse pressure gradient for a longer distance, boundary layer separation may be delayed or even avoided altogether. Delaying (or avoiding) the onset of separation reduces the pressure component of the drag (i.e., the form drag). For a blunt body or for a slender body at angle of attack, the reduction in form drag usually dominates the increase in skin friction drag.

When describing a turbulent flow, it is convenient to express the local velocity components as the sum of a mean motion plus a fluctuating, or eddying, motion. For example, as illustrated in Fig. 4.14,

$$u = \bar{u} + u' \tag{4.34}$$

where $\bar{u}$ is the time-averaged value of the u component of velocity, and u' is the velocity of fluctuation. The time-averaged value at a given point in space is calculated as

$$\bar{u} = \frac{1}{\Delta t} \int_{t_0}^{t_0 + \Delta t} u \, dt \tag{4.35}$$

The integration interval Δt should be much larger than any significant period of the fluctuation velocity u'. Thus, the mean value is independent of time. The integration interval depends on the physics and geometry of the problem. Referring to Eq. (4.35), we see that $\overline{u'} = 0$, by definition. The time average of any fluctuating parameter or its derivative is zero. The time average of products of fluctuating parameters and their derivatives is not zero. For example, $\overline{v'} = 0$, $[\partial(\overline{v'})]/\partial x = 0$; but $\overline{u'v'} \neq 0$. Of fundamental importance to turbulent motion is the way in which the fluctuations u', v', and w' influence the mean motion $\bar{u}$, $\bar{v}$, and $\bar{w}$.

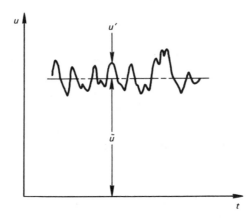

Figure 4.14 Histories of the mean component ($\bar{u}$) and the fluctuating component (u') of the streamwise velocity u for a turbulent boundary layer.

Derivation of the Momentum Equation for Turbulent Boundary Layer

Let us now derive the x (or streamwise) momentum equation for a steady, constant-property, two-dimensional, turbulent boundary layer. Since the density is constant, the continuity equation is

$$\frac{\partial(\bar{u} + u')}{\partial x} + \frac{\partial(\bar{v} + v')}{\partial y} = 0 \tag{4.36}$$

Expanding yields

$$\frac{\partial \bar{u}}{\partial x} + \frac{\partial \bar{v}}{\partial y} + \frac{\partial u'}{\partial x} + \frac{\partial v'}{\partial y} = 0 \tag{4.37}$$

Let us take the time-averaged value for each of these terms. The first two terms already are time-averaged values. As noted when discussing equation (4.35), the time-averaged value of a fluctuating component is zero; that is,

$$\frac{\partial \overline{u'}}{\partial x} = \frac{\partial \overline{v'}}{\partial y} = 0$$

Thus, for a turbulent flow, we learn from the continuity equation that

$$\frac{\partial \bar{u}}{\partial x} + \frac{\partial \bar{v}}{\partial y} = 0 \tag{4.38a}$$

and that

$$\frac{\partial u'}{\partial x} + \frac{\partial v'}{\partial y} = 0 \tag{4.38b}$$

Substituting the fluctuating descriptions for the velocity components into the x momentum equation (4.6), we have

$$\rho(\bar{u} + u')\frac{\partial(\bar{u} + u')}{\partial x} + \rho(\bar{v} + v')\frac{\partial(\bar{u} + u')}{\partial y} = \rho_e u_e \frac{du_e}{dx} + \mu \frac{\partial^2(\bar{u} + u')}{\partial y^2}$$

Expanding gives us

$$\rho\bar{u}\frac{\partial\bar{u}}{\partial x} + \rho\bar{v}\frac{\partial\bar{u}}{\partial y} + \rho u'\frac{\partial\bar{u}}{\partial x} + \rho v'\frac{\partial\bar{u}}{\partial y} + \rho u'\frac{\partial u'}{\partial x} + \rho v'\frac{\partial u'}{\partial y} + \rho\bar{u}\frac{\partial u'}{\partial x} + \rho\bar{v}\frac{\partial u'}{\partial y}$$

$$= \rho_e u_e\frac{du_e}{dx} + \mu\frac{\partial^2\bar{u}}{\partial y^2} + \mu\frac{\partial^2 u'}{\partial y}$$

Taking the time average of the terms in this equation, the terms that contain only one fluctuating parameter vanish, since their time-averaged value is zero. However, the time average of terms involving the product of fluctuating terms is not zero. Thus, we obtain

$$\rho\bar{u}\frac{\partial\bar{u}}{\partial x} + \rho\bar{v}\frac{\partial\bar{u}}{\partial y} + \overline{\rho u'\frac{\partial u'}{\partial x}} + \overline{\rho v'\frac{\partial u'}{\partial y}} = \rho_e u_e\frac{du_e}{dx} + \mu\frac{\partial^2\bar{u}}{\partial y^2} \qquad (4.39)$$

Let us now multiply the fluctuating portion of the continuity equation (4.38b) by $\rho(\bar{u} + u')$. We obtain

$$\rho\bar{u}\frac{\partial u'}{\partial x} + \rho u'\frac{\partial u'}{\partial x} + \rho\bar{u}\frac{\partial v'}{\partial y} + \rho u'\frac{\partial v'}{\partial y} = 0$$

Taking the time average of these terms, we find that

$$\overline{\rho u'\frac{\partial u'}{\partial x}} + \overline{\rho u'\frac{\partial v'}{\partial y}} = 0 \qquad (4.40)$$

Adding equation (4.40) to (4.39) and rearranging the terms, we obtain

$$\rho\bar{u}\frac{\partial\bar{u}}{\partial x} + \rho\bar{v}\frac{\partial\bar{u}}{\partial y} = \rho_e u_e\frac{du_e}{dx} + \mu\frac{\partial^2\bar{u}}{\partial y^2} - \rho\frac{\partial}{\partial y}(\overline{u'v'}) - \rho\frac{\partial}{\partial x}(\overline{u'})^2 \qquad (4.41)$$

We will neglect the streamwise gradient of the time-averaged value of the square of the fluctuating velocity component, that is, $(\partial/\partial x)\overline{(u')^2}$ as compared to the transverse gradient. Thus, the momentum equation becomes

$$\rho\bar{u}\frac{\partial\bar{u}}{\partial x} + \rho\bar{v}\frac{\partial\bar{u}}{\partial y} = \rho_e u_e\frac{du_e}{dx} + \mu\frac{\partial^2\bar{u}}{\partial y^2} - \rho\frac{\partial}{\partial y}(\overline{u'v'}) \qquad (4.42)$$

Let us further examine the last two terms,

$$\frac{\partial}{\partial y}\left(\mu\frac{\partial\bar{u}}{\partial y} - \rho\overline{u'v'}\right) \qquad (4.43)$$

Recall that the first term is the laminar shear stress. To evaluate the second term, let us consider a differential area dA such that the normal to dA is parallel to the y axis, and the directions x and z are in the plane of dA. The mass of fluid passing through this area in time dt is given by the product $(\rho v)(dA)(dt)$. The flux of momentum in the x direction is given by the product $(u)(\rho v)(dA)(dt)$. For a constant density flow, the time-averaged flux of momentum per unit time is

$$\overline{\rho u v}\, dA = \rho(\overline{u}\overline{v} + \overline{u'v'})\, dA$$

Since the flux of momentum per unit time through an area is equivalent to an equal-and-opposite force exerted on the area by the surroundings, we can treat the term $-\rho\overline{u'v'}$ as equivalent to a "turbulent" shear stress. This "apparent," or Reynolds, stress can be added to the stresses associated with the mean flow. Thus, we can write

$$\tau_{xy} = \mu\left(\frac{\partial\overline{u}}{\partial y}\right) - \rho\overline{u'v'} \tag{4.44}$$

Mathematically, then, the turbulent inertia terms behave as if the total stress on the system were composed of the Newtonian viscous stress plus an apparent turbulent stress.

The term $-\rho\overline{u'v'}$ is the source of considerable difficulties in the analysis of a turbulent boundary layer because its analytical form is not known a priori. It is related not only to physical properties of the fluid but also to the local flow conditions (velocity, geometry, surface roughness, upstream history, etc.). Furthermore, the magnitude of $-\rho\overline{u'v'}$ depends on the distance from the wall. Because the wall is a streamline, there is no flow through it. Thus, $\overline{v}$ and v' go to zero at the wall, and the flow for $y < 0.02\delta$ is basically laminar. At points away from the wall, $-\rho\overline{u'v'}$ is the dominant term and is called the *turbulent shear*. Experimental programs provide the information necessary to develop correlations for this stress.

The development of correlations in terms of known parameters is usually termed the *closure problem*. Closure procedures can be placed into one of three categories: (1) simple (or zeroth-order) methods, (2) first-order (or mean-field) closure methods, and (3) second-order (or mean-turbulence-field) closure methods. Zeroth-order techniques are characterized by a substantial amount of empiricism. For the mean-field closure approach, the turbulent-flux terms are related to mean-flow quantities. For the mean-turbulence-field closure technique, differential equations are derived from the Navier–Stokes equations, introducing additional unknowns and the need for additional information about the structure of the turbulent flow. There are numerous references providing detailed reviews (e.g., Ref. 4.7) for the reader who seeks additional information. The present discussion will be limited to simple methods.

Turbulent Boundary Layer for a Flat Plate

Since u_e is a constant for flat plate, the pressure gradient term is zero. Even with this simplification, there is no exact solution for the turbulent boundary layer. Very near the wall, the viscous shear dominates. Ludwig Prandtl deduced that the mean velocity in this region must depend on the wall shear stress, the fluid's physical properties, and the distance y from the wall. Thus, $\overline{u}$ is a function of (τ_w, ρ, μ, y). To a first order, the velocity profile is linear; that is, $\overline{u}$ is proportional to y. Thus,

$$\tau_w = \mu\frac{\partial\overline{u}}{\partial y} = \mu\frac{\overline{u}}{y} \tag{4.45}$$

Let us define

$$u^+ = \frac{\overline{u}}{u^*} \tag{4.46a}$$

and

$$y^+ = \frac{yu^*}{\nu} \tag{4.46b}$$

where u^* is called the *wall-friction velocity* and is defined as

$$u^* = \sqrt{\frac{\tau_w}{\rho}} \tag{4.46c}$$

Note that y^+ has the form of a Reynolds number.

Substituting these definitions into equation (4.45), we obtain

$$\tau_w = \mu \frac{u^+ u^*}{(y^+ \nu)/u^*} = (u^+/y^+)\,\rho u^{*2}$$

Introducing the definition of the wall-friction velocity, it is clear that

$$u^+ = y^+ \tag{4.47}$$

for the laminar sublayer. In the laminar sublayer, the velocities are so small that viscous forces dominate and there is no turbulence. The edge of the laminar sublayer corresponds to a y^+ of 5 to 10.

In 1930, Theodor von Kármán deduced that, in the outer region of a turbulent boundary layer, the mean velocity $\bar{u}$ is reduced below the free-stream value (u_e) in a manner that is independent of the viscosity but is dependent on the wall shear stress and the distance y over which its effect has diffused. Thus, the velocity defect ($u_e - \bar{u}$) for the outer region is a function of (τ_w, ρ, y, δ). For the outer region, the velocity-defect law is given by

$$\frac{u_e - \bar{u}}{u^*} = g\left(\frac{y}{\delta}\right) \tag{4.48}$$

The outer region of a turbulent boundary layer contains 80 to 90% of the boundary-layer thickness δ.

In 1933, Prandtl deduced that the mean velocity in the inner region must depend on the wall shear stress, the fluid physical properties, and the distance y from the wall. Thus, $\bar{u}$ is a function of (τ_w, ρ, μ, y). Specifically,

$$\frac{\bar{u}}{u^*} = f\left[\left(\frac{y}{\delta}\right)\left(\frac{\delta u^*}{\nu}\right)\right] \tag{4.49}$$

for the inner region.

Since the velocities of the two regions must match at their interface,

$$\frac{\bar{u}}{u^*} = f\left[\left(\frac{y}{\delta}\right)\left(\frac{\delta u^*}{\nu}\right)\right] = \frac{u_e}{u^*} - g\left(\frac{y}{\delta}\right)$$

As a result, the velocity in the inner region is given by

$$\frac{\bar{u}}{u^*} = \frac{1}{\kappa}\ln\frac{yu^*}{\nu} + B \tag{4.50a}$$

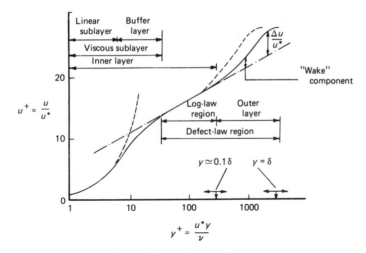

Figure 4.15 Turbulent boundary layer illustrating wall-layer nomenclature.

or, in terms of u^+, y^+, the equation can be written as

$$u^+ = \frac{1}{\kappa} \ln y^+ + B \qquad (4.50b)$$

This velocity correlation is valid only in regions where the laminar shear stress can be neglected in comparison with the turbulent stress. Thus, the flow in this region (i.e., $70 < y^+ < 400$), is fully turbulent.

The velocity in the outer region is given by

$$\frac{u_e - \bar{u}}{u^*} = -\frac{1}{\kappa} \ln \frac{y}{\delta} + A \qquad (4.51)$$

where κ, A, and B are dimensionless parameters. For incompressible flow past a flat plate,

$$\kappa \simeq 0.40 \text{ or } 0.41$$

$$A \simeq 2.35$$

$$B \simeq 5.0 \text{ to } 5.5$$

The resultant velocity profile is presented in Fig. 4.15.

EDDY VISCOSITY AND MIXING LENGTH CONCEPTS

In the late nineteenth century, Boussinesq introduced the concept of eddy viscosity to model the Reynolds shear stress. It was assumed that the Reynolds stresses act like the viscous (laminar) shear stresses and are proportional to the transverse gradient of the (mean) streamwise velocity component. The coefficient of the proportionality is called the *eddy viscosity* (ϵ_m) and is defined as

$$-\rho\overline{u'v'} = \rho\epsilon_m\left(\frac{\partial\overline{u}}{\partial y}\right) \tag{4.52}$$

Having introduced the concept of eddy viscosity, equation (4.44) for the total shear stress may be written

$$\tau = \tau_l + \tau_t = \rho\nu\frac{\partial\overline{u}}{\partial y} + \rho\epsilon_m\left(\frac{\partial\overline{u}}{\partial y}\right) \tag{4.53}$$

Like the kinematic viscosity ν, ϵ_m has the units of L^2/T. However, whereas ν is a property of the fluid and is defined once the pressure and the temperature are known, ϵ_m is a function of the flow field (including such factors as surface roughness, pressure gradients, etc.).

In an attempt to obtain a more generally applicable relation, Prandtl proposed the mixing length concept, whereby the shear stress is given as

$$-\rho\overline{u'v'} = \rho l^2\left|\frac{\partial\overline{u}}{\partial y}\right|\frac{\partial\overline{u}}{\partial y} \tag{4.54}$$

Equating the expressions for the Reynolds stress, given by equations (4.52) and (4.54), we can now write a relation between the eddy viscosity and the mixing length:

$$\epsilon_m = l^2\left|\frac{\partial\overline{u}}{\partial y}\right| \tag{4.55}$$

From this point on, we will use only the time-average (or mean-flow) properties. Thus, we will drop the overbar notation in the subsequent analysis.

The distributions of ϵ_m and of l across the boundary layer are based on experimental data. Because the eddy viscosity and the mixing length concepts are based on local equilibrium ideas, they provide only rough approximations to the actual flow and are said to lack generality. In fact, the original derivations included erroneous physical arguments. However, since they are relatively simple to use and provide reasonable values of the shear stress for many engineering applications, they are among the more popular and extensively used concepts.

A general conclusion that is drawn from the experimental evidence is that the turbulent boundary layer should be treated as a composite layer consisting of an inner region and an outer region. For the inner region, the mixing length is given as

$$l_i = \kappa y[1 - \exp{(-y/A)}] \tag{4.56}$$

where $\kappa = 0.41$ (as discussed earlier) and A, the van Driest damping parameter, is

$$A = \frac{26\nu}{Nu^*} \tag{4.57a}$$

where u^* is the wall friction velocity as defined by equation (4.46c),

$$N = (1 - 11.8p^+)^{0.5} \tag{4.57b}$$

and

$$p^+ = \frac{vu_e}{(u^*)^3} \frac{du_e}{dx} \tag{4.57c}$$

Thus, the eddy viscosity for the inner region becomes

$$(\epsilon_m)_i = (\kappa y)^2 [1 - \exp(-y/A)]^2 \left| \frac{\partial u}{\partial y} \right| \tag{4.58}$$

For the outer region, the eddy viscosity is given as

$$(\epsilon_m)_0 = \alpha u_e \delta^* \tag{4.59}$$

where δ^* is the displacement thickness as defined by equation (4.26),

$$\alpha = \frac{0.02604}{1 + \Pi} \tag{4.60a}$$

$$\Pi = 0.55 [1 - \exp(-0.243 \sqrt{z_1} - 0.298 z_1)] \tag{4.60b}$$

and

$$z_1 = \frac{\text{Re}_\theta}{425} - 1 \tag{4.60c}$$

The y coordinate of the interface between the inner region and the outer region is determined by the requirement that the y distribution of the eddy viscosity be continuous. Thus, the inner region expression, equation (4.58), is used to calculate the eddy viscosity until its value becomes equal to that given by the outer region expression, equation (4.59). The y coordinate, where the two expressions for the eddy viscosity are equal [i.e., when $(\epsilon_m)_i = (\epsilon_m)_0$], is the interface value, y_c. For $y \geq y_c$, the eddy viscosity is calculated using the outer region expression, equation (4.59).

Recall that the transition process occurs over a finite length; that is, the boundary layer does not instantaneously change from a laminar state to a fully turbulent profile. For most practical boundary-layer calculations, it is necessary to calculate the viscous flow along its entire length. That is, for a given pressure distribution (inviscid flow field) and for a given transition criterion, the boundary-layer calculation is started at the leading edge or at the forward stagnation point of the body (where the boundary layer is laminar) and proceeds downstream through the transitional flow into the fully turbulent region. To treat the boundary layer in the transition zone, the expressions for the eddy viscosity are multiplied by an intermittency factor, γ_{tr}. The expression for γ_{tr} is

$$\gamma_{tr} = 1 - \exp \left[-G(x - x_{tr}) \int_{x_{tr}}^{x} \frac{dx}{u_e} \right] \tag{4.61a}$$

where x_{tr} is the x coordinate for the onset of transition and

$$G = 8.35 \times 10^{-4} \left(\frac{u_e^3}{v^2} \right) (\text{Re}_{x,tr})^{-1.34} \tag{4.61b}$$

The intermittency factor varies from 0 (in the laminar region and at the onset of transition) to 1 (at the end of the transition zone and for fully turbulent flow). Thus, in the transition zone,

$$(\epsilon_m)_i = (\kappa y)^2 [1 - \exp(-y/A)]^2 \left| \frac{\partial u}{\partial y} \right| \gamma_{tr} \tag{4.62a}$$

and

$$(\epsilon_m)_0 = \alpha u_e \delta^* \gamma_{tr} \tag{4.62b}$$

Solutions have been obtained using these equations to describe the laminar, transitional, and turbulent boundary layer for flow past a flat plate. The results are presented in Fig. 4.16.

INTEGRAL EQUATIONS FOR A FLAT-PLATE BOUNDARY LAYER

The eddy viscosity concept or one of the higher-order closure methods are used in developing turbulent boundary-layer solutions using the differential equations of motion. Although approaches using the differential equations are most common in computational

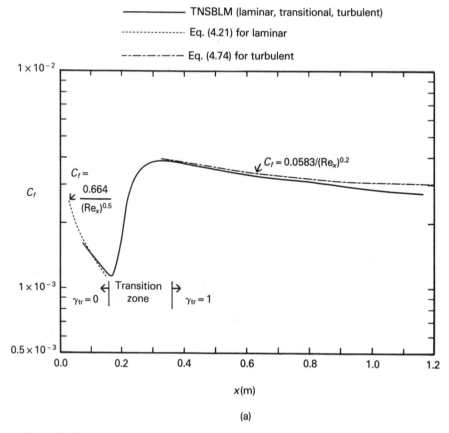

Figure 4.16 Sample of computed boundary layer for incompressible flow past a flat plate, $U_e = 114.1$ ft/s, $T_e = 542°R$, $P_e = 2101.5$ psf, $T_w = 540°R$: (a) skin-friction distribution; (b) turbulent viscosity/thermodynamic viscosity ratio.

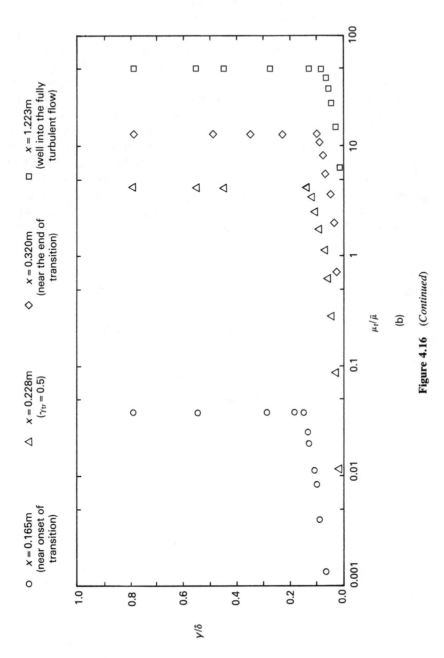

Figure 4.16 (*Continued*)

fluid dynamics, the integral approach can also be used to obtain approximate solutions for a turbulent boundary layer. Following a suggestion by Prandtl, the turbulent velocity profile will be represented by a power-law approximation.

We shall use the mean-flow properties in the integral form of the equation of motion to develop engineering correlations for the skin-friction coefficient and the boundary-layer thickness for an incompressible, turbulent boundary layer on a flat plate. Since we will use only the time-averaged (or mean-flow) properties in this section, we will drop the overbar notation. Consider the control volume shown in Fig. 4.17. The wall (which is, of course, a streamline) is the inner boundary of the control volume. A streamline outside the boundary layer is the outer boundary. Any streamline that is outside the boundary layer (and, therefore, has zero shear force acting across it) will do. Because the viscous action retards the flow near the surface, the outer boundary is not parallel to the wall. Thus, the streamline is a distance Y_0 away from the wall at the initial station and is a distance Y away from the wall at the downstream station, with $Y > Y_0$. Since $\vec{V} \cdot \hat{n}\, dA$ is zero for both boundary streamlines, the continuity equation (2.5) yields

$$\int_0^Y u\, dy - u_e Y_0 = 0 \tag{4.63}$$

But also,

$$\int_0^Y u\, dy = \int_0^Y [u_e + (u - u_e)]\, dy$$
$$= u_e Y + \int_0^Y (u - u_e)\, dy \tag{4.64}$$

Combining these two equations and introducing the definition for the displacement thickness,

$$\delta^* = \int_0^\delta \left(1 - \frac{u}{u_e}\right) dy$$

we find that

$$Y - Y_0 = \delta^* \tag{4.65}$$

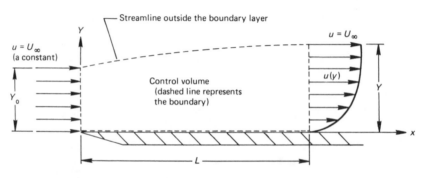

Figure 4.17 Control volume used to analyze the boundary layer on a flat plate.

Thus, we have derived the expected result that the outer streamline is deflected by the transverse distance δ^*. In developing this relation, we have used both δ and Y as the upper limit for the integration. Since the integrand goes to zero for $y \geq \delta$, the integral is independent of the upper limit of integration, provided that it is at, or beyond, the edge of the boundary layer.

Similarly, application of the integral form of the momentum equation (2.13) yields

$$-d = \int_0^Y u(\rho u\, dy) - \int_0^{Y_0} u_e(\rho u_e\, dy)$$

Note that the lowercase letter designates the drag per unit span (d). Thus,

$$d = \rho u_e^2 Y_0 - \int_0^Y (\rho u^2\, dy) \tag{4.66}$$

Using equation (4.63), we find that

$$d = \rho u_e \int_0^Y u\, dy - \int_0^Y \rho u^2\, dy$$

This equation can be rewritten in terms of the section drag coefficient as

$$\begin{aligned}
C_d &= \frac{d}{\frac{1}{2}\rho_\infty U_\infty^2 L} \\[2mm]
&= \frac{2}{L}\left(\int_0^Y \frac{u}{u_e}\, dy - \int_0^Y \frac{u^2}{u_e^2}\, dy\right)
\end{aligned} \tag{4.67}$$

Let us introduce the definition for the momentum thickness of an incompressible flow:

$$\theta = \int_0^\delta \frac{u}{u_e}\left(1 - \frac{u}{u_e}\right) dy \tag{4.68}$$

Note that the result is independent of the upper limit of integration provided that the upper limit is equal to or greater than the boundary-layer thickness. Thus, the drag coefficient (for one side of a flat plate of length L) is

$$C_d = \frac{2\theta}{L} \tag{4.69}$$

The equations developed in this section are valid for incompressible flow past a flat plate whether the boundary layer is laminar or tubulent. The value of the integral technique is that it requires only a "reasonable" approximation for the velocity profile [i.e., $u(y)$] in order to achieve "fairly accurate" drag predictions, because the integration often averages out positive and negative deviations in the assumed velocity function.

Application of the Integral Equations to a Turbulent, Flat-Plate Boundary Layer

Now let us apply these equations to develop correlations for a turbulent boundary layer on a flat plate. As discussed earlier, an analytical form for the turbulent shear is not known a priori. Therefore, we need some experimental information. Experimental measurements have shown that the time-averaged velocity may be represented by the power law,

$$\frac{u}{u_e} = \left(\frac{y}{\delta}\right)^{1/7} \tag{4.70}$$

when the local Reynolds number Re_x is in the range 5×10^5 to 1×10^7. However, note that the velocity gradient for this profile,

$$\frac{\partial u}{\partial y} = \frac{u_e}{7} \frac{1}{\delta^{1/7}} \frac{1}{y^{6/7}}$$

goes to infinity at the wall. Thus, although the correlation given in equation (4.70) provides a reasonable representation of the actual velocity profile, we need another piece of experimental data: a correlation for the shear at the wall. Blasius found that the skin friction coefficient for a turbulent boundary layer on a flat plate where the local Reynolds number is in the range 5×10^5 to 1×10^7 is given by

$$C_f = \frac{\tau}{\frac{1}{2}\rho u_e^2} = 0.0456\left(\frac{\nu}{u_e \delta}\right)^{0.25} \tag{4.71}$$

Differentiating equation (4.67) gives us

$$C_f = -2\frac{d}{dx}\left[\delta \int_0^1 \frac{u}{u_e}\left(\frac{u}{u_e} - 1\right) d\left(\frac{y}{\delta}\right)\right] \tag{4.72}$$

Substituting equations (4.70) and (4.71) into equation (4.72), we obtain

$$0.0456\left(\frac{\nu}{u_e\delta}\right)^{0.25} = -2\frac{d}{dx}\left\{\delta \int_0^1 \left[\left(\frac{y}{\delta}\right)^{2/7} - \left(\frac{y}{\delta}\right)^{1/7}\right] d\left(\frac{y}{\delta}\right)\right\}$$

which becomes

$$\delta^{0.25} \, d\delta = 0.2345\left(\frac{\nu}{u_e}\right)^{0.25} dx$$

If we assume that the boundary-layer thickness is zero when $x = 0$, we find that

$$\delta = 0.3747\left(\frac{\nu}{u_e}\right)^{0.2} (x)^{0.8}$$

Rearranging, the thickness of a turbulent boundary layer on a flat plate is given by

$$\frac{\delta}{x} = \frac{0.3747}{(Re_x)^{0.2}} \tag{4.73}$$

Comparing the turbulent correlation given by equation (4.73) with the laminar correlation given by equation (4.23), we see that a turbulent boundary layer grows at a faster rate than a laminar boundary layer subject to the same conditions. Furthermore, at a given x station, a turbulent boundary layer is thicker than a laminar boundary layer for the same stream conditions.

Substitution of equation (4.73) into equation (4.71) yields

$$C_f = \frac{0.0583}{(\text{Re}_x)^{0.2}} \tag{4.74}$$

The skin-friction coefficient correlations for laminar boundary layers and for turbulent boundary layers are compared in Fig. 4.18. Included is an approximate transition criteria for flat-plate boundary layers,

$$\text{Re}_{x,\text{tr}} \simeq 500{,}000$$

When the local Reynolds number is significantly above this value, the boundary layer is probably turbulent.

Example 4.4

Air at atmospheric pressure and 5°C flows at 200 km/h across a flat plate. Compare the velocity distribution for a laminar boundary layer and for a turbulent boundary layer at the transition point, assuming that the transition process is completed instantaneously at that location.

Solution For air at atmospheric pressure and 5°C,

$$\rho_\infty = \frac{1.01325 \times 10^5 \text{ N/m}^2}{(287.05 \text{ N} \cdot \text{m/kg} \cdot \text{K})\,(278.15 \text{ K})} = 1.2691 \text{ kg/m}^3$$

$$\mu_\infty = 1.458 \times 10^{-6}\frac{(278.15)^{1.5}}{278.15 + 110.4} = 1.7404 \times 10^{-5} \text{ kg/s} \cdot \text{m}$$

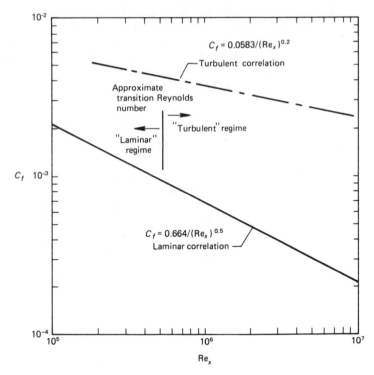

Figure 4.18 Skin-friction correlation for flat-plate boundary layers.

and

$$U_\infty = \frac{(200 \text{ km/h})(1000 \text{ m/km})}{3600 \text{ s/h}} = 55.556 \text{ m/s}$$

We will assume that the transition Reynolds number for this incompressible flow past a flat plate is 500,000. Thus,

$$x_{tr} = \frac{\text{Re}_{x,tr}}{\rho u_e / \mu} = 0.12344 \text{ m}$$

The thickness of a laminar boundary layer at this point is

$$\delta_{\text{lam}} = \frac{5.0x}{\sqrt{\text{Re}_x}} = 8.729 \times 10^{-4} \text{ m}$$

For comparison, we will calculate the thickness of the turbulent boundary layer at this point for this Reynolds number, assuming that the boundary layer is turbulent all the way from the leading edge. Thus,

$$\delta_{\text{turb}} = \frac{0.3747x}{(\text{Re}_x)^{0.2}} = 3.353 \times 10^{-3} \text{ m}$$

In reality, the flow is continuous at the transition location and the boundary layer thickness does not change instantaneously. Furthermore, since we are at the transition location, it is not realistic to use the assumption that the boundary layer is turbulent all the way from the leading edge. (This assumption would be reasonable far downstream of the transition location so that $x \gg x_{tr}$.) Nevertheless, the object of these calculations is to illustrate the characteristics of the turbulent boundary layer relative to a laminar boundary layer at the same conditions.

The resultant velocity profiles are compared in Table 4.5 and Fig. 4.19. Note that the streamwise velocity component u increases much more rapidly with y near the wall for the turbulent boundary layer. Thus, the shear at the wall is greater for the turbulent boundary layer even though this layer is much thicker than the laminar boundary layer for the same conditions at a given x station. The macroscopic transport of fluid in the y direction causes both increased shear and increased thickness of the boundary layer.

TABLE 4.5 VELOCITY PROFILES FOR EXAMPLE 4.4

y (m)	u_{lam} (m/s)	u_{turb} (m/s)
0.00000	0.00	0.00
0.00017	17.78	36.21
0.00034	33.33	39.98
0.00067	52.50	44.14
0.00101	Inviscid flow	46.78
0.00134		48.74
0.00168		50.32
0.00201		51.64
0.00235		52.79
0.00268		53.81
0.00302		54.71
0.00335		55.56

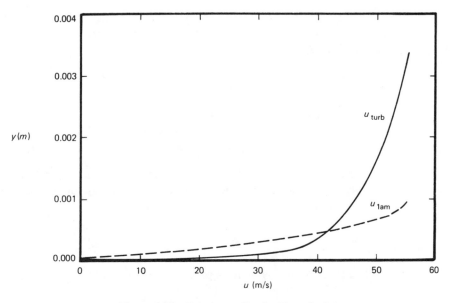

Figure 4.19   Velocity profiles for Example 4.4.

Integral Solutions for a Turbulent Boundary Layer with a Pressure Gradient

If we apply the integral equations of motion to a flow with a velocity gradient external to the boundary layer, we obtain

$$\frac{d\theta}{dx} + (2 + H)\frac{\theta}{u_e}\frac{du_e}{dx} = \frac{C_f}{2} \tag{4.75}$$

where θ, the momentum thickness, was defined in equation (4.68). H, the momentum shape factor, is defined as

$$H = \frac{\delta^*}{\theta} \tag{4.76}$$

where δ^*, the displacement thickness, is defined in equation (4.26). Equation (4.75) contains three unknown parameters, θ, H, and C_f, for a given external velocity distribution. For a turbulent boundary layer, these parameters are interrelated in a complex way. Head (Ref. 4.8) assumed that the rate of entrainment is given by

$$\frac{d}{dx}(u_e\theta H_1) = u_e F \tag{4.77}$$

where H_1 is defined as

$$H_1 = \frac{\delta - \delta^*}{\theta} \tag{4.78}$$

Head also assumed that H_1 is a function of the shape factor H, that is, $H_1 = G(H)$. Correlations of several sets of experimental data that were developed by Cebeci and Bradshaw (Ref. 4.9) yielded

$$F = 0.0306(H_1 - 3.0)^{-0.6169} \tag{4.79}$$

and

$$G = \begin{cases} 0.8234(H - 1.1)^{-1.287} + 3.3 & \text{for } H \leqslant 1.6 \\ 1.5501(H - 0.6778)^{-3.064} + 3.3 & \text{for } H \geqslant 1.6 \end{cases} \tag{4.80}$$

Equations (4.77) through (4.80) provide a relationship between θ and H. A relation between C_f, θ, and H is needed to complete our system of equations. A curve-fit formula given in Ref. 4.3 is

$$C_f = \frac{0.3e^{-1.33H}}{(\log \text{Re}_\theta)^{(1.74 + 0.31H)}} \tag{4.81}$$

where Re_θ is the Reynolds number based on the momentum thickness:

$$\text{Re}_\theta = \frac{\rho u_e \theta}{\mu} \tag{4.82}$$

We can numerically solve this system of equations for a given inviscid flow field. To start the calculations at some initial streamwise station, such as the transition location, values for two of the three parameters, θ, H, and C_f, must be specified at this station. The third parameter is then calculated using equation (4.81). Using this method, the shape factor H can be used as a criterion for separation. Although it is not possible to define an exact value of H corresponding to the separation, the value of H for separation is usually in the range 1.8 to 2.8.

THERMAL BOUNDARY LAYER FOR CONSTANT-PROPERTY FLOWS

As noted earlier, there are many constant-property flows for which we are interested in calculating the convective heat transfer. Thus, the temperature variations in the flow field are sufficiently large that there is heat transfer to or from a body in the flow but are small enough that the corresponding variations in density and viscosity can be neglected. Let us examine one such flow, the thermal boundary layer for a steady, low-speed flow past a flat plate. We will consider flows where the boundary layer is laminar. The solution for the velocity field for this flow has been described earlier in this chapter; see equations (4.6) through (4.27).

We will now solve the energy equation (2.32), in order to determine the temperature distribution. For a low-speed, constant-property, laminar boundary layer, the viscous dissipation is negligible (i.e., $\phi = 0$). For flow past a flat plate, $dp/dt = 0$. Thus, for a perfect gas, equation (2.32) becomes

$$\rho u c_p \frac{\partial T}{\partial x} + \rho v c_p \frac{\partial T}{\partial y} = k \frac{\partial^2 T}{\partial y^2} \tag{4.83}$$

Note that we have already neglected $k\,(\partial^2 T/\partial x^2)$ since it is small compared to $k\,(\partial^2 T/\partial y^2)$. We made a similar assumption about the corresponding velocity gradients when working with the momentum equation; see equation (4.6).

Let us now change the dependent variable from T to the dimensionless parameter θ, where

$$\theta = \frac{T - T_w}{T_e - T_w}$$

Note that $\theta = 0$ at the wall (i.e., at $y = 0$), and $\theta = 1$ at the edge of the thermal boundary layer. Using θ as the dependent variable, the energy equation becomes

$$\rho u \frac{\partial \theta}{\partial x} + \rho v \frac{\partial \theta}{\partial y} = \frac{k}{c_p} \frac{\partial^2 \theta}{\partial y^2} \tag{4.84}$$

Since the pressure is constant along the flat plate, the velocity at the edge of the boundary layer (u_e) is constant, and the momentum equation becomes

$$\rho u \frac{\partial u}{\partial x} + \rho v \frac{\partial u}{\partial y} = \mu \frac{\partial^2 u}{\partial y^2} \tag{4.85}$$

Let us replace u in the derivatives by the dimensionless parameter, u^*, where

$$u^* = \frac{u}{u_e}$$

Thus, equation (4.85) becomes

$$\rho u \frac{\partial u^*}{\partial x} + \rho v \frac{\partial u^*}{\partial y} = \mu \frac{\partial^2 u^*}{\partial y^2} \tag{4.86}$$

Note that $u^* = 0$ at the wall (i.e., at $y = 0$), and $u^* = 1$ at the edge of the velocity boundary layer.

Compare equations (4.84) and (4.86). Note that the equations are identical if $k/c_p = \mu$. Furthermore, the boundary conditions are identical: $\theta = 0$ and $u^* = 0$ at the wall, and $\theta = 1$ and $u^* = 1$ at the edge of the boundary layer. Thus, if

$$\frac{\mu c_p}{k} = 1$$

the velocity and the thermal boundary layers are identical. This ratio is called the Prandtl number (Pr) in honor of the German scientist:

$$\mathrm{Pr} = \frac{\mu c_p}{k} \tag{4.87}$$

The Prandtl number is an important dimensionless parameter for problems involving convective heat transfer where one encounters both fluid motion and heat conduction.

Reynolds Analogy

The shear at the wall is defined as

$$\tau = \left(\mu \frac{\partial u}{\partial y} \right)_{y=0}$$

Therefore, the skin-friction coefficient for a flat plate is

$$C_f = \frac{\tau}{\frac{1}{2}\rho u_e^2} = \frac{2\mu}{\rho u_e}\frac{\partial u^*}{\partial y} \tag{4.88}$$

The rate at which heat is transferred to the surface $(\dot{q})$ is defined as

$$\dot{q} = \left(k\frac{\partial T}{\partial y}\right)_{y=0} \tag{4.89}$$

The Stanton number (designated by the symbols St or C_h), which is a dimensionless heat-transfer coefficient, is defined as

$$St \equiv C_h = \frac{\dot{q}}{\rho u_e c_p(T_e - T_w)} \tag{4.90}$$

Combining these last two expressions, we find that the Stanton number is

$$St = \frac{k}{\rho u_e c_p}\frac{\partial\theta}{\partial y} \tag{4.91}$$

Relating the Stanton number, as given by equation (4.91), to the skin-friction coefficient, as defined by equation (4.88), we obtain the ratio

$$\frac{C_f}{St} = \frac{2\mu c_p}{k}\frac{\partial u^*/\partial y}{\partial\theta/\partial y} \tag{4.92}$$

Note that if $\mu c_p/k \equiv Pr = 1$, then

$$\frac{\partial u^*}{\partial y} = \frac{\partial\theta}{\partial y}$$

Thus, if the Prandtl number is 1,

$$St = \frac{C_f}{2} \tag{4.93}$$

This relation between the heat-transfer coefficient and the skin-friction coefficient is known as the Reynolds analogy.

Example 4.5

The thermal conductivity of air can be calculated using the relation

$$k = 4.76 \times 10^{-6}\frac{T^{1.5}}{T + 112}\ \text{cal/cm} \cdot \text{s} \cdot \text{K} \tag{4.94}$$

over the range of temperatures below those for which oxygen dissociates, that is, approximately 2000 K at atmospheric pressure. What is the Prandtl number for air at 15°C, that is, at 288.15 K?

Solution Using the results from Example 1.3, the viscosity is 1.7894×10^{-5} kg/s $\cdot$ m. The specific heat is 1004.7 J/kg $\cdot$ K. Using the equation to calculate the thermal conductivity,

$$k = 4.76 \times 10^{-6}\frac{(288.15)^{1.5}}{400.15} = 5.819 \times 10^{-5}\ \text{cal/cm} \cdot \text{s} \cdot \text{K}$$

Noting that there are 4.187 J/cal, the thermal conductivity is

$$k = 2.436 \times 10^{-2} \text{ J/m} \cdot \text{s} \cdot \text{K}$$

Thus,

$$\text{Pr} = \frac{\mu c_p}{k} = \frac{(1.7894 \times 10^{-5} \text{ kg/s} \cdot \text{m}) (1004.7 \text{ J/kg} \cdot \text{K})}{2.436 \times 10^{-2} \text{ J/m} \cdot \text{s} \cdot \text{K}} = 0.738$$

Note that, as a rule of thumb, the Prandtl number for air is essentially constant (approximately 0.7) over a wide range of flow conditions.

Thermal Boundary Layer for Pr ≠ 1

To solve for the temperature distribution for the laminar, flat-plate boundary layer, let us introduce the transformation of equation (4.10):

$$\eta = y \sqrt{\frac{u_e}{2\nu x}}$$

Using the transformed stream function f, as defined by equation (4.12b), equation (4.84) becomes

$$\theta'' + (\text{Pr})f\,\theta' = 0 \tag{4.95a}$$

where the ′ denotes differentiation with respect to η. But we have already obtained the solution for the stream function. Referring to equation (4.16) for a flat plate (i.e., $\beta = 0$), we have

$$f = -\frac{f'''}{f''} \tag{4.95b}$$

Combining equations (4.95a) and (4.95b) and rearranging, we obtain

$$\frac{\theta''}{\theta'} = (\text{Pr})\frac{f'''}{f''}$$

Integrating twice yields

$$\theta = C \int_0^\eta (f'')^{\text{Pr}}\, d\eta + \theta_0 \tag{4.96}$$

where C and θ_0 are constants of integration. They can be evaluated by applying the boundary conditions (1) at $\eta = 0$, $\theta = 0$, and (2) for $\eta \to$ large, $\theta = 1$.

$$\theta = \frac{T - T_w}{T_e - T_w} = 1 - \frac{\displaystyle\int_\eta^\infty (f'')^{\text{Pr}}\, d\eta}{\displaystyle\int_0^\infty (f'')^{\text{Pr}}\, d\eta} \tag{4.97}$$

The rate at which heat is transferred to the wall ($\dot{q}$) can be calculated using

$$\dot{q} = \left(k \frac{\partial T}{\partial y} \right)_{y=0} = k(T_e - T_w) \frac{\partial \eta}{\partial y} \left(\frac{\partial \theta}{\partial \eta} \right)_{\eta=0}$$

Using the values of Pohlhausen, we find that

$$\left(\frac{\partial \theta}{\partial \eta} \right)_{\eta=0} = 0.4696(Pr)^{0.333}$$

Combining these two relations, the rate at which heat is transferred from a laminar boundary layer to the wall is given by the relation

$$\dot{q} = 0.332k(T_e - T_w)(Pr)^{0.333} \sqrt{\frac{u_e}{vx}} \qquad (4.98)$$

The heat transfer can be expressed in terms of the Stanton number using equation (4.90). Thus,

$$St = 0.332 \frac{k}{\mu c_p} \sqrt{\frac{\mu}{\rho u_e x}} (Pr)^{0.333}$$

Using the definitions for the Reynolds number and Prandtl number, the Stanton number is

$$St = \frac{0.332}{(Pr)^{0.667}(Re_x)^{0.5}} \qquad (4.99)$$

Another popular dimensionless heat-transfer parameter is the Nusselt number. The Nusselt number is defined as

$$Nu_x = \frac{hx}{k} \qquad (4.100a)$$

In this equation, h is the local heat-transfer coefficient, which is defined as

$$h = \frac{\dot{q}}{T_e - T_w} \qquad (4.100b)$$

Combining this definition with equations (4.98) and (4.100a) gives us

$$Nu_x = 0.332(Re_x)^{0.5}(Pr)^{0.333} \qquad (4.101)$$

By dividing the expression for the Stanton number, equation (4.99), by that for the skin-friction coefficient, equation (4.21), we obtain

$$St = \frac{C_f}{2(Pr)^{0.667}} \qquad (4.102)$$

Because of the similarity between this equation and equation (4.93), we shall call this the *modified Reynolds analogy*.

Example 4.6

Using the modified Reynolds analogy, develop relations for the dimensionless heat-transfer parameters, St and Nu_x, for a turbulent flat-plate boundary layer.

Solution Referring to the discussion of turbulent boundary layers, we note that

$$C_f = \frac{0.0583}{(\text{Re}_x)^{0.2}} \tag{4.74}$$

Thus, using equation (4.102) for the modified Reynolds analogy, we can approximate the Stanton number as

$$\text{St} = \frac{0.0292}{(\text{Re}_x)^{0.2}(\text{Pr})^{0.667}} \tag{4.103}$$

Comparing equations (4.99) and (4.101), we can see that the Nu_x is given as

$$\text{Nu}_x = (\text{St})(\text{Pr})(\text{Re}_x)$$

Thus, the Nusselt number for turbulent flow past a flat plate can be approximated as

$$\text{Nu}_x = 0.0292(\text{Re}_x)^{0.8}(\text{Pr})^{0.333} \tag{4.104}$$

Example 4.7

The radiator systems on many of the early racing aircraft were flush mounted on the external surface of the airplane. Let us assume that the local heat-transfer rate can be estimated using the flat-plate relations. What is the local heating rate for $x = 3.0$ m when the airplane is flying at 468 km/h at an altitude of 3 km? The surface temperature is 330 K.

Solution Using Table 1.2 to find the free-stream flow properties,

$$p_\infty = 7.012 \times 10^4 \text{ N/m}^2 \qquad T_\infty = 268.659 \text{ K}$$
$$\rho_\infty = 0.9092 \text{ kg/m}^3 \qquad \mu_\infty = 1.6938 \times 10^{-5} \text{ kg/s} \cdot \text{m}$$

Since we have assumed that the flow corresponds to that for a flat plate, these values are also the local properties at the edge of the boundary layer at $x = 3.0$ m. Note that

$$u_e = U_\infty = 468 \text{ km/h} = 130 \text{ m/s}$$

To determine whether the boundary layer is laminar or turbulent, let us calculate the local Reynolds number:

$$\text{Re}_x = \frac{\rho_e u_e x}{\mu_e} = \frac{(0.9092)(130)(3.0)}{1.6938 \times 10^{-5}} = 2.093 \times 10^7$$

This is well above the transition value. In fact, if the transition Reynolds number is assumed to be 500,000, transition would occur at a point

$$x_{tr} = \frac{500,000}{(\rho_e u_e)/\mu_e} = 0.072 \text{ m}$$

from the leading edge. Thus, the calculation of the heating will be based on the assumption that the boundary layer is turbulent over its entire length.

Combining equations (4.100a) and (4.100b),

$$\dot{q} = \frac{\mathrm{Nu}_x k(T_e - T_w)}{x}$$

where the Nu_x is given by equation (4.104). Thus,

$$\dot{q} = \frac{0.0292(\mathrm{Re}_x)^{0.8}(\mathrm{Pr})^{0.333}k(T_e - T_w)}{x}$$

To calculate the thermal conductivity for air,

$$k = 4.76 \times 10^{-6} \frac{T^{1.5}}{T + 112} = 5.506 \times 10^{-5} \, \mathrm{cal/cm \cdot s \cdot K}$$
$$= 2.306 \times 10^{-2} \, \mathrm{J/m \cdot s \cdot K}$$

Since $1 \, \mathrm{W} = 1 \, \mathrm{J/s}$,

$$k = 2.306 \times 10^{-2} \, \mathrm{W/m \cdot K}$$

Furthermore, the Prandtl number is

$$\mathrm{Pr} = \frac{\mu c_p}{k} = 0.738$$

Thus,

$$\dot{q} = \frac{(0.0292)(2.093 \times 10^7)^{0.8}(0.738)^{0.333}(2.306 \times 10^{-2} \, \mathrm{W \cdot K})(268.659 - 330) \, \mathrm{K}}{3.0 \, \mathrm{m}}$$

$$= -8.944 \times 10^3 \, \mathrm{W/m^2} = -8.944 \, \mathrm{kW/m^2}$$

The minus sign indicates that heat is transferred from the surface to the air flowing past the aircraft. This is as it should be, since the surface is hotter than the adjacent air. Furthermore, since the problem discusses a radiator, proper performance would produce cooling. Since there are 1.341 hp/kW, the heat transfer rate is equivalent to 1.114 hp/ft².

SUMMARY

In this chapter we have developed techniques by which we can obtain solutions for a thin, viscous boundary layer near the surface. Techniques have been developed both for a laminar and for a turbulent boundary layer using both integral and differential approaches. We now have reviewed the basic concepts of fluid mechanics in Chapters 1 through 4 and are ready to apply them specifically to aerodynamic problems.

PROBLEMS

4.1. A very thin, "flat-plate" wing of a model airplane moves through the air at standard sea-level conditions at a velocity of 15 m/s. The dimensions of the plate are such that its chord (streamwise dimension) is 0.5 m and its span (length perpendicular to the flow direction) is 5 m. What is the Reynolds number at the trailing edge ($x = 0.5$ m)? Assume that the boundary layer is laminar in answering the remaining questions. What are the boundary-layer thickness and the displacement thickness at the trailing edge? What are the local shear at the wall and skin-friction coefficient at $x = 0.5$ m? Calculate the total drag on the wing (both sides). Prepare a graph of $\bar{u}$ as a function of y, where $\bar{u}$ designates the x component of velocity relative to a point on the ground, at $x = 0.5$ m.

4.2. Assume that the inviscid external flow over a configuration is given by

$$u_e = Ax$$

Thus, the stagnation point occurs at the leading edge of the configuration (i.e., at $x = 0$). Obtain the expression for β. Using Fig. 4.4 and assuming that the boundary layer is laminar, determine the value of $f''(0)$, that is, the value of the shear function at the wall. What is the relation between the shear at a given value of x for this flow and that for a flat plate?

4.3. Consider two-dimensional, incompressible flow over a cylinder. For ease of use with the nomenclature of the current chapter, we will assume that the windward plane of symmetry (i.e., the stagnation point) is $\theta = 0$ and that θ increases in the streamwise direction. Thus,

$$u_e = 2U_\infty \sin \theta \quad \text{and} \quad x = R\theta$$

Determine the values of β at $\theta = 30°$, at $\theta = 45°$, and at $\theta = 90°$.

4.4. Assume that the wall is porous so that there can be flow through the wall; that is, $v(x, 0) = v_w \neq 0$. Using equation (4.14), show that

$$\frac{v_w}{u_e} = -\frac{f(0)}{\sqrt{2\mathrm{Re}_x}}$$

in order to have similarity solutions; that is $(\partial f/\partial s)_\eta = 0$ for steady, incompressible flow past a flat plate.

4.5. We plan to use suction through a porous wall as a means of boundary-layer control. Using the equation developed in Problem 4.4, determine $f(0)$ if $v_w = -0.001u_e$ for steady flow past a flat plate where $u_e = 10$ m/s at standard sea-level conditions. What are the remaining two boundary conditions?

4.6. Transpiration (or injecting gas through a porous wall into the boundary layer) is to be used to reduce the skin-friction drag for steady, laminar flow past a flat plate. Using the equation developed in Problem 4.4, determine v_w if $f(0) = -0.25$. The inviscid velocity (u_e) is 50 ft/s with standard atmospheric conditions.

4.7. When we derived the integral equations for a flat-plate boundary layer, the outer boundary of our control volume was a streamline outside the boundary layer (see Fig. 4.17). Let us now apply the integral equations to a rectangular control volume to calculate the sectional drag coefficient for incompressible flow past a flat plate of length L. Thus, as shown in Fig. P4.7, the outer boundary is a line parallel to the wall and outside the boundary layer at all x stations. Owing to the growth of the boundary layer, fluid flows through the upper boundary with a velocity v_e which is a function of x. How does the resultant expression compare with equation (4.69)?

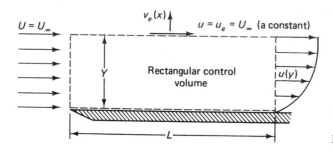

Figure P4.7

4.8. Use the integral momentum analysis and the assumed velocity profile for a laminar boundary layer:

$$\frac{u}{u_e} = \frac{3}{2}\left(\frac{y}{\delta}\right) - \frac{1}{2}\left(\frac{y}{\delta}\right)^3$$

where δ is the boundary-layer thickness, to describe the incompressible flow past a flat plate. For this profile, compute: **(a)** $(\delta/x)\sqrt{Re_x}$, **(b)** $(\delta^*/x)\sqrt{Re_x}$, **(c)** $(v_e/u_e)\sqrt{Re_x}$, **(d)** $C_f\sqrt{Re_x}$, and **(e)** $C_d\sqrt{Re_x}$. Compare these values with those presented in the text, which were obtained using the more exact differential technique, e.g., $(\delta/x)\sqrt{Re_x} = 5.0$. Prepare a graph comparing this approximate velocity profile and that given in Table 4.3. For the differential solution, use $\eta = 3.5$ to define δ when calculating y/δ.

4.9. Use the integral momentum analysis and a linear velocity profile for a laminar boundary layer,

$$\frac{u}{u_e} = \frac{y}{\delta}$$

where δ is the boundary layer thickness. If the viscous flow is incompressible, calculate $(\delta/x)\sqrt{Re_x}$, $(\delta^*/x)\sqrt{Re_x}$, and $C_f\sqrt{Re_x}$. Compare these values with those presented in the chapter that were obtained using the more exact differential technique [e.g., $(\delta/x)\sqrt{Re_x} = 5.0$].

4.10. Let us represent the wing of an airplane by a flat plate. The airplane is flying at standard sea-level conditions at 170 mph. The dimensions of the wing are chord = 4 ft and span = 28 ft. What is the total friction drag acting on the wing? What is the drag coefficient?

4.11. A flat plate at zero angle of attack is mounted in a wind tunnel where

$$p_\infty = 1.01325 \times 10^5 \text{ N/m}^2 \qquad U_\infty = 100 \text{ m/s}$$

$$\mu_\infty = 1.7894 \times 10^{-5} \text{ kg/m} \cdot \text{s} \qquad \rho_\infty = 1.2250 \text{ kg/m}^3$$

A pitot probe is to be used to determine the velocity profile at a station 1.0 m from the leading edge (Fig. P4.11).

(a) Using a transition criterion that $Re_{x,tr} = 500,000$, where does transition occur?

(b) Use equation (4.73) to calculate the thickness of the turbulent boundary layer at a point 1.00 m from the leading edge.

(c) If the streamwise velocity varies as the $\frac{1}{7}$th power law [i.e., $u/u_e = (y/\delta)^{1/7}$], calculate the pressure you should expect to measure with the pitot probe (p_t) as a function of y. Present the predicted values as:

(1) The difference between that sensed by the pitot probe and that sensed by the static port in the wall [i.e., y versus $p_t(y) - p_{static}$].

(2) As the pressure coefficient

$$y \text{ versus } C_p(y) = \frac{p_t(y) - p_\infty}{\frac{1}{2}\rho_\infty U_\infty^2}$$

Note that for part (c) we can use Bernoulli's equation to relate the static pressure and the velocity on the streamline just ahead of the probe and the stagnation pressure sensed by the probe. Even though this is in the boundary layer, we can use Bernoulli's equation, since we relate properties on a streamline and since we calculate these properties at a "point" not letting viscous forces affect the properties.

(d) Is the flow described by this velocity function rotational or irrotational?

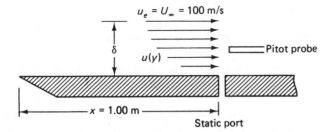

$u_e = U_\infty = 100$ m/s

δ

$u(y)$

Pitot probe

$x = 1.00$ m

Static port

Figure P4.11

4.12. Air at atmospheric pressure and 100°C flows at 100 km/h across a flat plate. Compare the streamwise velocity as a function of y for a laminar boundary layer and for a turbulent boundary layer at the transition point, assuming that the transition process is completed instantaneously at that location. Use Table 4.3 to define the laminar profile and the one-seventh power law to describe the turbulent profile.

4.13. A thin symmetric airfoil section is mounted at zero angle of attack in a low-speed wind tunnel. A pitot probe is used to determine the velocity profile in the viscous region downstream of the airfoil, as shown in Fig. P4.13. The resultant velocity distribution in the region $-w \leq z \leq +w$ is

$$u(z) = U_\infty - \frac{U_\infty}{2} \cos \frac{\pi z}{2w}$$

If we apply the integral form of the momentum equation [equation (2.13)] to the flow between the two streamlines bounding this wake, we can calculate the drag force acting on the airfoil section. The integral continuity equation [equation (2.5)] can be used to

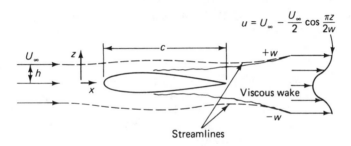

$u = U_\infty - \dfrac{U_\infty}{2} \cos \dfrac{\pi z}{2w}$

U_∞

z

h

c

$+w$

x

Viscous wake

$-w$

Streamlines

Figure 4P.13

relate the spacing between the streamlines in the undisturbed flow ($2h$) to their spacing ($2w$) at the x location where the pitot profile was obtained. If $w = 0.009c$, what is the section drag coefficient, C_d?

4.14. Using equation (4.94), calculate the thermal conductivity of air at 2000 K. What is the Prandtl number of perfect air at this temperature?

4.15. The boundary conditions that were used in developing the equation for the laminar thermal boundary layer were that the temperature is known at the two limits (1) $\theta = 0$ at $\eta = 0$, and (2) $\theta = 1$ at $\eta \to$ large. What would be the temperature distribution if the boundary conditions were (1) an adiabatic wall (i.e., $\theta' = 0$ at $\eta = 0$), and (2) $\theta = 1$ at $\eta \to$ large. *Hint*: From equation (4.96),

$$\theta' = C(f'')^{\text{Pr}}$$

4.16. Represent the wing of an airplane by a flat plate. The airplane is flying at standard sea-level conditions at 170 mi/h. The dimensions of the wing are chord = 4 ft and span = 28 ft. What is the total heat transferred to the wing if the temperature of the wing is 50°F?

4.17. A wind tunnel has a 1-m², 6-m-long test section in which air at standard sea-level conditions moves at 70 m/s. It is planned to let the walls diverge slightly (slant outward) to compensate for the growth in boundary-layer displacement thickness, and thus maintain a constant area for the inviscid flow. This allows the free-stream velocity to remain constant. At what angle should the walls diverge to maintain a constant velocity between $x = 1.5$ m and $x = 6$ m?

REFERENCES

4.1. Brune, G. W., P. W. Rubbert, and T. C. Nark, Jr., "A New Approach to Inviscid Flow/Boundary Layer Matching," *AIAA Paper 74-601*, presented at the 7th Fluid and Plasma Dynamics Conference, Palo Alto, Calif., 1974.

4.2. Schlichting, H., *Boundary Layer Theory*, 7th ed., McGraw-Hill Book Company, New York, 1979.

4.3. White, F. M., *Viscous Fluid Flow*, McGraw-Hill Book Company, New York, 1974.

4.4. Dorrance, W. H., *Viscous Hypersonic Flow*, McGraw-Hill Book Company, New York, 1962.

4.5 Abbott, I. H., and A. E. von Doenhoff, *Theory of Wing Sections*, Dover Publications, New York, 1949.

4.6 Cebeci, T., and A. M. O. Smith, *Analysis of Turbulent Boundary Layers*, Academic Press, Orlando, Fla., 1974.

4.7. Bushnell, D. M., A. M. Cary, Jr., and J. E. Harris, "Calculation Methods for Compressible Turbulent Boundary Layers—1976," *SP-422*, NASA, Langley Research Center, 1977.

4.8. Head, M. R., "Cambridge Work on Entrainment, in *Proceedings, Computation of Turbulent Boundary Layers—1968 AFOSR-IFP-Stanford Conference*, Vol. 1, Stanford University Press, Stanford, Calif., 1969.

4.9. Cebeci, T., and P. Bradshaw, *Momentum Transfer in Boundary Layers*, McGraw-Hill Book Company, New York, 1979.

5

CHARACTERISTIC PARAMETERS FOR AIRFOIL AND WING AERODYNAMICS

AIRFOIL GEOMETRY PARAMETERS

If a horizontal wing is cut by a vertical plane parallel to the centerline of the vehicle, the resultant section is called the *airfoil section*. The generated lift and the stall characteristics of the wing depend strongly on the geometry of the airfoil sections that make up the wing. Geometric parameters that have an important effect on the aerodynamic characteristics of an airfoil section include (1) the leading-edge radius, (2) the mean camber line, (3) the maximum thickness and the thickness distribution of the profile, and (4) the trailing-edge angle. The effect of these parameters, which are illustrated in Fig. 5.1 will be discussed after a brief introduction to airfoil-section nomenclature.

Airfoil-Section Nomenclature

Quoting from Ref. 5.1: "The gradual development of wing theory tended to isolate the wing-section problems from the effects of planform and led to a more systematic experimental approach. The tests made at Göttingen during World War I contributed much to the development of modern types of wing sections. Up to about World War II, most wing sections in common use were derived from more or less direct extensions of the work at Göttingen. During this period, many families of wing sections were tested in the laboratories of various countries, but the work of the NACA was out-

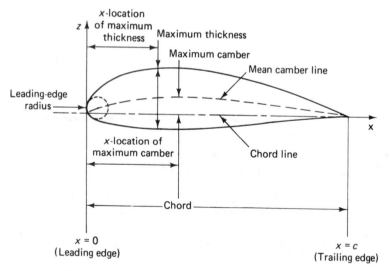

Figure 5.1   Airfoil-section geometry and its nomenclature.

standing. The NACA investigations were further systematized by separation of the effects of camber and thickness distribution, and the experimental work was performed at higher Reynolds number than were generally obtained elsewhere.''

As a result, the geometry of many airfoil sections is uniquely defined by the NACA designation for the airfoil. There are a variety of classifications, including NACA four-digit wing sections, NACA five-digit wing sections, and NACA 6 series wing sections. As an example, consider the NACA four-digit wing sections. The first integer indicates the maximum value of the mean camber-line ordinate (see Fig. 5.1) in percent of the chord. The second integer indicates the distance from the leading edge to the maximum camber in tenths of the chord. The last two integers indicate the maximum section thickness in percent of the chord. Thus, the NACA 0010 is a symmetric airfoil section whose maximum thickness is 10% of the chord. The NACA 4412 airfoil section is a 12% thick airfoil which has a 4% maximum camber located at 40% of the chord.

A series of "standard" modifications are designated by a suffix consisting of a dash followed by two digits. These modifications consist essentially of (1) changes of the leading-edge radius from the normal value and (2) changes of the position of maximum thickness from the normal position (which is at $0.3c$). Thus,

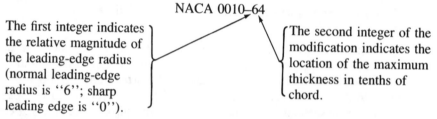

NACA 0010–64

The first integer indicates the relative magnitude of the leading-edge radius (normal leading-edge radius is "6"; sharp leading edge is "0").

The second integer of the modification indicates the location of the maximum thickness in tenths of chord.

However, because of the rapid improvements, both in computer hardware and computer software, and because of the broad use of sophisticated numerical codes, one often encounters airfoil sections being developed that are not described by the standard NACA geometries.

Leading-Edge Radius and Chord Line

The *chord line* is defined as the straight line connecting the leading and trailing edges. The leading edge of airfoils used in subsonic applications is rounded, with a radius that is on the order of 1% of the chord length. The leading-edge radius of the airfoil section is the radius of a circle centered on a line tangent to the leading-edge camber connecting tangency points of the upper and the lower surfaces with the leading edge. The center of the leading-edge radius is located such that the cambered section projects slightly forward of the leading-edge point. The magnitude of the leading-edge radius has a significant effect on the stall (or boundary-layer separation) characteristics of the airfoil section.

The geometric angle of attack is the angle between the chord line and the direction of the undisturbed, "free-stream" flow. For many airplanes the chord lines of the airfoil sections are inclined relative to the vehicle axis (see Table 5.1).

Mean Camber Line

The locus of the points midway between the upper surface and the lower surface, as measured perpendicular to the chord line, defines the *mean camber line*. The shape of the mean camber line is very important in determining the aerodynamic characteristics of an airfoil section. As will be seen in the theoretical solutions and in the experimental data that will be presented in this book, cambered airfoils in a subsonic flow generate lift even when the section angle of attack is zero. Thus, an effect of camber is a change in the zero-lift angle of attack, α_{0l}. While the symmetric sections have zero lift at zero angle of attack, zero lift results for sections with positive camber when they are at negative angles of attack.

Furthermore, camber has a beneficial effect on the maximum value of the section lift coefficient. If the maximum lift coefficient is high, the stall speed will be low, all other factors being the same. It should be noted, however, that the high thickness and camber necessary for high maximum values for the section lift coefficient produce low critical Mach numbers (see Chapter 9) and high twisting moments at high speeds. Thus, one needs to consider the trade-offs in selecting a design value for a particular parameter.

Maximum Thickness and Thickness Distribution

The maximum thickness and the thickness distribution strongly influence the aerodynamic characteristics of the airfoil section. The maximum lift coefficient for an airfoil section increases as the maximum thickness of the airfoil increases. In addition,

TABLE 5.1 WING-GEOMETRY PARAMETERS

Type	Wing span [m(ft)]	Aspect ratio, AR	Sweep angle	Dihedral	Airfoil section	Speed [km/h (mi/h)]
a. Four-Place Single-Engine Aircraft						
Socata Rallye (France)	9.61 (31.52)	7.57	None	7°	63A414 (mod), 63A416, inc. 4°	173–245 (108–152)
Ambrosini NF 15 (Italy)	9.90 (32.5)	7.37	None	6°	64–215, inc. 4°	325 (202)
Beechcraft Bonanza V35B	10.20 (33.46)	6.30	None	6°	23016.5 at root, 23012 at tip, inc. 4° at root, 1° at tip	298–322 (185–200)
Beechcraft Sierra	9.98 (32.75)	7.35	None	6°30'	63₂A415, inc. 3° at root, 1° at tip	211–281 (131–162)
Cessna 172	10.92 (35.83)	7.32	None	1°44'	NACA 2412, inc. 1°30' at root, −1°30' at tip	211 (131)
Piper Commanche	10.97 (36.0)	7.28	2°30' for-ward	5°	64₂A215, inc. 2°	298 (185)
Bellanca, Model 25	10.67 (35.0)	6.70	None	2°	NACA 63₂–215, inc. 2°	458–499 (285–310)
Piper Warrior II	10.67 (35.0)	7.24	None	7°	NACA 65₂–415, inc. 2° at root, −1° at tip	191–235 (119–146)
b. Commercial Jetliners and Transports						
Caravelle 210 (France)	34.3 (112.5)	8.02	20° at c/4	3°	NACA 65,212	790 (490)
BAC 111 (UK)	26.97 (88.5)	8.00	20° at c/4	2°	NACA cambered section (mod.), t/c = 0.125 at root, 0.11 at tip, inc. 2°30'	815 (507)
Tupolev 154 (USSR)	37.55 (123.2)	7.03	35° at c/4	—	—	975 (605)
McDonnell-Douglas DC9	27.25 (89.42)	8.25	24° at c/4	—	t/c = 0.116 (av.)	903 (561)

TABLE 5.1 *(Continued)*

Type	Wing span [m(ft)]	Aspect ratio, AR	Sweep angle	Dihedral	Airfoil section	Speed [km/h (mi/h)]
Boeing 727	32.92 (108.0)	7.67	32° at c/4	3°	Special Boeing sections $t/c = 0.09$ to 0.13, inc. 2°	975 (605)
Boeing 737	28.35 (93.0)	8.83	25° at c/4	6°	$t/c = 0.129$ (av.)	848 (527)
Boeing 747	59.64 (195.7)	6.95	37°30' at c/4	7°	$t/c = 0.134$ (inboard), 0.078 (midspan), 0.080 (outboard), inc. 2°	958 (595)
Lockheed C-5A	67.88 (222.8)	7.75	25° at c/4	Anhedral 5°30'	NACA 0011 (mod.) near midspan, inc. 3°30'	815 (507)
Airbus A310 (International)	43.89 (144.0)	8.8	28° at c/4	11°8' (inboard at the trailing edge)	$t/c = 0.152$ at root, $t/c = 0.108$ at tip, inc. 5°3' at root	667–895 (414–556)
Ilyushin IL76 (USSR)	50.50 (165.7)	8.5	25° at c/4	Anhedral with wing mounted above the fuselage		750–850 (466–528)
c. High-Speed Military Aircraft						
SAAB-35 Draken (Sweden)	9.40 (30.8)	1.77	Central: leading edge 80°, outer: leading edge 57°	—	$t/c = 0.05$	Mach 1.4–2.0
Dassault Mirage III (France)	8.22 (27.0)	1.94	Leading edge 60°34'	Anhedral 1°	$t/c = 0.045$–0.035	Mach 2.2

TABLE 5.1 *(Continued)*

Type	Wing span [m(ft)]	Aspect ratio, AR	Sweep angle	Dihedral	Airfoil section	Speed [km/h (mi/h)]
Northrop F-5E	8.13 (26.67)	3.82	24° at c/4	None	65A004.8 (mod.), $t/c = 0.048$	Mach 1.23
McDonnell-Douglas F4	11.70 (38.4)	2.78	45°	Outer panel 12°	$t/c = 0.051$ (av.)	Over Mach 2.0
LTV F-8	10.81 (35.7)	3.39	35°	Anhedral 5°	Thin, laminar flow section	Nearly Mach 2
LTV A-7	11.80 (38.75)	4.0	35° at c/4	Anhedral 5°	65A007, inc. −1°	1123 (698)
Mitsubishi T2 (Japan)	7.88 (25.85)	2.93	35° 47′ at c/4	Anhedral 9°	NACA 65 series (mod.), $t/c = 0.0466$	Mach 1.6
General Dynamics F-16	10.0 (32.83)	3.2	40° on leading edges	—	NACA 64A-204	Mach 2.0+

Source: Data from Refs. 5.2, 5.3, and 5.4.

the thicker airfoils benefit more from the use of high lift devices but have a lower critical Mach number.

The maximum local velocity to which a fluid particle accelerates as it flows around an airfoil section increases as the maximum thickness increases (see the discussion associated with Fig. 4.10). Thus, the minimum pressure value is smallest for the thickest airfoil. As a result, the adverse pressure gradient associated with the deceleration of the flow from the location of this pressure minimum to the trailing edge is greatest for the thickest airfoil. As the adverse pressure gradient becomes larger, the boundary layer becomes thicker (and is more likely to separate producing relatively large values for the form drag). Thus, the beneficial effects of increasing the maximum thickness are limited.

The thickness distribution for an airfoil affects the pressure distribution and the character of the boundary layer. As the location of the maximum thickness moves aft, the velocity gradient (and hence the pressure gradient) in the midchord region decreases. The resultant favorable pressure gradient in the midchord region promotes boundary-layer stability and increases the possibility that the boundary layer remains laminar. Laminar boundary layers produce less skin friction drag than turbulent boundary layers but are also more likely to separate under the influence of an adverse pressure gradient. This will be discussed in more detail later in this chapter.

Trailing-Edge Angle

The trailing-edge angle affects the location of the aerodynamic center (which is defined later in this chapter). The aerodynamic center of thin airfoil sections in a subsonic stream is theoretically located at the quarter-chord.

WING-GEOMETRY PARAMETERS

By placing the airfoil sections discussed in the preceding section in spanwise combinations, wings, horizontal tails, vertical tails, canards, and/or other lifting surfaces are formed. When the parameters that characterize the wing planform are introduced, attention must be directed to the existence of flow components in the spanwise direction. In other words, airfoil section properties deal with flow in two dimensions while planform properties relate to the resultant flow in three dimensions.

In order to fully describe the planform of a wing, several terms are required. The terms that are pertinent to defining the aerodynamic characteristics of a wing are illustrated in Fig. 5.2.

1. The *wing area*, S, is simply the plan surface area of the wing. Although a portion of the area may be covered by fuselage or nacelles, the pressure carryover on these surfaces allows legitimate consideration of the entire plan area.
2. The *wing span*, b, is measured tip to tip.
3. The *average chord*, $\bar{c}$, is the geometric average. The product of the span and the average chord is the wing area ($b \times \bar{c} = S$).

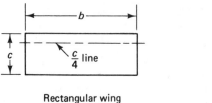
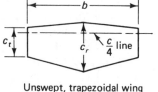

Rectangular wing   Unswept, trapezoidal wing

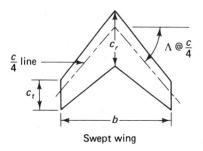
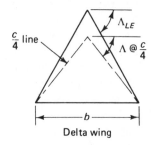

Swept wing   Delta wing

Figure 5.2   Geometric characteristics of the wing planform.

4. The *aspect ratio*, AR, is the ratio of the span and the average chord. For a rectangular wing, the aspect ratio is simply

$$AR = \frac{b}{c}$$

For a nonrectangular wing,

$$AR = \frac{b^2}{S}$$

The aspect ratio is a fineness ratio of the wing and is useful in determining the aerodynamic characteristics and structural weight. Typical aspect ratios vary from 35 for a high-performance sailplane to 2 for a supersonic jet fighter.

5. The *root chord*, c_r, is the chord at the wing centerline, and the *tip chord*, c_t, is measured at the tip.

6. Considering the wing planform to have straight lines for the leading and trailing edges, the *taper ratio*, λ, is the ratio of the tip chord to the root chord:

$$\lambda = \frac{c_t}{c_r}$$

The taper ratio affects the lift distribution and the structural weight of the wing. A rectangular wing has a taper ratio of 1.0 while the pointed tip delta wing has a taper ratio of 0.0.

7. The *sweep angle* Λ is usually measured as the angle between the line of 25 percent chord and a perpendicular to the root chord. Sweep angles of the leading

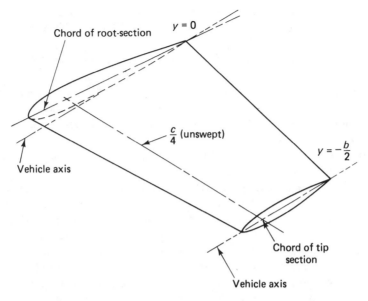

Figure 5.3   Unswept, tapered wing with geometric twist (wash-out).

edge or of the trailing edge are often presented with the parameters, since they are of interest for many applictions. The sweep of a wing causes definite changes in the maximum lift, in the stall characteristics, and in the effects of compressibility.

8. The *mean aerodynamic chord* (m.a.c) is used together with S to nondimensionalize the pitching moments. Thus, the mean aerodynamic chord represents an average chord which, when multiplied by the product of the average section moment coefficient, the dynamic pressure, and the wing area, gives the moment for the entire wing. The mean aerodynamic chord is given by

$$\bar{c} = \frac{1}{S} \int_{-0.5b}^{+0.5b} [c(y)]^2 \, dy$$

9. The *dihedral angle* is the angle between a horizontal plane containing the root chord and a plane midway between the upper and lower surfaces of the wing. If the wing lies below the horizontal plane, it is termed an *anhedral angle*. The dihedral angle affects the lateral stability characteristics of the airplane.

10. *Geometric twist* defines the situation where the chord lines for the spanwise distribution of airfoil sections do not all lie in the same plane. Thus, there is a spanwise variation in the geometric angle of incidence for the sections. The chord of the root section of the wing shown in the sketch of Fig. 5.3 is inclined 4° relative to the vehicle axis. The chord of the tip section, however, is parallel to the vehicle axis. In this case, where the incidence of the airfoil sections relative to the vehicle axis decrease toward the tip, the wing has "wash out." The wings of numerous subsonic aircraft have wash out to control the spanwise lift distri-

bution and, hence, the boundary-layer separation (i.e., stall) characteristics. If the angle of incidence increases toward the tip, the wing has "wash in."

The aspect ratio, taper ratio, twist, and sweepback of a planform are the principal factors that determine the aerodynamic characteristics of a wing and have an important bearing on its stall properties. These same quantities also have a definite influence on the structural weight and stiffness of a wing. Values of these parameters for a variety of aircraft have been taken from Refs. 5.2, 5.3, and 5.4 and are summarized in Table 5.1. Data are presented for four-place single-engine aircraft, commercial jetliners and transports, and high-speed military aircraft. Note how the values of these parameters vary from one group of aircraft to another.

The data presented in Table 5.1 indicate that similar designs result for similar applications, regardless of the national origin of the specific design. For example, the aspect ratio for the four-seat, single-engine civil aviation designs is approximately 7, whereas the aspect ratio for the supersonic military aircraft is between 1.7 and 3.8. In fact, as noted in Ref. 5.5, which was a case study in aircraft design for the Northrop F-5, "The selection of wing aspect ratio represents an interplay between a large value for low drag-due-to-lift and a small value for reduced wing weight."

There are other trends exhibited in parameters relating to aircraft performance. Notice the grouping by generic classes of aircraft for the correlation between the power-to-weight ratio and the wing loading that is presented in Fig. 5.4. There is a tendency for airplanes to get larger, faster, heavier, and more complex with technological improvements. The trend toward larger, heavier aircraft is evident in this figure. Note that a fully loaded B-17G Flying Fortess, a heavy bomber of World War II, weighed 29,700 kg (65,500 lb) with a wing span of 31.62 m (103.75 ft), whereas the F15, a

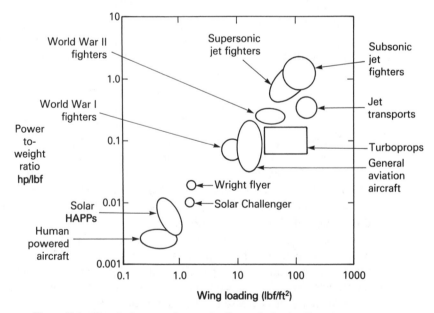

Figure 5.4 Historical ranges of power loading and wing loading. (From Ref. 5.6.)

modern fighter aircraft, has a maximum takeoff weight of 30,845 kg (68,000 lb) with a wing span of 13.05 m (42.81 ft). However, the successful human-powered aircraft fall in the lower left corner of Fig. 5.4, with wing loadings (the ratio of takeoff gross weight to wing area) less than 1 lbf/ft². It is in this same region that Lockheed's solar high-altitude powered platform (Solar HAPP) operates.

CHARACTERIZATION OF AERODYNAMIC FORCES

The motion of the air around the vehicle produces pressure and velocity variations which produce the aerodynamic forces and moments. Although viscosity is a fluid property, and therefore acts throughout the flow field, the viscous forces depend on the velocity gradients as well as on the viscosity. For many aerodynamics problems, the viscous forces are significant only within a narrow region near the wall where the velocity gradients are large (the boundary layer) or within the viscous wake should boundary-layer separation occur. When the viscous effects are confined to the boundary layer, it is possible to use the equations of motion for an inviscid flow to determine the pressure distribution around the vehicle and hence the velocity of the air particles at the edge of the boundary layer. With the inviscid flow field known, the velocity distribution across the viscous boundary layer, and hence the tangential shear forces at the wall, can be calculated. The normal and tangential forces, which act on the surface due to the fluid motion around the vehicle, are shown in Fig. 5.5.

It is conventional to resolve the sum of the normal, or pressure, forces and the tangential, or viscous shear, forces into three components along axes parallel and perpendicular to the free-stream direction. The three components are designated as lift, drag, and side force. The relation of the lift and the drag forces with respect to the free-stream velocity is shown in Fig. 5.6.

1. *Lift.* Lift is the force component acting upward, perpendicular to the direction of flight, or to the undisturbed free-stream velocity. The aerodynamic lift is produced primarily by the pressure forces acting on the vehicle surface.
2. *Drag.* Drag is the net aerodynamic force acting in the same direction as the undisturbed free-stream velocity. The aerodynamic drag is produced by the pressure forces and by skin friction forces that act on the surface.
3. *Side force.* Side force is the component of force in a direction perpendicular to both the lift and the drag. The side force is positive when acting toward the starboard wing (i.e., the pilot's right).

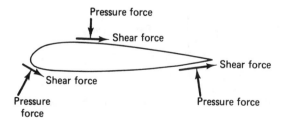

Figure 5.5 Normal (or pressure) and tangential (or shear) forces on an airfoil surface.

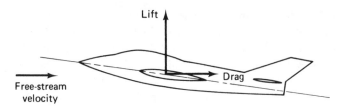

Figure 5.6   Aerodynamic forces in the plane of symmetry.

The resulting aerodynamic force usually will not act through the origin of the airplane's axis system (i.e., the center of gravity). The moment due to the resultant force acting at a distance from the origin may be divided into three components, referred to the airplane's reference axes. The three moment components are the pitching moment, the rolling moment, and the yawing moment, as shown in Fig. 5.7.

1. *Pitching moment.* The moment about the lateral axis (the y axis of the airplane-fixed-coordinate system) is the pitching moment. The pitching moment is the result of the lift and the drag forces acting on the vehicle. A positive pitching moment is in the nose-up direction.
2. *Rolling moment.* The moment about the longitudinal axis of the airplane (the x axis) is the rolling moment. A rolling moment is often created by a differential lift, generated by some type of ailerons or spoilers. A positive rolling moment causes the right, or starboard, wingtip to move downward.

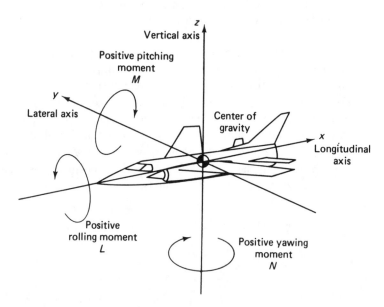

Figure 5.7   Reference axes of the airplane and the corresponding aerodynamic moments.

3. *Yawing moment.* The moment about the vertical axis of the airplane (the z axis) is the yawing moment. A positive yawing moment tends to rotate the nose to the pilot's right.

Parameters That Govern Aerodynamic Forces

The magnitude of the forces and the moments that act on a vehicle depend on the combined effects of many different variables. Personal observations of the aerodynamic forces acting on an arm extended from a car window or on a ball in flight demonstrate the effect of velocity and of configuration. Pilot manuals advise that a longer length of runway is required if the ambient temperature is relatively high or if the airport elevation is high (i.e., the ambient density is relatively low). The parameters that govern the magnitude of aerodynamic forces and moments include:

1. Configuration geometry
2. Angle of attack
3. Vehicle size or model scale
4. Free-stream velocity
5. Density of the undisturbed air
6. Reynolds number (as it relates to viscous effects)
7. Mach number (as it relates to compressibility effects)

The calculation of the aerodynamic forces and moments acting on a vehicle often requires that the engineer be able to relate data obtained at other flow conditions to the conditions of interest. Thus, the engineer often uses data from the wind tunnel, where scale models are exposed to flow conditions that simulate the design environment or data from flight tests at other flow conditions. So that one can correlate the data for various stream conditions and configuration scales, the measurements are usually presented in dimensionless form. Ideally, once in dimensionless form, the results would be independent of all but the first two parameters listed above, configuration geometry and angle of attack. In practice, flow phenomena, such as boundary-layer separation, shock-wave/boundary-layer interactions, and compressibility effects, limit the range of flow conditions over which the dimensionless force and moment coefficients remain constant.

AERODYNAMIC FORCE AND MOMENT COEFFICIENTS

Lift Coefficient

Let us develop the equation for the normal force coefficient to illustrate the physical significance of a dimensionless force coefficient. We choose the normal (or z) component of the resultant force since it is relatively simple to calculate and it has the same relation to the pressure and the shear forces as does the lift. For a relatively thin airfoil section at a relatively low angle of attack, it is clear from Fig. 5.5 that the lift (and

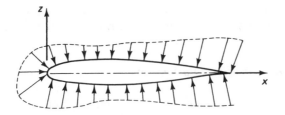

Figure 5.8   Pressure distribution for a lifting airfoil section.

similarly the normal force) results primarily from the action of the pressure forces. The shear forces act primarily in the chordwise direction (i.e., contribute primarily to the drag). Therefore, to calculate the force in the z direction, we need consider only the pressure contribution, which is illustrated in Fig. 5.8. The pressure force acting on a differential area of the vehicle surface is $dF = p\, ds\, dy$, as shown in Fig. 5.9. The elemental surface area is the product of ds, the wetted length of the element in the plane of the cross section, times dy, the element's length in the direction perpendicular to the plane of the cross section (or spanwise direction). Since the pressure force acts normal to the surface, the force component in the z direction is the product of the pressure times the projected planform area:

$$dF_z = p\, dx\, dy \qquad (5.1)$$

Integrating over the entire wing surface (including the upper and the lower surfaces), the net force in the z direction is given by

$$F_z = \oiint p\, dx\, dy \qquad (5.2)$$

Note that the resultant force in any direction due to a constant pressure over a closed surface is zero:

$$\oiint p_\infty\, dx\, dy = 0 \qquad (5.3)$$

Combining equations (5.2) and (5.3), the resultant force component is

$$F_z = \oiint (p - p_\infty)\, dx\, dy \qquad (5.4)$$

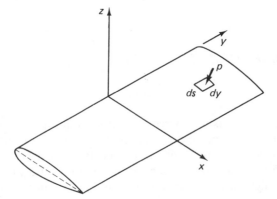

Figure 5.9   Pressure acting on an elemental surface area.

To nondimensionalize the factors on the right-hand side of equation (5.4), divide by the product $q_\infty cb$, which has the units of force.

$$\frac{F_z}{q_\infty cb} = \oiint \frac{p - p_\infty}{q_\infty} d\left(\frac{x}{c}\right) d\left(\frac{y}{b}\right)$$

Since the product cb represents the planform area S of the rectangular wing of Fig. 5.9,

$$\frac{F_z}{q_\infty S} = \oiint C_p d\left(\frac{x}{c}\right) d\left(\frac{y}{b}\right) \tag{5.5}$$

When the boundary layer is thin, the pressure distribution around the airfoil is essentially that of an inviscid flow. Thus, the pressure distribution is independent of Reynolds number and does not depend on whether the boundary layer is laminar or turbulent. When the boundary layer is thin, the pressure coefficient at a particular location on the surface given by the dimensionless coordinates $(x/c, y/b)$ is independent of vehicle scale and of the flow conditions. Over the range of flow conditions for which the pressure coefficient is a unique function of the dimensionless coordinates $(x/c, y/b)$, the value of the integral in equation (5.5) depends only on the configuration geometry and on the angle of attack. Thus, the resulting dimensionless force parameter, or force coefficient (in this case, the normal force coefficient), is independent of model scale and of flow conditions.

A similar analysis can be used to calculate the lift coefficient, which is defined as

$$C_L = \frac{L}{q_\infty S} \tag{5.6}$$

Data are presented in Fig. 5.10 for a NACA 23012 airfoil; that is, the wind-tunnel model represented a wing of infinite span. Thus, the lift measurements are presented in terms of the section lift coefficient C_l. The section lift coefficient is the lift per unit span (l) divided by the product of the dynamic pressure times the plan area per unit span, which is the chord length (c).

$$C_l = \frac{l}{q_\infty c} \tag{5.7}$$

The data from Ref. 5.1 were obtained in a wind tunnel that could be operated at pressures up to 10 atm. As a result, the Reynolds number ranged from 3×10^6 to 9×10^6 at Mach numbers less than 0.17. In addition to the measurements obtained with a smooth model, data are presented for a model that had "standard" surface roughness applied near the leading edge. Additional comments will be made about surface roughness later in this chapter.

The experimental section lift coefficient is independent of Reynolds number over the range tested and is a linear function of the angle of attack from $-10°$ to $+10°$.

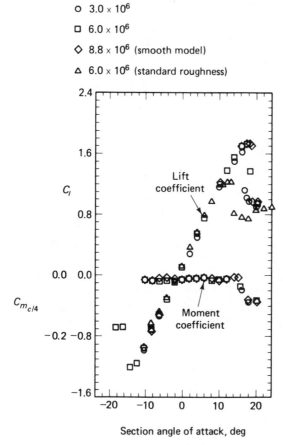

Re_c

○ 3.0×10^6

□ 6.0×10^6

◇ 8.8×10^6 (smooth model)

△ 6.0×10^6 (standard roughness)

Figure 5.10 Section lift coefficient and section moment coefficient (with respect to $c/4$) for an NACA 23012 airfoil. (Data from Ref. 5.1.)

The slope of this linear portion of the curve is called the *two-dimensional lift curve slope*. For this airfoil,

$$\frac{dC_l}{d\alpha} = C_{l,\alpha} = 0.104 \text{ per degree} \tag{5.8}$$

Since the NACA 23012 airfoil section is cambered (the maximum camber is approximately 2% of the chord length), lift is generated at zero angle of attack. In fact, zero lift is obtained at $-1.2°$, which is designated α_{0l} or the section angle of attack for zero lift. As the angle of attack is increased above 10°, the section lift coefficient continues to increase (but not linearly with angle or attack) until a maximum value, $C_{l,\max}$, is reached. Referring to Fig. 5.10, $C_{l,\max}$ is 1.79 and occurs at an angle of attack of 18°. Partly because of this relatively high value of $C_{l,\max}$, the NACA 23012 section has been used on many aircraft (e.g., the Beechcraft Bonanza; see Table 5.1, and the Brewster Buffalo).

At angles of attack in excess of 10°, the section lift coefficients exhibit a Reynolds-number dependence. Note that the adverse pressure gradient which the air particles encounter as they move toward the trailing edge of the upper surface increases as the angle of attack increases. At these angles of attack, the air particles which have been slowed by the viscous forces cannot overcome the relatively large adverse pressure gradient, and the boundary layer separates. The separation location depends on the character (laminar or turbulent) of the boundary layer and its thickness, and therefore on the Reynolds number. As will be discussed, boundary-layer separation has a profound effect on the drag acting on the airfoil.

The study of airfoil lift as a function of incidence has shown that, in many instances, the presence of a separation bubble near the leading edge of the airfoil results in laminar section stall. Experimental data on two-dimensional "peaky" airfoil sections indicates that $C_{l,\max}$, as limited by laminar stall, is strongly dependent on the leading-edge shape and on the Reynolds number. If the laminar boundary layer, which develops near the leading edge of the upper surface of the airfoil, is subjected to a relatively high adverse pressure gradient, it will separate because the relatively low kinetic energy level of the air particles near the wall is insufficient to surmount the "pressure hill" of the adverse pressure gradient. The separated shear layer that is formed may curve back onto the surface within a very short distance. This is known as *short bubble separation*. The separated viscous layer near the pressure peak may not reattach to the surface at all, or it may reattach within 0.3 chord length or more downstream. This extended separation region is known as *long bubble separation*.

The separation bubble may not appear at all on relatively thick, strongly cambered profiles operating at high Reynolds numbers. The reason for this is that the Reynolds number is then large enough for a natural transition to a turbulent boundary layer to occur upstream of the strong pressure rise. The relatively high kinetic energy of the air particles in a turbulent boundary layer can climb the "pressure hill," and boundary-layer separation occurs only a short distance upstream of the trailing edge (trailing-edge stall). The separation point moves upstream continuously with increasing angle of attack, and the lift does not drop abruptly after $C_{l,\max}$ but decreases gradually.

Moment Coefficients

The moment created by the aerodynamic forces acting on a wing (or airfoil) is determined about a particular reference axis. The reference axis may be the leading edge, the quarter-chord location, the aerodynamic center, and so on. The significance of these reference axes in relation to the coefficients for thin airfoils will be discussed in subsequent chapters.

The procedure used to nondimensionalize the moments created by the aerodynamic forces is similar to that used to nondimensionalize the lift. To demonstrate this nondimensionalization, the pitching moment about the leading edge due to the pressures acting on the surface will be calculated (refer again to Fig. 5.9). The contribution of the chordwise component of the pressure force to the moment is small and is neglected. Thus, the pitching moment about the leading edge due to the pressure force acting on the surface element whose area is ds times dy and which is located at a distance x from the leading edge is

$$dM_0 = p \, dx \, dy \, x \qquad (5.9)$$

where $dx\ dy$ is the projected area. Integrating over the entire wing surface, the net pitching moment is given by

$$M_0 = \oiint px\ dx\ dy \tag{5.10}$$

When a uniform pressure acts on any closed surface, the resultant pitching moment due to this constant pressure is zero. Thus,

$$\oiint p_\infty x\ dx\ dy = 0 \tag{5.11}$$

Combining equations (5.10) and (5.11), the resulting pitching moment about the leading edge is

$$M_0 = \oiint (p - p_\infty)x\ dx\ dy \tag{5.12}$$

To nondimensionalize the factors on the right-hand side of equation (5.12), divide by $q_\infty c^2 b$, which has the units of force times length:

$$\frac{M_0}{q_\infty c^2 b} = \oiint \frac{p - p_\infty}{q_\infty}\frac{x}{c}\ d\left(\frac{x}{c}\right)\ d\left(\frac{y}{b}\right)$$

Since the product of cb represents the planform area of the wing S,

$$\frac{M_0}{q_\infty Sc} = \oiint C_p \frac{x}{c}\ d\left(\frac{x}{c}\right)\ d\left(\frac{y}{b}\right) \tag{5.13}$$

Thus, the dimensionless moment coefficient is

$$C_{M0} = \frac{M_0}{q_\infty Sc} \tag{5.14}$$

Since the derivation of equation (5.14) was for the rectangular wing of Fig. 5.9, the chord c is used. However, as noted previously in this chapter, the mean aerodynamic chord is used together with S to nondimensionalize the pitching moment.

The section moment coefficient is used to represent the dimensionless moment per unit span (m_0):

$$C_{m0} = \frac{m_0}{q_\infty cc} \tag{5.15}$$

since the surface area per unit span is the chord length c. The section pitching moment coefficient depends on the camber and on the thickness ratio. Section pitching moment coefficients for a NACA 23012 airfoil section with respect to the quarter chord and with respect to the aerodynamic center are presented in Figs. 5.10 and 5.11, respectively. The *aerodynamic center* is that point about which the section moment coefficient is independent of the angle of attack. Thus, the aerodynamic center is that point along the chord where all changes in lift effectively take place. Since the moment about the aerodynamic center is the product of a force (the lift that acts at the center of pressure)

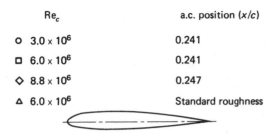

Re$_c$	a.c. position (x/c)
o 3.0 × 10^6	0.241
□ 6.0 × 10^6	0.241
◇ 8.8 × 10^6	0.247
△ 6.0 × 10^6	Standard roughness

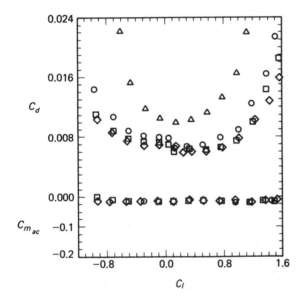

Figure 5.11 Section drag coefficient and section moment coefficient (with respect to the ac) for an NACA 23012 airfoil. (Data from Ref. 5.1.)

and a lever arm (the distance from the aerodynamic center to the center of pressure), the center of pressure must move toward the aerodynamic center as the lift increases. The quarter-chord location is significant, since it is the theoretical aerodynamic center for incompressible flow about a two-dimensional airfoil.

Note that the pitching moment coefficient is independent of the Reynolds number (for those angles of attack where the lift coefficient is independent of the Reynolds number), since the pressure coefficient depends only on the dimensionless space coordinates $(x/c, y/b)$ [see equation (5.13)]. One of the features of the NACA 23012 section is a relatively high $C_{l,\max}$ with only a small $C_{m_{ac}}$.

The characteristic length (or moment arm) for the rolling moment and for the yawing moment is the wing span b (instead of the chord). Therefore, the rolling moment coefficient is

$$C_{\mathscr{L}} = \frac{\mathscr{L}}{q_\infty Sb} \tag{5.16}$$

and the yawing moment coefficient is

$$C_N = \frac{N}{q_\infty Sb} \tag{5.17}$$

Drag Coefficient

The drag force on a wing is due in part to skin friction and in part to the integrated effect of pressure. If $\vec{\tau}$ denotes the tangential shear stress at a point on the body surface, p the static pressure, and $\hat{n}$ the external normal to the element of surface dS, the drag can be formally expressed as

$$D = \oiint \vec{\tau} \cdot \hat{e}_\infty \, dS - \oiint p\hat{n} \cdot \hat{e}_\infty \, dS \tag{5.18}$$

where $\hat{e}_\infty$ is a unit vector parallel to the free stream and the integration takes place over the entire wetted surface. The first integral represents the friction component and the second integral represents the pressure drag.

The most straightforward approach to calculating the pressure drag is to perform the numerical integration indicated by the second term in equation (5.18). This approach is known as the near-field method of drag computation. Unfortunately, this can be a relatively inaccurate procedure for streamlined configurations at small angles of attack. The inaccuracy results because the pressure drag integral is the difference between the integration on forward-facing and rearward-facing surface elements, this difference being a second-order quantity for slender bodies. This difficulty can be circumvented by applying the integral form of the momentum equation, as demonstrated in Chapter 2 and in the homework problems of Chapters 2 and 4. This approach is known as the far-field method of drag computation.

In Chapter 3 we learned that zero drag results for irrotational, steady, incompressible flow past a two-dimensional body. For an airfoil section (i.e., a two-dimensional geometry) which is at a relatively low angle of attack so that the boundary layer is thin and does not separate, the pressure distribution is essentially that for an inviscid flow. Thus, skin friction is a major component of the chordwise force per unit span (f_x). Referring to Figs. 5.5 and 5.8, sf_x can be approximated as

$$sf_x \approx \oint \tau \, dx \tag{5.19}$$

where sf_x is the chordwise force per unit span due to skin friction. Dividing both sides of equation (5.19) by the product $q_\infty c$ yields an expression for the dimensionless force coefficient:

$$\frac{sf_x}{q_\infty c} \approx \oint C_f \, d\left(\frac{x}{c}\right) \tag{5.20}$$

C_f, the skin friction coefficient, is defined as

$$C_f = \frac{\tau}{\frac{1}{2}\rho_\infty U_\infty^2} \tag{5.21}$$

As was stated in the general discussion of the boundary-layer characteristics in Chapter 4 (see Fig. 4.18) skin friction for a turbulent boundary layer is much greater than that for a laminar boundary layer for given flow conditions. Equations for calculating the skin friction coefficient were developed in Chapter 4. However, let us introduce the correlations for the skin friction coefficient for incompressible flow past a flat plate to gain insight into the force coefficient of equation (5.20). Of course, the results for a flat plate only approximate those for an airfoil. The potential function given by equation (3.35) shows that the velocity at the edge of the boundary layer and, therefore, the local static pressure, is constant along the plate. Such is not the case for an airfoil section, for which the flow accelerates from a forward stagnation point to a maximum velocity, then decelerates to the trailing edge. Nevertheless, the analysis will provide useful insights into the section drag coefficient,

$$C_d = \frac{d}{q_\infty c} \tag{5.22}$$

for an airfoil at relatively low angles of attack.

Referring to Chapter 4, when the boundary layer is laminar,

$$C_f = \frac{0.664}{(\mathrm{Re}_x)^{0.5}} \tag{5.23}$$

If the boundary layer is turbulent,

$$C_f = \frac{0.0583}{(\mathrm{Re}_x)^{0.2}} \tag{5.24}$$

For equations (5.23) and (5.24), the local Reynolds number is defined as

$$\mathrm{Re}_x = \frac{\rho_\infty U_\infty x}{\mu_\infty} \tag{5.25}$$

Example 5.1

Calculate the local skin friction at a point 0.5 m from the leading edge of a flat-plate airfoil flying at 60 m/s at a height of 6 km.

Solution   Refer to Table 1.2 to obtain the static properties of undisturbed air at 6 km:

$$\rho_\infty = 0.6601 \text{ kg/m}^3$$

$$\mu_\infty = 1.5949 \times 10^{-5} \text{ kg/s·m}$$

Thus, using equation (5.25),

$$\mathrm{Re}_x = \frac{(0.6601 \text{ kg/m}^3)(60 \text{ m/s})(0.5 \text{ m})}{1.5949 \times 10^{-5} \text{ kg/s·m}}$$

$$= 1.242 \times 10^6$$

If the boundary layer is laminar,

$$C_f = \frac{0.664}{(\mathrm{Re}_x)^{0.5}} = 5.959 \times 10^{-4}$$

$$\tau = C_f(\tfrac{1}{2}\rho_\infty U_\infty^2) = 0.708 \text{ N/m}^2$$

If the boundary layer is turbulent,

$$C_f = \frac{0.0583}{(\mathrm{Re}_x)^{0.2}} = 3.522 \times 10^{-3}$$

$$\tau = C_f(\tfrac{1}{2}\rho_\infty U_\infty^2) = 4.185 \text{ N/m}^2$$

It is obvious that the force coefficient of equation (5.20) depends on the Reynolds number. The Reynolds number not only affects the magnitude of C_f, but it is also used as an indicator of whether the boundary layer is laminar or turbulent. Since the calculation of the force coefficient requires integration of C_f over the chord length, we must know at what point, if any, the boundary layer becomes turbulent (i.e., where transition occurs).

Boundary-Layer Transition

Near the forward stagnation point on a wing or near the leading edge of a flat plate, the boundary layer is laminar. As the flow proceeds downstream, the boundary layer thickens and the viscous forces continue to dissipate the energy of the airstream. Disturbances to the flow in the growing viscous layer may be caused by surface roughness, a temperature variation in the surface, pressure pulses, and so on. If the Reynolds number is low, the disturbances will be damped by viscosity and the boundary layer will remain laminar. At higher Reynolds numbers, the disturbances grow. In such cases, the boundary layer may become unstable and, eventually, turbulent (i.e., transition will occur). The details of the transition process are quite complex and depend on many parameters.

The engineer who must develop a transition criterion for design purposes usually uses the Reynolds number. For instance, if the surface of a flat plate is smooth and if the external airstream has no turbulence, transition "occurs" at a Reynolds number (Re_x) of approximately 500,000. However, experience has shown that the Reynolds number at which the disturbances will grow and the length over which the transition process takes place depends on the magnitude of the disturbances and on the flow field. Let us consider briefly the effect of surface roughness, of the surface temperature, of a pressure gradient in the inviscid flow, and of the local Mach number on transition.

1. *Surface roughness.* Since transition is the amplification of disturbances to the flow, the presence of surface roughness significantly promotes transition (i.e., causes transition to occur at relatively low Reynolds numbers).

2. *Surface temperature.* The boundary-layer thickness decreases as the surface temperature is decreased. Cooling the surface usually delays transition.

3. *Pressure gradient.* A favorable pressure gradient (i.e., the static pressure decreases in the streamwise direction or, equivalently, the inviscid flow is accelerating) delays transition. Conversely, an adverse pressure gradient promotes transition.

4. *Mach number*. The transition Reynolds number is usually higher when the flow is compressible (i.e., as the Mach number is increased).

Stability theory (e.g., Ref. 5.7) can be an important tool that provides insights into the importance of individual parameters without introducing spurious effects that might be due to flow disturbances peculiar to a test facility (e.g., "noise" in a wind tunnel). For a more detailed discussion of transition, the reader is referred to a textbook on boundary-layer theory (e.g., Ref. 5.8).

If the skin friction is the dominant component of the drag, transition should be delayed as long as possible to obtain a low-drag section. To delay transition on a low-speed airfoil section, the point of maximum thickness should be moved aft so that the boundary layer is subjected to a favorable pressure gradient over a longer run. Consider the NACA 0009 section and the NACA 66–009 section. Both are symmetric, having a maximum thickness of $0.09c$. The maximum thickness for the NACA 66–009 section is at $0.45c$, while that for the NACA 0009 section is at $0.3c$ (see Fig. 5.12). As a

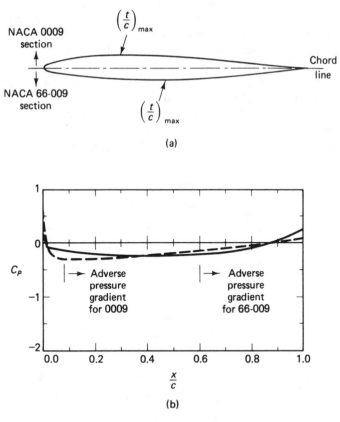

(a)

(b)

Figure 5.12 Comparison of geometries and resultant pressure distributions for a "standard" airfoil section (NACA 0009) and for a laminar airfoil section (NACA 66-009): (a) comparison of cross section for an NACA 0009 airfoil with that for an NACA 66-009 airfoil; (b) static pressure distribution.

result, the minimum pressure coefficient occurs at $x = 0.6c$ for the NACA 66–009 and a favorable pressure gradient acts to stabilize the boundary layer to this point. For the NACA 0009, the minimum pressure occurs near $x = 0.1c$. The lower local velocities near the leading edge and the extended region of favorable pressure gradient cause transition to be farther aft on the NACA 66–009. Since the drag for a streamlined airfoil at low angles of attack is primarily due to skin friction, use of equation (5.20) would indicate that the drag is lower for the NACA 66–009. This is verified by the data from Ref. 5.1, which are reproduced in Fig. 5.13. The subsequent reduction in the friction drag creates a *drag bucket* for the NACA 66–009 section. Note that the section drag curve varies only slightly with C_l for moderate excursions in angle of attack, since the skin friction coefficient varies little with angle of attack. At the very high Reynolds numbers that occur at some flight conditions, it is difficult to maintain a long run of laminar boundary layer, especially if surface roughnesses develop during the flight operations. However, a *laminar-flow section*, such as the NACA 66–009, offers additional benefits. Comparing the cross sections presented in Fig. 5.12, the cross section of the NACA 66-009 airfoil provides more flexibility for carrying fuel and for accommodating the load-carrying structure.

For larger angles of attack, the section drag coefficient depends both on Reynolds number and on angle of attack. As the angle of attack and the section lift coefficient increase, the minimum pressure coefficient decreases. The adverse pressure gradient that results as the flow decelerates toward the trailing edge increases. When the air particles in the boundary layer, already slowed by viscous action, encounter the relatively strong adverse pressure gradient, the boundary layer thickens and separates. Because the thickening boundary layer and its separation from the surface cause the pressure distribution to be significantly different from the inviscid model at the higher angles of attack, form drag dominates. Note that, at the higher angles of attack (where form

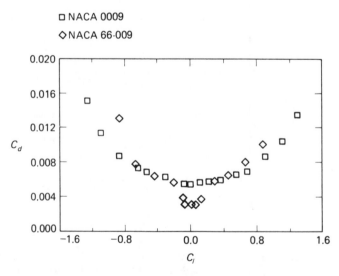

Figure 5.13 Section drag coefficients for NACA 0009 airfoil and the NACA 66-009 airfoil, $Re_c = 6 \times 10^6$. (Data from Ref. 5.1.)

drag is important), the drag coefficient for the NACA 66–009 is greater than that for the NACA 0009.

Thus, we see that the lift coefficient and the moment coefficient depend only on vehicle geometry and angle of attack for low-speed flows where viscous effects are secondary. However, the drag coefficient exhibits a Reynolds number dependence both at the low angles of attack, where the boundary layer is thin (and the transition location is important), and at high angles of attack, where extensive regions of separated flow exist.

The section drag coefficient for a NACA 23012 airfoil is presented as a function of the section lift coefficient in Fig. 5.11. The data illustrate the dependence on Reynolds number and on angle of attack, which has been discussed. Note that measurements, which are taken from Ref. 5.1, include data for a *standard roughness*.

Effect of Surface Roughness on the Aerodynamic Forces

As discussed in Chapter 2, the Reynolds number is an important parameter when comparing the viscous character of two flow fields. If one desires to reproduce the Reynolds number for a flight condition in the wind tunnel, then

$$\left(\frac{\rho_\infty U_\infty c}{\mu_\infty}\right)_{\text{wt}} = \left(\frac{\rho_\infty U_\infty c}{\mu_\infty}\right)_{\text{ft}} \tag{5.26}$$

where the subscripts wt and ft designate wind-tunnel and flight conditions, respectively. In many low-speed wind tunnels, the free-stream values for density and viscosity are roughly equal to the atmospheric values. Thus,

$$(U_\infty c)_{\text{wt}} \simeq (U_\infty c)_{\text{ft}} \tag{5.27}$$

If the wind-tunnel model is 0.2 scale, the wind-tunnel value for the free-stream velocity would have to be 5 times the flight value. As a result, the tunnel flow would be transonic or supersonic, which obviously would not be a reasonable simulation. Thus, since the maximum Reynolds number for this "equal density" subsonic wind-tunnel simulation is much less than the flight value, controlled surface roughness is often added to the model to fix boundary-layer transition at the location at which it would occur naturally in flight.

Abbott and von Doenhoff (Ref. 5.1) present data on the effect of surface condition on the aerodynamic forces. "The standard leading-edge roughness selected by the NACA for 24-in chord models consisted of 0.011-in carborundum grains applied to the surface of the model at the leading edge over a surface length of $0.08c$ measured from the leading edge on both surfaces. The grains were thinly spread to cover 5 to 10% of the area. This standard roughness is considerably more severe than that caused by the usual manufacturing irregularities or deterioration in service, but it is considerably less severe than that likely to be encountered in service as a result of accumulation of ice, mud, or damage in military combat." The data for the NACA 23012 airfoil (Fig. 5.10) indicate that the angle of zero lift and the lift-curve slope are practically unaffected by the standard leading-edge roughness. However, the maximum lift coefficient is affected by surface roughness. This is further illustrated by the data presented in Fig. 5.14.

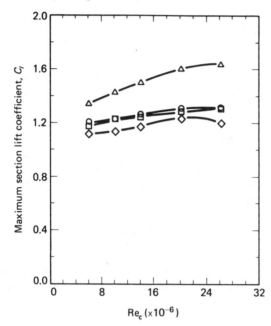

△ Smooth

○ 0.002 roughness

□ 0.004 roughness

◇ 0.011 roughness

Figure 5.14   Effect of roughness near the leading edge on the maximum section lift for the NACA 63(420)-422 airfoil. (Data from Ref. 5.1.)

When there is no appreciable separation of the flow, the drag on the airfoil is caused primarily by skin friction. Thus, the value of the drag coefficient depends on the relative extent of the laminar boundary layer. A sharp increase in the drag coefficient results when transition is suddenly shifted forward. If the wing surface is sufficiently rough to cause transition near the wing leading edge, large increases in drag are observed, as is evident in the data of Fig. 5.11 for the NACA 23012 airfoil section. In other test results presented in Ref. 5.1, the location of the roughness strip was systematically varied. The minimum drag increased progressively with forward movement of the roughness strip.

Scaling effects between model simulations and flight applications (as they relate to the viscous parameters) are especially important when the flow field includes an interaction between a shock wave and the boundary layer. The transonic flow field for an airfoil may include a shock-induced separation, a subsequent reattachment to the airfoil surface, and another boundary-layer separation near the trailing edge. According to Ref. 5.9, the prime requirements for correct simulation of these transonic shock/boundary-layer interactions include that the boundary layer is turbulent at the point of interaction and that the thickness of the turbulent boundary layer for the model flow is not so large in relation to the full-scale flow that a rear separation would occur in the simulation that would not occur in the full-scale flow.

Reference 5.10 provides some general guidelines for the use of grit-type boundary-layer transition trips. Whereas it is possible to fix boundary-layer transition far forward on wind-tunnel models at subsonic speeds using grit-type transition trips having little or no grit drag, the roughness configurations that are required to fix transition in a supersonic flow often cause grit drag. Fixing transition on wind-tunnel models becomes increasingly difficult as the Mach number is increased. Since roughness heights several times larger than the boundary-layer thickness can be required to fix transition in a hypersonic flow, the required roughness often produces undesirable distortions of the flow. Reference 5.11 provides some general guidelines for the use of boundary-layer trips for hypersonic flows. Data are presented to illustrate problems that can arise from poor trip designs.

The comments regarding the effects of surface roughness were presented for two reasons. (1) The reader should note that, since it is impossible to match the Reynolds number in many scale-model simulations, surface roughness (in the form of boundary-layer trips) is often used to fix transition and therefore the relative extent of the laminar boundary layer. (2) When surface roughness is used, considerable care should be taken to properly size and locate the roughness elements in order to properly simulate the desired flow.

Wing-section profile-drag measurements have been made for the Boeing 727 in flight using the Boeing Airborne Traversing Probe, examples of which are presented in Fig. 5.15. The probe consists of four main components: (1) flow sensors, (2) a rotating arm, (3) the drive unit, and (4) the mounting base. As reported by Bowes (Ref. 5.12): ''The measured minimum section profile drag at $M = .73$ was about 15 percent higher than predicted from wind-tunnel test data for a smooth airfoil. The wind-tunnel data used in this correlation were also from wake surveys on the 727 wing. The data were adjusted to fully turbulent flow and extrapolated to flight Reynolds numbers. This quite sizeable difference between the measured and extrapolated values of $C_{d,\min}$ has been attributed to surface roughness and excrescences on the airplane wing, although the 15-percent increase in wing-section profile drag is larger than traditionally allotted in airplane drag estimates. The wing section where this survey was performed was inspected and had numerous steps and bumps due to control devices and manufacturing tolerances which would account for this local level of excrescence drag. This is not representative of the entire wing surface.''

Wings of Finite Span

The drag coefficients presented thus far are for two-dimensional airfoils (i.e., configurations of infinite span). For a wing of finite span that is generating lift, the pressure differential between the lower surface and the upper surface causes a spanwise flow. The resultant three-dimensional flow field produces a drag component which, as we show in Chapter 7, is proportional to

$$C_{D_v} \propto \frac{C_L^2}{\pi \cdot \text{AR}} \tag{5.28}$$

This is termed the *vortex drag*, the *induced drag*, or the *drag due to lift*. Let us introduce another term, designated ΔC_{D_M}, which will include compressibility effects

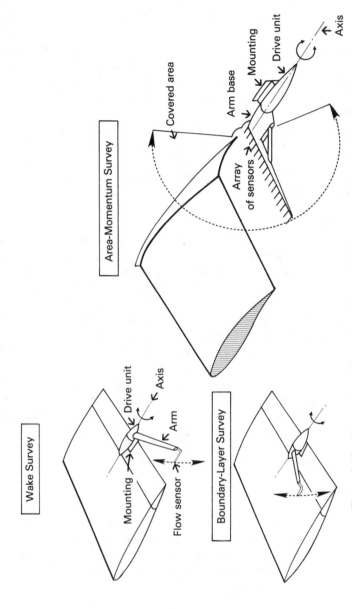

Figure 5.15 Airborne traversing probe concept and configurations. (From Ref. 5.12.) (The original version of this material was first published by the Advisory Group for Aerospace Research and Development, North Atlantic Treaty Organization [AGARD, NATO] in Lecture Series 67, May 1974, Ref. 5.12.)

and may, possibly, include the changes in the airflow accompanying thrust variations (according to Ref. 5.12). Thus, the total drag for an airplane may be written as the sum of (1) the profile drag, which is independent of the lift coefficient and therefore exists when the configuration generates zero lift (C_{D_0} or ΔC_{D_p}), (2) the drag associated with lift (kC_L^2), and (3) the compressibility-related effects that are correlated in terms of the Mach number and are known as wave drag:

$$C_D = C_{D_0} + kC_L^2 + \Delta C_{D_M} \tag{5.29}$$

Although the relative importance of the different drag sources depends on the aircraft type and the mission to be flown, approximate breakdowns (by category) for large, subsonic transports are presented in Figs. 5.16 and 5.17. According to Thomas (Ref. 5.13), the greatest contribution arises from turbulent skin-friction drag. The next most significant contribution arises from lift-induced drag, which, when added to skin-friction drag, accounts for about 85% of the total drag of a typical transport aircraft (see Fig. 5.16). Thomas cited the pressure drag due to open separation in the afterbody and other regions, interference effects between aerodynamic components, wave drag due to the compressibility effect at near-sonic flight conditions, and miscellaneous effects, such as roughness effects and leakage, as constituting the remaining 15%.

Using slightly different division of the drag-contributing sources, Bowes (Ref. 5.12) presented the lift/drag polar which is reproduced in Fig. 5.17. The majority of the lift-related drag is the vortex drag for an elliptic load distribution at subcritical speeds (i.e., the entire flow is subsonic). Bowes notes that a good wing design should approach an elliptic loading at the design condition.

1. *Zero-lift drag.* Skin friction and form drag components can be calculated for the wing, tail, fuselage, nacelles, and so on. When evaluating the zero-lift drag, one must consider interactions such as how the growth of the boundary layer on the fuselage reduces the boundary-layer velocities on the wing-root surface. Because of the interaction, the wing-root boundary layer is more easily separated in the presence of an adverse pressure gradient. Since the upper wing surface

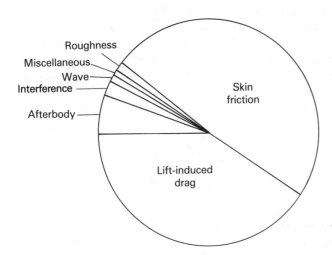

Figure 5.16 Contributions of different drag sources for a typical transport aircraft. (From Ref. 5.13.)

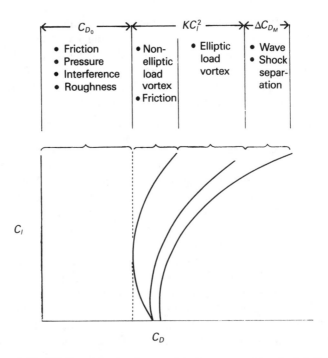

Figure 5.17 Lift/drag polar for a large, subsonic transport. (From Ref. 5.12.) (The original version of this material was first published by the Advisory Group for Aerospace Research and Development, North Atlantic Treaty Organization [AGARD, NATO] in Lecture Series 67, May 1974, Ref. 5.12.)

has the more critical pressure gradients, a low wing position on a circular fuselage would be sensitive to this interaction. Adequate filleting and control of the local pressure gradients can minimize the interaction effects.

A representative value of C_{D0} would be 0.020, of which the wings may account for 50%, the fuselage and the nacelles 40%, and the tail 10%. Since the wing constitutes such a large fraction of the drag, reducing the wing area would reduce C_{D0} if all other factors are unchanged. Although other factors must be considered, the reasoning above implies that an optimum airplane configuration would have a minimum wing surface area and, therefore, highest practical wing loading (N/m²)

2. *Drag due to lift.* The drag due to lift may be represented as:

$$kC_L^2 = \frac{C_L^2}{\pi(AR)e} \tag{5.30}$$

where e is the airplane efficiency factor. Typical values of the airplane efficiency factor range from 0.6 to 0.9. At high lift coefficients (near $C_{L,\max}$), e should be changed to account for increased form drag. The deviation of the actual airplane drag from the quadratic correlation where e is a constant is significant for airplanes with low aspect ratios and sweepback.

Compressibility

Another factor to consider is the effect of compressibility. When the flight Mach number is equal to or greater than 0.5 (approximately), the drag increases very rapidly with speed, as we discuss in Chapter 9. The correlation of equation (5.29) includes wave drag (i.e., the compressibility effects) as the third term. Low aspect ratios and sweep-back are factors that delay and reduce the compressibility drag rise.

Lift/Drag Ratio

The configuration and the application of an airplane is closely related to the lift/drag ratio. Many important items of airplane performance are obtained in flight at $(L/D)_{max}$. Performance conditions that occur at $(L/D)_{max}$ include:

1. Maximum range of propeller-driven airplanes
2. Maximum climb angle for jet-powered airplanes
3. Maximum power-off glide ratio (for jet-powered or propeller-driven airplanes)

   Representative values of the maximum lift/drag ratios for subsonic flight speeds are:

Type of airplane	$(L/D)_{max}$
High-performance sailplane	25–40
Commercial transport	12–20
Supersonic fighter	4–9

PROBLEMS

5.1. Using equation (5.29) and treating C_{D0} and k as constants, show that the lift coefficient for maximum lift/drag ratio, and the maximum lift/drag ratio for an incompressible flow are given by

$$C_{L_{(L/D)max}} = \sqrt{\frac{C_{D0}}{k}}$$

$$\left(\frac{L}{D}\right)_{max} = \frac{1}{2\sqrt{kC_{D0}}}$$

5.2. A Cessna 172 is cruising at 10,000 ft on a standard day ($\rho_\infty = 0.001756$ slug/ft³) at 130 mi/h. If the airplane weighs 2300 lb, what C_L is required to maintain level flight?

5.3. Assume that the lift coefficient is a linear function of α for its operating range. Assume further that the wing has a positive camber so that its zero-lift angle of attack (α_{0L}) is negative, and that the slope of the straight-line portion of the C_L versus α curve is $C_{L,\alpha}$. Using the results of Problem 5.1, derive an expression for $\alpha_{(L/D)max}$.

5.4. Using the results of Problem 5.1, what is $C_{D_{(L/D)max}}$?

5.5. Consider a flat plate at zero angle of attack in a uniform flow where $U_\infty = 35$ m/s in the standard sea-level atmosphere (Fig. P5.5). Assume that $Re_{x,tr} = 500,000$ defines the transition point. Determine the section drag coefficient, C_d. Neglect plate edge effects (i.e., assume two-dimensional flow). What error would be incurred if it is assumed that the boundary layer is turbulent along the entire length of the plate?

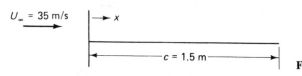

$U_\infty = 35$ m/s $\rightarrow x$

$c = 1.5$ m

Figure P5.5

5.6. An airplane that weighs 50,000 N and has a wing area of 21.5 m² is landing at an airport.
  (a) Graph C_L as a function of the true airspeed over the airspeed range 300 to 180 km/h, if the airport is at sea level.
  (b) Repeat the problem for an airport that is at an altitude of 1600 m. For the purposes of this problem, assume that the airplane is in steady, level flight to calculate the required lift coefficients.

5.7. Pressure distribution measurements from Ref. 5.14 are presented in Table 5.2 for the midspan section of a 12.7- by 76.2-cm model which had a NACA 4412 section. Graph C_p as a function of x/c for these two angles of attack. Comment on the movement of the stagnation point. Comment on changes in the magnitude of the adverse pressure gradient toward the trailing edge of the upper surface. How does this relate to possible boundary-layer separation (or stall)?
  Use equation (3.13),

$$\frac{U}{U_\infty} = \sqrt{1 - C_p}$$

to calculate the maximum value of the local velocity both on the upper surface and on the lower surface for both angles of attack. If these velocities are representative of the changes with angle of attack, how does the circulation (or lift) change with the angle of attack?

5.8. Using the small-angle approximations for the local surface inclinations, integrate the experimental chordwise pressure distributions of Table 5.2 to obtain values of the section lift coefficient for $\alpha = 2°$ and $\alpha = 16°$. Assuming that the section lift coefficient is a linear function of α, calculate

$$C_{l,\alpha} = \frac{dC_l}{d\alpha}$$

5.9. Using the small-angle approximations for the local surface inclinations, integrate the experimental chordwise pressure distributions of Table 5.2 to obtain values of the section pitching moment coefficient about the quarter chord for $\alpha = 2°$ and $\alpha = 16°$. Thus, in equation (5.9), x is replaced by $(x - 0.25c)$. Recall that a positive pitching moment is in the nose-up direction.

5.10. The lift/drag ratio of a sailplane is 30. The sailplane has a wing area of 10.0 m² and weighs 3150 N. What is C_D when the aircraft is in steady level flight at 170 km/h at an altitude of 1.0 km?

TABLE 5.2 EXPERIMENTAL PRESSURE DISTRIBUTIONS FOR AN
NACA 4412 AIRFOIL[a]

Orifice location		Value of the pressure coefficient, C_p	
x Station (percent c from leading edge)	z Ordinate (percent c above chord)	$\alpha = 2°$	$\alpha = 16°$
100.00	0	0.181	0.010
97.92	−0.10	0.164	0.121
94.86	−0.16	0.154	0.179
89.90	−0.22	0.152	0.231
84.94	−0.28	0.148	0.257
74.92	−0.52	0.136	0.322
64.94	−0.84	0.120	0.374
54.48	−1.24	0.100	0.414
49.98	−1.44	0.091	0.426
44.90	−1.64	0.088	0.459
39.98	−1.86	0.071	0.485
34.90	−2.10	0.066	0.516
29.96	−2.30	0.048	0.551
24.90	−2.54	0.025	0.589
19.98	−2.76	−0.011	0.627
14.94	−2.90	−0.053	0.713
9.96	−2.86	−0.111	0.818
7.38	−2.72	−0.131	0.896
4.94	−2.46	−0.150	0.980
2.92	−2.06	−0.098	0.993
1.66	−1.60	0.028	0.791
0.92	−1.20	0.254	0.264
0.36	−0.70	0.639	−1.379
0	0	0.989	−3.648
0	0.68	0.854	−6.230
0.44	1.56	0.336	−5.961
0.94	2.16	0.055	−5.210
1.70	2.78	−0.148	−4.478
2.94	3.64	−0.336	−3.765
4.90	4.68	−0.485	−3.190
7.50	5.74	−0.568	−2.709
9.96	6.56	−0.623	−2.440
12.58	7.34	−0.676	−2.240
14.92	7.88	−0.700	−2.149
17.44	8.40	−0.721	−1.952
19.96	8.80	−0.740	−1.841
22.44	9.16	−0.769	−1.758
24.92	9.52	−0.746	−1.640
27.44	9.62	−0.742	−1.535
29.88	9.76	−0.722	−1.438
34.98	9.90	−0.693	−1.269
39.90	9.84	−0.635	−1.099
44.80	9.64	−0.609	−0.961
49.92	9.22	−0.525	−0.786
54.92	8.76	−0.471	−0.649
59.94	8.16	−0.438	−0.551

TABLE 5.2 (*Continued*)

Orifice location		Value of the pressure coefficient, C_p	
x Station (percent c from leading edge)	z Ordinate (percent c above chord)	$\alpha = 2°$	$\alpha = 16°$
64.90	7.54	−0.378	−0.414
69.86	6.76	−0.319	−0.316
74.90	5.88	−0.252	−0.212
79.92	4.92	−0.191	−0.147
84.88	3.88	−0.116	−0.082
89.88	2.74	−0.026	−0.043
94.90	1.48	0.076	−0.016
98.00	0.68	0.143	−0.004

[a] Average pressure (standard atmospheres): 21; average Reynolds number: 3,100,000.

Source: Ref. 5.14.

REFERENCES

5.1. Abbott, I. H., and A. E. von Doenhoff, *Theory of Wing Sections*, Dover Publications, New York, 1949.

5.2. Taylor, J. W. R. (Ed.), *Jane's All the World's Aircraft, 1973–1974*, Jane's Yearbooks, London, 1973.

5.3. Taylor, J. W. R. (Ed.), *Jane's All the World's Aircraft, 1966–1967*, Jane's Yearbooks, London, 1966.

5.4. Taylor, J. W. R. (Ed.), *Jane's All the World's Aircraft, 1984–1985*, Jane's Yearbooks, London, 1984.

5.5. Stuart, W. G., *Northrop F-5 Case Study in Aircraft Design*, AIAA Professional Study Series, Sept. 1978.

5.6. Hall, D. W., "To Fly on the Wings of the Sun," *Lockheed Horizons*, Issue 18, Lockheed Corporation, Burbank, Calif., June 1985, pp. 60–68.

5.7. Mack, L. M., "Boundary-Layer Linear Stability Theory," *Special Course on Stability and Transition of Laminar Flow*, edited by R. Michel, *AGARD Report 709*, pp. 3–1 to 3–81, 1984.

5.8. Schlichting, H., *Boundary-Layer Theory*, McGraw-Hill Book Company, New York, 1968.

5.9. Pearcey, H. H., J. Osborne, and A. B. Haines, "The Interaction Between Local Effects at the Shock and Rear Separation—A Source of Significant Scale Effects in Wind-Tunnel Tests on Aerofoils and Wings," Paper 11 in "Transonic Aerodynamics," *AGARD Conference Proceedings*, Vol. 35, Sept. 1968.

5.10. Braslow, A. L., R. M. Hicks, and R. V. Harris, Jr., "Use of Grit-Type Boundary-Layer Transition Trips on Wind-Tunnel Models," *TN D-3579*, NASA, Sept. 1966.

5.11. Sterret, J. R., E. L. Morrisette, A. H. Whitehead, Jr., and R. M. Hicks, "Transition Fixing for Hypersonic Flow," *TN D-4129*, NASA, Oct. 1966.

5.12. Bowes, G. M., *Aircraft Lift and Drag Prediction and Measurement*, Boeing Commercial

Airplane Company, prepared as Lecture Series 67, for AGARD (Advisory Group for Aerospace Research and Development), May 1974.

5.13. Thomas, A. S. W., "Aircraft Viscous Drag Reduction Technology," *Lockheed Horizons*, Oct. 1985, Issue 19, pp. 22–32.

5.14. Pinkerton, R. M., "Calculated and Measured Pressure Distributions over the Midspan Section of the NACA 4412 Airfoil," *Report 563*, NACA, 1936.

6

TWO-DIMENSIONAL, INCOMPRESSIBLE FLOWS AROUND THIN AIRFOILS

Theoretical relations that describe an inviscid, low-speed flow around a thin airfoil will be developed in this chapter. To obtain the governing equations, it is assumed that the airfoil extends to infinity in both directions from the plane of symmetry (i.e., that the airfoil is a wing of infinite aspect ratio). Thus, the flow field around the airfoil is the same for any cross section perpendicular to the wing and the flow is two-dimensional. The flow around a two-dimensional airfoil can be idealized by super-imposing a translational flow past the airfoil section, a distortion of the stream that is due to the airfoil thickness, and a circulatory flow that is related to the lifting characteristics of the airfoil. Since it is a two-dimensional configuration, an airfoil in an incompressible stream experiences no drag force, if one neglects the effects of viscosity. However, as discussed in Chapters 4 and 5, the viscous forces produce a velocity gradient near the surface of the airfoil, and hence drag due to skin friction. Furthermore, the presence of the viscous flow near the surface modifies the inviscid flow field and may produce a significant drag force due to the integrated effect of the pressure field (i.e., form drag).

For a lifting airfoil, the pressure on the lower surface of the airfoil is, on the average, greater than the pressure on the upper surface. Bernoulli's equation for steady, incompressible flow leads to the conclusion that the velocity over the upper surface is on the average, greater than that past the lower surface. Thus, the flow around the airfoil can be represented by the combination of a translational flow from left to right

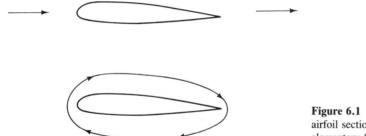

Figure 6.1 Flow around the lifting airfoil section, as represented by two elementary flows.

and a circulating flow in a clockwise direction, as shown in Fig. 6.1. This flow model assumes that the airfoil is sufficiently thin so that thickness effects may be neglected.

CIRCULATION AND THE GENERATION OF LIFT

If the fluid is initially at rest, the line integral of the velocity around any curve completely surrounding the airfoil is zero, because the velocity is zero for all fluid particles. Thus, the circulation around the line of Fig. 6.2a is zero. According to Kelvin's theorem for a frictionless flow, the circulation around this line of fluid particles remains zero if the fluid is suddenly given a uniform velocity with respect to the airfoil. Therefore, in Fig. 6.2b and c, the circulation around the line which encloses the lifting airfoil and which contains the same fluid particles as the line of Fig. 6.2a should be zero. However, circulation is necessary to produce lift. Thus, as explained in the following paragraphs, the circulation around each of the component curves of Fig. 6.2b and c is not zero but equal in magnitude and opposite in sign to the others.

Starting Vortex

When an airfoil is accelerated from rest, the circulation around it, and therefore the lift is not produced instantaneously. At the instant of starting, the flow is a potential flow without circulation, and the streamlines are as shown in Fig. 6.3a, with a stagnation point occurring on the rear upper surface. At the sharp trailing edge, the air is required to change direction suddenly. However, because of viscosity, the large velocity gradients produce large viscous forces, and the air is unable to flow around the sharp trailing edge. Instead, a surface of discontinuity emanating from the sharp trailing edge is rolled up into a vortex, which is called the *starting vortex*. The stagnation point moves toward the trailing edge, as the circulation around the airfoil and, therefore, the lift increase progressively. The circulation around the airfoil increases in intensity until the flows from the upper surface and the lower surface join smoothly at the trailing edge, as shown in Fig. 6.3b. Thus, the generation of circulation around the wing and the resultant lift are necessarily accompanied by a starting vortex, which results because of the effects of viscosity.

A line which encloses both the airfoil and the starting vortex and which always contains the same fluid particles is presented in Fig. 6.2. The total circulation around this line remains zero, since the circulation around the airfoil is equal in strength but

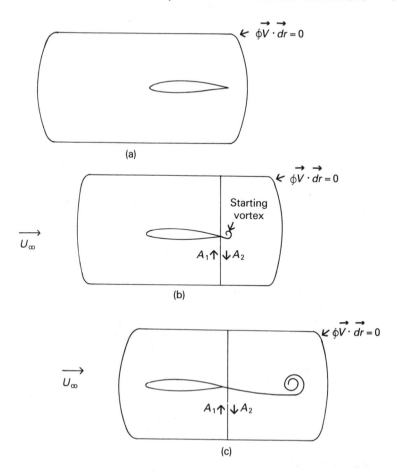

Figure 6.2 Circulation around a fluid line containing the airfoil remains zero: (a) fluid at rest; (b) fluid at time t; (c) fluid at time $t + \Delta t$.

opposite in direction to that of the starting vortex. Thus, the existence of circulation is not in contradiction to Kelvin's theorem. Referring to Fig. 6.2, the line integral of the tangential component of the velocity around the curve that encloses area A_1 must be equal and opposite to the corresponding integral for area A_2.

If either the free-stream velocity or the angle of attack of the airfoil is increased, another vortex is shed which has the same direction as the starting vortex. However, if the velocity or the angle of attack is decreased, a vortex is shed which has the opposite direction of rotation relative to the initial vortex.

A simple experiment that can be used to visualize the starting vortex requires only a pan of water and a small, thin board. Place the board upright in the water so that it cuts the surface. If the board is accelerated suddenly at moderate incidence, the starting vortex will be seen leaving the trailing edge of the "airfoil." If the board is stopped suddenly, another vortex of equal strength but of opposite rotation is generated.

Figure 6.3   Streamlines around the airfoil section: (a) zero circulation, stagnation point on the rear upper surface; (b) full circulation, stagnation point on the trailing edge.

GENERAL THIN-AIRFOIL THEORY

The essential assumptions of thin-airfoil theory are (1) that the lifting characteristics of an airfoil below stall are negligibly affected by viscosity, (2) that the airfoil is operating at a small angle of attack, and (3) that the resultant of the pressure forces (magnitude, direction, and line of action) is only slightly influenced by the airfoil thickness, since the maximum mean camber is small and the ratio of maximum thickness to chord is small. Typically, airfoil sections have a maximum thickness of approximately 12% of the chord and a maximum mean camber of approximately 2% of the chord. For thin-airfoil theory, the airfoil will be represented by its mean camber line in order to calculate the section aerodynamic characteristics.

A velocity difference across the infinitely thin profile which represents the airfoil section is required to produce the lift-generating pressure difference. A vortex sheet produces a velocity distribution that exhibits the required velocity jump. Therefore, the desired flow is obtained by superimposing on a uniform field of flow a field induced by a series of line vortices of infinitesimal strength which are located along the camber line as shown in Fig. 6.4. The total circulation is the sum of the circulations of the vortex filaments

$$\Gamma = \int_0^c \gamma(s)\, ds \qquad (6.1)$$

where $\gamma(s)$ is the distribution of vorticity for the line vortices. The length of an arbitrary element of the camber line is ds and the circulation is in the clockwise direction.

The velocity field around the sheet is the sum of the free-stream velocity and the velocity induced by all the vortex filaments that make up the vortex sheet. For the vortex sheet to be a streamline of the flow, it is necessary that the resultant velocity be tangent to the mean camber line at each point. Thus, the sum of the components normal to the surface for these two velocities is zero. In addition, the condition that the flows from the upper surface and the lower surface join smoothly at the trailing edge (i.e., the Kutta condition) requires that $\gamma = 0$ at the trailing edge.

The portion of the vortex sheet designated ds in Fig. 6.5 produces a velocity at point P which is perpendicular to the line whose length is r and which joins the element

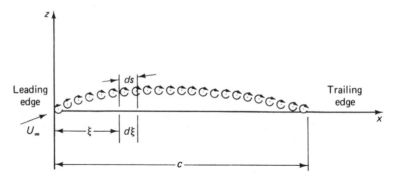

Figure 6.4 Representation of the mean camber line by a vortex sheet whose filaments are of variable strength $\gamma(s)$.

ds and the point P. The induced velocity component normal to the camber line at P due to the vortex element ds is

$$dV_{s,n} = - \frac{\gamma \, ds \cos \delta_3}{2\pi r}$$

where the negative sign results because the circulation induces a clockwise velocity and the normal to the upper surface is positive outward. To calculate the resultant vortex-induced velocity at a particular point P, one must integrate over all the vortex filaments from the leading edge to the trailing edge. The chordwise location of the point of interest P will be designated in terms of its x coordinate. The chordwise location of a given element of vortex sheet ds will be given in terms of its ξ coordinate. Thus, to calculate the cumulative effect of all the vortex elements, it is necessary to integrate over the ξ coordinate from the leading edge ($\xi = 0$) to the trailing edge ($\xi = c$). Noting that

$$\cos \delta_2 = \frac{x - \xi}{r} \quad \text{and} \quad ds = \frac{d\xi}{\cos \delta_1}$$

the resultant vortex-induced velocity at any point P (which has the chordwise location x) is given by

$$V_{s,n}(x) = - \frac{1}{2\pi} \int_0^c \frac{\gamma(\xi) \cos \delta_2 \cos \delta_3 \, d\xi}{(x - \xi) \cos \delta_1} \tag{6.2}$$

The component of the free-stream velocity normal to the mean camber line at P is given by

$$U_{\infty,n}(x) = U_\infty \sin (\alpha - \delta_P)$$

where α is the angle of attack and δ_P is the slope of the camber line at the point of interest P. Thus,

$$\delta_P = \tan^{-1} \frac{dz}{dx}$$

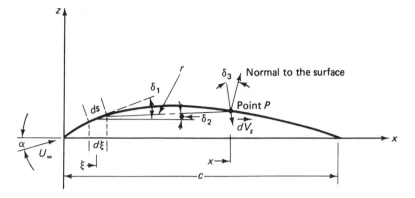

Figure 6.5 Thin-airfoil geometry parameters.

where $z(x)$ describes the mean camber line. As a result,

$$U_{\infty,n}(x) = U_\infty \sin\left(\alpha - \tan^{-1}\frac{dz}{dx}\right) \tag{6.3}$$

Since the sum of the velocity components normal to the surface must be zero at all points along the vortex sheet,

$$\frac{1}{2\pi}\int_0^c \frac{\gamma(\xi)\cos\delta_2\cos\delta_3\,d\xi}{(x-\xi)\cos\delta_1} = U_\infty \sin\left(\alpha - \tan^{-1}\frac{dz}{dx}\right) \tag{6.4}$$

The vorticity distribution $\gamma(\xi)$ that satisfies this integral equation makes the vortex sheet a streamline of the flow. The desired vorticity distribution must also satisfy the Kutta condition that $\gamma(c) = 0$.

Within the assumptions of thin-airfoil theory, the angles δ_1, δ_2, δ_3, and α are small. Using the approximate trigonometric relations for small angles, equation (6.4) becomes

$$\frac{1}{2\pi}\int_0^c \frac{\gamma(\xi)\,d\xi}{x-\xi} = U_\infty\left(\alpha - \frac{dz}{dx}\right) \tag{6.5}$$

THIN, SYMMETRIC FLAT-PLATE AIRFOIL

The mean camber line of a symmetric airfoil is coincident with the chord line. Thus, when the profile is replaced by its mean camber line, a flat plate with a sharp leading edge is obtained. For subsonic flow past a flat plate even at small angles of attack, a region of dead air (or stalled flow) will exist over the upper surface. For the actual airfoil, the rounded nose allows the flow to accelerate from the stagnation point onto the upper surface without separation. Of course, when the angle of attack is sufficiently large (the value depends on the cross-section geometry), stall will occur for the actual profile. The approximate theoretical solution for a thin airfoil with two sharp edges represents an irrotational flow with finite velocity at the trailing edge but with infinite velocity at the leading edge. The approximate solution does not describe the chordwise variation of the flow around the actual airfoil. However, as will be discussed, the theoretical values of the lift coefficient (obtained by integrating the circulation distribution along the airfoil) are in reasonable agreement with the experimental values.

For the camber line of the symmetric airfoil, dz/dx is everywhere zero. Thus, equation (6.5) becomes

$$\frac{1}{2\pi} \int_0^c \frac{\gamma(\xi)}{x - \xi} \, d\xi = U_\infty \alpha \tag{6.6}$$

It is convenient to introduce the coordinate transformation

$$\xi = \frac{c}{2} (1 - \cos \theta) \tag{6.7}$$

Similarly, the x coordinate transforms to θ_0 using

$$x = \frac{c}{2} (1 - \cos \theta_0)$$

The corresponding limits of integration are

$$\xi = 0, \quad \theta = 0 \qquad \text{and} \qquad \xi = c, \quad \theta = \pi$$

Equation (6.6) becomes

$$\frac{1}{2\pi} \int_0^\pi \frac{\gamma(\theta) \sin \theta \, d\theta}{\cos \theta - \cos \theta_0} = U_\infty \alpha \tag{6.8}$$

The required vorticity distribution, $\gamma(\theta)$, must not only satisfy this integral equation, but must satisfy the Kutta condition, $\gamma(\pi) = 0$. The solution is

$$\gamma(\theta) = 2\alpha U_\infty \frac{1 + \cos \theta}{\sin \theta} \tag{6.9}$$

That this is a valid solution can be seen by substituting the expression for $\gamma(\theta)$ given by equation (6.9) into equation (6.8). The resulting equation,

$$\frac{\alpha U_\infty}{\pi} \int_0^\pi \frac{(1 + \cos \theta) \, d\theta}{\cos \theta - \cos \theta_0} = U_\infty \alpha$$

can be reduced to an identity using the relation (Ref. 6.1)

$$\int_0^\pi \frac{\cos n\theta \, d\theta}{\cos \theta - \cos \theta_0} = \frac{\pi \sin n\theta_0}{\sin \theta_0} \tag{6.10}$$

where n assumes only integer values. Using l'Hospital's rule, it can be readily shown that the expression for $\gamma(\theta)$ also satisfies the Kutta condition.

The Kutta-Joukowski theorem for steady flow about a two-dimensional body of any cross section shows that the force per unit span is equal to $\rho_\infty U_\infty \Gamma$ and acts perpendicular to U_∞. Thus, for two-dimensional inviscid flow, an airfoil has no drag but experiences a lift per unit span equal to the product of free-stream density, the free-stream velocity, and the total circulation. The lift per unit span is

$$l = \int_0^c \rho_\infty U_\infty \gamma(\xi) \, d\xi \tag{6.11}$$

Using the circulation distribution of equation (6.9) and the coordinate transformation of equation (6.7), the lift per unit span is

$$l = \rho_\infty U_\infty^2 \alpha c \int_0^\pi (1 + \cos\theta) \, d\theta$$
$$l = \pi \rho_\infty U_\infty^2 \alpha c \tag{6.12}$$

To determine the section lift coefficient of the airfoil, the reference area is the chord times the unit span. The section lift coefficient is

$$C_l = \frac{l}{0.5\rho_\infty U_\infty^2 c} = 2\pi\alpha \tag{6.13}$$

Thus, thin-airfoil theory yields a section lift coefficient for a symmetric airfoil that is directly proportional to the geometric angle of attack. The geometric angle of attack is the angle between the free-stream velocity and the chord line of the airfoil. The theoretical relation is independent of the airfoil thickness.

The pressure distribution also produces a pitching moment about the leading edge (per unit span), which is given by

$$m_0 = -\int_0^c \rho_\infty U_\infty \gamma(\xi) \, \xi \, d\xi \tag{6.14}$$

The lift-generating circulation of an element $d\xi$ produces an upward force that acts a distance ξ downstream of the leading edge. The lift force, therefore, produces a nose-down pitching moment about the leading edge. Thus, the negative sign is used in equation (6.14) because nose-up pitching moments are considered positive. Again, using the coordinate transformation [equation (6.7)] and the circulation distribution [equation (6.9)], the pitching moment (per unit span) about the leading edge is

$$m_0 = -0.5\rho_\infty U_\infty^2 \alpha c^2 \int_0^\pi (1 - \cos^2\theta) \, d\theta$$
$$= -\frac{\pi}{4} \rho_\infty U_\infty^2 \alpha c^2 \tag{6.15}$$

The section moment coefficient is given by

$$C_{m0} = \frac{m_0}{0.5\rho_\infty U_\infty^2 cc} \tag{6.16}$$

Note that the reference area for the airfoil is the chord times the unit span and the reference length is the chord. For the symmetric airfoil,

$$C_{m0} = -\frac{\pi}{2}\alpha = -\frac{C_l}{4} \tag{6.17}$$

The center of pressure x_{cp} is the x coordinate, where the resultant lift force could be placed to produce the pitching moment m_0. Equating the moment about the leading edge [equation (6.15)] to the product of the lift [equation (6.12)] and the center of pressure yields

$$-\frac{\pi}{4}\rho_\infty U_\infty^2 \alpha c^2 = -\pi\rho_\infty U_\infty^2 \alpha c x_{cp}$$

Solving for x_{cp}, one obtains

$$x_{cp} = \frac{c}{4} \tag{6.18}$$

The result is independent of the angle of attack and is therefore independent of the section lift coefficient.

Example 6.1

The theoretical aerodynamic coefficients calculated using the thin-airfoil relations are compared with the data of Abbott and von Doenhoff (Ref. 6.2) in Fig. 6.6. Data are presented for two different airfoil sections. One, the NACA 0009, has a maximum thickness

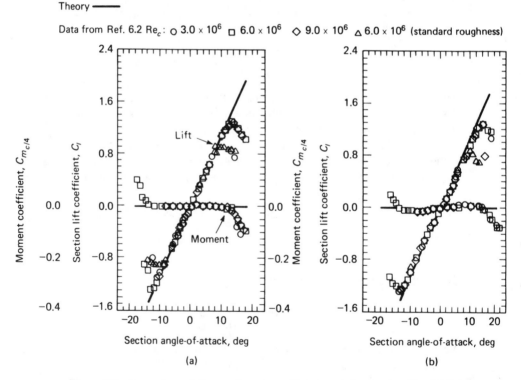

Theory ———

Data from Ref. 6.2 Re_c: ○ 3.0×10^6 □ 6.0×10^6 ◇ 9.0×10^6 △ 6.0×10^6 (standard roughness)

Figure 6.6 Comparison of the aerodynamic coefficients calculated using thin-airfoil theory for symmetric airfoils with the data of Ref. 6.2: (a) NACA 0009 wing section; (b) NACA 0012-64 wing section.

which is 9% of chord and is located at $x = 0.3c$. The theoretical lift coefficient calculated using equation (6.13) is in excellent agreement with the data for the NACA 0009 airfoil up to an angle of attack of 12°. At higher angles of attack, the viscous effects significantly alter the flow field and hence the experimental lift coefficients. Thus, theoretical values would not be expected to agree with the data at high angles of attack. Since the theory presumes that viscous effects are small, it is valid only for angles of attack below stall. According to thin-airfoil theory, the moment about the quarter chord is zero. The measured moments for the NACA 0009 are also in excellent agreement with theory prior to stall. The correlation between the theoretical values and the experimental values is not as good for the NACA 0012–64 airfoil section. The difference in the correlation between theory and data for these two airfoil sections is attributed to viscous effects. The maximum thickness of the NACA 0012–64 is greater and located farther aft. Thus, the adverse pressure gradients that cause separation of the viscous boundary layer and thereby alter the forces would be greater for the NACA 0012–64.

CAMBERED AIRFOIL

The method of determining the aerodynamic characteristics for a cambered airfoil is similar to that followed for the symmetric airfoil. Thus, a vorticity distribution is sought which satisfies both the condition that the mean camber line is a streamline [equation (6.5)], and the Kutta condition. However, because of camber, the actual computations are more involved. Again, the coordinate transformation

$$\xi = \frac{c}{2}(1 - \cos \theta) \tag{6.7}$$

is used, so the integral equation to be solved is

$$\frac{1}{2\pi} \int_0^\infty \frac{\gamma(\theta) \sin \theta \, d\theta}{\cos \theta - \cos \theta_0} = U_\infty \left(\alpha - \frac{dz}{dx}\right) \tag{6.19}$$

Recall that this integral equation expresses the requirement that the resultant velocity for the inviscid flow is parallel to the mean camber line (which represents the airfoil). The vorticity distribution $\gamma(\theta)$ which satisfies this integral equation makes the vortex sheet a streamline of the flow.

Vorticity Distribution

The desired vorticity distribution, which satisfies equation (6.19) and the Kutta condition, may be represented by the series involving:

1. A term of the form for the vorticity distribution for a symmetric airfoil,

$$2U_\infty A_0 \frac{1 + \cos \theta}{\sin \theta}$$

2. A Fourier sine series whose terms automatically satisfy the Kutta condition,

$$2U_\infty \sum_{n=1}^\infty A_n \sin n\theta$$

The coefficients A_n of the Fourier series depend on the shape of the camber line. Thus,

$$\gamma(\theta) = 2U_\infty \left(A_0 \frac{1 + \cos\theta}{\sin\theta} + \sum_{n=1}^{\infty} A_n \sin n\theta \right) \tag{6.20}$$

Since each term is zero when $\theta = \pi$, the Kutta condition is satisfied [i.e., $\gamma(\pi) = 0$].

Substituting the vorticity distribution [equation (6.20)] into equation (6.19) yields

$$\frac{1}{\pi} \int_0^\pi \frac{A_0(1 + \cos\theta)\, d\theta}{\cos\theta - \cos\theta_0} + \frac{1}{\pi} \int_0^\pi \sum_{n=1}^{\infty} \frac{A_n \sin n\theta \sin\theta\, d\theta}{\cos\theta - \cos\theta_0} = \alpha - \frac{dz}{dx} \tag{6.21}$$

This integral equation can be used to evaluate the coefficients $A_0, A_1, A_2, \ldots, A_n$ in terms of the angle of attack and the mean camber-line slope, which is known for a given airfoil section. The first integral on the left-hand side of equation (6.21) can be readily evaluated using equation (6.10). To evaluate the series of integrals represented by the second term, one must use equation (6.10) and the trigonometric identity:

$$(\sin n\theta)(\sin\theta) = 0.5[\cos(n-1)\theta - \cos(n+1)\theta]$$

Equation (6.21) becomes

$$\frac{dz}{dx} = \alpha - A_0 + \sum_{n=1}^{\infty} A_n \cos n\theta \tag{6.22}$$

which applies to any chordwise station. Since we are evaluating both dz/dx and $\cos n\theta_0$ at the general point θ_0 (i.e., x), we have dropped the subscript 0 from equation (6.22) and from all subsequent equations. Thus, the coefficients $A_0, A_1, A_2, \ldots,$ A_n must satisfy equation (6.22) if equation (6.20) is to represent the vorticity distribution which satisfies the condition that the mean camber line is a streamline. Since the geometry of the mean camber line would be known for the airfoil of interest, the slope is a known function of θ. One can, therefore, determine the values of the coefficients. To evaluate A_0, note that

$$\int_0^\pi A_n \cos n\theta\, d\theta = 0$$

for any value of n. Thus, by algebraic manipulation of equation (6.22),

$$A_0 = \alpha - \frac{1}{\pi} \int_0^\pi \frac{dz}{dx}\, d\theta \tag{6.23}$$

Multiplying both sides of equation (6.22) by $\cos m\theta$, where m is an unspecified integer, and integrating from 0 to π, one obtains

$$\int_0^\pi \frac{dz}{dx} \cos m\theta\, d\theta = \int_0^\pi (\alpha - A_0) \cos m\theta\, d\theta + \int_0^\pi \sum_{n=1}^{\infty} A_n \cos n\theta \cos m\theta\, d\theta$$

The first term on the right-hand side is zero for any value of m. Note that

$$\int_0^\pi A_n \cos n\theta \cos m\theta\, d\theta = 0 \qquad \text{when } n \neq m$$

but

$$\int_0^\pi A_n \cos n\theta \cos m\theta \, d\theta = \frac{\pi}{2} A_n \qquad \text{when } n = m$$

Thus,

$$A_n = \frac{2}{\pi} \int_0^\pi \frac{dz}{dx} \cos n\theta \, d\theta \qquad\qquad (6.24)$$

Using equations (6.23) and (6.24) to define the coefficients, equation (6.20) can be used to evaluate the vorticity distribution for a cambered airfoil in terms of the geometric angle of attack and the shape of the mean camber line. Note that, for a symmetric airfoil, $A_0 = \alpha$, $A_1 = A_2 = \cdots = A_n = 0$. Thus, the vorticity distribution for a symmetric airfoil, as determined using equation (6.20), is

$$\gamma(\theta) = 2\alpha U_\infty \frac{1 + \cos \theta}{\sin \theta}$$

This is identical to equation (6.9). Therefore, the general expression for the cambered airfoil includes the symmetric airfoil as a special case.

Aerodynamic Coefficients for a Cambered Airfoil

The lift and the moment coefficients for a cambered airfoil are found in the same manner as for the symmetric airfoil. The section lift coefficient is given by

$$C_l = \frac{1}{0.5\rho_\infty U_\infty^2 c} \int_0^c \rho_\infty U_\infty \gamma(\xi) \, d\xi$$

Using the coordinate transformation [equation (6.7)] and the expression for γ [equation (6.20)] gives us

$$C_l = 2 \left[\int_0^\pi A_0(1 + \cos \theta) \, d\theta + \int_0^\pi \sum_{n=1}^\infty A_n \sin n\theta \sin \theta \, d\theta \right]$$

Note that $\int_0^\pi A_n \sin n\theta \sin \theta \, d\theta = 0$ for any value of n other than unity. Thus, upon integration, one obtains

$$C_l = \pi(2A_0 + A_1) \qquad\qquad (6.25)$$

The section moment coefficient for the pitching moment about the leading edge is given by

$$C_{m0} = \frac{-1}{0.5\rho_\infty U_\infty^2 c^2} \int_0^c \rho_\infty U_\infty \gamma(\xi) \, \xi \, d\xi$$

Again, using the coodinate transformation and the vorticity distribution, one obtains, upon integration,

$$C_{m0} = -\frac{\pi}{2} \left(A_0 + A_1 - \frac{A_2}{2} \right) \qquad\qquad (6.26)$$

The center of pressure relative to the leading edge is found by dividing the moment about the leading edge (per unit span) by the lift per unit span.

$$x_{cp} = -\frac{m_0}{l}$$

The negative sign is used since a positive lift force with a positive moment arm x_{cp} results in a nose-down, or negative moment, as shown in the sketch of Fig. 6.7. Thus,

$$x_{cp} = \frac{c}{4}\left(\frac{2A_0 + 2A_1 - A_2}{2A_0 + A_1}\right)$$

Noting that $C_l = \pi(2A_0 + A_1)$, the expression for the center of pressure becomes

$$x_{cp} = \frac{c}{4}\left[1 + \frac{\pi}{C_l}(A_1 - A_2)\right] \tag{6.27}$$

Thus, for the cambered airfoil, the position of the center of pressure depends on the lift coefficient and hence the angle of attack. The line of action for the lift, as well as the magnitude, must be specified for each angle of attack.

If the pitching moment per unit span produced by the pressure distribution is referred to a point $0.25c$ downstream of the leading edge (i.e., the quarter-chord), the moment is given by

$$m_{0.25c} = \int_0^{0.25c} \rho_\infty U_\infty \gamma(\xi)\left(\frac{c}{4} - \xi\right) d\xi - \int_{0.25c}^{c} \rho_\infty U_\infty \gamma(\xi)\left(\xi - \frac{c}{4}\right) d\xi$$

Again, the signs are chosen so that a nose-up moment is positive. Rearranging the relation yields

$$m_{0.25c} = \frac{c}{4}\int_0^{c} \rho_\infty U_\infty \gamma(\xi) \, d\xi - \int_0^{c} \rho_\infty U_\infty \gamma(\xi) \, \xi \, d\xi$$

The first integral on the right-hand side of this equation represents the lift per unit span, while the second integral represents the moment per unit span about the leading edge. Thus,

$$m_{0.25c} = \frac{c}{4}l + m_0 \tag{6.28}$$

The section moment coefficient about the quarter-chord point is given by

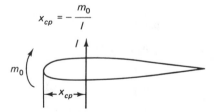

Figure 6.7 Center of pressure for a thin, cambered airfoil.

$$C_{m_{0.25c}} = \frac{1}{4} C_l + C_m = \frac{\pi}{4} (A_2 - A_1) \qquad (6.29)$$

Since A_1 and A_2 depend on the camber only, the section moment coefficient about the quarter-chord point is independent of the angle of attack. The point about which the section moment coefficient is independent of the angle of attack is called the *aerodynamic center of the section*. Thus, according to the theoretical relations for a thin-airfoil section, the aerodynamic center is at the quarter-chord point.

In order for the section pitching moment to remain constant as the angle of attack is increased, the product of the moment arm (relative to the aerodynamic center) and C_l must remain constant. Thus, the moment arm (relative to the aerodynamic center) decreases as the lift increases. This is evident in the expression for the center of pressure which is given in equation (6.27). Alternatively, the aerodynamic center is the point at which all changes in lift effectively take place. Because of these factors, the center of gravity is usually located near the aerodynamic center. If we include the effects of viscosity on the flow around the airfoil, the lift due to angle of attack would not necessarily be concentrated at the exact quarter-chord point. However, for angles of attack below the onset of stall, the actual location of the aerodynamic center for the various sections is usually between the 23% chord point and the 27% chord point. Thus, the moment coefficient about the aerodynamic center, which is given the symbol C_{mac}, is also given by equation (6.29). If equation (6.24) is used to define A_1 and A_2, then C_{mac} becomes

$$C_{mac} = \frac{1}{2} \int_0^\pi \frac{dz}{dx} (\cos 2\theta - \cos \theta) \, d\theta \qquad (6.30)$$

Note, as discussed when comparing theory with data in the preceding section, the C_{mac} is zero for a symmetric airfoil.

Example 6.2

The relations developed in this section will now be used to calculate the aerodynamic coefficients for a representative airfoil section. The airfoil section selected for use in this sample problem is the NACA 2412. As discussed in Ref. 6.2, the first digit defines the maximum camber in percent of chord, the second digit defines the location of the maximum camber in tenths of chord, and the last two digits represent the thickness ratio (i.e., the maximum thickness in percent of chord). The equation for the mean camber line is defined in terms of the maximum camber and its location. Forward of the maximum camber position, the equation of the mean camber line is

$$\left(\frac{z}{c}\right)_{fore} = 0.125 \left[0.8 \left(\frac{x}{c}\right) - \left(\frac{x}{c}\right)^2 \right]$$

while aft of the maximum camber position,

$$\left(\frac{z}{c}\right)_{aft} = 0.0555 \left[0.2 + 0.8 \left(\frac{x}{c}\right) - \left(\frac{x}{c}\right)^2 \right]$$

To calculate the section lift coefficient and the section moment coefficient, it is necessary to evaluate the coefficients A_0, A_1, and A_2. To evaluate these coefficients it is necessary to integrate an expression involving the function which defines the slope of the mean camber line. Therefore, the slope of the mean camber line will be expressed in terms of the θ coordinate, which is given in equation (6.7). Forward of the maximum camber location, the slope is given by

$$\left(\frac{dz}{dx}\right)_{fore} = 0.1 - 0.25\frac{x}{c} = 0.125 \cos\theta - 0.025$$

Aft of the maximum camber location, the slope is given by

$$\left(\frac{dz}{dx}\right)_{aft} = 0.0444 - 0.1110\frac{x}{c} = 0.0555 \cos\theta - 0.0111$$

Since the maximum camber location serves as a limit for the integrals, it is necessary to convert the x coodinate, which is $0.4c$, to the corresponding θ coordinate. To do this,

$$\frac{c}{2}(1 - \cos\theta) = 0.4c$$

Thus, the location of the maximum camber is $\theta = 78.463° = 1.3694$ rad. Referring to equations (6.23) and (6.24) the necessary coefficients are

$$A_0 = \alpha - \frac{1}{\pi}\left[\int_0^{1.3694} (0.125 \cos\theta - 0.025)\, d\theta + \int_{1.3694}^{\pi} (0.0555 \cos\theta \right.$$

$$\left. - 0.0111)\, d\theta\right] = \alpha - 0.004517$$

$$A_1 = \frac{2}{\pi}\left[\int_0^{1.3694} (0.125 \cos^2\theta - 0.025 \cos\theta)\, d\theta \right.$$

$$\left. + \int_{1.3694}^{\pi} (0.0555 \cos^2\theta - 0.0111 \cos\theta)\, d\theta\right]$$

$$= 0.08146$$

$$A_2 = \frac{2}{\pi}\left[\int_0^{1.3694} (0.125 \cos\theta \cos 2\theta - 0.025 \cos 2\theta)\, d\theta \right.$$

$$\left. + \int_{1.3694}^{\pi} (0.0555 \cos\theta \cos 2\theta - 0.0111 \cos 2\theta)\, d\theta\right]$$

$$= 0.01387$$

The section lift coefficient is

$$C_l = 2\pi\left(A_0 + \frac{A_1}{2}\right) = 2\pi\alpha + 0.2297$$

Solving for the angle of attack for zero lift, we obtain

$$\alpha_{ol} = -\frac{0.2297}{2\pi} \text{ rad} = -2.095°$$

According to thin-airfoil theory, the aerodynamic center is at the quarter-chord point. Thus, the section moment coefficient for the moment about the quarter chord is equal to that about the aerodynamic center. The two coefficients are given by

$$C_{m_{ac}} = C_{m0.25c} = \frac{\pi}{4}(A_2 - A_1) = -0.05309$$

The theoretical values of the section lift coefficient and of the section moment coefficients are compared with the measured values from Ref. 6.2 in Figs. 6.8 and 6.9, respectively. Since the theoretical coefficients do not depend on the airfoil section thickness, they will be compared with data from Ref. 6.2 for a NACA 2418 airfoil as well as for a NACA 2412 airfoil. For both airfoil sections, the maximum camber is 2% of the chord length and is located at $x = 0.4c$. The maximum thickness is 12% of chord for the NACA 2412 airfoil section and is 18% of the chord for the NACA 2418 airfoil section.

The correlation between the theoretical and the experimental values of lift is

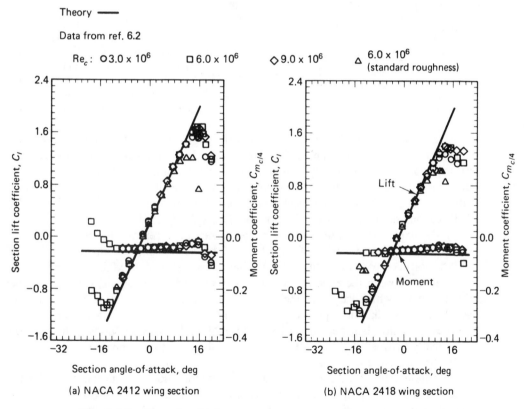

(a) NACA 2412 wing section   (b) NACA 2418 wing section

Figure 6.8 Comparison of the aerodynamic coefficients calculated using thin airfoil theory for cambered airfoils with the data of Ref. 6.2: (a) NACA 2412 wing section; (b) NACA 2418 wing section.

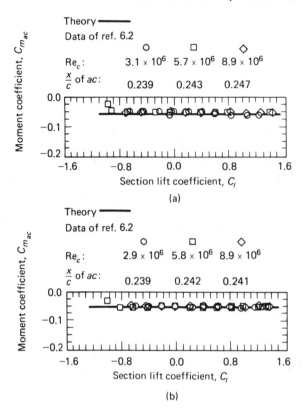

Figure 6.9 Comparison of the theoretical and the experimental section moment coefficient (about the aerodynamic center) for two cambered airfoils: (a) NACA 2412 wing section; (b) NACA 2418 wing section.

satisfactory for both airfoils (Fig. 6.8) until the angle of attack becomes so large that viscous phenomena significantly affect the flow field. The theoretical value for the zero lift angle of attack agrees very well with the measured values for the two airfoils. The theoretical value of $C_{l,\alpha}$ is 2π. Based on the measured lift coefficients for angles of attack for 0° to 10°, the experimental value of $C_{l,\alpha}$ is approximately 6.0 for the NACA 2412 airfoil and approximately 5.9 for the NACA 2418 airfoil.

The experimental values of the moment coefficient referred to the aerodynamic center (approximately -0.045 for the NACA 2412 section and -0.050 for the NACA 2418 airfoil) compare favorably with the theoretical value of -0.053 (Fig. 6.9). The correlation between the experimental values of the moment coefficient referred to the quarter chord, which vary with the angle of attack, and the theoretical value is not as good. Note also that the experimentally determined location of the aerodynamic center for these two airfoils is between $0.239c$ and $0.247c$. As noted previously, the location is normally between $0.23c$ and $0.27c$ for a real fluid flow, as compared with the value of $0.25c$ calculated using thin-airfoil theory.

Although the thickness ratio of the airfoil section does not enter into the theory, except as an implied limit to its applicability, the data of Figs. 6.8 and 6.9 show thickness-related variations. Note that the maximum value of the experimental lift coefficient is consistently greater for the NACA 2412 and that it occurs at a higher angle of attack. Also note that, as the angle of attack increases beyond the maximum lift value, the measured lift coefficients decrease more sharply for the NACA 2412.

Thus, the thickness ratio influences the interaction between the adverse pressure gradient and the viscous boundary layer. The interaction, in turn, affects the aerodynamic coefficients. In Fig. 6.10, $C_{l,max}$ is presented as a function of the thickness ratio for the NACA 2400 series airfoils. The data of Ref. 6.2 and the results of McCormick (Ref. 6.3) are presented. McCormick notes that below a thickness ratio of approximately 12%, $C_{l,max}$ decreases rapidly with decreasing thickness. Above a thickness ratio of 12%, the variation in $C_{l,max}$ is less pronounced.

The correlations presented in Figs. 6.6 and 6.8 indicate that at low angles of attack, the theoretical lift coefficients based on thin-airfoil theory are in good agreement with the measured values from Ref. 6.2. However, to compute the airfoil lift and pitching moment coefficients for various configurations exposed to a wide range of flow environments, especially if knowledge of the maximum section lift coefficient $C_{l,max}$ is important, it is necessary to include the effects of the boundary layer and of the separated wake. Using repeated application of a panel method (see Chapter 3) to solve for the separated wake displacement surface, Henderson (Ref. 6.4) discussed the relative importance of separation effects. These effects are illustrated in Fig. 6.11. The lift coefficient calculated using potential flow analysis with no attempt to account for the effects either of the boundary layer or of separation is compared with wind-tunnel data in Fig. 6.11a. At low angles of attack where the boundary layer is thin and there is little, if any, separation, potential flow analysis of the surface alone is a fair approximation to the data; but as the angle of attack is increased, the correlation degrades.

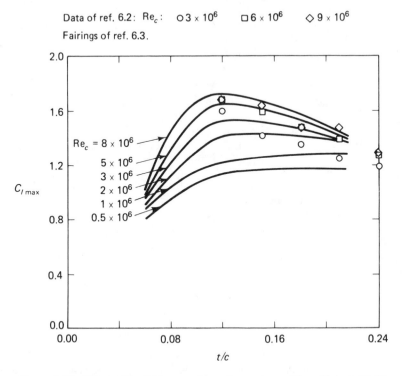

Figure 6.10 Effect of the thickness ratio on the maximum lift coefficient, NACA 24XX series airfoil sections.

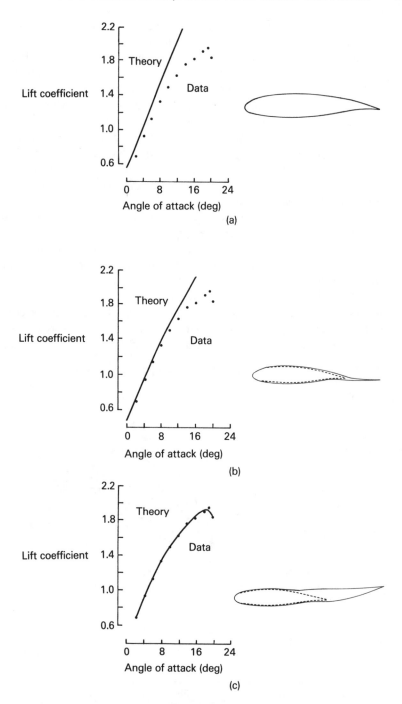

Figure 6.11 Relative importance of separation effects: (a) analysis of geometry alone; (b) analysis with boundary layer modeled; (c) analysis with boundary layer and separated wake modeled. (From Ref. 6.4.)

Henderson (Ref. 6.4) notes: "Rarely will the boundary layers be thin enough that potential flow analysis of the bare geometry will be sufficiently accurate." By including the effect of the boundary layer but not the separated wake in the computational flow model, the agreement between the theoretical lift coefficients and the wind-tunnel values is good at low angles of attack, as shown in Fig. 6.11b. When the angle of attack is increased and separation becomes important, the predicted and the measured lift coefficients begin to diverge.

Separation effects must be modeled in order to predict the maximum lift coefficient. As shown in Fig. 6.11c, when one accounts for the boundary layer and the separated wake, there is good agreement between theoretical values and experimental values through $C_{l,\max}$. This will be the case for any gradually separating section, such as the GAW-1, used in the example of Fig. 6.11.

HIGH-LIFT AIRFOIL SECTIONS

As noted by Smith (Ref. 6.5): "The problem of obtaining high lift is that of developing the lift in the presence of boundary layers—getting all the lift possible without causing separation. Provided that boundary-layer control is not used, our only means of obtaining higher lift is to modify the geometry of the airfoil. For a one-piece airfoil, there are several possible means for improvement—changed leading-edge radius, a flap, changed camber, a nose flap, a variable-camber leading edge, and changes in detail shape of a pressure distribution."

Thus, if more lift is to be generated, the circulation around the airfoil section must be increased, or, equivalently, the velocity over the upper surface must be increased relative to the velocity over the lower surface. However, once the effect of the boundary layer is included, the Kutta condition at the trailing edge requires that the upper-surface and the lower-surface velocities assume a value slightly less than the free-stream velocity. Hence, when the higher velocities over the upper surface of the airfoil are produced in order to get more lift, larger pressure gradients are required to decelerate the flow from the maximum velocity to the trailing-edge velocity. Again, referring to (Ref. 6.5): "The process of deceleration is critical, for if it is too severe, separation develops. The science of developing high lift, therefore, has two components: (1) analysis of the boundary layer, prediction of separation, and determination of the kinds of flows that are most favorable with respect to separation; and (2) analysis of the inviscid flow about a given shape with the purpose of finding shapes that put the least stress on a boundary layer."

Stratford (Ref. 6.6) has developed a formula for predicting the point of separation in an arbitrary decelerating flow. The resultant Stratford pressure distribution, which recovers a given pressure distribution in the shortest distance, has been used in the work of Liebeck (Ref. 6.7). To develop a class of high-lift airfoil sections, Liebeck used a velocity distribution that satisfied "three criteria: (1) the boundary layer does not separate; (2) the corresponding airfoil shape is practical and realistic; and (3) maximum possible C_l is obtained." The optimized form of the airfoil velocity distribution is markedly different than that for a typical airfoil section (which is presented in Fig. 6.12). The velocity distribution is presented as a function of s, the distance

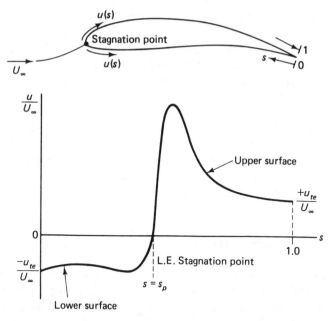

Figure 6.12 General form of the velocity distribution around a typical airfoil. (From Ref. 6.7.)

along the airfoil surface, where s begins at the lower-surface trailing edge and proceeds clockwise around the airfoil surface to the upper-surface trailing edge. In the s-coordinate system, the velocities are negative on the lower surface and positive on the upper surface. The "optimum" velocity distribution, modified to obtain a realistic airfoil, is presented in Fig. 6.13. The lower-surface velocity is as low as possible in the interest of obtaining the maximum lift and increases continuously from the leading-edge stagnation point to the trailing-edge velocity. The upper-surface acceleration region is shaped to provide good off-design peformance. A short boundary-layer transition ramp (the region where the flow decelerates, since an adverse pressure gradient promotes transition) is used to ease the boundary layer's introduction to the severe initial Stratford gradient.

Once the desired airfoil velocity distribution has been defined, there are two options available for calculating the potential flow. One method uses conformal mapping of the flow to a unit circle domain to generate the airfoil (e.g., Refs. 6.8 and 6.9). A second method uses the panel method for the airfoil analysis (e.g., Ref. 6.10). Olsen et al. (Ref. 6.11) note that, in the potential flow analysis, the airfoil section is represented by a closed polygon of planar panels connecting the input coordinate pairs. The boundary condition for the inviscid flow—that there be no flow through the airfoil surface—is applied at each of the panel centers. An additional equation, used to close the system, specifies that the upper- and lower-surface velocities have a common limit at the trailing edge (i.e., the Kutta condition). The effect of boundary-layer displacement is simulated by piecewise linear source distributions on the panels describing the airfoil contour. Thus, instead of modifying the airfoil geometry by an appropriate displacement

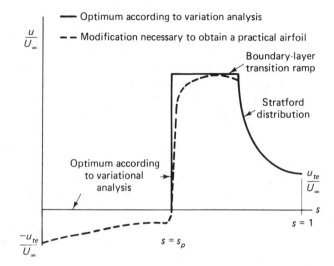

Figure 6.13 "Optimized" velocity distribution for a high-lift, single-element airfoil section. (From Ref. 6.7.)

thickness to account for the boundary layer, the boundary condition is modified by introducing surface transpiration. Miranda (Ref. 6.12) notes that "The latter approach is more satisfactory because the surface geometry and the computational grid are not affected by the boundary layer. This means that, for panel methods, the aerodynamic influence coefficients and, for finite difference methods, the computational grid do not

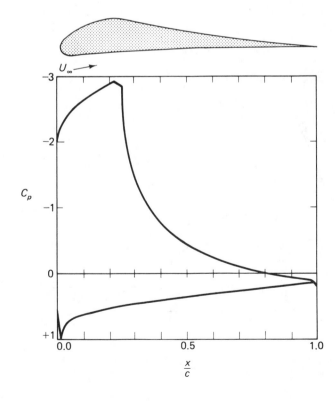

Figure 6.14 Theoretical pressure distribution for high-lift, single-element airfoil, $Re_c = 3 \times 10^6$, $t_{max} = 0.125c$, $C_l = 1.35$. (From Ref. 6.7.)

have to be recomputed at each iteration.'' The boundary condition on the suface panels requires that the velocity normal to the surface equals the strength of the known source distribution.

Liebeck has developed airfoil sections which, although they ''do not appear to be very useful'' (the quotes are from Ref. 6.7), develop an L/D of 600. The airfoil section, theoretical pressure distribution, the experimental lift curve and drag polar, and the experimental pressure distributions for a more practical, high-lift section are presented in Figs. 6.14 through 6.16. The pressure distributions indicate that the flow remained attached all the way to the trailing edge. The flow remained completely attached until the stalling angle was reached, at which point the entire recovery region separated instantaneously. Reducing the angle of attack less than $\frac{1}{2}°$ resulted in an instantaneous and complete reattachment, indicating a total lack of hysteresis effect on stall recovery.

Improvements of a less spectacular nature have been obtained for airfoil sections

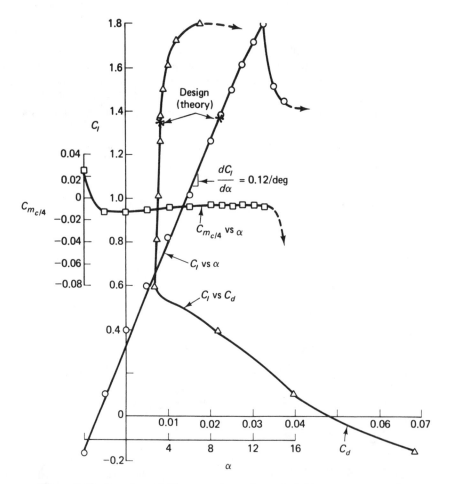

Figure 6.15 Experimental life curve, drag polar, and pitching moment curve for a high-lift, single-element airfoil, $Re_c = 3 \times 10^6$. (From Ref. 6.7.)

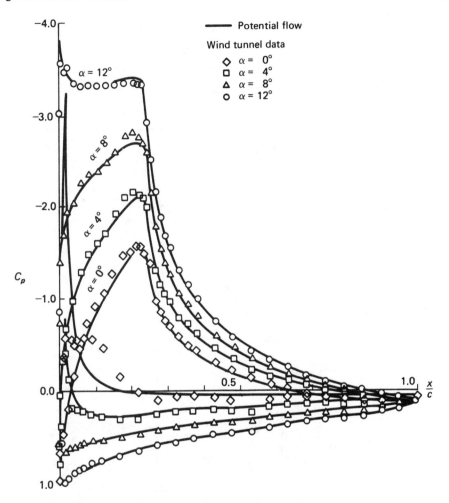

Figure 6.16 Comparison of the theoretical potential-flow and the experimental pressure distribution for a high-lift, single-element airfoil, $Re_c = 3 \times 10^6$. (From Ref. 6.7.)

being developed by NASA for light airplanes. One such airfoil section is the General Aviation (Whitcomb) number 1 airfoil, GA(W)-1, which is 17% thick with a blunt nose and a cusped lower surface near the trailing edge. The geometry of the GA(W)-1 section is similar to that of the supercritical airfoil, which is discussed in Chapter 9. Experimentally determined lift coefficients, drag coefficients, and pitching moment coefficients, which are taken from Ref. 6.13, are presented in Fig. 6.17. Included for comparison are the corresponding correlations for the NACA 65_2–415 and the NACA 65_3–418 airfoil sections. Both the GA(W)-1 and the NACA 65_3–418 airfoils have the same design lift coefficient (0.40), and both have roughly the same mean thickness distribution in the region of the structural box ($0.15c$ to $0.60c$). However, the experimental value of the maximum section lift coefficient for the GA(W)-1 was

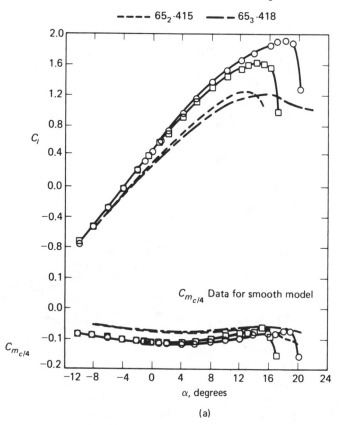

NASA GA(W)-1 airfoil

○ NASA standard roughness

□ NACA standard roughness

NACA airfoils, NACA standard roughness

---- 65_2-415 ——— 65_3-418

(a)

Figure 6.17 Aerodynamic coefficients for a NASA GA(W)-1 airfoil, for a NACA 65_2-415 airfoil, and for a NACA 65_3-418 airfoil; $M_\infty = 0.20$, Re_c 6 × 10⁶: (a) lift-coefficient and pitching-moment coefficient curves; (b) drag polars. (Data from Ref. 6.13.)

approximately 30 percent greater than that for the NACA 65 series airfoil for a Reynolds number of 6 × 10⁶. Since the section drag coefficient remains approximately constant to higher lift coefficients for the GA(W)-1, significant increases in the lift/drag ratio are obtained. At a lift coefficient of 0.90, the lift/drag ratio for the GA(W)-1 was approximately 70, which is 50% greater than that for the NACA 65_3–418 section. This is of particular importance from a safety standpoint for light general aviation airplanes, where large values of section lift/drag ratio at high lift coefficients result in improved climb performance.

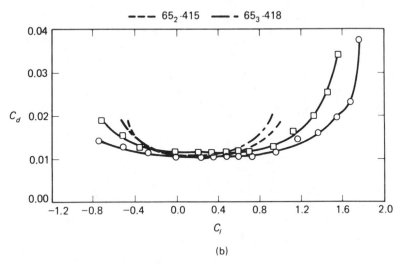

NASA GA(W)-1 airfoil

○   NASA standard roughness

□   NACA standard roughness

NACA airfoils, NACA standard roughness

--- 65_2-415   — - 65_3-418

(b)

Figure 6.17   (*Continued*)

MULTIELEMENT AIRFOILS

The problem of computing the aerodynamic characteristics of multielement airfoils can be subdivided into the following broad topical areas, each requiring models for the computer program (Ref. 6.10):

1. Geometry definition
2. Solution for the inviscid, potential flow
3. Solution for the conventional boundary layer
4. Solution for the viscous wakes and slot-flow characteristics
5. Combined inviscid/viscous solution

Stevens et al. (Ref. 6.10) note that the geometric modeling of the complete airfoil, including slots, slats, vanes, and flaps, requires a highly flexible indexing system to ensure that conventional arrangements of these components can be readily adapted to the code.

To compute the inviscid, potential flow, Stevens et al. (Ref. 6.10) and Olson et al. (Ref. 6.11) use distributed vortex singularities as the fundamental solution to the

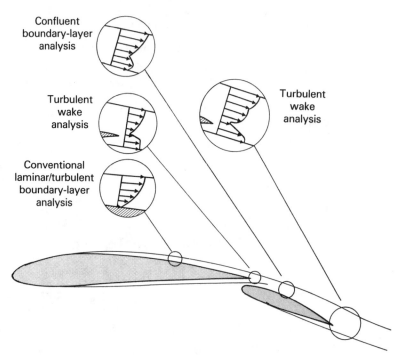

Figure 6.18   Theoretical flow models for the various viscous regions. (From Ref. 6.11.)

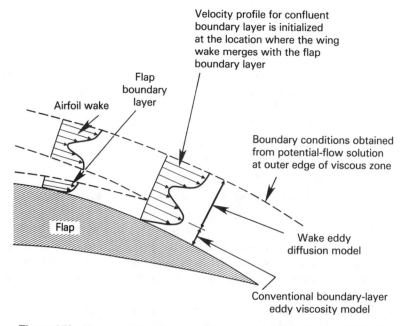

Figure 6.19   Flow model for merging of the wake from the principal airfoil with the boundary layer on the flap to form the confluent boundary layer on the upper surface of the flap. (From Ref. 6.11.)

Laplace equation. Olson et al. note that viscous calculations can be separated into three types of flows: conventional boundary layers, turbulent wakes, and confluent boundary layers (i.e., wakes merging with conventional boundary layers). These are illustrated in Fig. 6.18. To obtain a complete viscous calculation, the conventional boundary layers on the upper and lower surfaces of the main airfoil are first analyzed. These calculations provide the initial conditions to start the turbulent-wake analysis at the trailing edge of the principal airfoil. The calculations proceed downstream until the wake merges with the outer edge of the boundary layer on the upper surface of the flap, as shown in Fig. 6.19. The wake from the principal airfoil and the boundary layer of the flap combine into a single viscous layer at this point, a so-called confluent boundary layer. The calculation procedure continues stepwise downstream to the flap trailing edge. At the flap trailing edge, this confluent boundary-layer solution merges with the boundary layer from the lower surface of the flap. The calculation then continues downstream into the wake along a potential-flow streamline.

Although the techniques used to calculate the viscous effects differ from those

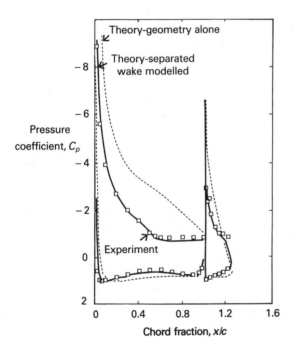

NASA GAW-1 airfoil, 30% Fowler flap,
Angle of attack = 12.5°, flap angle = 40°

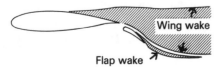

Figure 6.20 Comparison of experimental and calculated pressure distributions on a two-element airfoil with separation from both surfaces. (From Ref. 6.4.)

described in the preceding paragraph, the importance of including the viscous effects is illustrated in Fig. 6.20. Using a repeated application of a panel method to solve for the separated wake displacement surface, Henderson found a significant effect on the pressure distributions both on the principal airfoil and the flap for a GAW-1 airfoil for which α was 12.5° and the flap angle was 40°. Although a separation wake occurred for both models, the agreement between the calculated pressures and the experimental values was quite good.

PROBLEMS

6.1. Using the identity given in equation (6.10), show that the vorticity distribution

$$\gamma(\theta) = 2\alpha U_\infty \frac{1 + \cos\theta}{\sin\theta}$$

satisfies the condition that flow is parallel to the surface [i.e., equation (6.8)]. Show that the Kutta condition is satisfied. Sketch the $2\gamma/U_\infty$ distribution as a function of x/c for a section lift coefficient of 0.5. What is the physical significance of $2\gamma/U_\infty$? What angle of attack is required for a symmetric airfoil to develop a section lift coefficient of 0.5?

Using the vorticity distribution, calculate the section pitching moment about a point 0.75 chord from the leading edge. Verify your answer, using the fact that the center of pressure (x_{cp}) is at the quarter-chord for all angles of attack and the definition for lift.

6.2. Calculate C_l and $C_{m0.25c}$ for a NACA 0009 airfoil that has a plain flap whose length is $0.2c$ and which is deflected 25°. When the geometric angle of attack is 4°, what is the section lift coefficient? Where is the center of pressure?

6.3. The mean camber line of an airfoil is formed by a segment of a circular arc (having a constant radius of curvature). The maximum mean camber (which occurs at midchord) is equal to kc, where k is a constant and c is a chord length. Develop an expression for the γ distribution in terms of the free-stream velocity U_∞ and the angle of attack α. Since kc is small, you can neglect the higher-order terms in kc in order to simplify the mathematics. What is the angle of attack for zero lift (α_{0l}) for this airfoil section? What is the section moment coefficient about the aerodynamic center (C_{mac})?

6.4. The numbering system for wing sections of the NACA five-digit series is based on a combination of theoretical aerodynamic characteristics and geometric characteristics. The first integer indicates the amount of camber in terms of the relative magnitude of the design lift coefficient; the design lift coefficient in tenths is three-halves of the first integer. The second and third integers together indicate the distance from the leading edge to the location of the maximum camber; this distance in percent of the chord is one-half the number represented by these integers. The last two integers indicate the section thickness in percent of the chord. The NACA 23012 wing section thus has a design lift coefficient of 0.3, has its maximum camber at 15% of the chord, and has a maximum thickness of $0.12c$. The equation for the mean camber line is

$$\frac{z}{c} = 2.6595 \left[\left(\frac{x}{c}\right)^3 - 0.6075 \left(\frac{x}{c}\right)^2 + 0.11471 \left(\frac{x}{c}\right) \right]$$

NACA 23012 Airfoil Section
Data from ref. 6.2
Re_c: ○ 3.0×10^6 □ 6.0×10^6 ◇ 8.8×10^6

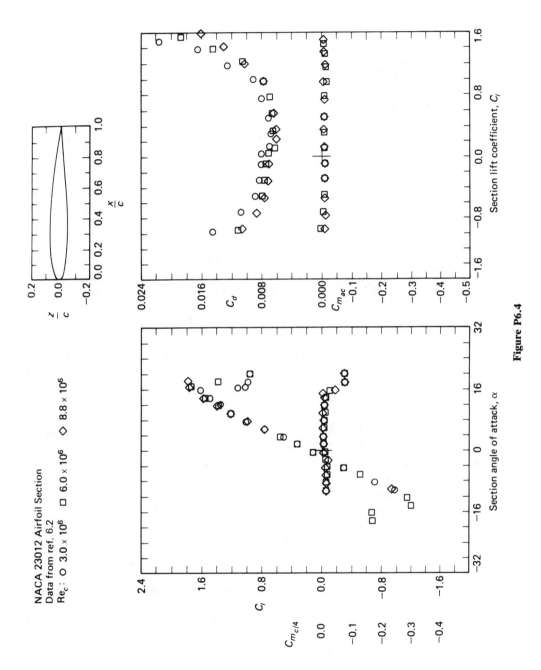

Figure P6.4

for the region $0.0c \leq x \leq 0.2025c$ and

$$\frac{z}{c} = 0.022083 \left(1 - \frac{x}{c} \right)$$

for the region $0.2025c \leq x \leq 1.000c$.

    Calculate A_0, A_1, and A_2 for this airfoil section. What is the section lift coefficient, C_l? What is the angle of attack for zero lift, α_{0l}? What angle of attack is required to develop the design lift coefficient of 0.3? Calculate the section moment coefficient about the theoretical aerodynamic center. Compare your theoretical values with the experimental values in Fig. P6.4 that are reproduced from the work of Abbott and von Doenhoff (Ref. 6.2). When the geometric angle of attack is 3°, what is the section lift coefficient? What is the x/c location of the center of pressure?

REFERENCES

6.1. Glauert, H., *Elements of Aerofoil and Airscrew Theory*, Macmillan Publishing Co., New York, 1943.

6.2. Abbott, I. H., and A. E. von Doenhoff, *Theory of Wing Sections*, Dover Publications, New York, 1949.

6.3. McCormick, B. W., Jr., *Aerodynamics of V/STOL Flight*, Academic Press, New York, 1967.

6.4. Henderson, M. L., "A Solution to the 2-D Separated Wake Modeling Problem and Its Use to Predict C_{Lmax} of Arbitrary Airfoil Sections," *AIAA Paper 78–156*, presented at the AIAA 16th Aerospace Sciences Meeting, Huntsville, Ala., Jan. 1978.

6.5. Smith, A.M. O., "High-Lift Aerodynamics," *Journal of Aircraft*, June 1975, Vol. 12, No. 6, pp. 501–530.

6.6. Stratford, B. S., "The Prediction of Separation of the Turbulent Boundary Layer," *Journal of Fluid Mechanics*, Jan. 1959, Vol. 5, Part 1, pp. 1–16.

6.7. Liebeck, R. H., "A Class of Airfoils Designed for High Lift in Incompressible Flow," *Journal of Aircraft*, Oct. 1973, Vol. 10, No. 10, pp. 610–617.

6.8. Eppler, R., and D. M. Somers, "A Computer Program for the Design and Analysis of Low-Speed Airfoils," *Technical Memorandum 80210*, NASA, Aug. 1980.

6.9. Liebeck, R. H., "On the Design of Subsonic Airfoils for High Lift," *AIAA Paper 76–406*, presented at the AIAA 9th Fluid and Plasma Dynamics Conference, San Diego, Calif., July 1976.

6.10. Stevens, W. A., S. H. Goradia, and J. A. Braden, "Mathematical Model for Two-Dimensional Multi-component Airfoils in Viscous Flow," *CR-1843*, NASA, July 1971.

6.11. Olson, L. E., W. D. James, and P. R. McGowan, "Theoretical and Experimental Study of the Drag of Multielement Airfoils," *AIAA Paper 78–1223*, presented at the AIAA 11th Fluid and Plasma Dynamics Conference, Seattle, Wash., July 1978.

6.12. Miranda, L. R., "Application of Computational Aerodynamics to Airplane Design," *Journal of Aircraft*, June 1984, Vol. 21, No. 6, pp. 355–369.

6.13. McGhee, R. J., and W. D. Beasley, "Low-Speed Aerodynamic Characteristics of a 17-Percent-Thick Section Designed for General Aviation Applications," *TN D-7428*, NASA, Dec. 1973.

7

INCOMPRESSIBLE FLOW
ABOUT WINGS
OF FINITE SPAN

The aerodynamic characteristics for subsonic flow about an unswept airfoil have been discussed in Chapters 5 and 6. Since the span of an airfoil is infinite, the flow is identical for each spanwise station (i.e., the flow is two-dimensional). The lift produced by the pressure differences between the lower surface and the upper surface of the wing, and therefore the circulation (integrated along the chord length of the section) does not vary along the span. For a wing of finite span, the high-pressure air beneath the wing spills out around the wing tips toward the low-pressure regions above the wing. As a consequence of the tendency of the pressures acting on the top surface near the tip of the wing to equalize with those on the bottom surface, the lift force per unit span decreases toward the tips. A sketch of a representative aerodynamic load distribution is presented in Fig. 7.1. As indicated in Fig. 7.1a, there is a chordwise variation in the pressure differential between the lower surface and the upper surface. The resultant lift force acting on a section (i.e., a unit span) is obtained by integrating the pressure distribution over the chord length. A procedure that can be used to determine the sectional lift coefficient has been discussed in Chapter 6.

As indicated in the sketch of Fig. 7.1b, there is a spanwise variation in the lift force. As a result of the spanwise pressure variation, the air on the upper surface flows inboard toward the root. Similarly, on the lower surface, air will tend to flow outward toward the tips. The resultant flow around a wing of finite span is three-dimensional, having both chordwise and spanwise velocity components. Where the flows from the

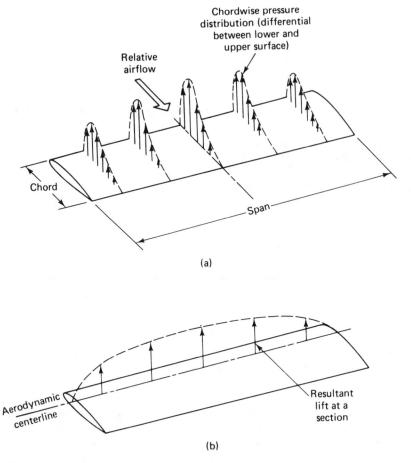

Figure 7.1　Aerodynamic load distribution for a rectangular wing in subsonic airstream: (a) differential pressure distribution along the chord for several spanwise stations; (b) spanwise lift distribution.

upper surface and the lower surface join at the trailing edge, the difference in spanwise velocity components will cause the air to roll up into a number of streamwise vortices, distributed along the span. These small vortices roll up into two large vortices just inboard of the wing tips (see Fig. 7.2). At this point, it is customary to assume (1) that the vortex wake, which is of finite thickness, may be replaced by an infinitesimally thin surface of discontinuity, designated the trailing vortex sheet, and (2) that the trailing vortex sheet remains flat as it extends downstream from the wing. Spreiter and Sacks (Ref. 7.1) note that "it has been firmly established that these assumptions are sufficiently valid for the prediction of the forces and moments on finite-span wings."

Thus, an important difference in the three-dimensional flow field around a wing (as compared with the two-dimensional flow around an airfoil) is the spanwise variation in lift. Since the lift force acting on the wing section at a given spanwise location is related to the strength of the circulation, there is a corresponding spanwise variation

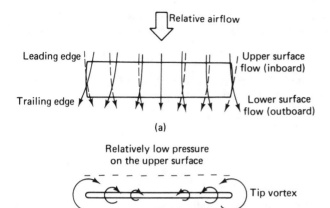

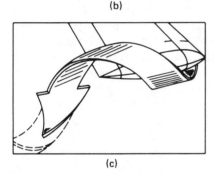

Figure 7.2 Generation of the trailing vortices due to the spanwise load distribution: (a) view from bottom; (b) view from trailing edge; (c) formation of the tip vortex.

in circulation, such that the circulation at the wing tip is zero. Procedures that can be used to determine the vortex-strength distribution produced by the flow field around a three-dimensional lifting wing are presented in this chapter.

VORTEX SYSTEM

A solution is sought for the vortex system which would impart to the surrounding air a motion similar to that produced by a lifting wing. A suitable distribution of vortices would represent the physical wing in every way except that of thickness. The vortex system consists of:

1. The bound vortex system
2. The trailing vortex system
3. The "starting" vortex

As stated in Chapter 6, the "starting" vortex is associated with a change in circulation and would, therefore, relate to changes in lift that might occur at some time.

The representation of the wing by a bound vortex system is not to be interpreted as a rigorous flow model. However, the idea allows a relation to be established between:

1. The physical load distribution for the wing (which depends on the wing geometry and on the aerodynamic characteristics of the wing sections)
2. The trailing vortex system

UNSWEPT WINGS

If the wing of interest is unswept (or is only slightly swept) and has an aspect ratio greater than about 4.0, we can represent the spanwise lift distribution by a single bound vortex system, the axis of which is normal to the plane of symmetry and passes through the aerodynamic center of the lifting surface. Since the theoretical relations developed in Chapter 6 for inviscid flow past a thin airfoil showed that the aerodynamic center is at the quarter chord, we shall place the bound vortex system at the quarter-chord line. The single vortex has a circulation Γ whose strength varies along the span (i.e., is a function of y). As discussed in Chapter 3, the vortex theorems of Helmholtz state that a vortex filament cannot end in the fluid. Thus, as the strength of the bound vortex system changes to represent the spanwise change in lift, vortex filaments equal in strength to the change in circulation leave the bound vortex, as shown in Fig. 7.3. If the strength of the bound vortex decreases by $\Delta\Gamma$ over the spanwise segment Δy, a trailing vortex of strength $\Delta\Gamma$ must be shed in the x direction. Therefore, although the filaments of the bound vortex system are of varying length (in the wing), they do not "end" in the wing but turn backward at each end to form a pair of the vortex elements in the trailing system. For steady flight conditions, the "starting" vortex is left behind so that the trailing vortex pair effectively stretches to infinity. Thus, for practical purposes, the system consists of the bound vortex system and the related system of trailing vortices. A sketch of the three-sided vortex system, which is termed a *horseshoe vortex*, is presented in Fig. 7.3. Also included is a sketch of a symmetrical lift distribution which the vortex system is to represent.

Prandtl and Tietjens (Ref. 7.2) hypothesized that each section of the wing acts as though it is an isolated two-dimensional section, provided that the spanwise flow is not great. The lateral pressure gradients associated with a swept-back wing produce significant spanwise flow violating the condition of Prandtl's hypothesis. Procedures for a swept wing will be discussed later in this chapter.

Trailing Vortices and Downwash

A consequence of the vortex theorems of Helmholtz is that a vortex tube does not change strength between two sections unless a vortex filament equal in strength to the change joins or leaves the vortex tube. If $\Gamma\,(y)$ denotes the strength of the circulation along the y axis (the spanwise coordinate), a semi-infinite vortex of strength $\Delta\Gamma$ trails

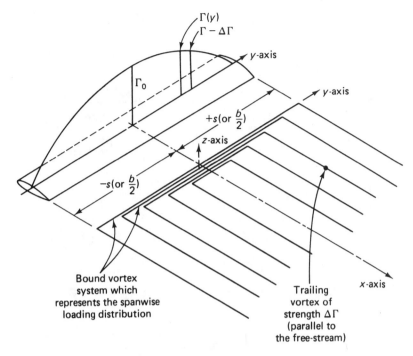

Figure 7.3 Trailing vortex system.

from the segment Δy, as shown in Fig. 7.4. The strength of the trailing vortex is given by

$$\Delta \Gamma = \frac{d\Gamma}{dy} \Delta y$$

It is assumed that each spanwise strip of the wing (Δy) behaves as if the flow were locally two-dimensional. To calculate the influence of a trailing vortex filament located at y, consider the semi-infinite vortex line, parallel to the x axis (which is parallel to the free-stream flow) and extending downstream to infinity from the line through the aerodynamic center of the wing (i.e., the y axis). The vortex at y induces a velocity at a general point y_1 on the aerodynamic centerline which is one-half the velocity that would be induced by an infinitely long vortex filament of the same strength:

$$\delta w_{y_1} = \frac{1}{2} \left[+ \frac{d\Gamma}{dy} dy \frac{1}{2\pi(y - y_1)} \right]$$

The positive sign results because, when both $(y - y_1)$ and $d\Gamma/dy$ are negative, the trailing vortex at y induces an upward component of velocity, as shown in Fig. 7.4.

To calculate the resultant induced velocity at any point y_1 due to the cumulative effect of all the trailing vortices, the expression above is integrated with respect to y from the left wing tip ($-s$) to the right wing tip ($+s$):

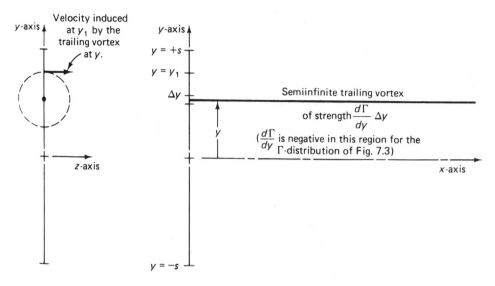

Figure 7.4 Geometry for the calculation of the induced velocity at $y = y_1$.

$$w_{y_1} = +\frac{1}{4\pi}\int_{-s}^{+s}\frac{d\Gamma/dy}{y - y_1}\,dy \qquad (7.1)$$

The resultant induced velocity at y_1 is, in general, in a downward direction (i.e, negative) and is called the *downwash*. As shown in the sketch of Fig. 7.5, the downwash angle is

$$\epsilon = \tan^{-1}\left(\frac{w_{y_1}}{U_\infty}\right) \approx -\frac{w_{y_1}}{U_\infty} \qquad (7.2)$$

The downwash has the effect of "tilting" the undisturbed air, so the effective angle of attack at the aerodynamic center (i.e., the quarter-chord) is

$$\alpha_e = \alpha - \epsilon \qquad (7.3)$$

Note that, if the wing has a geometric twist, both the angle of attack (α) and the downwash angle (ϵ) would be a function of the spanwise position. Since the direction of the resultant velocity at the aerodynamic center is inclined downward relative to the direction of the undisturbed free-stream air, the effective lift of the section of interest is inclined aft by the same amount. Thus, the effective lift on the wing has a component of force parallel to the undisturbed free-stream air (refer to Fig. 7.5). This drag force is a consequence of the lift developed by a finite wing and is termed *vortex drag* (or *induced drag*). Thus, for subsonic flow past a finite-span wing, in addition to the skin-friction drag and the form (or pressure) drag, there is a drag component due to lift. As a result of the induced downwash velocity, the lift generated by the airfoil section of a finite-span wing which is at the geometric angle of attack α is less than that for the same airfoil section of an infinite-span airfoil at the same angle of attack.

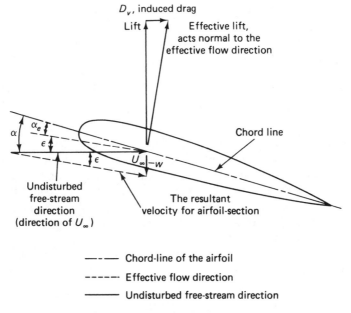

Figure 7.5 Induced flow.

Based on the Kutta-Joukowski theorem, the lift on an elemental airfoil section of the wing is

$$l(y) = \rho_\infty U_\infty \Gamma(y) \qquad (7.4)$$

while the vortex drag is

$$d_v(y) = -\rho_\infty w(y)\Gamma(y) \qquad (7.5)$$

The minus sign results because a downward (or negative) value of w produces a positive drag force. Integrating over the entire span of the wing, the total lift is given by

$$L = \int_{-s}^{+s} \rho_\infty U_\infty \Gamma(y) \, dy \qquad (7.6)$$

and the total vortex drag is given by

$$D_v = -\int_{-s}^{+s} \rho_\infty w(y)\Gamma(y) \, dy \qquad (7.7)$$

Note that for the two-dimensional airfoil (i.e., a wing of infinite span) the circulation strength Γ is constant across the span (i.e., it is independent of y) and the induced downwash velocity is zero at all points. Thus, $D_v = 0$, as discussed in Chapter 6. As a consequence of the trailing vortex system, the aerodynamic characteristics are modified significantly from those of a two-dimensional airfoil of the same section.

Case of Elliptic Spanwise Circulation Distribution

An especially simple circulation distribution, which also has significant practical implications, is given by the elliptic relation (see Fig. 7.6)

$$\Gamma(y) = \Gamma_0 \sqrt{1 - \left(\frac{y}{s}\right)^2} \qquad (7.8)$$

Since the lift is a function only of the free-stream density, the free-stream velocity, and the circulation, an elliptic distribution for the circulation would produce an elliptic distribution for the lift. However, to calculate the section lift coefficient, the section lift force is divided by the product of q_∞ and the local chord length at the section of interest. Hence, only when the wing has a rectangular planform (and c is, therefore, constant) is the spanwise lift coefficient distribution (C_l) elliptic when the spanwise lift distribution is elliptic.

For the elliptic spanwise circulation distribution of equation (7.8), the induced downwash velocity is

$$w_{y_1} = - \frac{\Gamma_0}{4\pi s} \int_{-s}^{+s} \frac{y}{\sqrt{s^2 - y^2}(y - y_1)} \, dy$$

which can be rewritten as

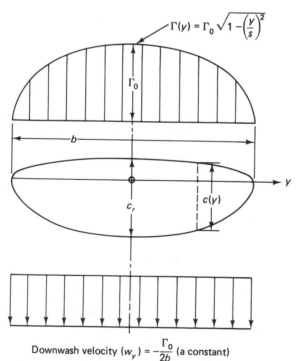

Downwash velocity $(w_y) = -\dfrac{\Gamma_0}{2b}$ (a constant)

Figure 7.6 Elliptic circulation distribution and the resultant downwash velocity.

$$w_{y_1} = -\frac{\Gamma_0}{4\pi s}\left[\int_{-s}^{+s}\frac{(y - y_1)\,dy}{\sqrt{s^2 - y^2}(y - y_1)} + \int_{-s}^{+s}\frac{y_1\,dy}{\sqrt{s^2 - y^2}(y - y_1)}\right]$$

Integration yields

$$w_{y_1} = -\frac{\Gamma_0}{4\pi s}(\pi + y_1 I) \qquad (7.9)$$

where

$$I = \int_{-s}^{+s}\frac{dy}{\sqrt{s^2 - y^2}(y - y_1)}$$

Since the elliptic loading is symmetric about the pitch plane of the vehicle (i.e., $y = 0$), the velocity induced at a point $y_1 = +a$ should be equal to the velocity at a point $y_1 = -a$. Referring to equation (7.9), this can be true only if $I = 0$. Thus, for the elliptic load distribution

$$w_{y_1} = w(y) = -\frac{\Gamma_0}{4s} \qquad (7.10)$$

The induced velocity is independent of the spanwise coordinate.

The total lift for the wing is

$$L = \int_{-s}^{+s}\rho_\infty U_\infty \Gamma_0\sqrt{1 - \left(\frac{y}{s}\right)^2}\,dy$$

Using the coordinate transformation

$$y = -s\cos\phi$$

the equation for lift becomes

$$L = \int_0^\pi \rho_\infty U_\infty \Gamma_0\sqrt{1 - \cos^2\phi}\;s\sin\phi\,d\phi$$

Thus,

$$L = \rho_\infty U_\infty \Gamma_0 s\frac{\pi}{2} = \frac{\pi}{4}b\rho_\infty U_\infty \Gamma_0 \qquad (7.11)$$

The lift coefficient for the wing is

$$C_L = \frac{L}{\frac{1}{2}\rho_\infty U_\infty^2 S} = \frac{\pi b\Gamma_0}{2U_\infty S} \qquad (7.12)$$

Similarly, we can calculate the total vortex (or induced) drag for the wing.

$$D_v = \int_{-s}^{+s}\frac{\rho_\infty\Gamma_0}{4s}\Gamma_0\sqrt{1 - \left(\frac{y}{s}\right)^2}\,dy$$

Introducing the coordinate transformation again, we have

$$D_v = \frac{\rho_\infty \Gamma_0^2}{4s} \int_0^\pi \sqrt{1 - \cos^2\phi} \; s \sin \phi \; d\phi \tag{7.13}$$

$$= \frac{\pi}{8} \rho_\infty \Gamma_0^2$$

and the drag coefficient for the induced component is

$$C_{D_v} = \frac{D_v}{\frac{1}{2}\rho_\infty U_\infty^2 S} = \frac{\pi \Gamma_0^2}{4 U_\infty^2 S} \tag{7.14}$$

Rearranging equation (7.12) to solve for Γ_0 gives

$$\Gamma_0 = \frac{2 C_L U_\infty S}{\pi b} \tag{7.15}$$

Thus,

$$C_{D_v} = \frac{\pi}{4 U_\infty^2 S} \left(\frac{2 C_L U_\infty S}{\pi b} \right)^2$$

or

$$C_{D_v} = \frac{C_L^2}{\pi} \left(\frac{S}{b^2} \right)$$

Since the aspect ratio is defined as

$$\text{AR} = \frac{b^2}{S} \tag{7.16}$$

$$C_{D_v} = \frac{C_L^2}{\pi \cdot \text{AR}}$$

Note that once again we see that the induced drag is zero for a two-dimensional airfoil (i.e., a wing with an aspect ratio of infinity). Note also that the trailing vortex drag for an inviscid flow around a wing is not zero but is proportional to C_L^2.

The induced drag coefficient given by equation (7.16) and the measurements for a wing whose aspect ratio is 5 are compared in Fig. 7.7. The experimental values of the induced drag coefficient, which were presented in Ref. 7.3, closely follow the theoretical values up to an angle of attack of 20°. The relatively constant difference between the measured values and the theoretical values is due to the influence of skin friction, which was not included in the development of equation (7.16). Therefore, as noted in Chapter 5, the drag coefficient is typically written as

$$C_D = C_{D0} + k C_L^2 \tag{7.17}$$

where C_{D0} is the drag coefficient at zero lift and $k C_L^2$ is the lift-dependent drag coefficient. The lift-dependent drag coefficient includes that part of the viscous drag and of the form drag, which results as the angle of attack changes from α_{0l}.

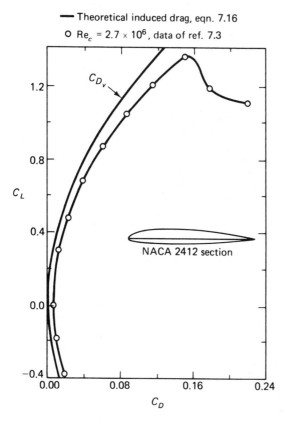

Figure 7.7 Experimental drag polar for a wing with an aspect ratio of 5 compared with the theoretical induced drag.

These relations describing the influence of the aspect ratio on the lift and the drag have been verified experimentally by Prandtl and Betz. If one compares the drag polars for two wings which have aspect ratios of AR_1 and AR_2, respectively, then for a given value of the lift coefficient,

$$C_{D,2} = C_{D,1} + \frac{C_L^2}{\pi}\left(\frac{1}{AR_2} - \frac{1}{AR_1}\right) \tag{7.18}$$

where $C_{D0,1}$ has been assumed to be equal to $C_{D0,2}$. The data from Ref. 7.4 for a series of rectangular wings are reproduced in Fig. 7.8. The experimentally determined drag polars are presented in Fig. 7.8a. Equation (7.18) has been used to convert the drag polars for the different aspect ratio wings to the equivalent drag polar for a wing whose aspect ratio is 5 (i.e., $AR_2 = 5$). These converted drag polars, which are presented in Fig. 7.8b, collapse quite well to a single curve. Thus, the correlation of the measurements confirms the validity of equation (7.18).

A similar analysis can be used to examine the effect of aspect ratio on the lift. Combining the definition for the downwash [equation (7.2)], the downwash velocity for the elliptic load distribution [equation (7.10)], and the correlation between the lift coefficient and Γ_0 [equation (7.15)], we obtain

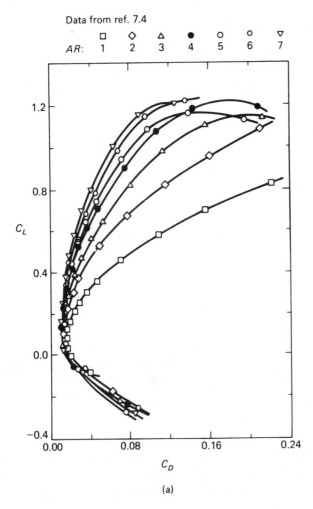

Data from ref. 7.4

| AR: | □ 1 | ◇ 2 | △ 3 | ● 4 | ○ 5 | ○ 6 | ▽ 7 |

Figure 7.8 Effect of the aspect ratio on the drag polar for rectangular wings (AR from 1 to 7): (a) measured drag polars.

(a)

$$\epsilon = \frac{C_L}{\pi \cdot AR} \tag{7.19}$$

One can determine the effect of the aspect ratio on the correlation between the lift coefficient and the geometric angle of attack. To calculate the geometric angle of attack α_2 required to generate a particular lift coefficient for a wing of AR_2, if a wing with an aspect ratio of AR_1 generates the same lift coefficient at α_1, we use the equation

$$\alpha_2 = \alpha_1 + \frac{C_L}{\pi} \left(\frac{1}{AR_2} - \frac{1}{AR_1} \right) \tag{7.20}$$

Experimentally determined lift coefficients (from Ref. 7.4) are presented in Fig. 7.9. The data presented in Fig. 7.9a are for the same rectangular wings of Fig. 7.8. The results of converting the coefficient-of-lift measurements using equation (7.20) in terms of a wing whose aspect ratio is 5 (i.e., $AR_2 = 5$) are presented in Fig. 7.9b. Again,

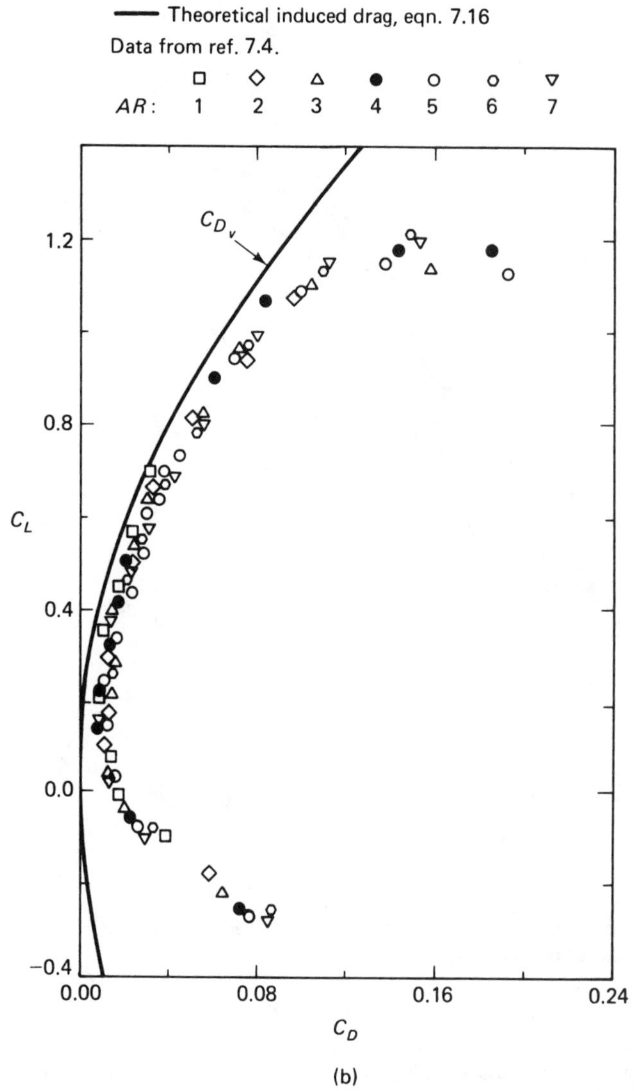

Figure 7.8 (*Continued*) (b) drag polars converted to AR = 5.

(b)

the converted curves collapse into a single correlation. Therefore, the validity of equation (7.20) is experimentally verified.

Technique for General Spanwise Circulation Distribution

Consider a spanwise circulation distribution that can be represented by a Fourier sine series consisting of N terms:

$$\Gamma(\phi) = 4sU_\infty \sum_1^N A_n \sin n\phi \qquad (7.21)$$

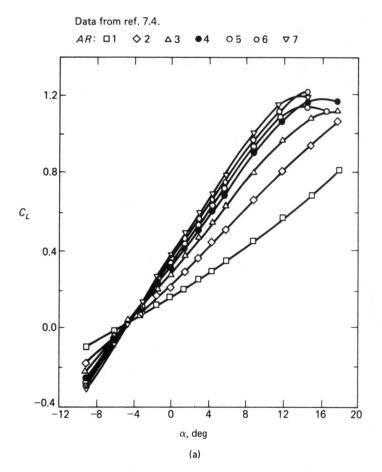

Data from ref. 7.4.

AR: □ 1   ◇ 2   △ 3   ● 4   ○ 5   ○ 6   ▽ 7

(a)

Figure 7.9 Effect of aspect ratio on the lift coefficient for rectangular wings (AR from 1 to 7): (a) measured lift correlations.

As was done previously, the physical spanwise coordinate (y) has been replaced by the ϕ coordinate:

$$\frac{y}{s} = -\cos \phi$$

A sketch of one such Fourier series is presented in Fig. 7.10. Since the spanwise lift distribution represented by the circulation of Fig. 7.10 is symmetrical, only the odd terms remain.

The section lift force [i.e., the lift acting on that spanwise section for which the circulation is $\Gamma(\phi)$] is given by

$$l(\phi) = \rho_\infty U_\infty \Gamma(\phi) = 4\rho_\infty U_\infty^2 s \sum_1^N A_n \sin n\phi \qquad (7.22)$$

Data from ref. 7.4.

AR: □ 1   ◇ 2   △ 3   ● 4   ○ 5   ○ 6   ▽ 7

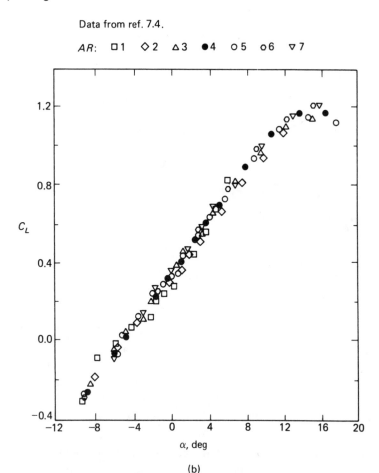

(b)

Figure 7.9   (*Continued*)   (b) lift correlations converted to AR = 5.

To evaluate the coefficients $A_1, A_2, A_3, \ldots, A_N$, it is necessary to determine the circulation at N spanwise locations. Once this is done, the N-resultant linear equations can be solved for the A_n coefficients.

Recall that the section lift coefficient is defined as

$$C_l(\phi) = \frac{\text{lift per unit span}}{\frac{1}{2}\rho_\infty U_\infty^2 c}$$

Using the local circulation to determine the local lift per unit span, we obtain

$$C_l(\phi) = \frac{\rho_\infty U_\infty \Gamma(\phi)}{\frac{1}{2}\rho_\infty U_\infty^2 c} = \frac{2\Gamma(\phi)}{U_\infty c} \tag{7.23}$$

It is also possible to evaluate the section lift coefficient by using the linear correlation between the lift and the angle of attack for the equivalent two-dimensional flow. Thus, referring to Fig. 7.11 for the nomenclature, we have

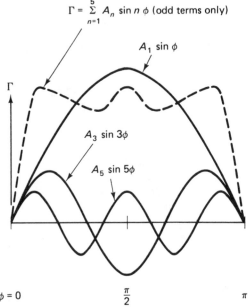

$$\Gamma = \sum_{n=1}^{5} A_n \sin n\phi \text{ (odd terms only)}$$

$A_1 \sin \phi$

Γ

$A_3 \sin 3\phi$

$A_5 \sin 5\phi$

$\phi = 0$ $\dfrac{\pi}{2}$ π

$y = -s$ 0 $+s$

Figure 7.10 Symmetric spanwise lift distribution as represented by a sine series.

$$C_l = \left(\frac{dC_l}{d\alpha}\right)_e (\alpha_e - \alpha_{0l}) \qquad (7.24)$$

Let the equivalent lift-curve slope $(dC_l/d\alpha)_e$ be designated by the symbol a_e. Note that since $\alpha_e = \alpha - \epsilon$, equations (7.23) and (7.24) can be combined to yield the relation

$$\frac{2\Gamma(\phi)}{c(\phi)a_e} = U_\infty[\alpha(\phi) - \alpha_{0l}(\phi)] - U_\infty\epsilon(\phi) \qquad (7.25)$$

For the present analysis, five parameters in equation (7.25) may depend on the spanwise location ϕ (or equivalently y) at which we will evaluate the terms. The five parameters are (1) Γ, the local circulation; (2) ϵ, the downwash angle, which depends on the circulation distribution; (3) c, the chord length, which varies with ϕ for a tapered wing planform; (4) α, the local geometric angle of attack, which varies with ϕ when the wing is twisted (i.e., geometric twist, which is illustrated in Fig. 5.3); and (5) α_{0l}, the zero lift angle of attack, which varies with ϕ when the airfoil section varies in the spanwise direction (which is known as *aerodynamic twist*). Note that

$$U_\infty\epsilon = -w = -\frac{1}{4\pi}\int_{-s}^{+s} \frac{d\Gamma/dy}{y - y_1}\, dy$$

Using the Fourier series representation for Γ and the coordinate transformation, we obtain

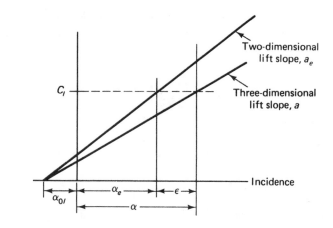

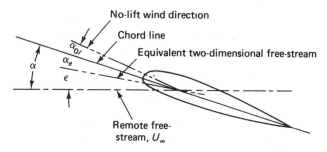

Figure 7.11 Nomenclature for wing/airfoil lift.

$$-w = U_\infty \frac{\sum n A_n \sin n\phi}{\sin \phi}$$

Equation (7.25) can be rewritten as

$$\frac{2\Gamma}{ca_e} = U_\infty(\alpha - \alpha_{0l}) - U_\infty \frac{\sum n A_n \sin n\phi}{\sin \phi}$$

Since $\Gamma = 4sU_\infty \sum A_n \sin n\phi$, the equation becomes

$$\frac{8s}{ca_e} \sum A_n \sin n\phi = (\alpha - \alpha_{0l}) - \frac{\sum n A_n \sin n\phi}{\sin \phi}$$

Defining $\mu = ca_e/8s$, the resultant governing equation is

$$\mu(\alpha - \alpha_{0l}) \sin \phi = \sum A_n \sin n\phi(\mu n + \sin \phi) \qquad (7.26)$$

which is known as the *monoplane equation*. If we consider only symmetrical loading distributions, only the odd terms of the series need be considered. That is, as shown in the sketch of Fig. 7.10,

$$\Gamma(\phi) = 4sU_\infty(A_1 \sin \phi + A_3 \sin 3\phi + A_5 \sin 5\phi + \cdots)$$

Lift on the Wing

$$L = \int_{-s}^{+s} \rho_\infty U_\infty \Gamma(y)\, dy = \int_0^\pi \rho_\infty U_\infty s \Gamma(\phi) \sin \phi\, d\phi$$

Using the Fourier series for $\Gamma(\phi)$ gives us

$$L = 4\rho_\infty U_\infty^2 s^2 \int_0^\pi \sum A_n \sin n\phi \sin \phi\, d\phi$$

Noting that $\sin A \sin B = \frac{1}{2} \cos (A - B) - \frac{1}{2} \cos (A + B)$, the integration yields

$$L = 4\rho_\infty U_\infty^2 s^2 \left\{ A_1 \left[\frac{\phi}{2} + \frac{\sin 2\phi}{4} \right] \Big|_0^\pi + \sum_3^N \frac{1}{2} A_n \left[\frac{\sin (n - 1)\phi}{n - 1} - \frac{\sin (n + 1)\phi}{n + 1} \right] \Big|_0^\pi \right\}$$

The summation represented by the second term on the right-hand side of the equation is zero, since each of the terms is zero for $n \neq 1$. Thus, the integral expression for the lift becomes

$$L = (4s^2)(\tfrac{1}{2}\rho_\infty U_\infty^2) A_1 \pi = C_L(\tfrac{1}{2}\rho_\infty U_\infty^2)(S) \tag{7.27}$$

$$C_L = A_1 \pi \cdot \text{AR}$$

The lift depends only on the magnitude of the first coefficient, no matter how many terms may be present in the series describing the distribution.

Vortex-Induced Drag

$$D_v = -\int_{-s}^{+s} \rho_\infty w \Gamma\, dy$$

$$= \rho_\infty \int_0^\pi \underbrace{\frac{U_\infty \sum n A_n \sin n\phi}{\sin \phi}}_{-w} \underbrace{4sU_\infty \sum A_n \sin n\phi}_{\Gamma}\, \underbrace{s \sin \phi\, d\phi}_{dy}$$

$$= 4\rho_\infty s^2 U_\infty^2 \int_0^\pi \sum n A_n \sin n\phi \sum A_n \sin n\phi\, d\phi$$

The integral

$$\int_0^\pi \sum n A_n \sin n\phi \sum A_n \sin n\phi\, d\phi = \frac{\pi}{2} \sum n A_n^2$$

Thus, the coefficient for the vortex-induced drag

$$C_{D_v} = \pi \cdot \text{AR} \sum n A_n^2 \tag{7.28}$$

Since $A_1 = C_L/(\pi \cdot \text{AR})$,

$$C_{D_v} = \frac{C_L^2}{\pi \cdot \text{AR}} \sum n \left(\frac{A_n}{A_1} \right)^2$$

where only the odd terms in the series are considered for the symmetric load distribution.

$$C_{D_v} = \frac{C_L^2}{\pi \cdot AR}\left[1 + \left(\frac{3A_3^2}{A_1^2} + \frac{5A_5^2}{A_1^2} + \frac{7A_7^2}{A_1^2} + \cdots\right)\right]$$

or

$$C_{D_v} = \frac{C_L^2}{\pi \cdot AR}(1 + \delta) \tag{7.29}$$

where

$$\delta = \frac{3A_3^2}{A_1^2} + \frac{5A_5^2}{A_1^2} + \frac{7A_7^2}{A_1^2} + \cdots$$

Since $\delta \geq 0$, the drag is a minimum when $\delta = 0$. In this case, the only term in the series representing the circulation distribution is the first term:

$$\Gamma(\phi) = 4sU_\infty A_1 \sin \phi$$

which is the elliptic distribution.

Example 7.1

The monoplane equation [i.e., equation (7.26)] will be used to compute the aerodynamic coefficients of a wing for which aerodynamic data are available. The geometry of the wing to be studied is illustrated in Fig. 7.12. The wing, which is unswept at the quarter chord, is composed of NACA 65–210 airfoil sections. Referring to the data of Ref. 7.5, the zero-lift angle of attack (α_{ol}) is approximately $-1.2°$ across the span. Since the wing is untwisted, the geometric angle of attack is the same at all spanwise positions. The aspect ratio (AR) is 9.00. The taper ratio λ (i.e., c_t/c_r) is 0.40. Since the wing planform is trapezoidal,

$$S = 0.5(c_r + c_t)b = 0.5c_r(1 + \lambda)b$$

and

$$AR = \frac{2b}{c_r + c_t}$$

Thus, the parameter μ in equation (7.26) becomes

$$\mu = \frac{ca_e}{4b} = \frac{ca_e}{2(AR) \cdot c_r(1 + \lambda)}$$

Since the terms are to be evaluated at spanwise stations for which $0 \leq \phi \leq \pi/2$ [i.e., $-s \leq y \leq 0$ (which corresponds to the port wing or left side of the wing)],

$$\mu = \frac{a_e}{2(1 + \lambda)AR}[1 + (\lambda - 1)\cos\phi]$$
$$= 0.24933(1 - 0.6\cos\phi) \tag{7.30}$$

where the equivalent lift-curve slope, a_e, has been assumed to be equal to 2π. It might be noted that numerical solutions for the lift and the vortex-drag coefficients were essentially the same for this geometry whether the series representing the spanwise circulation

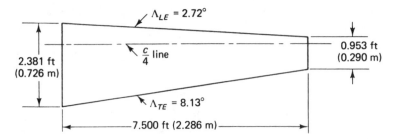

Figure 7.12 Planform for an unswept wing, AR = 9.00. λ = 0.40, airfoil section NACA 65-210.

distribution included four terms or ten terms. Therefore, so that the reader can perform the required calculations with a pocket calculator, a four-term series will be used to represent the spanwise loading. Equation (7.26) is

$$\mu(\alpha - \alpha_{ol}) \sin \phi = A_1 \sin \phi(\mu + \sin \phi) + A_3 \sin 3\phi(3\mu + \sin \phi) \\ + A_5 \sin 5\phi(5\mu + \sin \phi) + A_7 \sin 7\phi(7\mu + \sin \phi) \tag{7.31}$$

Since there are four coefficients (i.e., A_1, A_3, A_5, and A_7) to be evaluated, equation (7.31) must be evaluated at four spanwise locations. The resultant values for the factors are summarized in Table 7.1. Note that, since we are considering the left side of the wing, the y coordinate is negative.

TABLE 7.1 VALUES OF THE FACTOR FOR EQUATION (7.31)

Station	ϕ	$-\dfrac{y}{s}$ $(= \cos \phi)$	$\sin \phi$	$\sin 3\phi$	$\sin 5\phi$	$\sin 7\phi$	μ
1	22.5°	0.92388	0.38268	0.92388	0.92388	0.38268	0.11112
2	45.0°	0.70711	0.70711	0.70711	−0.70711	−0.70711	0.14355
3	67.5°	0.38268	0.92388	−0.38268	−0.38268	0.92388	0.19208
4	90.0°	0.00000	1.00000	−1.00000	1.00000	−1.00000	0.24933

For a geometric angle of attack of 4°, equation (7.31) becomes

$$0.00386 = 0.18897A_1 + 0.66154A_3 + 0.86686A_5 + 0.44411A_7$$

for $\phi = 22.5°$ (i.e., $y = -0.92388s$). For the other stations, the equation becomes

$$0.00921 = 0.60150A_1 + 0.80451A_3 - 1.00752A_5 - 1.21053A_7$$
$$0.01611 = 1.03101A_1 - 0.57407A_3 - 0.72109A_5 + 2.09577A_7$$
$$0.02263 = 1.24933A_1 - 1.74799A_3 + 2.24665A_5 - 2.74531A_7$$

The solution of this system of linear equations yields

$$A_1 = 1.6459 \times 10^{-2}$$
$$A_3 = 7.3218 \times 10^{-5}$$
$$A_5 = 8.5787 \times 10^{-4}$$
$$A_7 = -9.6964 \times 10^{-5}$$

Using equation (7.27), the lift coefficient for an angle of attack of 4° is

$$C_L = A_1\pi \cdot AR = 0.4654$$

The theoretically determined lift coefficients are compared in Fig. 7.13 with data for this wing. In addition to the geometric characteristics already described, the wing had a dihedral angle of 3°. The measurements reported in Ref. 7.6 were obtained at a Reynolds number of approximately 4.4×10^6 and a Mach number of approximately 0.17. The agreement between the theoretical values and the experimental values is very good.

The spanwise distribution for the local lift coefficient of this wing is presented in Fig. 7.14. As noted in Ref. 7.6, the variation of the section lift coefficient can be used to determine the spanwise position of initial stall. The local lift coefficient is given by

$$C_l = \frac{\rho_\infty U_\infty \Gamma}{0.5\, \rho_\infty U_\infty^2 c}$$

which for the trapezoidal wing under consideration is

$$C_l = 2(AR)(1 + \lambda)\frac{c_r}{c}\sum A_{2n-1} \sin(2n-1)\phi \tag{7.32}$$

The theoretical value of the induced drag coefficient for an angle of attack of 4°, as determined using equation (7.29), is

$$C_{Dv} = \frac{C_L^2}{\pi \cdot AR}\left(1 + \frac{3A_3^2}{A_1^2} + \frac{5A_5^2}{A_1^2} + \frac{7A_7^2}{A_1^2}\right)$$

$$= 0.00766(1.0136) = 0.00776$$

—— Theory

o Data of ref. 7.6

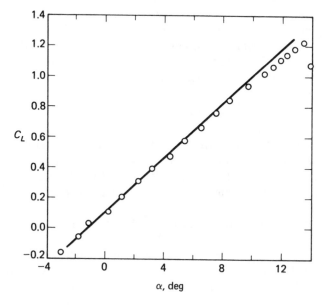

Figure 7.13 Comparison of the theoretical and the experimental lift coefficients for an unswept wing in a subsonic stream. (Wing is that of Fig. 7.12.)

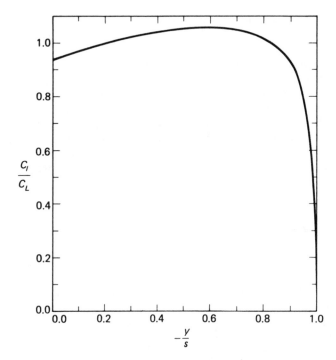

Figure 7.14 Spanwise distribution of the local lift coefficient, AR $= 9$, $\lambda = 0.4$, untwisted wing composed of NACA 65-210 airfoil sections.

The theoretically determined induced drag coefficients are compared in Fig. 7.15 with the measured drag coefficients for this wing. As has been noted earlier, the theoretical relations developed in this chapter do not include the effects of skin friction. The relatively constant difference between the measured values and the theoretical values is due to the influence of skin friction.

The effect of the taper ratio on the spanwise variation of the lift coefficient is illustrated in Fig. 7.16. Theoretical solutions are presented for untwisted wings having taper ratios from 0 to 1. The wings, which were composed of NACA 2412 airfoil sections, all had an aspect ratio of 7.28. Again, the local lift coefficient has been divided by the overall lift coefficient for the wings. Thus,

$$\frac{C_l}{C_L} = \frac{2(1 + \lambda)}{\pi A_1} \frac{c_r}{c} \sum A_{2n-1} \sin (2n - 1)\phi$$

The values of the local (or section) lift coefficient near the tip of the highly tapered wings are significantly greater than the overall lift coefficient for that planform. As noted earlier, this result is important relative to the separation (or stall) of the boundary layer for a particular planform when it is operating at a relatively high angle of attack.

Sketches of stall patterns are presented in Fig. 7.17. The desirable stall pattern for a wing is a stall which begins at the root sections so that the ailerons remain effective at high angles of attack.

The spanwise load distribution for a rectangular wing indicates stall will begin at the root and proceed outward. Thus, the stall pattern is favorable. The spanwise load distribution for a wing with a moderate taper ratio ($\lambda = 0.4$) approximates that

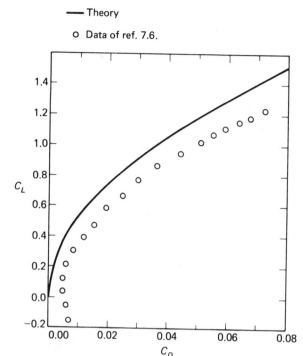

— Theory

o Data of ref. 7.6.

Figure 7.15 Comparison of the theoretical induced drag coefficients and the measured drag coefficients for an unswept wing in a subsonic stream. (Wing is that of Fig. 7.12.)

of an elliptical wing (i.e., the local lift coefficient is roughly constant across the span). As a result, all sections will reach stall at essentially the same angle of attack.

Tapering of the wing reduces the wing-root bending moments, since the inboard portion of the wing carries more of the wing's lift than the tip. Furthermore, the longer wing-root chord makes it possible to increase the actual thickness of the wing while maintaining a low thickness ratio which is needed if the airplane is to operate at high speeds also. While taper reduces the actual loads carried outboard,the lift coefficients near the tip are higher than those near the root for a tapered wing. Therefore, there is a strong tendency to stall near (or at) the tip for the highly tapered (or pointed) wings.

In order to prevent the stall pattern from beginning in the region of the ailerons, the wing may be given a geometric twist, or washout, to decrease the local angles of attack at the tip (refer to Table 5.1). The addition of leading edge slots or slats toward the tip increases the stall angle of attack and is useful in avoiding tip stall and the loss of aileron effectiveness.

PANEL METHODS

Although lifting-line theory (i.e., the monoplane equation) provides a reasonable estimate of the lift and of the induced drag for an unswept, thin wing of relatively high aspect ratio in a subsonic stream, an improved flow model is needed to calculate

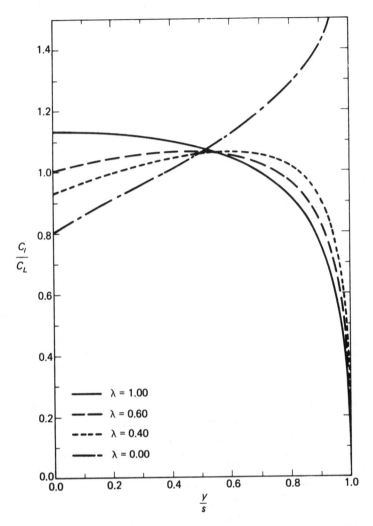

Figure 7.16 Effect of taper ratio on the spanwise variation of the lift coefficient for an untwisted wing.

the lifting flow field about a highly swept wing or a delta wing. A variety of methods have been developed to compute the flow about a thin wing which is operating at a small angle of attack so that the resultant flow may be assumed to be steady, inviscid, irrotational, and incompressible.

The basic concept of panel methods is illustrated in Fig. 7.18. The configuration is modeled by a large number of elementary quadrilateral panels lying either on the actual aircraft surface, or on some mean surface, or a combination thereof. To each elementary panel, there is attached one or more types of singularity distributions, such as sources, vortices, and doublets. These singularities are determined by specifying some functional variation across the panel (e.g., constant, linear, quadratic, etc.), whose actual value is set by corresponding strength parameters. These strength pa-

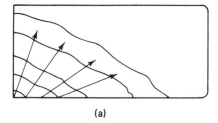

(a)

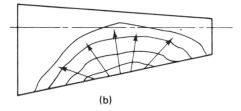

(b)

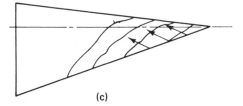

(c)

Figure 7.17 Typical stall patterns:
(a) rectangular wing, $\lambda = 1.0$;
(b) moderately tapered wing, $\lambda = 0.4$;
(c) pointed wing, $\lambda = 0.0$.

rameters are determined by solving the appropriate boundary condition equations. Once the singularity strengths have been determined, the velocity field and the pressure field can be computed.

Numerous codes using panel-method techniques have been developed (e.g., Ref. 7.7), the variations depending mainly on the choice of type and form of singularity distribution, the geometric layout of the elementary panels, and the type of boundary condition imposed. The choice of combinations is not a trivial matter. Although many different combinations are in principle mathematically equivalent, their numerical implementation may yield significantly different results from the point of view of numerical stability, computational economy, accuracy, and overall code robustness. Bristow and Grose (Ref. 7.7) note that there is an important equivalence between surface doublet distributions and vorticity distributions. A surface doublet distribution can be replaced by an equivalent surface vortex distribution. The theoretical equivalency between vorticity distributions and doublet distributions does not imply equivalent simplicity in a numerical formulation.

For each control point (and there is one control point per panel), the velocities induced at that control point by the singularities associated with each of the panels of the configuration are summed, resulting in a set of linear algebraic equations that express the exact boundary condition of flow tangency on the surface. For many

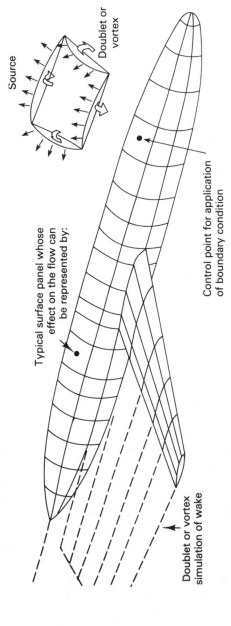

Figure 7.18   Representation of an airplane flowfield by panel (or singularity) methods.

applications, the aerodynamic coefficients computed using panel methods are reasonably accurate. Bristow and Grose (Ref. 7.7) discuss some problems with the source panel class of methods when used on thin, highly loaded surfaces such as the aft portion of a supercritical airfoil. In such cases, source strengths with strong gradients can degrade the local velocity calculations.

Margason et al. (Ref. 7.8) compare computed aerodynamic coefficients using one vortex lattice method (VLM), one source panel method, two low-order surface potential distributions, and two high-order surface potential distributions. The computed values of C_L are presented as a function of α for a 45° swept-back and a 45° swept-forward wing in Figs. 7.19a and b, respectively. The five surface panel methods consistently overpredict the experimental data with little difference between the lift coefficients predicted by the various surface panel methods. As Margason et al. (Ref. 7.8) note: ''The VLM predicts the experimental data very well, due to the fact that vortex lattice methods neglect both thickness and viscosity effects. For most cases, the effect of viscosity offsets the effect of thickness, fortuitously yielding good agreement between the VLM and experiment.''

VORTEX LATTICE METHOD

The vortex lattice method is the simplest of the methods reviewed by Margason et al. (Ref. 7.8). The VLM represents the wing as a planar surface on which a grid of horseshoe vortices is superimposed. The velocities induced by each horseshoe vortex at a specified control point are calculated using the law of Biot-Savart. A summation is performed for all control points on the wing to produce a set of linear algebraic equations for the horseshoe vortex strengths that satisfy the boundary condition of no flow through the wing. The vortex strengths are related to the wing circulation and the pressure differential between the upper and lower wing surfaces. The pressure differentials are integrated to yield the total forces and moments.

In our approach to solving the governing equation, the continuous distribution of bound vorticity over the wing surface is approximated by a finite number of discrete horseshoe vortices, as shown in Fig. 7.20. The individual horseshoe vortices are placed in trapezoidal panels (also called *finite elements* or *lattices*). This procedure for obtaining a numerical solution to the flow is termed the vortex lattice method.

The bound vortex coincides with the quarter-chord line of the panel (or element) and is, therefore, aligned with the local sweepback angle. In a rigorous theoretical analysis, the vortex lattice panels are located on the mean camber surface of the wing and, when the trailing vortices leave the wing, they follow a curved path. However, for many engineering applications, suitable accuracy can be obtained using linearized theory in which straight-line trailing vortices extend downstream to infinity. In the linearized approach the trailing vortices are aligned either parallel to the free stream or parallel to the vehicle axis. Both orientations provide similar accuracy within the assumptions of linearized theory. In this text we shall assume that the trailing vortices are parallel to the axis of the vehicle, as shown in Fig. 7.21. This orientation of the trailing vortices is chosen because the computation of the influences of the various vortices (which we will call the *influence coefficients*) is simpler. Furthermore, these

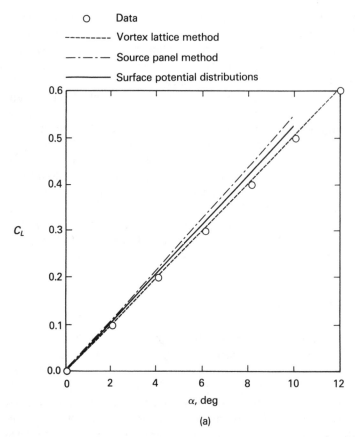

Figure 7.19 Comparison of the lift coefficient as a function of angle of attack: (a) $\Lambda_{C/4} = 45°$, NACA 64A010 section, AR = 3.0, $\lambda = 0.5$, aft swept wing. (From Ref. 7.8.)

geometric coefficients do not change as the angle of attack is changed. Application of the boundary condition that the flow is tangent to the wing surface at "the" control point of each of the 2N panels (i.e., there is no flow through the surface) provides a set of simultaneous equations in the unknown vortex circulation strengths. The control point of each panel is centered spanwise on the three-quarter-chord line midway between the trailing vortex legs.

An indication of why the three-quarter-chord location is used as the control point may be seen by referring to Fig. 7.22. A vortex filament whose strength Γ represents the lifting character of the section is placed at the quarter-chord location. It induces a velocity,

$$U = \frac{\Gamma}{2\pi r}$$

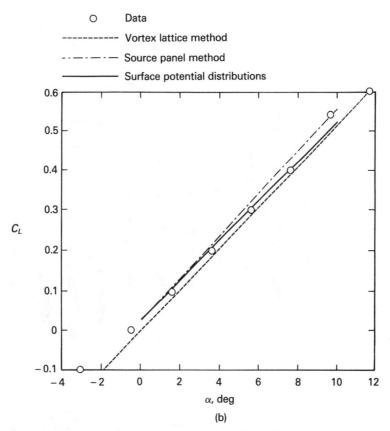

Figure 7.19 (*Continued*) (b) $\Lambda_{C/4} = -45°$, NACA 64A112 section, AR = 3.55, $\lambda = 0.5$, forward swept wing. (From Ref. 7.8.)

at the point c, the control point which is a distance r from the vortex filament. If the flow is to be parallel to the surface at the control point, the incidence of the surface relative to the free stream is given by

$$\alpha \approx \sin \alpha = \frac{U}{U_\infty} = \frac{\Gamma}{2\pi r U_\infty}$$

But, as was discussed in equations (6.11) and (6.12),

$$l = \tfrac{1}{2}\rho_\infty U_\infty^2 c 2\pi\alpha = \rho_\infty U_\infty \Gamma$$

Combining the relations above gives us

$$\pi\rho_\infty U_\infty^2 c \frac{\Gamma}{2\pi r U_\infty} = \rho_\infty U_\infty \Gamma$$

Solving for r yields

$$r = \frac{c}{2}$$

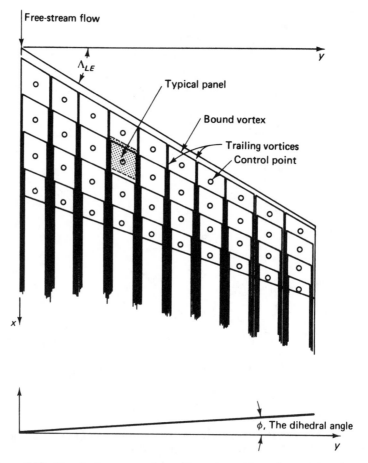

Figure 7.20 Coordinate system, elemental panels, and horseshoe vortices for a typical wing planform in the vortex lattice method.

Thus, we see that the control point is at the three-quarter-chord location for this two-dimensional geometry. The use of the chordwise slope at the 0.75-chord location to define the effective incidence of a panel in a finite-span wing has long been in use (e.g., Refs. 7.9 and 7.10).

Velocity Induced by a General Horseshoe Vortex

The velocity induced by a vortex filament of strength Γ_n and a length of dl is given by the *law of Biot and Savart* (see Ref. 7.11):

$$\overrightarrow{dV} = \frac{\Gamma_n(\overrightarrow{dl} \times \overrightarrow{r})}{4\pi r^3} \tag{7.33}$$

Referring to the sketch of Fig. 7.23, the magnitude of the induced velocity is

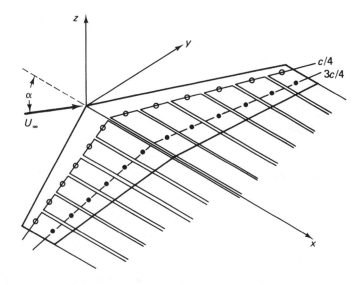

Filled circles represent the
control points

Figure 7.21 Distributed horseshoe vortices representing the lifting flow field over a swept wing.

$$dV = \frac{\Gamma_n \sin \theta \, dl}{4\pi r^2} \qquad (7.34)$$

Since we are interested in the flow field induced by a horseshoe vortex which consists of three straight segments, let us use equation (7.33) to calculate the effect of each segment separately. Let AB be such a segment, with the vorticity vector directed from A to B. Let C be a point in space whose normal distance from the line AB is r_p. Since

$$r = \frac{r_p}{\sin \theta} \quad \text{and} \quad dl = r_p(\csc^2 \theta) \, d\theta$$

we can integrate between A and B to find the magnitude of the induced velocity:

$$V = \frac{\Gamma_n}{4\pi r_p} \int_{\theta_1}^{\theta_2} \sin \theta \, d\theta = \frac{\Gamma_n}{4\pi r_p} (\cos \theta_1 - \cos \theta_2) \qquad (7.35)$$

Note that, if the vortex filament extends to infinity in both directions, then $\theta_1 = 0$ and $\theta_2 = \pi$. In this case

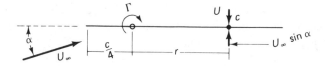

Figure 7.22 Planar airfoil section indicating location of control point where flow is parallel to the surface.

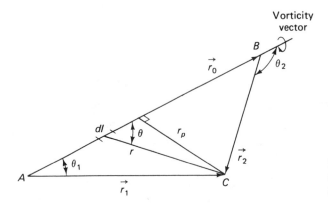

Figure 7.23 Nomenclature for calculating the velocity induced by a finite length vortex segment.

$$V = \frac{\Gamma_n}{2\pi r_p}$$

which is the result used in Chapter 6 for the infinite-span airfoils. Let $\vec{r}_0$, $\vec{r}_1$, and $\vec{r}_2$ designate the vectors $\vec{AB}$, $\vec{AC}$, and $\vec{BC}$, respectively, as shown in Fig. 7.23. Then

$$r_p = \frac{|\vec{r}_1 \times \vec{r}_2|}{r_0} \qquad \cos\theta_1 = \frac{\vec{r}_0 \cdot \vec{r}_1}{r_0 r_1} \qquad \cos\theta_2 = \frac{\vec{r}_0 \cdot \vec{r}_2}{r_0 r_2}$$

In these equations, if a vector quantity (such as $\vec{r}_0$) is written without a superscript arrow, the symbol represents the magnitude of the parameter. Thus, r_0 is the magnitude of the vector $\vec{r}_0$. Also note that $|\vec{r}_1 \times \vec{r}_2|$ represents the magnitude of the vector cross product. Substituting these expressions into equation (7.35) and noting that the direction of the induced velocity is given by the unit vector

$$\frac{\vec{r}_1 \times \vec{r}_2}{|\vec{r}_1 \times \vec{r}_2|}$$

yields

$$\vec{V} = \frac{\Gamma_n}{4\pi} \frac{\vec{r}_1 \times \vec{r}_2}{|\vec{r}_1 \times \vec{r}_2|^2} \left[\vec{r}_0 \cdot \left(\frac{\vec{r}_1}{r_1} - \frac{\vec{r}_2}{r_2} \right) \right] \tag{7.36}$$

This is the basic expression for the calculation of the induced velocity by the horseshoe vortices in the VLM. It can be used regardless of the assumed orientation of the vortices.

We shall now use equation (7.36) to calculate the velocity that is induced at a general point in space (x, y, z) by the horseshoe vortex shown in Fig. 7.24. This horseshoe vortex may be assumed to represent that for a typical wing panel (e.g., the nth panel) in Fig. 7.20. Segment AB represents the bound vortex portion of the horseshoe system and coincides with the quarter-chord line of the panel element. The trailing vortices are parallel to the x axis. The resultant induced velocity vector will be calculated by considering the influence of each of the elements.

For the bound vortex, segment $\vec{AB}$:

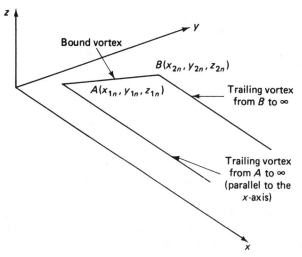

Figure 7.24 "Typical" horseshoe vortex.

$$\vec{r}_0 = \overrightarrow{AB} = (x_{2n} - x_{1n})\hat{i} + (y_{2n} - y_{1n})\hat{j} + (z_{2n} - z_{1n})\hat{k}$$
$$\vec{r}_1 = (x - x_{1n})\hat{i} + (y - y_{1n})\hat{j} + (z - z_{1n})\hat{k}$$
$$\vec{r}_2 = (x - x_{2n})\hat{i} + (y - y_{2n})\hat{j} + (z - z_{2n})\hat{k}$$

Using equation (7.36) to calculate the velocity induced at some point $C(x, y, z)$ by the vortex filament AB (shown in Figs. 7.24 and 7.25),

$$\vec{V}_{AB} = \frac{\Gamma_n}{4\pi}\{Fac1_{AB}\}\{Fac2_{AB}\} \tag{7.37a}$$

where

$$\{Fac1_{AB}\} = \frac{\vec{r}_1 \times \vec{r}_2}{|\vec{r}_1 \times \vec{r}_2|^2}$$

$$= \{[(y - y_{1n})(z - z_{2n}) - (y - y_{2n})(z - z_{1n})]\hat{i}$$
$$- [(x - x_{1n})(z - z_{2n}) - (x - x_{2n})(z - z_{1n})]\hat{j}$$
$$+ [(x - x_{1n})(y - y_{2n}) - (x - x_{2n})(y - y_{1n})]\hat{k}\}/$$
$$\{[(y - y_{1n})(z - z_{2n}) - (y - y_{2n})(z - z_{1n})]^2$$
$$+ [(x - x_{1n})(z - z_{2n}) - (x - x_{2n})(z - z_{1n})]^2$$
$$+ [(x - x_{1n})(y - y_{2n}) - (x - x_{2n})(y - y_{1n})]^2\}$$

and

$$\{Fac2_{AB}\} = \left(\vec{r}_0 \cdot \frac{\vec{r}_1}{r^1} - \vec{r}_0 \cdot \frac{\vec{r}_2}{r^2}\right)$$

$$= \{[(x_{2n} - x_{1n})(x - x_{1n}) + (y_{2n} - y_{1n})(y - y_{1n}) + (z_{2n} - z_{1n})(z - z_{1n})]/$$
$$\sqrt{(x - x_{1n})^2 + (y - y_{1n})^2 + (z - z_{1n})^2}$$
$$- [(x_{2n} - x_{1n})(x - x_{2n}) + (y_{2n} - y_{1n})(y - y_{2n}) + (z_{2n} - z_{1n})(z - z_{2n})]/$$
$$\sqrt{(x - x_{2n})^2 + (y - y_{2n})^2 + (z - z_{2n})^2}\}$$

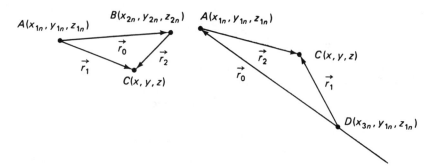

Figure 7.25 Vector elements for the calculation of the induced velocities.

To calculate the velocity induced by the filament that extends from A to ∞, let us first calculate the velocity induced by the collinear, finite-length filament that extends from A to D. Since $\vec{r}_0$ is in the direction of the vorticity vector,

$$\vec{r}_0 = \overrightarrow{DA} = (x_{1n} - x_{3n})\hat{i}$$
$$\vec{r}_1 = (x - x_{3n})\hat{i} + (y - y_{1n})\hat{j} + (z - z_{1n})\hat{k}$$
$$\vec{r}_2 = (x - x_{1n})\hat{i} + (y - y_{1n})\hat{j} + (z - z_{1n})\hat{k}$$

as shown in Fig. 7.25. Thus, the induced velocity is

$$\overrightarrow{V_{AD}} = \frac{\Gamma_n}{4\pi}\{\text{Fac1}_{AD}\}\{\text{Fac2}_{AD}\}$$

where

$$\{\text{Fac1}_{AD}\} = \frac{(z - z_{1n})\hat{j} + (y_{1n} - y)\hat{k}}{[(z - z_{1n})^2 + (y_{1n} - y)^2](x_{3n} - x_{1n})}$$

and

$$\{\text{Fac2}_{AD}\} = (x_{3n} - x_{1n})\left\{\frac{x_{3n} - x}{\sqrt{(x - x_{3n})^2 + (y - y_{1n})^2 + (z - z_{1n})^2}}\right.$$
$$\left. + \frac{x - x_{1n}}{\sqrt{(x - x_{1n})^2 + (y - y_{1n})^2 + (z - z_{1n})^2}}\right\}$$

Letting x_3 go to ∞, the first term of $\{\text{Fac2}_{AD}\}$ goes to 1.0. Therefore, the velocity induced by the vortex filament which extends from A to ∞ in a positive direction parallel to the x axis is given by

$$\overrightarrow{V_{A\infty}} = \frac{\Gamma_n}{4\pi}\left\{\frac{(z - z_{1n})\hat{j} + (y_{1n} - y)\hat{k}}{[(z - z_{1n})^2 + (y_{1n} - y)^2]}\right\}\left[1.0\right.$$
$$\left. + \frac{x - x_{1n}}{\sqrt{(x - x_{1n})^2 + (y - y_{1n})^2 + (z - z_{1n})^2}}\right] \qquad (7.37b)$$

Similarly, the velocity induced by the vortex filament that extends from B to ∞ in a positive direction parallel to the x axis is given by

$$\overrightarrow{V_{B\infty}} = -\frac{\Gamma_n}{4\pi} \left\{ \frac{(z - z_{2n})\hat{j} + (y_{2n} - y)\hat{k}}{[(z - z_{2n})^2 + (y_{2n} - y)^2]} \right\} \left[1.0 \right.$$

$$\left. + \frac{x - x_{2n}}{\sqrt{(x - x_{2n})^2 + (y - y_{2n})^2 + (z - z_{2n})^2}} \right] \qquad (7.37c)$$

The total velocity induced at some point (x, y, z) by the horseshoe vortex representing one of the surface elements (i.e., that for the nth panel) is the sum of the components given in equation (7.37). Let the point (x, y, z) be the control point of the mth panel, which we will designate by the coordinates (x_m, y_m, z_m). The velocity induced at the mth control point by the vortex representing the nth panel will be designated as $\overrightarrow{V_{m,n}}$. Examining equation (7.37), we see that

$$\overrightarrow{V_{m,n}} = \overrightarrow{C_{m,n}}\Gamma_n \qquad (7.38)$$

where the influence coefficient $\overrightarrow{C_{m,n}}$ depends on the geometry of the nth horseshoe vortex and its distance from the control point of the mth panel. Since the governing equation is linear, the velocities induced by the $2N$ vortices are added together to obtain an expression for the total induced velocity at the mth control point:

$$\overrightarrow{V_m} = \sum_{n=1}^{2N} \overrightarrow{C_{m,n}}\Gamma_n \qquad (7.39)$$

We have $2N$ of these equations, one for each of the control points.

Application of the Boundary Conditions

Thus, it is possible to determine the resultant induced velocity at any point in space, if the strengths of the $2N$ horseshoe vortices are known. However, their strengths are not known a priori. To compute the strengths of the vortices, Γ_n, which represent the lifting flow field of the wing, we use the boundary condition that the surface is a streamline. That is, the resultant flow is tangent to the wing at each and every control point (which is located at the midspan of the three-quarter-chord line of each elemental panel). If the flow is tangent to the wing, the component of the induced velocity normal to the wing at the control point balances the normal component of the free-stream velocity. To evaluate the induced velocity components, we must introduce at this point our convention that the trailing vortices are parallel to the vehicle axis [i.e., the x axis for equation (7.37) is the vehicle axis]. Referring to Fig. 7.26, the tangency requirement yields the relation

$$-u_m \sin \delta \cos \phi - v_m \cos \delta \sin \phi + w_m \cos \phi \cos \delta$$
$$+ U_\infty \sin (\alpha - \delta) \cos \phi = 0 \qquad (7.40)$$

where ϕ is the dihedral angle, as shown in Fig. 7.20 and δ is the slope of the mean camber line at the control point. Thus,

$$\delta = \tan^{-1} \left(\frac{dz}{dx} \right)_m$$

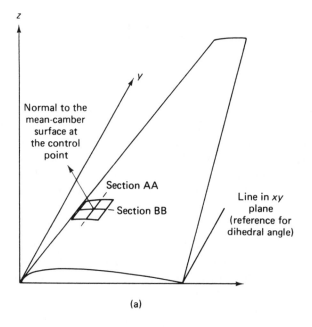

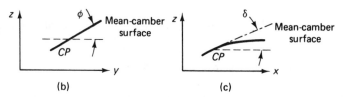

Figure 7.26 Nomenclature for the tangency requirement: (a) normal to element of the mean-camber surface; (b) section AA; (c) section BB.

For wings where the slope of the mean camber line is small and which are at small angles of attack, equation (7.40) can be replaced by the approximation

$$w_m - v_m \tan \phi + U_\infty \left[\alpha - \left(\frac{dz}{dx} \right)_m \right] = 0 \qquad (7.41)$$

This approximation is consistent with the assumptions of linearized theory. The unknown circulation strengths (Γ_n) required to satisfy these tangent flow boundary conditions are determined by solving the system of simultaneous equations represented by equation (7.39). The solution involves the inversion of a matrix.

Relations for a Planar Wing

Equations (7.37) through (7.41) are those for the VLM where the trailing vortices are parallel to the x axis. As such, they can be solved to determine the lifting flow for a twisted wing with dihedral. Let us apply these equations to a relatively simple geometry, a planar wing (i.e., one that lies in the xy plane), so that we can learn the significance of the various operations using a geometry which we can readily visualize. For a planar wing, $z_{1n} = z_{2n} = 0$ for all the bound vortices. Furthermore, $z_m = 0$ for all the control points. Thus, for our planar wing,

$$\overrightarrow{V_{AB}} = \frac{\Gamma_n}{4\pi} \frac{\hat{k}}{(x_m - x_{1n})(y_m - y_{2n}) - (x_m - x_{2n})(y_m - y_{1n})}$$

$$\left[\frac{(x_{2n} - x_{1n})(x_m - x_{1n}) + (y_{2n} - y_{1n})(y_m - y_{1n})}{\sqrt{(x_m - x_{1n})^2 + (y_m - y_{1n})^2}} \right.$$

$$\left. - \frac{(x_{2n} - x_{1n})(x_m - x_{2n}) + (y_{2n} - y_{1n})(y_m - y_{2n})}{\sqrt{(x_m - x_{2n})^2 + (y_m - y_{2n})^2}} \right] \qquad (7.42a)$$

$$\overrightarrow{V_{A\infty}} = \frac{\Gamma_n}{4\pi} \frac{\hat{k}}{y_{1n} - y_m} \left[1.0 + \frac{x_m - x_{1n}}{\sqrt{(x_m - x_{1n})^2 + (y_m - y_{1n})^2}} \right] \qquad (7.42b)$$

$$\overrightarrow{V_{B\infty}} = -\frac{\Gamma_n}{4\pi} \frac{\hat{k}}{y_{2n} - y_m} \left[1.0 + \frac{x_m - x_{2n}}{\sqrt{(x_m - x_{2n})^2 + (y_m - y_{2n})^2}} \right] \qquad (7.42c)$$

Note that, for the planar wing, all three components of the vortex representing the nth panel induce a velocity at the control point of the mth panel which is in the z direction (i.e., a downwash). Therefore, we can simplify equation (7.42) by combining the components into one expression:

$$w_{m,n} = \frac{\Gamma_n}{4\pi} \left\{ \frac{1}{(x_m - x_{1n})(y_m - y_{2n}) - (x_m - x_{2n})(y_m - y_{1n})} \right.$$

$$\left[\frac{(x_{2n} - x_{1n})(x_m - x_{1n}) + (y_{2n} - y_{1n})(y_m - y_{1n})}{\sqrt{(x_m - x_{1n})^2 + (y_m - y_{1n})^2}} \right.$$

$$\left. - \frac{(x_{2n} - x_{1n})(x_m - x_{2n}) + (y_{2n} - y_{1n})(y_m - y_{2n})}{\sqrt{(x_m - x_{2n})^2 + (y_m - y_{2n})^2}} \right]$$

$$+ \frac{1.0}{y_{1n} - y_m} \left[1.0 + \frac{x_m - x_{1n}}{\sqrt{(x_m - x_{1n})^2 + (y_m - y_{1n})^2}} \right]$$

$$\left. - \frac{1.0}{y_{2n} - y_m} \left[1.0 + \frac{x_m - x_{2n}}{\sqrt{(x_m - x_{2n})^2 + (y_m - y_{2n})^2}} \right] \right\} \qquad (7.43)$$

Summing the contributions of all the vortices to the downwash at the control point of the mth panel:

$$w_m = \sum_{n=1}^{2N} w_{m,n} \qquad (7.44)$$

Let us now apply the tangency requirement defined by equations (7.40) and (7.41). Since we are considering a planar wing in this section, $(dz/dx)_m = 0$ everywhere and $\phi = 0$. The component of the free-stream velocity perpendicular to the wing is

$U_\infty \sin \alpha$ at any point on the wing. Thus, the resultant flow will be tangent to the wing if the total vortex-induced downwash at the control point of the mth panel, which is calculated using equation (7.44), balances the normal component of the free-stream velocity:

$$w_m + U_\infty \sin \alpha = 0 \qquad (7.45)$$

For small angles of attack,

$$w_m = -U_\infty \alpha \qquad (7.46)$$

Example 7.2

Let us use the relations developed in this section to calculate the lift coefficient for a swept wing. So that the calculation procedures can be easily followed, let us consider a wing that has a relatively simple geometry (i.e., that illustrated in Fig. 7.27). The wing

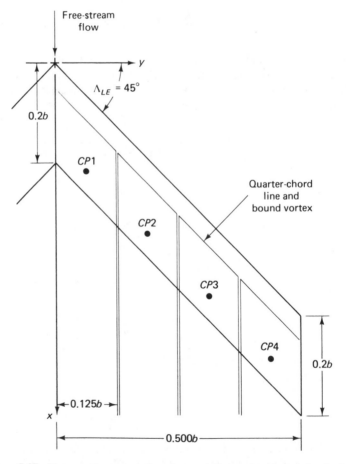

Figure 7.27 Four-panel representation of a swept planar wing, taper ratio of unity, AR = 5, $\Lambda = 45°$.

has an aspect ratio of 5, a taper ratio of unity (i.e., $c_r = c_t$), and an uncambered section (i.e., it is a flat plate). Since the taper ratio is unity, the leading edge, the quarter-chord line, the three-quarter-chord line, and the trailing edge all have the same sweep, 45°. Since

$$\text{AR} = 5 = \frac{b^2}{S}$$

and since for a swept, untapered wing

$$S = bc$$

it is clear that $b = 5c$. Using this relation, it is possible to calculate all of the necessary coordinates in terms of the parameter b. Therefore, the solution does not require that we know the physical dimensions of the configuration.

The flow field under consideration is symmetric with respect to the $y = 0$ plane (xz plane); that is, there is no yaw. Thus, the lift force acting at a point on the starboard wing ($+y$) is equal to that at the corresponding point on the port wing ($-y$). Because of symmetry, we need only to solve for the strengths of the vortices of the starboard wing. Furthermore, we need to apply the tangency condition [i.e., equation (7.46)] only at the control points of the starboard wing. However, we must remember to include the contributions of the horseshoe vortices of the port wing to the velocities induced at these control points (of the starboard wing). Thus, for this planar symmetric flow, equation (7.44) becomes

$$w_m = \sum_{n=1}^{N} w_{m,ns} + \sum_{n=1}^{N} w_{m,np}$$

where the symbols s and p represent the starboard and port wings, respectively.

The planform of the starboard wing is divided into four panels, each panel extending from the leading edge to the trailing edge. By limiting ourselves to only four spanwise panels, we can calculate the strengths of the horseshoe vortices using only a pocket electronic calculator. Thus, we can more easily see how the terms are to be evaluated. As before, the bound portion of each horseshoe vortex coincides with the quarter-chord line of its panel and the trailing vortices are in the plane of the wing, parallel to the x axis. The control points are designated by the solid symbols in Fig. 7.27. Recall that (x_m, y_m, 0) are the coordinates of a given control point and that (x_{1n}, y_{1n}, 0) and (x_{2n}, y_{2n}, 0) are the coordinates of the "ends" of the bound vortex filament AB. The coordinates for a 4×1 lattice (four spanwise divisions and one chordwise division) for the starboard (right) wing are summarized in Table 7.2.

TABLE 7.2 COORDINATES OF THE BOUND VORTICES AND OF THE CONTROL POINTS OF THE STARBOARD (RIGHT) WING

Panel	x_m	y_m	x_{1n}	y_{1n}	x_{2n}	y_{2n}
1	0.2125b	0.0625b	0.0500b	0.0000b	0.1750b	0.1250b
2	0.3375b	0.1875b	0.1750b	0.1250b	0.3000b	0.2500b
3	0.4625b	0.3125b	0.3000b	0.2500b	0.4250b	0.3750b
4	0.5875b	0.4375b	0.4250b	0.3750b	0.5500b	0.5000b

Using equation (7.43) to calculate the downwash velocity at the CP of panel 1 induced by the horseshoe vortex of panel 1 of the starboard wing:

$$
\begin{aligned}
w_{1,1s} = \frac{\Gamma_1}{4\pi} \Bigg\{ & \frac{1.0}{(0.1625b)(-0.0625b) - (0.0375b)(0.0625b)} \\
& \left[\frac{(0.1250b)(0.1625b) + (0.1250b)(0.0625b)}{\sqrt{(0.1625b)^2 + (0.0625b)^2}} \right. \\
& \left. - \frac{(0.1250b)(0.0375b) + (0.1250b)(-0.0625b)}{\sqrt{(0.0375b)^2 + (-0.0625b)^2}} \right] \\
& + \frac{1.0}{-0.0625b} \left[1.0 + \frac{0.1625b}{\sqrt{(0.1625b)^2 + (0.0625b)^2}} \right] \\
& - \frac{1.0}{0.0625b} \left[1.0 + \frac{0.0375b}{\sqrt{(0.0375b)^2 + (0.0625b)^2}} \right] \Bigg\} \\
= \frac{\Gamma_1}{4\pi b} & (-16.3533 - 30.9335 - 24.2319)
\end{aligned}
$$

Note that, as one would expect, each of the vortex elements induces a negative (downward) component of velocity at the control point. The student should visualize the flow induced by each segment of the horseshoe vortex to verify that a negative value for each of the components is intuitively correct. In addition, the velocity induced by the vortex trailing from A to ∞ is greatest in magnitude. Adding the components together, we find

$$
w_{1,1s} = \frac{\Gamma_1}{4\pi b} (-71.5187)
$$

The downwash velocity at the CP of panel 1 (of the starboard wing) induced by the horseshoe vortex of panel 1 of the port wing:

$$
\begin{aligned}
w_{1,1p} = \frac{\Gamma_1}{4\pi} \Bigg\{ & \frac{1.0}{(0.0375b)(0.0625b) - (0.1625b)(0.1875b)} \\
& \left[\frac{(-0.1250b)(0.0375b) + (0.1250b)(0.1875b)}{\sqrt{(0.0375b)^2 + (0.1875b)^2}} \right. \\
& \left. - \frac{(-0.1250b)(0.1625b) + (0.1250b)(0.0625b)}{\sqrt{(0.1625b)^2 + (0.0625b)^2}} \right] \\
& + \frac{1.0}{-0.1875b} \left[1.0 + \frac{0.0375b}{\sqrt{(0.0375b)^2 + (0.1875b)^2}} \right] \\
& - \frac{1.0}{-0.0625b} \left[1.0 + \frac{0.1625b}{\sqrt{(0.1625b)^2 + (0.0625b)^2}} \right] \Bigg\} \\
= \frac{\Gamma_1}{4\pi b} & (18.5150)
\end{aligned}
$$

Similarly, using equation (7.43) to calculate the downwash velocity at the CP of panel 2 induced by the horseshoe vortex of panel 4 of the starboard wing, we obtain

$$w_{2,4s} = \frac{\Gamma_4}{4\pi} \left\{ \frac{1.0}{(-0.0875b)(-0.3125b) - (-0.2125b)(-0.1875b)} \right.$$

$$\left[\frac{(0.1250b)(-0.0875b) + (0.1250b)(-0.1875b)}{\sqrt{(-0.0875b)^2 + (-0.1875b)^2}} \right.$$

$$\left. - \frac{(0.1250b)(-0.2125b) + (0.1250b)(-0.3125b)}{\sqrt{(-0.2125b)^2 + (-0.3125b)^2}} \right]$$

$$+ \frac{1.0}{0.1875b} \left[1.0 + \frac{-0.0875b}{\sqrt{(-0.0875b)^2 + (-0.1875b)^2}} \right]$$

$$\left. - \frac{1.0}{0.3125b} \left[1.0 + \frac{-0.2125b}{\sqrt{(-0.2125b)^2 + (-0.3125b)^2}} \right] \right\}$$

$$= \frac{\Gamma_4}{4\pi b} (-0.60167 + 3.07795 - 1.40061)$$

$$= \frac{\Gamma_4}{4\pi b} (1.0757)$$

Again, the student should visualize the flow induced by each segment to verify that the signs and the relative magnitudes of the components are individually correct.

Evaluating all of the various components (or influence coefficients), we find that at control point 1:

$$w_1 = \frac{1}{4\pi b} [(-71.5187\Gamma_1 + 11.2933\Gamma_2 + 1.0757\Gamma_3 + 0.3775\Gamma_4)_s$$
$$+ (+18.5150\Gamma_1 + 2.0504\Gamma_2 + 0.5887\Gamma_3 + 0.2659\Gamma_4)_p]$$

At CP 2:

$$w_2 = \frac{1}{4\pi b} [(+20.2174\Gamma_1 - 71.5187\Gamma_2 + 11.2933\Gamma_3 + 1.0757\Gamma_4)_s$$
$$+ (+3.6144\Gamma_1 + 1.1742\Gamma_2 + 0.4903\Gamma_3 + 0.2503\Gamma_4)_p]$$

At CP 3:

$$w_3 = \frac{1}{4\pi b} [(+3.8792\Gamma_1 + 20.2174\Gamma_2 - 71.5187\Gamma_3 + 11.2933\Gamma_4)_s$$
$$+ (+1.5480\Gamma_1 + 0.7227\Gamma_2 + 0.3776\Gamma_3 + 0.2179\Gamma_4)_p]$$

At CP 4:

$$w_4 = \frac{1}{4\pi b} [(+1.6334\Gamma_1 + 3.8792\Gamma_2 + 20.2174\Gamma_3 - 71.5187\Gamma_4)_s$$
$$+ (+0.8609\Gamma_1 + 0.4834\Gamma_2 + 0.2895\Gamma_3 + 0.1836\Gamma_4)_p]$$

Since it is a planar wing with no dihedral, the no-flow condition of equation (7.46) requires that

$$w_1 = w_2 = w_3 = w_4 = -U_\infty \alpha$$

Thus

$$-53.0037\Gamma_1 + 13.3437\Gamma_2 + 1.6644\Gamma_3 + 0.6434\Gamma_4 = -4\pi b U_\infty \alpha$$
$$+23.8318\Gamma_1 - 70.3445\Gamma_2 + 11.7836\Gamma_3 + 1.3260\Gamma_4 = -4\pi b U_\infty \alpha$$
$$+5.4272\Gamma_1 + 20.9401\Gamma_2 - 71.1411\Gamma_3 + 11.5112\Gamma_4 = -4\pi b U_\infty \alpha$$
$$+2.4943\Gamma_1 + 4.3626\Gamma_2 + 20.5069\Gamma_3 - 71.3351\Gamma_4 = -4\pi b U_\infty \alpha$$

Solving for Γ_1, Γ_2, Γ_3, and Γ_4, we find that

$$\Gamma_1 = +0.02728(4\pi bU_\infty\alpha) \tag{7.47a}$$

$$\Gamma_2 = +0.02869(4\pi bU_\infty\alpha) \tag{7.47b}$$

$$\Gamma_3 = +0.02841(4\pi bU_\infty\alpha) \tag{7.47c}$$

$$\Gamma_4 = +0.02490(4\pi bU_\infty\alpha) \tag{7.47d}$$

Having determined the strength of each of the vortices by satisfying the boundary conditions that the flow is tangent to the surface at each of the control points, the lift of the wing may be calculated. For wings that have no dihedral over any portion of the wing, all the lift is generated by the free-stream velocity crossing the spanwise vortex filament, since there are no sidewash or backwash velocities. Furthermore, since the panels extend from the leading edge to the trailing edge, the lift acting on the nth panel is

$$\Delta l_n = l = \rho_\infty U_\infty \Gamma_n \tag{7.48}$$

which is also the lift per unit span. Since the flow is symmetric, the total lift for the wing is

$$L = 2 \int_0^{0.5b} \rho_\infty U_\infty \Gamma(y)\, dy \tag{7.49a}$$

or, in terms of the finite-element panels,

$$L = 2\rho_\infty U_\infty \sum_{n=1}^{4} \Gamma_n\, \Delta y_n \tag{7.49b}$$

Since $\Delta y_n = 0.1250b$ for each panel,

$$L = 2\rho_\infty U_\infty 4\pi bU_\infty\alpha(0.02728 + 0.02869 + 0.02841 + 0.02490)0.1250b$$
$$= \rho_\infty U_\infty^2 b^2 \pi\alpha(0.10928)$$

To calculate the lift coefficient, recall that $S = bc$ and $b = 5c$ for this wing. Therefore,

$$C_L = \frac{L}{q_\infty S} = 1.0928\pi\alpha$$

Furthermore,

$$C_{L_\alpha} = \frac{dC_L}{d\alpha} = 3.43314 \text{ per radian} = 0.05992 \text{ per degree}$$

Comparing this value of C_{L_α} with that for an unswept wing (such as the results presented in Fig. 7.13), it is apparent that an effect of sweepback is the reduction in the lift-curve slope.

The theoretical lift curve generated using the VLM is compared in Fig. 7.28 with experimental results reported in Ref. 7.12. The experimentally determined values of the lift coefficient are for a wing of constant chord and of constant section, which was swept $45°$ and which had an aspect ratio of 5. The theoretical lift coefficients are in good agreement with the experimental values.

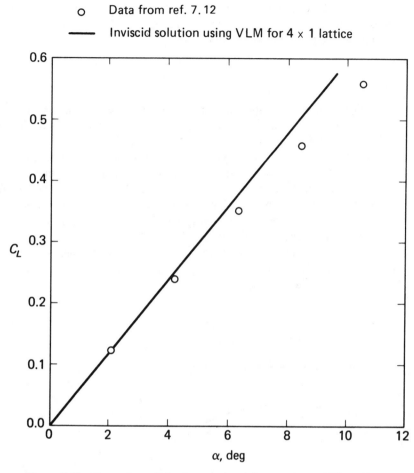

○   Data from ref. 7.12

──   Inviscid solution using VLM for 4 × 1 lattice

Figure 7.28 Comparison of the theoretical and the experimental lift coefficients for the swept wing of Fig. 7.27 in a subsonic stream.

Since the lift per unit span is given by equation (7.48), the section lift coefficient is

$$C_l = \frac{l}{\frac{1}{2}\rho_\infty U_\infty^2 c_{av}} = \frac{2\Gamma}{U_\infty c_{av}} \tag{7.50}$$

When the panels extend from the leading edge to the trailing edge, such as is the case for the 4 × 1 lattice shown in Fig. 7.27, the value of Γ given in equation (7.47) is used in equation (7.50). When there are a number of discrete panels in the chordwise direction, such as the 10 × 4 lattice shown in Fig. 7.20, you should sum (from the leading edge to the trailing edge) the values of Γ for those bound-vortex filaments at the spanwise location (i.e., in the chordwise strip) of interest. For a chordwise row,

$$\frac{C_l c}{c_{av}} = \sum_{j=1}^{J_{max}} \left(\frac{l}{q_\infty c_{av}}\right)_j \tag{7.51}$$

where c_{av} is the average chord (and is equal to S/b), c is the local chord, and j is the index for an elemental panel in the chordwise row. The total lift coefficient is obtained by integrating the lift over the span

$$C_L = \int_0^1 \frac{C_l c}{c_{av}} \, d\left(\frac{2y}{b}\right) \tag{7.52}$$

The spanwise variation in the section lift coefficient is presented in Fig. 7.29. The theoretical distribution is compared with the experimentally determined spanwise load distribution for an angle of attack of 4.2°, which was presented in Ref. 7.12. The increased loading of the outer wing sections promotes premature boundary-layer separation there. This unfavorable behavior is amplified by the fact that the spanwise velocity component causes the already decelerated fluid particles in the boundary layer to move toward the wing tips. This transverse flow results in a large increase in the boundary-layer thickness near the wing tips. Thus, at large angles of attack, premature separation may occur on the suction side of the wing near the tip. If significant tip stall occurs on the swept wing, there is a loss of the effectiveness of the control surfaces and a forward shift in the wing center of pressure that creates an unstable, nose-up increase in the pitching moment.

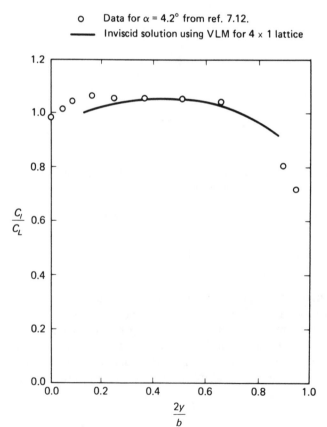

Figure 7.29 Comparison of the theoretical and the experimental spanwise lift distribution for the wing of Fig. 7.27.

Boundary-layer fences are often used to break up the spanwise flow on swept wings. The spanwise distribution of the local-lift coefficient (taken from Ref. 7.13) without and with a boundary-layer fence is presented in Fig. 7.30. The essential effect of the boundary-layer fence does not so much consist in the prevention of the transverse flow but, much more important, in that the fence divides each wing into an inner and an outer portion. Both transverse flow and boundary-layer separation may be present, but to a reduced extent. Boundary-layer fences are evident on the swept wings of the Trident, shown in Fig. 7.31.

Once we have obtained the solution for the section lift coefficient (i.e., that for a chordwise strip of the wing), the induced drag coefficient may be calculated using the relation given in Ref. 7.14:

$$C_{D_v} = \frac{1}{S} \int_{-0.5b}^{+0.5b} C_l c \alpha_i \, dy \tag{7.53}$$

where α_i, which is the induced incidence, given by

$$\alpha_i = -\frac{1}{8\pi} \int_{-0.5b}^{+0.5b} \frac{C_l c}{(y - \eta)^2} \, d\eta \tag{7.54}$$

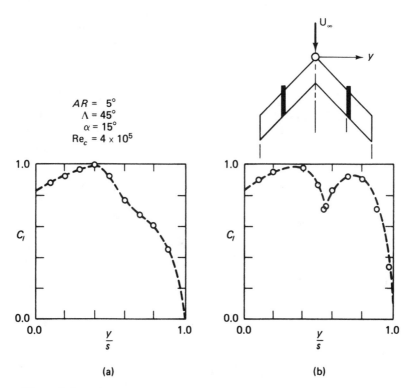

Figure 7.30 Effect of a boundary-layer fence on the spanwise distribution of the local lift coefficient: (a) without fence; (b) with fence. (Data from Ref. 7.13.)

Figure 7.31   Trident illustrating boundary-layer fences on the wing. (Courtesy of British Aerospace.)

For a symmetrical loading, equation (7.54) may be written

$$\alpha_i = -\frac{1}{8\pi} \int_0^{0.5b} \left[\frac{C_l c}{(y - \eta)^2} + \frac{C_l c}{(y + \eta)^2} \right] d\eta \tag{7.55}$$

Following the approach of Ref. 7.15, we consider the mth chordwise strip, which has a semiwidth of e_m and whose centerline is located at $\eta = y_m$. Let us approximate the spanwise lift distribution across the strip by a parabolic function:

$$\left(\frac{C_l c}{C_L \bar{c}} \right)_m = a_m \eta^2 + b_m \eta + c_m \tag{7.56}$$

To solve for the coefficients a_m, b_m, and c_m, note that

$$y_{m+1} = y_m + (e_m + e_{m+1})$$
$$y_{m-1} = y_m - (e_m + e_{m-1})$$

Thus,

$$c_m = \left(\frac{C_l c}{C_L \bar{c}} \right)_m - a_m \eta_m^2 - b_m \eta_m$$

$$a_m = \frac{1}{d_{mi} d_{mo} (d_{mi} + d_{mo})} \left\{ d_{mi} \left(\frac{C_l c}{C_L \bar{c}} \right)_{m+1} - (d_{mi} + d_{mo}) \left(\frac{C_l c}{C_L \bar{c}} \right)_m + d_{mo} \left(\frac{C_l c}{C_L \bar{c}} \right)_{m-1} \right\}$$

and

$$b_m = \frac{1}{d_{mi}d_{mo}(d_{mi} + d_{mo})}\left\{ d_{mo}(2\eta_m - d_{mo})\left[\left(\frac{C_l c}{C_L \bar{c}}\right)_m - \left(\frac{C_l c}{C_L \bar{c}}\right)_{m-1}\right]\right.$$

$$\left. - d_{mi}(2\eta_m - d_{mi})\left[\left(\frac{C_l c}{C_L \bar{c}}\right)_{m+1} - \left(\frac{C_l c}{C_L \bar{c}}\right)_m\right]\right\}$$

where

$$d_{mi} = e_m + e_{m-1}$$

and

$$d_{mo} = e_m + e_{m+1}$$

For a symmetric load distribution, we let

$$\left(\frac{C_l c}{C_L \bar{c}}\right)_{m-1} = \left(\frac{C_l c}{C_L \bar{c}}\right)_m$$

and

$$e_{m-1} = e_m$$

at the root. Similarly, we let

$$\left(\frac{C_l c}{C_L \bar{c}}\right)_{m+1} = 0$$

and

$$e_{m+1} = 0$$

at the tip. Substituting these expressions into equations (7.55) and (7.56), we then obtain the numerical form for the induced incidence:

$$\frac{\alpha_i(y)}{C_L \bar{c}} = -\frac{1}{4\pi}\sum_{m=1}^{N}\left\{\frac{y^2(y_m + e_m)a_m + y^2 b_m + (y_m + e_m)c_m}{y^2 - (y_m + e_m)^2}\right.$$

$$- \frac{y^2(y_m - e_m)a_m + y^2 b_m + (y_m - e_m)c_m}{y^2 - (y_m - e_m)^2}$$

$$+ \frac{1}{2}y a_m \log\left[\frac{(y - e_m)^2 - y_m^2}{(y + e_m)^2 - y_m^2}\right]^2$$

$$\left. + \frac{1}{4}b_m \log\left[\frac{y^2 - (y_m + e_m)^2}{y^2 - (y_m - e_m)^2}\right]^2 + 2e_m a_m\right\}$$

(7.57)

We then assume that the product $C_l c \alpha_i$ also has a parabolic variation across the strip.

$$\left[\left(\frac{C_l c}{C_L \bar{c}}\right)\left(\frac{\alpha_i}{C_L \bar{c}}\right)\right]_n = a_n y^2 + b_n y + c_n$$

(7.58)

The coefficients a_n, b_n, and c_n can be obtained using an approach identical to that employed to find a_m, b_m, and c_m. The numerical form of equation (7.53) is then a generalization of Simpson's rule:

$$\frac{C_{D_v}}{C_L^2} = \frac{4}{\text{AR}} \sum_{n=1}^{N} e_n \left\{ \left[y_n^2 + \left(\frac{1}{3}\right) e_n^2 \right] a_n + y_n b_n + c_n \right\} \qquad (7.59)$$

The lift developed along the chordwise bound vortices in a chordwise row of horseshoe vortices varies from the leading edge to the trailing edge of the wing because of the longitudinal variation of both the sidewash velocity and the local value of the vortex strength for planforms that have a nonzero dihedral angle. For techniques to compute the lift, the pitching moment, and the rolling moment for more general wings, the reader is referred to Ref. 7.16.

Numerous investigators have studied methods for improving the convergence and the rigor of the VLM. Therefore, there is an ever-expanding body of relevant literature with which the reader should be familiar before attempting to analyze complex flow fields. Furthermore, as noted in the introduction to this section, VLM is only one (and, indeed, a relatively simple one) of the methods used to compute the aerodynamic coefficients.

DELTA WINGS

As has been discussed, a major aerodynamic consideration in wing design is the prediction and the control of flow separation. However, as the sweep angle is increased and the section thickness is decreased in order to avoid undesirable compressibility effects, it becomes increasingly more difficult to prevent boundary-layer separation. Although many techniques have been discussed to alleviate these problems, it is often necessary to employ rather complicated variable-geometry devices in order to satisfy a wide range of conflicting design requirements which result due to the flow-field variations for the flight envelope of high-speed aircraft. Beginning with the delta-wing design of Alexander Lippisch in Germany during World War II, supersonic aircraft designs have often used thin, highly swept wings of low aspect ratio to minimize the wave drag at the supersonic cruise conditions. It is interesting to note that during the design of the world's first operational jet fighter, the Me 262, the outer panels of the wing were swept to resolve difficulties arising when increasingly heavier turbojets caused the center of gravity to move. Thus, the introduction of sweepback did not reflect an attempt to reduce the effects of compressibility (Ref. 7.17). This historical note is included here to remind the reader that many parameters enter into the design of an airplane; aerodynamics is only one of them. The final configuration will reflect design priorities and trade-offs.

At subsonic speeds, however, delta-wing planforms have aerodynamic characteristics which are substantially different from those of the relatively straight, high-aspect-ratio wings designed for subsonic flight. Because they operate at relatively high angles of attack, the boundary layer on the lower surface flows outward and separates as it goes over the leading edge, forming a free shear layer. The shear layer curves upward and inboard, eventually rolling up into a core of high vorticity, as shown in Fig. 7.32. There is an appreciable axial component of motion and the fluid spirals around and along the axis. A spanwise outflow is induced on the upper surface, beneath the coiled vortex sheet, and the flow separates again as it approaches the leading edge (Ref. 7.18). The size and the strength of the coiled vortex sheets increase with increasing incidence and they become a dominant feature of the flow, which remains steady throughout the range of practical flight attitudes of the wing. The formation of these vortices is responsible for the nonlinear aerodynamic characteristics that exist over the angle-of-attack range.

The leading-edge suction analogy developed by Polhamus (e.g., Refs. 7.19 and 7.20) can be used to calculate the lift and the drag-due-to-lift characteristics which

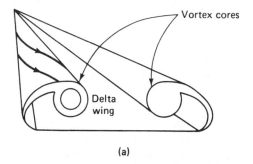

(a)

(b)

Figure 7.32   (a) Vortex core that develops for flow over a delta wing. (b) Water vapor condenses due to the pressure drop revealing the vortex core for a General Dynamic F-16. (Photograph courtesy of General Dynamics.)

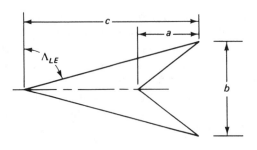

Figure 7.33 Wing geometry nomenclature.

arise when the separated flow around sharp-edge delta wings reattaches on the upper surface. Thus, the correlations apply to thin wings having neither camber nor twist. Furthermore, the method is applicable to wings for which the leading edges are of sufficient sharpness that separation is fixed at the leading edge. Since the vortex flow induces reattachment and since the Kutta condition must, therefore, still be satisfied at the trailing edge, the total lift coefficient consists of a potential-flow term and a vortex-lift term. The total lift coefficient can be represented by the sum

$$C_L = K_p \sin \alpha \cos^2 \alpha + K_v \sin^2 \alpha \cos \alpha \qquad (7.60)$$

The constant K_p is simply the normal-force slope calculated using the potential-flow lift-curve slope. The constant K_v can be estimated from the potential flow leading-edge suction calculations. Using the nomenclature for arrow-, delta-, and diamond-planform wings illustrated in Fig. 7.33, K_p and K_v are presented as a function of the planform parameters in Figs. 7.34 and 7.35, respectively.

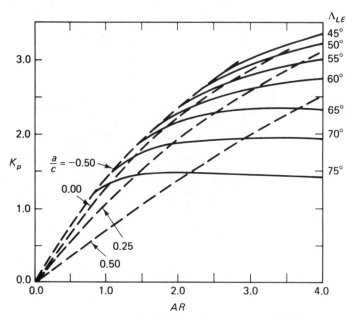

Figure 7.34 Variation of potential-flow lift constant with planform parameters. (From Ref. 7.20.)

Values of the lift coefficient calculated using equation (7.60) are compared in Fig. 7.36 with experimental values for uncambered delta wings that have sharp leading edges. Data are presented for wings of aspect ratio from 1.0 (Ref. 7.21) to 2.0 (Ref. 7.22). Since the analytical method is based on an analogy with potential-flow leading-edge suction, which requires that flow reattaches on the upper surface inboard of the vortex, the correlation between theory and data breaks down as flow reattachment fails to occur. The lift coefficients calculated using equation (7.60) for $AR = 1.0$ and 1.5 are in good agreement with the measured values up to angles of attack in excess of 20°. However, for the delta wing, whose aspect ratio is 2.0, significant deviations between the calculated values and the experimental values exist for angles of attack above 15°.

If the leading edges of the wing are rounded, separation occurs well inboard on the wing's upper surface. The result is that, outboard of the loci of the separation points, the flow pattern is essentially that of potential flow and the peak negative pressures at the section extremity are preserved. However, as noted in Ref. 7.21, increasing the thickness causes a reduction in net lift. The combined effect of the

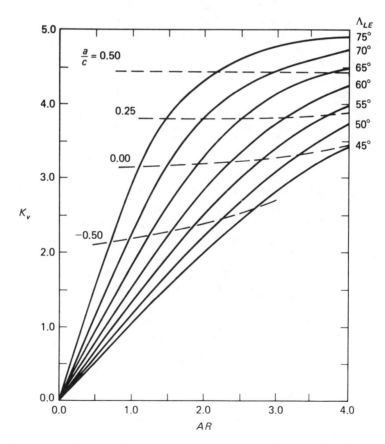

Figure 7.35   Variation of vortex-lift constant with planform parameters. (From Ref. 7.20.)

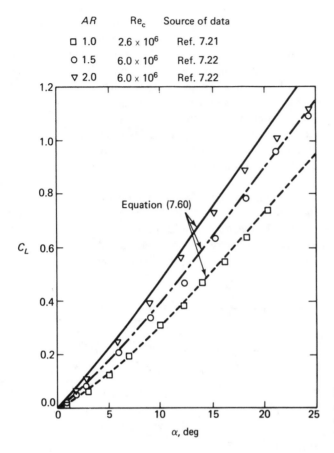

AR	Re_c	Source of data
□ 1.0	2.6×10^6	Ref. 7.21
○ 1.5	6.0×10^6	Ref. 7.22
▽ 2.0	6.0×10^6	Ref. 7.22

Equation (7.60)

C_L

α, deg

Figure 7.36 Comparison of the calculated and the experimental lift coefficients for thin, flat delta wings with sharp leading-edges.

thickness ratio and of the shape of the leading edges is illustrated in the experimental lift coefficients presented in Fig. 7.37, which are taken from Ref. 7.22. The lift coefficients are less for the thicker wings, which also have rounded leading edges.

The lift coefficients for a series of delta wings are presented as a function of the angle of attack in Fig. 7.38. The lift-curve slope, $dC_L/d\alpha$, becomes progressively smaller as the aspect ratio decreases. However, for all but the smallest aspect ratio wing, the maximum value of the lift coefficient and the angle of attack at which it occurs increase as the aspect ratio decreases.

For a thin flat-plate model, the resultant force acts normal to the surface. Thus, the induced drag ΔC_D for a flat-plate wing would be

$$\Delta C_D = C_D - C_{D0} = C_L \tan \alpha \qquad (7.61)$$

Using equation (7.60) to evaluate C_L,

$$C_D = C_{D0} + K_p \sin^2 \alpha \cos \alpha + K_v \sin^3 \alpha \qquad (7.62)$$

Experimental values of the drag coefficient (Ref. 7.22) are compared with the correlation of equation (7.61) in Fig. 7.39. The experimental drag coefficient increases with angle

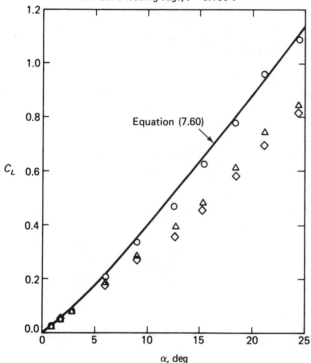

O Beveled (sharp) leading-edge, $t = 0.050\,c$
◇ Elliptical leading-edge, $t = 0.075\,c$
△ Round leading-edge, $t = 0.100\,c$

Equation (7.60)

C_L

α, deg

Figure 7.37 Effect of the leading-edge shape on the measured lift coefficient for thin, flat delta wings for which AR = 1.5, $Re_c = 6 \times 10^6$. (Data from Ref. 7.22.)

of attack (see Fig. 7.37). The correlation is best for the higher values of the lift coefficient.

The flow field over a delta wing is such that the resultant pressure distribution produces a large nose-down (negative) pitching moment about the apex, as illustrated by the experimental values from Ref. 7.22 presented in Fig. 7.40. The magnitude of the negative pitching moment increases as the angle of attack is increased. The resultant aerodynamic coefficients present a problem relating to the low-speed performance of a delta-wing aircraft which is designed to cruise at supersonic speeds, since the location of the aerodynamic center for subsonic flows differs from that for supersonic flows. At low speeds (and, therefore, at relatively low values of the dynamic pressure), delta wings must be operated at relatively high angles of attack in order to generate sufficient lift, since

$$L = \tfrac{1}{2}\rho_\infty U_\infty^2 \, S C_L$$

However, if the wing is at an angle of attack that is high enough to produce the desired C_L, a large nose-down pitching moment results. Thus, the basic delta configuration is often augmented by a lifting surface in the nose region (called a canard) which provides a nose-up trimming moment. The canards may be fixed, such as those shown in Fig. 7.41 on North American's XB-70 Valkyrie, or retractable, such as those on the Dassault

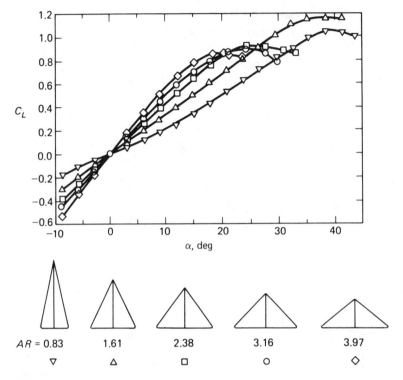

Figure 7.38 Lift coefficients for delta wings of various aspect ratios; $t = 0.12c$, $\mathrm{Re}_c \approx 7 \times 10^5$. (Data from Ref. 7.3.)

Mirage. An alternative design approach, which is used on the Space Shuttle Orbiter, uses a reflexed wing trailing edge (i.e., negative camber) to provide the required trimming moment.

As noted in Ref. 7.23: "the proper use of canard surfaces on a maneuvering aircraft can offer several attractive features such as potentially higher trimmed-lift capability, improved pitching moment characteristics, and reduced trimmed drag. In addition, the geometric characteristics of close-coupled canard configurations offer a potential for improved longitudinal progression of cross-sectional area which could result in reduced wave drag at low supersonic speeds and placement of the horizontal control surfaces out of the high wing downwash and jet exhaust." These benefits are primarily associated with the additional lift developed by the canard and with the beneficial interaction between the canard flow-field and that of the wing. These benefits may be accompanied by a longitudinal instability (or pitch up) at the higher angles of attack because of the vortex lift developed on the forward canards.

LEADING-EDGE EXTENSIONS

The designs of lightweight fighters that can cruise supersonically and maneuver transonically employ additional, highly swept areas ahead of the main wing. These forward areas are called strakes, gloves, fillets, apex regions, and leading-edge extensions. The

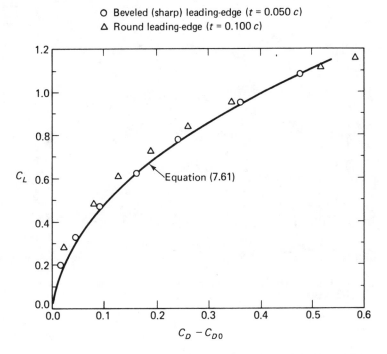

O Beveled (sharp) leading-edge ($t = 0.050\,c$)
△ Round leading-edge ($t = 0.100\,c$)

Figure 7.39 Drag correlation for thin, flat delta wings for which AR $=$ 1.5, Re$_c$ $= 6 \times 10^6$. (Data from Ref. 7.22.)

O Beveled (sharp) leading-edge, $t = 0.050\,c$
△ Round leading-edge, $t = 0.100\,c$

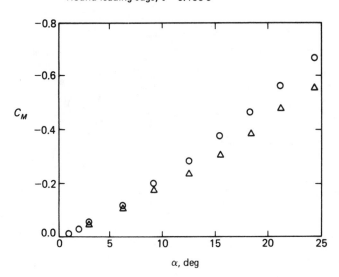

Figure 7.40 Moment coefficient (about the apex) for thin, flat delta wings for which AR $=$ 1.5, Re$_c$ $= 6 \times 10^6$. (Data from Ref. 7.22.)

Figure 7.41 North American XB-70 illustrating the use of canards. (Courtesy of NASA.)

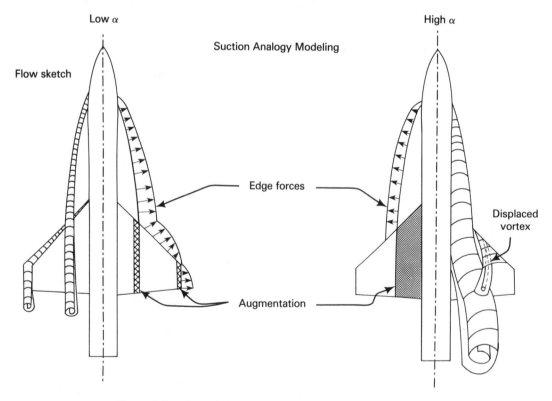

Figure 7.42 Theoretical vortex model for strake/wing configuration. (From Ref. 7.24.)

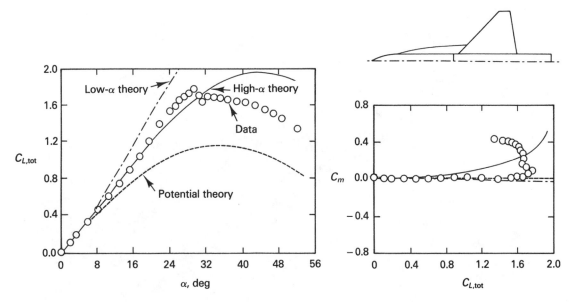

Figure 7.43 Predicted and measured $C_{L,tot}$ and C_m for strake/wing configuration at $M_\infty = 0.2$. (From Ref. 7.24.)

strake/wing combinations are also known by different names (e.g., ogee and double delta).

Lamar and Frink (Ref. 7.24) note that the mutual benefits derived from strake/ wing configurations "include for the wing: (1) minimal interference at or below the cruise angle of attack, (2) energizing of the upper surface boundary layer with the resulting flow reattachment on the outer wing panel at moderate to high angle of attack due to the strake vortex, and (3) reduced area required for maneuver lift. Benefits for the strake are (1) upwash from the main wing strengthens the strake vortex and (2) the need for only a small area (hence wetted area and comparatively lightweight structure) to generate its significant contribution to the total lift, because the strake provides large amounts of vortex lift. It should be mentioned that these vortex-induced benefits are realized when the strake vortex is stable and maintaining a well-organized vortex system over the wing. Once the angle of attack becomes sufficiently large that strake-vortex breakdown progresses ahead of the wing trailing edge, these favorable effects deteriorate significantly."

A method for estimating the aerodynamic forces and moments for strake/wing/ body configurations estimates the vortex flow effects with the suction analogy and the basic potential flow effects with the vortex-lattice method (Ref. 7.25). As noted by Lamar and Frink (Ref. 7.24) and illustrated in Fig. 7.42, "at low angle of α, it was concluded that the configuration had attached flow away from the leading and side edges, and downstream from the component tips, vortical flow existed. However, at the higher values of α, the strake vortex becomes much larger and tends to displace the wing-vortex-flow system off the wing, so that this system can no longer cause flow reattachment to occur on the wing. This lack of reattachment causes a large portion of theoretically available aerodynamic effects to be effectively lost to the configuration."

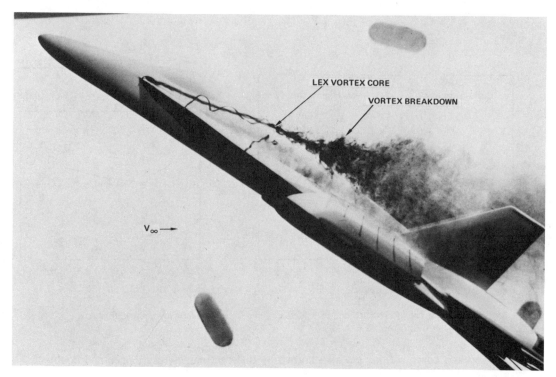

Figure 7.44   Flow visualization of a LEX vortex on a Subscale Advanced Fighter Model; $\alpha = 32°$, Reynolds number = 10^4 from the Northrop Water Tunnel. (From Ref. 7.26.)

The two theoretical solutions along with a potential flow solution are compared with data (as taken from Ref. 7.24) in Fig. 7.43. The C_L data are better estimated by the high-α theory. However, there may be an angle-of-attack range below the α for $C_{L,\max}$ for which the C_L data are underpredicted. This is most likely associated with the exclusion of any vortex lift from the wing in the high-α theory.

A flow visualization photograph of the vortices shed from the sharp leading edges of slender wings and wing leading-edge extensions (LEXs) or wing-body strakes at high angles of attack is presented in Fig. 7.44. The data were obtained in the Northrop Water Tunnel. Although the Reynolds number was relatively low, the separation point does not vary with Reynolds number, provided that flow separation occurs from a sharp leading edge. Thus, Erickson (Ref. 7.26) concludes that "the water tunnel provides a good representation of the wake shed from a wing."

As noted in Ref. 7.27, the original T-38 wing planform did not include a leading-edge extension (LEX) at the wing root. The effects of the LEX in terms of maximum trimmed angle of attack and trimmed $C_{L_{\max}}$ are illustrated in Fig. 7.45. The F-5E LEX with an area increase of 4.4% of the wing reference area provides a $C_{L_{\max}}$ increment of 38% of the no-LEX value. Northrop studies suggest a practical LEX upper size limit for any given configuration and the F-5E LEX is close to this value. The higher

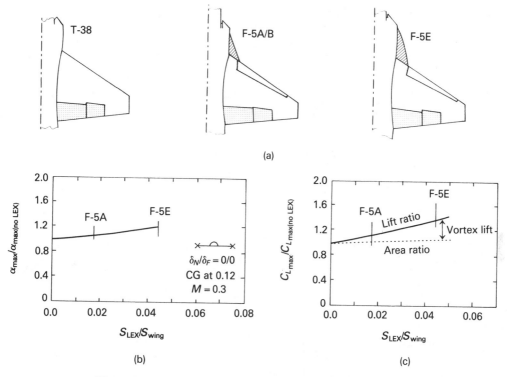

Figure 7.45 Effect of wing leading-edge extentions on the maximum angle of attack and the minimum lift for the F-5 family of aircraft: (a) configurations; (b) maximum angle of attack; (c) maximum lift. (From Ref. 7.27.)

trimmed angle of attack capability increases air combat effectiveness. The higher drag associated with the higher angle of attack is very useful in producing overshoots of the attacking aircraft. Although strakes (or leading-edge extensions) produce significant increases in the maximum angle of attack and in the maximum lift coefficient, the load distribution may concurrently produce large nose-up (positive) pitching moments at the high angles of attack.

ASYMMETRIC LOADS ON THE FUSELAGE AT HIGH ANGLES OF ATTACK

Ericsson and Reding (Ref. 7.28) noted that it has long been recognized that asymmetric vortex shedding can occur on bodies of revolution at high angles of attack. Large asymmetric loads can be induced on the body itself, even at zero sideslip. They note that experimental results have shown that the vortex-induced side force can exceed the normal force.

As a slender body of revolution is pitched through the angle-of-attack range from 0 to 90°, there are four distinct flow patterns that reflect the diminishing influence of

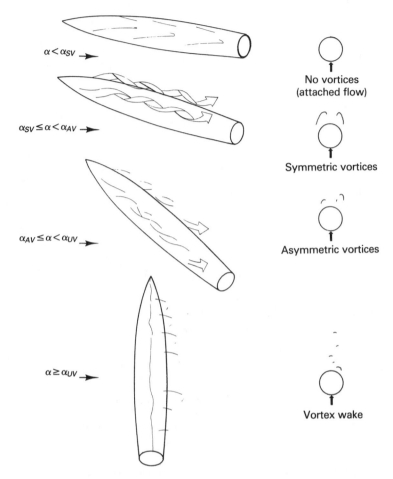

Figure 7.46 Effect of angle of attack on the leeside flowfield. (From Ref. 7.28.)

the axial flow component. Sketches of these four patterns are presented in Fig. 7.46. At low angles of attack ($0° \leq \alpha \leq \alpha_{SV}$), the axial-flow component dominates and the flow is attached, although the cross-flow effects will generate a thick viscous layer on the leeside. At intermediate angles of attack ($\alpha_{SV} \leq \alpha \leq \alpha_{AV}$) [i.e., the angle of attack is greater than the minimum for symmetric vortices (α_{SV}) but less than that for asymmetric vortices (α_{AV})], the cross-flow separates and generates a symmetric pair of vortices. In the angle-of-attack range where the vortex shedding is asymmetric ($\alpha_{AV} \leq \alpha \leq \alpha_{UV}$), as shown in Fig. 7.46c, the axial-flow component is still sufficient to produce steady vortices. However, the vortex pattern is asymmetric, producing a side force and a yawing moment, even at zero sideslip. Finally, at very high angles of attack ($\alpha_{UV} \leq \alpha \leq 90°$), the axial-flow component has less and less influence, so that the vortex shedding becomes unsteady, starting on the aft body and progressing toward the nose with increasing angle of attack. That is, the leeside flow resembles the wake of a two-dimensional cylinder normal to the flow.

Ericsson and Reding (Ref. 7.28) note that "the seriousness of the problem is illustrated by the wind-tunnel test results for the F-111, which show the vortex-induced yawing moment to exceed, by an order of magnitude, the available control capability through full rudder deflection. The problem can be cured by use of nose strakes or other types of body reshaping that cause the forebody vortices to be shed symmetrically."

SUMMARY

The shape of the wing is a dominant parameter in determining the performance and handling characteristics of an airplane. The planform of a wing is defined by the aspect ratio, sweep, and taper ratio. At subsonic speeds, the lift-dependent drag can be reduced by increasing the aspect ratio. A high aspect ratio produces a high-lift-curve slope, which is useful for takeoff and landing but has a weight penalty since, for a given wing area, more massive root sections are needed to handle the higher bending moments. Furthermore, a swept wing has a greater structural span than a straight wing of the same area and aspect ratio. Thus, the use of sweepback may also lead to a growth in wing weight. As a result, the choice of aspect ratio requires a trade study in which the designer may chose to emphasize one criterion at the expense of another (see the discussion relating to transport aircraft in Chapter 14).

Tapering the wing reduces the wing-root bending moments, since the inboard portion of the wing carries more of the wing's lift than the portion near the tip. While taper reduces the actual loads carried by the outboard sections, the wingtip sections are subjected to higher lift coefficients and tend to stall first (see Fig. 7.17). Thus, progressively reducing the incidence of the local section through the use of geometric twist from the root to the tip (i.e., washout) may be used either to reduce local loading or to prevent tip stalling.

PROBLEMS

7.1. Consider an airplane that weighs 14,700 N and cruises in level flight at 300 km/h at an altitude of 3000 m. The wing has a surface area of 17.0 m² and an aspect ratio of 6.2. Assume that the lift coefficient is a linear function of the angle of attack and that $\alpha_{0L} = -1.2°$. If the load distribution is elliptic, calculate the value of the circulation in the plane of symmetry (Γ_0), the downwash velocity (w_{y1}), the induced drag coefficient (C_{Dv}), the geometric angle of attack (α), and the effective angle of attack (α_e).

7.2. Consider the case where the spanwise circulation distribution for a wing is parabolic,

$$\Gamma(y) = \Gamma_0 \left(1 - \frac{y^2}{s^2} \right)$$

If the total lift generated by the wing with the parabolic circulation distribution is to be equal to the total lift generated by a wing with an elliptic circulation distribution, what is the relation between the Γ_0 values for the two distributions? What is the relation between the induced downwash velocities at the plane of symmetry for the two configurations?

7.3. When a GA(W)-1 airfoil section (i.e., a wing of infinite span) is at an angle of attack of 4°, the lift coefficient is 1.0. Using equation (7.20), calculate the angle of attack at which a wing whose aspect ratio is 7.5 would have to operate to generate the same lift coefficient. What would the angle of attack have to be to generate this lift coefficient for a wing whose aspect ratio is 6.0?

7.4. Consider a planar wing (i.e., no geometric twist) which has a NACA 0012 section and an aspect ratio of 7.0. Following Example 7.1, use a four-term series to represent the load distribution. Compare the lift coefficient, induced drag coefficient, and the spanwise lift distribution for taper ratios of (a) 0.4, (b) 0.5, (c) 0.6, and (d) 1.0.

7.5. Consider an airplane that weighs 10,000 N and cruises in level flight at 185 km/h at an altitude of 3.0 km. The wing has a surface area of 16.3 m², an aspect ratio of 7.52, and a taper ratio of 0.69. Assume that the lift coefficient is a linear function of the angle of attack and the airfoil section is a NACA 2412 (see Chapter 6 for the characteristics of this section). The incidence of the root section is $+1.5°$; the incidence of the tip section is $-1.5°$. Thus, there is a geometric twist of $-3°$ (washout). Following Example 7.1, use a four-term series to represent the load distribution and calculate:
  (a) The lift coefficient (C_L).
  (b) The spanwise load distribution ($C_l(y)/C_L$).
  (c) The induced drag coefficient (C_{D_v})
  (d) The geometric angle of attack (α).

7.6. Use equation (7.36) to calculate the velocity induced at some point $C(x, y, z)$ by the vortex filament AB (shown in Fig. 7.25); that is, derive equation (7.37a).

7.7. Use equation (7.36) to calculate the velocity induced at some point $C(x, y, 0)$ by the vortex filament AB in a planar wing; that is, derive equation (7.42a).

7.8. Calculate the downwash velocity at the CP of panel 4 induced by the horseshoe vortex of panel 1 of the starboard wing for the flow configuration of Example 7.2.

7.9. Following the VLM approach used in Example 7.2, calculate the lift coefficient for a swept wing. The wing has an aspect ratio of 8, a taper ratio of unity (i.e., $c_r = c_t$), and an uncambered section (i.e., it is a flat plate). Since the taper ratio is unity, the leading edge, the quarter-chord line, the three-quarter chord line, and the trailing edge all have the same sweep, 45°. How does the lift coefficient for this aspect ratio (8) compare with that for an aspect ratio of 5 (i.e., that computed in Example 7.2)? Is this consistent with our knowledge of the effect of aspect ratio (e.g., Fig. 7.9)?

7.10. Following the VLM approach used in Example 7.2, calculate the lift coefficient for a swept wing. The wing has an aspect ratio of 5, a taper ratio of 0.5 (i.e., $c_t = 0.5c_r$), an uncambered section, and the quarter-chord is swept 45°. Since the taper ratio is not unity, the leading edge, the quarter-chord line, the three-quarter chord line, and the trailing edge have different sweep angles. This should be taken into account when defining the coordinates of the horseshoe vortices and the control points.

7.11. Following the VLM approach used in Example 7.2, calculate the lift coefficient for the forward swept wing of Fig. 7.19b. The quarter chord is swept forward 45°, the aspect ratio is 3.55 and the taper ratio 0.5. The airfoil section (perpendicular to the quarter chord) is an NACA 64A112. For this airfoil section $\alpha_{ol} = -0.94°$ and $C_{l,\alpha} = 6.09$ per radian. For purposes of applying the no-flow boundary condition at the control points, assume that the wing is planar. Prepare a graph of the lift coefficient. How does this compare with that of Fig. 7.19?

7.12. Following the VLM approach used in Example 7.2, calculate the lift coefficient for a delta wing whose aspect ratio is 1.5. What is the sweep angle of the leading edge? The fact that the quarter-chord and the three-quarter chord lines have different sweeps should be taken into account when defining the coordinates of the horseshoe vortices and the control points. How do the calculated values for the lift coefficient compare with the experimental values presented in Fig. 7.36?

7.13. Use equation (7.60) to calculate the lift coefficient as a function of the angle of attack for a flat delta wing with sharp leading edges. The delta wing has an aspect ratio of 1.5. Compare the solution with the data of Fig. 7.36.

7.14. Use equation (7.61) to calculate the induced drag for a flat delta wing with sharp leading edges. The delta wing has an aspect ratio of 1.5. Compare the solution with the data of Fig. 7.39.

REFERENCES

7.1. Spreiter, J. R., and A. H. Sacks, "The Rolling Up of the Trailing Vortex Sheet and Its Effect on the Downwash Behind Wings," *Journal of the Aeronautical Sciences*, Jan. 1951, Vol. 18, No. 1, pp. 21–32.

7.2. Prandtl, L.,and O. G. Tietjens, *Applied Hydro- and Aeromechanics*, Dover Publications, New York, 1957.

7.3. Schlichting, H., and E. Truckenbrodt, *Aerodynamik des Flugzeuges*, Vol. 2, Chap. 7, Springer-Verlag, Berlin, 1969.

7.4. Prandtl, L., "Applications of Modern Hydrodynamics to Aeronautics," *Report 116*, NACA, 1921.

7.5. Abbott, I. H., and A. E. von Doenhoff, *Theory of Wing Sections*, Dover Publications, New York, 1949.

7.6. Sivells, J. C., "Experimental and Calculated Characteristics of Three Wings of NACA 64–210 and 65–210 Airfoil Sections with and without Washout," *TN 1422*, NACA, Aug. 1947.

7.7 Bristow, D. R., and G. G. Grose, "Modification of the Douglas Neumann Program to Improve the Efficiency of Predicting Component and High Lift Characteristics," *CR 3020*, NASA, Aug. 1978.

7.8. Margason, R. J., S. O. Kjelgaard, W. L. Sellers III, C. E. K. Morris, Jr., K. B. Walkey, and E. W. Shields, "Subsonic Panel Methods—A Comparison of Several Production Codes," *AIAA Paper 85–0280*, presented at the AIAA 23rd Aerospace Sciences Meeting, Reno, Nev., Jan. 1985.

7.9. Falkner, V. M., "The Calculation of Aerodynamic Loading on Surfaces of Any Shape,"*Reports and Memoranda 1910*, Aeronautical Research Committee, Aug. 1943.

7.10. Kalman, T. K., W. P. Rodden, and J. P. Giesing, "Application of the Doublet-Lattice Method to Nonplanar Configurations in Subsonic Flow," *Journal of Aircraft*, June 1971, Vol. 8, No. 6, pp. 406–413.

7.11. Robinson, A., and J. A. Laurmann, *Wing Theory*, Chap. 1, Cambridge University Press, Cambridge, 1956.

7.12. Weber, J., and G. G. Brebner, "Low-Speed Tests on 45-deg Swept-Back Wings, Part I: Pressure Measurements on Wings of Aspects Ratio 5," *Reports and Memoranda 2882*, Aeronautical Research Council, 1958.

7.13. Schlichting, H., "Some Developments in Boundary Layer Research in the Past Thirty

Years," *Journal of the Royal Aeronautical Society*, Feb. 1960, Vol. 64, No. 590, pp. 64–79.

7.14. Multhopp, H., "Method for Calculating the Lift Distribution of Wings (Subsonic Lifting Surface Theory)," *Reports and Memoranda 2884*, Aeronautical Research Council, Jan. 1950.

7.15. Kalman, T. P., J. P. Giesing, and W. P. Rodden, "Spanwise Distribution of Induced Drag in Subsonic Flow by the Vortex Lattice Method," *Journal of Aircraft*, Nov.–Dec. 1970, Vol. 7, No. 6, pp. 574–576.

7.16. Margason, R. J., and J. E. Lamar, "Vortex-Lattice Fortran Program for Estimating Subsonic Aerodynamic Characteristics of Complex Planforms," *TN D-6142*, NASA, Feb. 1971.

7.17. Voight, W., "Gestation of the Swallow," *Air International*, Mar. 1976, Vol. 10, No. 3, pp. 135–139, 153.

7.18. Stanbrook, A., and L. C. Squire, "Possible Types of Flow at Swept Leading Edges," *The Aeronautical Quarterly*, Feb. 1964, Vol. 15, pp. 72–82.

7.19. Polhamus, E. C., "Predictions of Vortex-Lift Characteristics by a Leading-Edge Suction Analogy," *Journal of Aircraft*, Apr. 1971, Vol. 8, No. 4, pp. 193–199.

7.20. Polhamus, E. C., "Charts for Predicting the Subsonic Vortex-Lift Characteristics of Arrow, Delta, and Diamond Wings," *TN D-6243*, NASA, Apr. 1971.

7.21. Peckham, D. H., "Low-Speed Wind-Tunnel Tests on a Series of Uncambered Slender Pointed Wings with Sharp Edges," *Reports and Memoranda 3186*, Aeronautical Research Council, Dec. 1958.

7.22. Bartlett, G. E., and R. J. Vidal, "Experimental Investigation of Influence of Edge Shape on the Aerodynamic Characteristics of Low Aspect Ratio Wings at Low Speeds," *Journal of the Aeronautical Sciences*, Aug. 1955, Vol. 22, No. 8, pp. 517–533, 588.

7.23. Gloss, B. B., and K. E. Washburn, "Load Distribution on a Close-Coupled Wing Canard at Transonic Speeds," *Journal of Aircraft*, Apr. 1978, Vol. 15, No. 4, pp. 234–239.

7.24. Lamar, J. E., and N. T. Frink, "Aerodynamic Features of Designed Strake-Wing Configurations," *Journal of Aircraft*, Aug. 1982, Vol. 19, No. 8, pp. 639–646.

7.25. Lamar, J. E., and B. B. Gloss, "Subsonic Aerodynamic Characteristics of Interacting Lifting Surfaces with Separated Flow Around Sharp Edges Predicted by a Vortex-Lattice Method," *TN D-7921*, NASA, Sept. 1975.

7.26. Erickson, G. E., "Vortex Flow Correlation," *ICAS Paper 82–6.61*, presented at the 13th Congress of the International Council of the Aeronautical Sciences, Aug. 1982. Also *NOR-82–158*, Northrop Corporation.

7.27. Stuart, W. G., "*Northrop F-5 Case Study in Aircraft Design*, AIAA Professional Study Series, Sept. 1978.

7.28. Ericsson, L. E. and J. P. Reding, "Asymmetric Vortex Shedding from Bodies of Revolution," in *Tactical Missile Aerodynamics*, edited by M. J. Hemsch and J. N. Nielsen, Progress in Aeronautics and Astronautics, Vol. 104 AIAA, 1986.

8

DYNAMICS
OF A COMPRESSIBLE,
INVISCID FLOW FIELD

Thus far, we have studied the aerodynamic forces for incompressible (constant density) flow past the vehicle. At low flight Mach numbers (e.g., below a free-stream Mach number M_∞ of approximately 0.5), Bernoulli's equation [equation (3.10)] provides the relation between the pressure distribution about an aircraft and the local velocity changes of the air as it flows around the various components of the vehicle. However, as the flight Mach number increases, changes in the local air density also affect the magnitude of the local static pressure. This leads to discrepancies between the actual aerodynamic forces and those predicted by incompressible flow theory. For our purposes, the Mach number is the parameter that determines the extent to which compressibility effects are important. The purpose of this chapter is to introduce those aspects of compressible flows (i.e., flows in which the density is not constant) that have applications to aerodynamics.

Throughout this chapter, air will be assumed to behave as a thermally perfect gas (i.e., the gas obeys the equation of state):

$$p = \rho RT \tag{1.3}$$

We will assume that the gas is also calorically perfect; that is, the specific heats, c_p and c_v, of the gas are constant. These specific heats are discussed further in the next section. The term *perfect gas* will be used to describe a gas that is both thermally and calorically perfect.

Even though we turn our attention in this and in the subsequent chapters to compressible flows, we may still divide the flow around the vehicle into (1) the boundary layer near the surface, where the effects of viscosity and heat conduction are important, and (2) the external flow, where the effects of viscosity and heat conduction can be neglected. As has been true in previous chapters, the inviscid flow is conservative.

THERMODYNAMIC CONCEPTS

Having reviewed the fundamentals of thermodynamics in Chapter 2 and having derived the energy equation therein, let us turn our attention to the related aerodynamic concepts.

Specific Heats

For simplicity (and without loss of generality), let us consider a system in which there are no changes in the kinetic and the potential energies. Equation (2.23) becomes

$$\delta q - p \, dv = du_e \tag{8.1}$$

As an extension of our discussion of fluid properties in Chapter 1, we note that, for any simple substance, the specific internal energy is a function of any other two independent fluid properties. Thus, consider $u_e = u_e(v, T)$. Then, by the chain rule of differentiation,

$$du_e = \left(\frac{\partial u_e}{\partial T}\right)_v dT + \left(\frac{\partial u_e}{\partial v}\right)_T dv \tag{8.2}$$

where the subscript is used to designate which variable is constant during the differentiation process.

From the principles of thermodynamics, it may be shown that for a thermally perfect gas, that is, one obeying equation (1.3),

$$\left(\frac{\partial u_e}{\partial v}\right)_T = 0$$

which is equivalent to saying that the internal energy of a perfect gas does not depend on the specific volume, or equivalently the density, and hence depends on the temperature alone. Thus, equation (8.2) becomes

$$du_e = \left(\frac{\partial u_e}{\partial T}\right)_v dT = c_v \, dT \tag{8.3}$$

In equation (8.3) we have introduced the definition that

$$c_v \equiv \left(\frac{\partial u_e}{\partial T}\right)_v$$

which is the *specific heat at constant volume*. It follows that

$$\Delta u_e = u_{e2} - u_{e1} = \int_1^2 c_v \, dT \tag{8.4}$$

Experimental evidence indicates that for most gases, c_v is constant over a wide range of conditions. For air below a temperature of approximately 850 K and over a wide range of pressure, c_v can be treated as a constant. The value for air is

$$c_v = 717.6 \frac{\text{N} \cdot \text{m}}{\text{kg} \cdot \text{K}} \left(\text{or} \frac{\text{J}}{\text{kg} \cdot \text{K}} \right) = 0.1717 \frac{\text{Btu}}{\text{lbm} \cdot {}^\circ\text{R}}$$

The assumption that c_v is constant is contained within the more general assumption that the gas is a perfect gas. Thus, for a perfect gas,

$$\Delta u_e = c_v \, \Delta T \qquad (8.5)$$

Since c_v and T are both properties of the fluid and since the change in a property between any two given states is independent of the process used in going between the two states, equation (8.5) is valid even if the process is not one of constant volume. Thus, equation (8.5) is valid for any simple substance undergoing any process where c_v can be treated as a constant.

Substituting equation (8.3) into equation (8.1), one sees that c_v is only directly related to the heat transfer if the process is one in which the volume remains constant (i.e., $dv = 0$). Thus, the name "specific heat" can be misleading. Physically, c_v is the proportionality constant between the amount of heat transferred to a substance and the temperature rise in the substance held at constant volume.

In analyzing many flow problems, the terms u_e and pv appear as a sum, and so it is convenient to define a symbol for this sum:

$$h \equiv u_e + pv \equiv u_e + \frac{p}{\rho} \qquad (8.6)$$

where h is called the *specific enthalpy*. Substituting the differential form of equation (8.6) into equation (8.1) and collecting terms yields

$$\delta q + v \, dp = dh \qquad (8.7)$$

which is the first law of thermodynamics expressed in terms of the enthalpy rather than the internal energy.

Since any property of a simple substance can be written as a function of any two other properties, we can write

$$h = h(p, T) \qquad (8.8)$$

Thus,

$$dh = \left(\frac{\partial h}{\partial p} \right)_T dp + \left(\frac{\partial h}{\partial T} \right)_p dT \qquad (8.9)$$

From the definition of the enthalpy, it follows that h is also a function of temperature only for a thermally perfect gas, since both u_e and p/ρ are functions of the temperature only. Thus,

$$\left(\frac{\partial h}{\partial p} \right)_T = 0$$

and equation (8.9) becomes

$$dh = \left(\frac{\partial h}{\partial T}\right)_p dT = c_p\, dT \tag{8.10}$$

We have introduced the definition

$$c_p \equiv \left(\frac{\partial h}{\partial T}\right)_p \tag{8.11}$$

which is the *specific heat at constant pressure*. In general, c_p depends on the composition of the substance and its pressure and temperature. It follows that

$$\Delta h = h_2 - h_1 = \int_1^2 c_p\, dT \tag{8.12}$$

Experimental evidence indicates that for most gases, c_p is essentially independent of temperature and of pressure over a wide range of conditions. Again, we conclude that, provided that the temperature extremes in a flow field are not too widely separated, c_p can be treated as a constant so that

$$\Delta h = c_p\, \Delta T \tag{8.13}$$

For air below a temperature of approximately 850 K, the value of c_p is 1004.7 N · m/ kg · K (or J/kg · K), which is equal to 0.2404 Btu/lbm · °R.

An argument parallel to that used for c_v shows that equation (8.13) is valid for any simple substance undergoing any process where c_p can be treated as a constant. Again, we note that the term ''specific heat'' is somewhat misleading, since c_p is only directly related to the heat transfer if the process is isobaric.

Additional Relations

A gas which is thermally and calorically perfect is one that obeys equations (1.3), (8.5), and (8.13). In such a case, there is a simple relation between c_p, c_v, and R. From the definitions of c_p and h, the perfect-gas law, and the knowledge that h depends upon T alone, we write

$$c_p \equiv \left(\frac{\partial h}{\partial T}\right)_p = \frac{dh}{dT} = \frac{du_e}{dT} + \frac{d}{dT}\left(\frac{p}{\rho}\right) = c_v + R \tag{8.14}$$

Let us introduce another definition, one for the ratio of specific heats:

$$\gamma \equiv \frac{c_p}{c_v} \tag{8.15}$$

For the most simple molecular model, the kinetic theory of gases shows that

$$\gamma = \frac{n + 2}{n}$$

where n is the number of degrees of freedom for the molecule. Thus, for a monatomic gas, such as helium, $n = 3$ and $\gamma = 1.667$. For a diatomic gas, such as nitrogen, oxygen, or air, $n = 5$ and $\gamma = 1.400$. Extremely complex molecules, such as Freon or tetrafluoromethane, have large values of n and values of γ which approach unity. In many treatments of air at high temperature and high pressure, approximate values of γ (e.g., 1.1 to 1.2) are used to approximate "real-gas" effects.

Combining equations (8.14) and (8.15), we can write

$$c_p = \frac{\gamma R}{\gamma - 1} \quad \text{and} \quad c_v = \frac{R}{\gamma - 1} \tag{8.16}$$

Second Law of Thermodynamics and Reversibility

The first law of thermodynamics does not place any constraints regarding what types of processes are physically possible and what types are not, providing that equation (2.22) is satisfied. However, we know from experience that not all processes permitted by the first law actually occur in nature. For instance, when one rubs sandpaper across a table, both the sandpaper and the table experience a rise in temperature. The first law is satisfied because the work done on the system by the sander's arm, which is part of the surroundings and which is, therefore, negative work for equation (2.23), is manifested as an increase in the internal energy of the system, which consists of the sandpaper and the table. Thus, the temperatures of the sandpaper and of the table increase. However, we do not expect that we can extract all the work back from the system and have the internal energy (and thus the temperature) decrease back to its original value, even though the first law would be satisfied. If this could occur, we would say the process was reversible, because the system and its surroundings would be restored to their original states.

The possibility of devising a reversible process, such as the type outlined above, cannot be settled by a theoretical proof. Experience shows that a truly reversible process has never been devised. This empirical observation is embodied in the *second law of thermodynamics*. For our purposes, an irreversible process is one that involves viscous (friction) effects, shock waves, or heat transfer through a finite temperature gradient. Thus, in regions outside boundary layers, viscous wakes, and planar shock waves, one can treat the flow as reversible. Note that the flow behind a curved shock wave can be treated as reversible only along a streamline.

The second law of thermodynamics provides a way to quantitatively determine the degree of reversibility (or irreversibility). Since the effects of irreversibility are dissipative and represent a loss of available energy (e.g., the kinetic energy of an aircraft wake, which is converted to internal energy by viscous stresses, is directly related to the aircraft's drag), the reversible process provides an ideal standard for comparison to real processes. Thus, the second law is a valuable tool available to the engineer.

There are several logically equivalent statements of the second law. In the remainder of this text we will usually be considering adiabatic processes, processes in which there is no heat transfer. This is not a restrictive assumption, since heat transfer in aerodynamic problems usually occurs only in the boundary layer and has a negligible effect on the flow in the inviscid region. The most convenient statement of the second law, for our purposes, is

$$ds \geq 0 \qquad (8.17)$$

for an adiabatic process. Thus, when a system is isolated from all heat exchange with its surroundings, s, the entropy of the system, either remains the same (if the process is reversible) or increases (if it is irreversible). It is not possible for a process to occur if the entropy of the system and its surroundings decreases. Thus, just as the first law led to the definition of internal energy as a property, the second law leads to the definition of entropy as a property.

The entropy change for a reversible process can be written as

$$\delta q = T \, ds$$

Thus, for a reversible process in which the only work done is that done at the moving boundary of the system,

$$T \, ds = du_e + p \, dv \qquad (8.18)$$

However, once we have written this equation, we see that it involves only changes in properties and does not involve any path-dependent functions. We conclude, therefore, that this equation is valid for all processes, both reversible and irreversible, and that it applies to the substance undergoing a change of state as the result of flow across the boundary of an open system (i.e., a control volume) as well as to the substance comprising a closed system (i.e., a control mass).

For a perfect gas, we can rewrite equation (8.18) as

$$ds = c_v \frac{dT}{T} + R \frac{dv}{v}$$

This equation can be integrated to give

$$s_2 - s_1 = c_v \ln \left\{ \left[\left(\frac{v_2}{v_1} \right)^{\gamma - 1} \right] \frac{T_2}{T_1} \right\} \qquad (8.19a)$$

Applying the equation of state for a perfect gas to the two states,

$$\frac{v_2}{v_1} = \frac{\rho_1}{\rho_2} = \frac{p_1}{p_2} \frac{T_2}{T_1}$$

equation (8.19a) can be written

$$s_2 - s_1 = R \ln \left\{ \left[\left(\frac{T_2}{T_1} \right)^{\gamma/(\gamma - 1)} \right] \frac{p_1}{p_2} \right\} \qquad (8.19b)$$

Equivalently,

$$s_2 - s_1 = c_v \ln \left\{ \left[\left(\frac{\rho_1}{\rho_2} \right)^{\gamma} \right] \frac{p_2}{p_1} \right\} \qquad (8.19c)$$

Using the various forms of equation (8.19), one can calculate the entropy change in terms of the properties of the end states.

In many compressible flow problems, the flow external to the boundary layer undergoes processes that are isentropic (i.e., adiabatic and reversible). If the entropy is constant at each step of the process, it follows from equation (8.19) that p, ρ, and T are interrelated. The following equations describe these relations for isentropic flow:

$$\frac{p}{\rho^{\gamma}} = \text{constant} \tag{8.20a}$$

$$\frac{T^{\gamma/(\gamma-1)}}{p} = \text{constant} \tag{8.20b}$$

and

$$Tv^{(\gamma-1)} = \text{constant} \tag{8.20c}$$

Speed of Sound

From experience, we know that the speed of sound in air is finite. To be specific, the *speed of sound* is defined as the rate at which infinitesimal disturbances are propagated from their source into an undisturbed medium. These disturbances can be thought of as small pressure pulses generated at a point and propagated in all directions. We shall learn later that finite disturbances such as shock waves propagate at a greater speed than that of sound waves.

Consider a motionless point source of disturbance in quiescent, homogeneous air (Fig. 8.1a). Small disturbances generated at the point move outward from the point in a spherically symmetric pattern. The distance between wave fronts is determined by the frequency of the disturbance. Since the disturbances are small, they leave the air behind them in essentially the same state it was before they arrived. The radius of a given wave front is given by

$$r = at \tag{8.21}$$

where a is the speed of propagation (speed of sound) of the wave front and t is the time since the particular disturbance was generated.

Now, suppose that the point source begins moving (Fig. 8.1b) from right to left at a constant speed U which is less than the speed of sound a. The wave-front pattern will now appear as shown in Fig. 8.1b. A stationary observer ahead of the source will detect an increase in frequency of the sound while one behind it will note a decrease. Still, however, each wave front is separate from its neighbors.

If the speed of the source reaches the speed of sound in the undisturbed medium, the situation will appear as shown in Fig. 8.1c. We note that the individual wave fronts are still separate, except at the point where they coalesce.

A further increase in source speed, such that $U > a$, leads to the situation depicted in Fig. 8.1d. The wave fronts now form a conical envelope, which is known as the *Mach cone*, within which the disturbances can be detected. Outside of this "zone of action" is the "zone of silence," where the pulses have not arrived.

We see that there is a fundamental difference between subsonic ($U < a$) and supersonic ($U > a$) flow. In subsonic flow, the effect of a disturbance propagates upstream of its location, and thus the upstream flow is "warned" of the approach of the disturbance. In supersonic flow, however, no such "warning" is possible. Stating

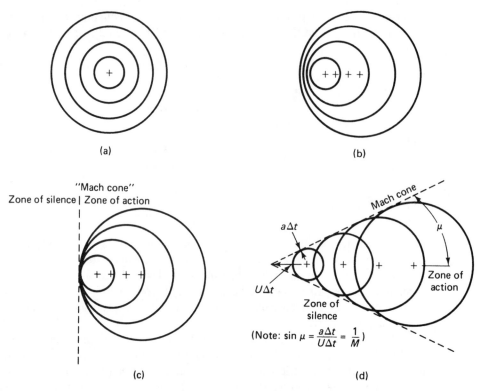

Figure 8.1 Wave pattern generated by pulsating disturbance of infinitesimal strength; (a) disturbance is stationary ($U = 0$); (b) disturbance moves to the left at subsonic speed ($U < a$); (c) disturbance moves to the left at sonic speed ($U = a$); (d) disturbance moves to the left at supersonic speed ($U > a$).

it another way: disturbances cannot propagate upstream in a supersonic flow relative to a source-fixed observer. This fundamental difference between the two types of flow has significant consequences on the flow field about a vehicle.

We note that the half-angle of the Mach cone dividing the zone of silence from the zone of action is given by

$$\sin \mu = \frac{1}{M} \tag{8.22}$$

where

$$M \equiv \frac{U}{a}$$

is the Mach number. At $M = 1$ (i.e., when $U = a$), $\mu = \pi/2$, and as $M \to \infty$, $\mu \to 0$.

To determine the speed of sound a, consider the wave front in Fig. 8.1a propagating into still air. A small portion of curved wave front can be treated as planar. To an observer attached to the wave, the situation appears as shown in Fig. 8.2. A control

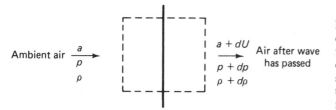

Ambient air $\dfrac{a}{p}$

ρ

$a + dU$ Air after wave
$\overrightarrow{\hspace{1cm}}$ has passed
$p + dp$
$\rho + d\rho$

Figure 8.2 Control volume used to determine the speed of sound. (A velocity of equal magnitude and of opposite direction has been superimposed on a wave of Fig. 8.1a so that the sound wave is stationary in this figure.)

volume is also shown attached to the wave. The boundaries of the volume are selected so that the flow is normal to faces parallel to the wave and tangent to the other faces. We make the key assumption (borne out by experiment) that, since the strength of the disturbance is infinitesimal, a fluid particle passing through the wave undergoes a process that is both reversible and adiabatic (i.e., isentropic).

The integral forms of the continuity and the momentum equations for a one-dimensional, steady, inviscid flow give

$$\rho a \, dA = (\rho + d\rho)(a + dU) \, dA \qquad (8.23)$$

$$p \, dA - (p + dp) \, dA = [(a + dU) - a]\rho a \, dA \qquad (8.24)$$

Simplifying equation (8.24), dividing equations (8.23) and (8.24) by dA, and combining the two relations gives

$$dp = a^2 \, d\rho \qquad (8.25)$$

However, since the process is isentropic,

$$a^2 = \left(\frac{\partial p}{\partial \rho}\right)_s \qquad (8.26)$$

where we have indicated that the derivative is taken with entropy fixed, as originally assumed.

For a perfect gas undergoing an isentropic process, equation (8.20a) gives

$$p = c\rho^\gamma$$

where c is a constant. Thus,

$$a^2 = \left(\frac{\partial p}{\partial \rho}\right)_s = \gamma c \rho^{\gamma - 1} = \frac{\gamma p}{\rho} \qquad (8.27)$$

Using the equation of state for a perfect gas, the speed of sound is

$$a = \sqrt{\gamma R T} \qquad (8.28)$$

ADIABATIC FLOW IN A VARIABLE-AREA STREAMTUBE

For purposes of derivation, consider the one-dimensional flow of a perfect gas through a variable-area streamtube (see Fig. 8.3). Let us apply the integral form of the energy equation [i.e., equation (2.35)] for steady, one-dimensional flow. Let us assume that

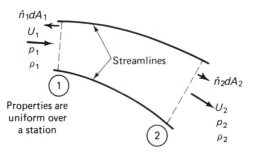

Figure 8.3 One-dimensional flow in a streamtube.

there is no heat transfer through the surface of the control volume (i.e., $\dot{Q} = 0$) and that only flow work (pressure-volume work) is done. Work is done on the system by the pressure forces acting at station 1 and is, therefore, negative. Work is done by the system at station 2. Thus,

$$+ p_1 U_1 A_1 - p_2 U_2 A_2 = -\left(u_{e1} \rho_1 U_1 A_1 + \frac{U_1^2}{2} \rho_1 U_1 A_1 \right)$$
$$+ \left(u_{e2} \rho_2 U_2 A_2 + \frac{U_2^2}{2} \rho_2 U_2 A_2 \right)$$

Rearranging, noting that $\rho_1 U_1 A_1 = \rho_2 U_2 A_2$ by continuity, and using the definition for the enthalpy, we obtain

$$H_t = h_1 + \frac{U_1^2}{2} = h_2 + \frac{U_2^2}{2} \tag{8.29}$$

where, as is usually the case in aerodynamics problems, changes in potential energy have been neglected. The assumption of one-dimensional flow is valid provided that the streamtube cross-sectional area varies smoothly and gradually in comparison to the axial distance along the streamtube.

For a perfect gas, equation (8.29) becomes

$$c_p T_1 + \frac{U_1^2}{2} = c_p T_2 + \frac{U_2^2}{2} \tag{8.30}$$

or

$$T_1 + \frac{U_1^2}{2c_p} = T_2 + \frac{U_2^2}{2c_p} \tag{8.31}$$

By definition, the *stagnation temperature* T_t is the temperature reached when the fluid is brought to rest adiabatically.

$$T_t = T + \frac{U^2}{2c_p} \tag{8.32}$$

Since the locations of stations 1 and 2 are arbitrary,

$$T_{t1} = T_{t2} \tag{8.33}$$

That is, the stagnation temperature is a constant for the adiabatic flow of a perfect gas and will be designated simply as T_t. Note that, whereas it is generally true that the stagnation enthalpy is constant for an adiabatic flow, as indicated by equation (8.29), the stagnation temperature is constant only if the additional condition that c_p is constant (i.e., the gas behaves as a perfect gas) is satisfied.

For any one-dimensional adiabatic flow,

$$\frac{T_t}{T} = 1 + \frac{U^2}{2[\gamma R/(\gamma - 1)]T} = 1 + \frac{\gamma - 1}{2} M^2 \tag{8.34}$$

Note that we have used the perfect-gas relations that $c_p = \gamma R/(\gamma - 1)$ and that $a^2 = \gamma RT$.

It is interesting to note that, when the flow is isentropic, Euler's equation for one-dimensional, steady flow (i.e., the inviscid-flow momentum equation) gives the same result as the energy equation for an adiabatic flow. To see this, let us write equation (3.2) for steady, one-dimensional, inviscid flow:

$$\rho u \frac{du}{ds} = -\frac{dp}{ds} \tag{8.35a}$$

$$\int \frac{dp}{\rho} + \int u \, du = 0 \tag{8.35b}$$

Note that for a one-dimensional flow $u = U$. For an isentropic process (which is of course adiabatic),

$$p = c\rho^\gamma$$

Differentiating, substituting the result into equation (8.35b), and integrating, we obtain

$$\frac{\gamma}{\gamma - 1} c\rho^{\gamma - 1} + \frac{U^2}{2} = \text{constant}$$

Thus,

$$\frac{1}{\gamma - 1} \frac{\gamma p}{\rho} + \frac{U^2}{2} = \text{constant}$$

Using the perfect-gas equation of state, we obtain

$$h + \frac{U^2}{2} = \text{constant}$$

which is equation (8.29).

Example 8.1

We are to design a supersonic wind tunnel using a large vacuum pump to draw air from the ambient atmosphere into our tunnel, as shown in Fig. 8.4. The air is continuously accelerated as it flows through a convergent/divergent nozzle so that flow in the test section is supersonic. If the ambient air is at the standard sea-level conditions, what is the maximum velocity that we can attain in the test section?

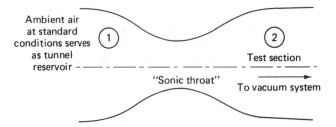

Ambient air
at standard
conditions serves
as tunnel
reservoir

"Sonic throat"

Test section

To vacuum system

Figure 8.4   Indraft, supersonic wind tunnel.

Solution   To calculate this maximum velocity, all we need is the energy equation for a steady, one-dimensional, adiabatic flow. Using equation (8.29), we have

$$h_1 + \tfrac{1}{2}U_1^2 = h_2 + \tfrac{1}{2}U_2^2$$

Since the ambient air (i.e., that which serves as the tunnel's "stagnation chamber" or "reservoir") is at rest, $U_1 = 0$. The maximum velocity in the test section occurs when $h_2 = 0$ (i.e., when the flow expands until the static temperature in the test section is zero).

$$(1004.7)288.15 \text{ J/kg} = \frac{U_2^2}{2}$$

$$U_2 = 760.9 \text{ m/s}$$

Of course, we realize that this limit is physically unattainable, since the gas would liquefy first. However, it does represent an upper limit for conceptual considerations.

Example 8.2

During a nominal reentry trajectory, the Space Shuttle Orbiter flies at 3964 ft/s at an altitude of 100,000 ft. The corresponding conditions at the stagnation point (point 2 in Fig. 8.5) are $p_2 = 490.2$ lbf/ft² and $T_2 = 1716.0°R$. The static pressures for two nearby locations (points 3 and 4 of Fig. 8.5) are $p_3 = 259.0$ lbf/ft² and $p_4 = 147.1$ lbf/ft². All three points are outside the boundary layer. What are the local static temperature, the local velocity, and the local Mach number at points 3 and 4.

Solution   At these conditions, the air can be assumed to behave approximately as a perfect gas. Furthermore, since all three points are outside the boundary layer and downstream of the bow shock wave, we will assume that the flow expands isentropically from point 2 to point 3 and then to point 4. (Note that because the shock wave is curved, the entropy will vary through the shock layer. For a further discussion of the rotational flow downstream of a curved shock wave, refer to Chapter 13. Thus, validity of the assumption that the expansion is isentropic should be verified for a given application.)

For an isentropic process, we can use equation (8.20b) to relate the temperature ratio to the pressure ratio between two points. Thus,

$$T_3 = T_2 \left(\frac{p_3}{p_2}\right)^{(\gamma-1)/\gamma} = (1716.0)(0.83337) = 1430.1°R$$

Similarly,

$$T_4 = (1716.0)(0.70899) = 1216.6°R$$

Using the energy equation for the adiabatic flow a perfect gas [i.e., equation (8.31)], and noting that $U_2 = 0$, since point 2 is a stagnation point,

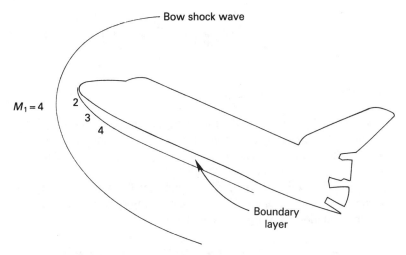

Figure 8.5   Shuttle flow field for Example 8.2.

$$U_3 = [2c_p(T_2 - T_3)]^{0.5}$$

$$= \left[2\left(0.2404 \, \frac{\text{Btu}}{\text{lbm} \cdot {}^\circ\text{R}} \right)\left(32.174 \, \frac{\text{ft} \cdot \text{lbm}}{\text{lbf} \cdot \text{s}^2} \right)\left(778.2 \, \frac{\text{ft} \cdot \text{lbf}}{\text{Btu}} \right)(285.9^\circ\text{R}) \right]^{0.5}$$

$$= 1855.2 \text{ ft/s}$$

Similarly,

$$U_4 = 2451.9 \text{ ft/s}$$

Using equation (1.7b) for the speed of sound in English units,

$$M_3 = \frac{U_3}{a_3} = \frac{1855.2}{49.02(1430.1)^{0.5}} = 1.001$$

$$M_4 = \frac{U_4}{a_4} = \frac{2451.9}{49.02(1216.6)^{0.5}} = 1.434$$

Thus, the flow accelerates from the stagnation conditions at point 2 to sonic conditions at point 3, becoming supersonic at point 4.

ISENTROPIC FLOW IN A VARIABLE-AREA STREAMTUBE

It is particularly useful to study the isentropic flow of a perfect gas in a variable-area streamtube, since it reveals many of the general characteristics of compressible flow. In addition, the assumption of constant entropy is not too restrictive, since the flow outside the boundary layer is essentially isentropic except while crossing linear shock waves or downstream of curved shock waves.

Using equations (8.20) and (8.34), we can write

$$\frac{p_{t1}}{p} = \left(1 + \frac{\gamma - 1}{2} M^2\right)^{\gamma/(\gamma - 1)} \tag{8.36}$$

$$\frac{\rho_{t1}}{\rho} = \left(1 + \frac{\gamma - 1}{2} M^2\right)^{1/(\gamma - 1)} \tag{8.37}$$

where p_{t1} and ρ_{t1} are the *stagnation pressure* and the *stagnation density*, respectively. Applying these equations between two streamwise stations shows that if T_t is constant and the flow is isentropic, the stagnation pressure p_{t1} is a constant. The equation of state requires that ρ_{t1} be constant also.

To get a feeling for the deviation between the pressure values calculated assuming incompressible flow and those calculated using the compressible flow relations, let us expand equation (8.36) in powers of M^2:

$$\frac{p_{t1}}{p} = 1 + \frac{\gamma}{2} M^2 + O(M^4) + \cdots \tag{8.38}$$

Since the flow is essentially incompressible when the Mach number is relatively low, we can neglect higher order terms. Retaining only terms of order M^2 yields

$$\frac{p_{t1}}{p} = 1 + \frac{\gamma}{2} M^2 \tag{8.39}$$

which for a perfect gas becomes

$$\frac{p_{t1}}{p} = 1 + \frac{\gamma}{2} \frac{U^2}{\gamma RT} = 1 + \frac{U^2}{2p/\rho} \tag{8.40}$$

Rearranging,

$$p_{t1} = p + \frac{\rho U^2}{2} \tag{8.41}$$

Thus, for low Mach numbers, the general relation, equation (8.36), reverts to Bernoulli's equation for incompressible flow. The static pressures predicted by equation (8.36) are compared with those of equation (8.41) in terms of percent error as a function of Mach number in Fig. 8.6. An error of less than 1% results when Bernoulli's equation is used if the local Mach number is less than or equal to 0.5 in air.

In deriving equations (8.34), (8.36), and (8.37), the respective stagnation properties have been used as references to nondimensionalize the static properties. Since the continuity equation for the one-dimensional steady flow requires that ρUA be a constant, the area becomes infinitely large as the velocity goes to zero. Let us choose the area where the flow is sonic (i.e., $M = 1$) as the reference area to relate to the streamtube area at a given station. Designating the sonic conditions by a * superscript, the continuity equation yields

$$\frac{A^*}{A} = \frac{\rho U}{\rho^* U^*} = \frac{\rho_{t1}(\rho/\rho_{t1})\sqrt{\gamma RT_t}\,\sqrt{T/T_t}\,M}{\rho_{t1}(\rho^*/\rho_{t1})\sqrt{\gamma RT_t}\,\sqrt{T^*/T_t}} \tag{8.42}$$

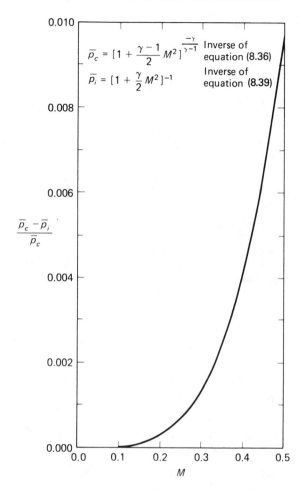

$$\bar{p}_c = [1 + \frac{\gamma - 1}{2} M^2]^{\frac{-\gamma}{\gamma-1}} \quad \text{Inverse of equation (8.36)}$$

$$\bar{p}_i = [1 + \frac{\gamma}{2} M^2]^{-1} \quad \text{Inverse of equation (8.39)}$$

Figure 8.6 Effect of compressibility on the theoretical value for the pressure ratio.

since $M^* = 1$. Noting that ρ^*/ρ_{t1} and T^*/T_t are to be evaluated at $M = M^* = 1$,

$$\frac{A^*}{A} = M\left[\frac{2}{\gamma + 1}\left(1 + \frac{\gamma - 1}{2} M^2\right)\right]^{-(\gamma + 1)/2(\gamma - 1)} \tag{8.43}$$

Given the area, A, and the Mach number, M, at any station one could compute an A^* for that station from equation (8.43). A^* is the area the streamtube would have to be if the flow were accelerated or decelerated to $M = 1$ isentropically. Equation (8.43) is especially useful in streamtube flows that are isentropic, and therefore where A^* is a constant.

In order to aid in the solution of isentropic flow problems, the temperature ratio [equation (8.34)], the pressure ratio [equation (8.36)], the density ratio [equation (8.37)], and the area ratio [equation (8.43)] are presented as a function of the Mach number in Table 8.1. A more complete tabulation of these data is given in Ref. 8.1. The results of Table 8.1 are summarized in Fig. 8.7.

TABLE 8.1   CORRELATIONS FOR A ONE-DIMENSIONAL, ISENTROPIC FLOW
OF PERFECT AIR($\gamma = 1.4$)

M	$\dfrac{A}{A^*}$	$\dfrac{p}{p_{t1}}$	$\dfrac{\rho}{\rho_{t1}}$	$\dfrac{T}{T_t}$	$\dfrac{A}{A^*}\dfrac{p}{p_{t1}}$
0	∞	1.00000	1.00000	1.00000	∞
0.05	11.592	0.99825	0.99875	0.99950	11.571
0.10	5.8218	0.99303	0.99502	0.99800	5.7812
0.15	3.9103	0.98441	0.98884	0.99552	3.8493
0.20	2.9635	0.97250	0.98027	0.99206	2.8820
0.25	2.4027	0.95745	0.96942	0.98765	2.3005
0.30	2.0351	0.93947	0.95638	0.98232	1.9119
0.35	1.7780	0.91877	0.94128	0.97608	1.6336
0.40	1.5901	0.89562	0.92428	0.96899	1.4241
0.45	1.4487	0.87027	0.90552	0.96108	1.2607
0.50	1.3398	0.84302	0.88517	0.95238	1.12951
0.55	1.2550	0.81416	0.86342	0.94295	1.02174
0.60	1.1882	0.78400	0.84045	0.93284	0.93155
0.65	1.1356	0.75283	0.81644	0.92208	0.85493
0.70	1.09437	0.72092	0.79158	0.91075	0.78896
0.75	1.06242	0.68857	0.76603	0.89888	0.73155
0.80	1.03823	0.65602	0.74000	0.88652	0.68110
0.85	1.02067	0.62351	0.71361	0.87374	0.63640
0.90	1.00886	0.59126	0.68704	0.86058	0.59650
0.95	1.00214	0.55946	0.66044	0.84710	0.56066
1.00	1.00000	0.52828	0.63394	0.83333	0.52828
1.05	1.00202	0.49787	0.60765	0.81933	0.49888
1.10	1.00793	0.46835	0.58169	0.80515	0.47206
1.15	1.01746	0.43983	0.55616	0.79083	0.44751
1.20	1.03044	0.41238	0.53114	0.77640	0.42493
1.25	1.04676	0.38606	0.50670	0.76190	0.40411
1.30	1.06631	0.36092	0.48291	0.74738	0.38484
1.35	1.08904	0.33697	0.45980	0.73287	0.36697
1.40	1.1149	0.31424	0.43742	0.71839	0.35036
1.45	1.1440	0.29272	0.41581	0.70397	0.33486
1.50	1.1762	0.27240	0.39498	0.68965	0.32039
1.55	1.2115	0.25326	0.37496	0.67545	0.30685
1.60	1.2502	0.23527	0.35573	0.66138	0.29414
1.65	1.2922	0.21839	0.33731	0.64746	0.28221
1.70	1.3376	0.20259	0.31969	0.63372	0.27099
1.75	1.3865	0.18782	0.30287	0.62016	0.26042
1.80	1.4390	0.17404	0.28682	0.60680	0.25044
1.85	1.4952	0.16120	0.27153	0.59365	0.24102
1.90	1.5555	0.14924	0.25699	0.58072	0.23211
1.95	1.6193	0.13813	0.24317	0.56802	0.22367
2.00	1.6875	0.12780	0.23005	0.55556	0.21567
2.05	1.7600	0.11823	0.21760	0.54333	0.20808
2.10	1.8369	0.10935	0.20580	0.53135	0.20087
2.15	1.9185	0.10113	0.19463	0.51962	0.19403
2.20	2.0050	0.09352	0.18405	0.50813	0.18751

TABLE 8.1 (Continued)

M	$\dfrac{A}{A^*}$	$\dfrac{p}{p_{t1}}$	$\dfrac{\rho}{\rho_{t1}}$	$\dfrac{T}{T_t}$	$\dfrac{A}{A^*}\dfrac{p}{p_{t1}}$
2.25	2.0964	0.08648	0.17404	0.49689	0.18130
2.30	2.1931	0.07997	0.16458	0.48591	0.17539
2.35	2.2953	0.07396	0.15564	0.47517	0.16975
2.40	2.4031	0.06840	0.14720	0.46468	0.16437
2.45	2.5168	0.06327	0.13922	0.45444	0.15923
2.50	2.6367	0.05853	0.13169	0.44444	0.15432
2.55	2.7630	0.05415	0.12458	0.43469	0.14963
2.60	2.8960	0.05012	0.11787	0.42517	0.14513
2.65	3.0359	0.04639	0.11154	0.41589	0.14083
2.70	3.1830	0.04295	0.10557	0.40684	0.13671
2.75	3.3376	0.03977	0.09994	0.39801	0.13276
2.80	3.5001	0.03685	0.09462	0.38941	0.12897
2.85	3.6707	0.03415	0.08962	0.38102	0.12534
2.90	3.8498	0.03165	0.08489	0.37286	0.12185
2.95	4.0376	0.02935	0.08043	0.36490	0.11850
3.00	4.2346	0.02722	0.07623	0.35714	0.11527
3.50	6.7896	0.01311	0.04523	0.28986	0.08902
4.00	10.719	0.00658	0.02766	0.23810	0.07059
4.50	16.562	0.00346	0.01745	0.19802	0.05723
5.00	25.000	$189(10)^{-5}$	0.01134	0.16667	0.04725
6.00	53.189	$633(10)^{-6}$	0.00519	0.12195	0.03368
7.00	104.143	$242(10)^{-6}$	0.00261	0.09259	0.02516
8.00	190.109	$102(10)^{-6}$	0.00141	0.07246	0.01947
9.00	327.189	$474(10)^{-7}$	0.000815	0.05814	0.01550
10.00	535.938	$236(10)^{-7}$	0.000495	0.04762	0.01263
∞	∞	0	0	0	0

In order to determine the mass-flow rate in a streamtube:

$$\dot{m} = \rho U A = \rho_{t1}\left(\frac{\rho}{\rho_{t1}}\right) M \sqrt{\gamma R T_t}\sqrt{\frac{T}{T_t}}\,A$$

$$\frac{\dot{m}}{A} = \sqrt{\frac{\gamma}{R}}\,\frac{p_{t1}}{\sqrt{T_t}}\,\frac{M}{\{1 + [(\gamma - 1)/2]M^2\}^{(\gamma + 1)/2(\gamma - 1)}} \tag{8.44}$$

Thus, the mass-flow rate is proportional to the stagnation pressure and inversely proportional to the square root of the stagnation temperature. To find the condition of maximum flow per unit area, one could compute the derivative of ($\dot{m}/A$) as given by equation (8.44) with respect to Mach number and set the derivative equal to zero. At this condition, one would find that $M = 1$. Thus, setting $M = 1$ in equation (8.44) yields

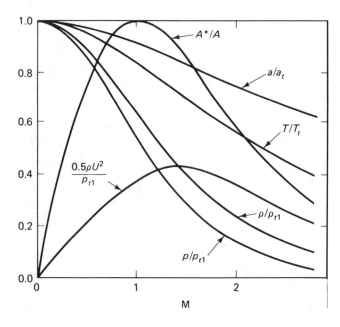

Figure 8.7   Property variations as functions of Mach number for isentropic flow for $\gamma = 1.4$.

$$\left(\frac{\dot{m}}{A}\right)_{max} = \frac{\dot{m}}{A^*} = \sqrt{\frac{\gamma}{R}\left(\frac{2}{\gamma + 1}\right)^{(\gamma + 1)/(\gamma - 1)}} \frac{p_{t1}}{\sqrt{T_t}}$$ (8.45)

The maximum flow rate per unit area occurs where the cross-section area is a minimum (designated A^*), where the Mach number is one. Thus, the maximum flow rate per unit area occurs at the sonic throat when the flow is choked.

Figure 8.7 shows that, for each value of A^*/A, there are two values of M: one subsonic, the other supersonic. Thus, from Fig. 8.7 we see that, while all static properties of the fluid monotonically decrease with Mach number, the area ratio does not. We conclude that to accelerate a subsonic flow, the streamtube must first converge until sonic conditions are reached, then diverge to achieve supersonic Mach numbers. Note that just because a streamtube is convergent/divergent, it does not necessarily follow that the flow is sonic at the narrowest cross section (or throat). However, if the Mach number is to be unity anywhere in the streamtube, it must be so at the throat.

For certain calculations (e.g., finding the true airspeed from Mach number and stagnation pressure) the ratio $0.5\,\rho U^2/p_{t1}$ is useful.

$$\frac{0.5\rho U^2}{p_{t1}} = \frac{\frac{1}{2}(p/RT)(\gamma/\gamma)U^2}{p_{t1}} = \frac{1}{2}\frac{\gamma p}{p_{t1}}\frac{U^2}{\gamma RT} = \frac{\gamma M^2}{2}\frac{p}{p_{t1}}$$ (8.46)

Thus,

$$\frac{0.5\rho U^2}{p_{t1}} = \frac{\gamma M^2}{2}\left(1 + \frac{\gamma - 1}{2}M^2\right)^{-\gamma/(\gamma - 1)}$$ (8.47)

The ratio of the local speed of sound to the speed of sound at the stagnation conditions is

$$\frac{a}{a_t} = \sqrt{\frac{\gamma R T}{\gamma R T_t}} = \left(\frac{T}{T_t}\right)^{0.5} = \left(1 + \frac{\gamma - 1}{2} M^2\right)^{-0.5} \qquad (8.48)$$

Note: The nomenclature used herein anticipates the discussion of shock waves. Since the stagnation pressure varies across a shock wave, the subscript $t1$ has been used to designate stagnation properties evaluated upstream of a shock wave (which correspond to the free-stream values in a flow with a single shock wave). Since the stagnation temperature is constant across a shock wave (for perfect-gas flows), it is designated by the simple subscript t.

Example 8.3

Assume that you are given the task of determining the aerodynamic forces and moments acting on a slender missile which flies at a Mach number of 3.5 at an altitude of 27,432 m (90,000 ft). Aerodynamic coefficients for the missile, which is 20.0 cm (7.874 in) in diameter and 10 diameters long, are required for angles of attack from 0° to 55°. The decision is made to obtain experimental values of the required coefficients in the Vought High-Speed Wind Tunnel (in Dallas, Texas). Upstream of the model shock system, the flow in the wind tunnel is isentropic and air, at these conditions, behaves as a perfect gas. Thus, the relations developed in this section can be used to calculate the wind-tunnel test conditions.

Solution

1. *Flight conditions.* Using the properties for a standard atmosphere (such as were presented in Chapter 1), the relevant parameters for the flight condition include

$$U_\infty = 1050 \text{ m/s}$$

$$p_\infty = 13.21 \text{ mm Hg}$$

$$T_\infty = 224 \text{ K}$$

$$\text{Re}_{\infty,d} = \frac{\rho_\infty U_\infty d}{\mu_\infty} = 3.936 \times 10^5$$

$$M_\infty = 3.5$$

2. *Wind-tunnel conditions.* Information about the operational characteristics of the Vought High-Speed Wind Tunnel is contained in the tunnel handbook (Ref. 8.2). To ensure that the model is not so large that its presence alters the flow in the tunnel (i.e., the model dimensions are within the allowable blockage area), the diameter of the wind-tunnel model, d_{wt}, will be 4.183 cm (1.6468 in).

Based on the discussion in Chapter 2, the Mach number and the Reynolds number are two parameters which we should try to simulate in the wind tunnel. The free-stream unit Reynolds number (U_∞/ν_∞) is presented as a function of the free-stream Mach number and of the stagnation pressure for a stagnation temperature of 311 K (100 °F) in Fig. 8.8 (which has been taken from Ref. 8.2 and has been left in English units). The student can use the equations of this section to verify the value for the unit Reynolds number given the conditions in the stagnation chamber and the free-stream Mach number. In order to match the flight Reynolds number of 3.936×10^5 in the wind tunnel,

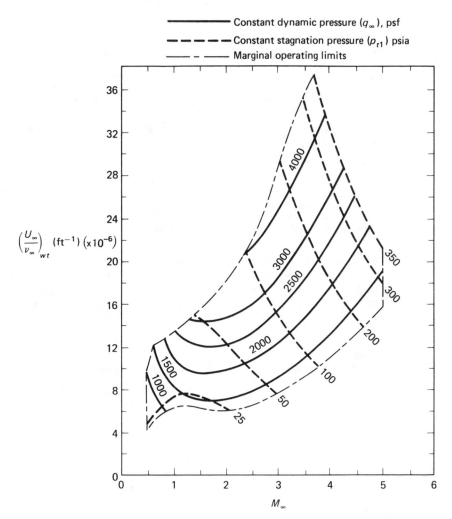

Figure 8.8 Free-stream unit Reynolds number as a function of the free-stream Mach number for $T_t = 100°F$ in the Vought high-speed wind tunnel. (From Ref. 8.2.)

$$\left(\frac{U_\infty}{v_\infty}\right)_{wt} = \frac{3.936 \times 10^5}{d_{wt}} = 2.868 \times 10^6/\text{ft}$$

But as indicated in Fig. 8.8, the lowest unit Reynolds number possible in this tunnel at $M_\infty = 3.5$ is approximately $9.0 \times 10^6/\text{ft}$. Thus, if the model is 4.183 cm in diameter, the lowest possible tunnel value of $Re_{\infty,d}$ is 1.235×10^6, which is greater than the flight value. This is much different than the typical subsonic flow, where (as discussed in Chapter 5) the maximum wind-tunnel Reynolds number is usually much less than the flight value. To obtain the appropriate Reynolds number for the current supersonic flow, we can choose to use a smaller model. Using a model that is 1.333 cm in diameter would yield a Reynolds number of 3.936×10^5 (as desired). If this model is too small, we cannot establish a tunnel condition which matches both the flight Mach number and the Reynolds number. In this case, the authors would choose to simulate the Mach number exactly rather than seek a compromise Mach-number/Reynolds-number test condition, since the pressure coefficients and the shock interaction phenomena on the control surfaces are Mach-number-dependent in this range of free-stream Mach number.

The conditions in the stagnation chamber of the tunnel are $T_t = 311$ K (560°R) and $p_{t1} = 4137$ mm Hg (80 psia). Thus, using either equations (8.34) and (8.36) or the values presented in Table 8.1, one finds that $T_\infty = 90.18$ K (162.32°R) and $p_\infty = 54.23$ mm Hg (1.049 psia). The cold free-stream temperature is typical of supersonic tunnels (e.g., most high-speed wind tunnels operate at temperatures near liquefaction of oxygen). Thus, the free-stream speed of sound is relatively low and, even though the free-stream Mach number is 3.5, the velocity in the test section U_∞ is only 665 m/s (2185 ft/s). In summary, the relevant parameters for the wind-tunnel conditions include

$$U_\infty = 665 \text{ m/s}$$

$$p_\infty = 54.23 \text{ mm Hg}$$

$$T_\infty = 90.18 \text{ K}$$

$$\text{Re}_{\infty,d} = 3.936 \times 10^5 \text{ if } d = 1.333 \text{ cm} \quad \text{or} \quad 1.235 \times 10^6 \text{ if } d = 4.183 \text{ cm}$$

$$M_\infty = 3.5$$

Because of the significant differences in the dimensional values of the flow parameters (such as U_∞, p_∞, and T_∞), we must again nondimensionalize the parameters so that correlations of wind-tunnel measurements can be related to the theoretical solutions or to the design flight conditions.

CHARACTERISTIC EQUATIONS AND PRANDTL-MEYER FLOW

Consider a two-dimensional flow around a slender airfoil shape. The deflection of the streamlines as flow encounters the airfoil are sufficiently small that shock waves are not generated. Thus, the flow may be considered as isentropic (excluding the boundary layer, of course). For the development of the equations for a more general flow, refer to Ref. 8.3. For the present type of flow, the equations of motion in a natural (or streamline) coordinate system, as shown in Fig. 8.9, are as follows. The continuity equation,

$$\frac{\partial(\rho U)}{\partial s} + \rho U \frac{\partial \theta}{\partial n} = 0 \tag{8.49}$$

the s-momentum equation,

$$\rho U \frac{\partial U}{\partial s} + \frac{\partial p}{\partial s} = 0 \tag{8.50}$$

and the n-momentum equation,

$$\rho U^2 \frac{\partial \theta}{\partial s} + \frac{\partial p}{\partial n} = 0 \tag{8.51}$$

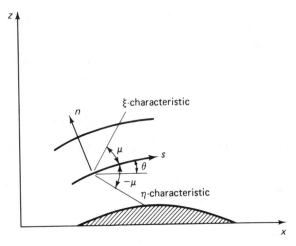

Figure 8.9 Supersonic flow around an airfoil in natural (streamline) coordinates.

Since the flow is isentropic, the energy equation provides no unique information and is, therefore, not used. However, since the flow is isentropic, the change in pressure with respect to the change in density is equal to the square of the speed of sound. Thus,

$$\frac{\partial p}{\partial \rho} = a^2 \tag{8.52}$$

and the continuity equation becomes

$$\frac{\partial p}{\partial s} \frac{M^2 - 1}{\rho U^2} + \frac{\partial \theta}{\partial n} = 0 \tag{8.53}$$

Combining equations (8.51) and (8.53) and introducing the concept of the directional derivative (e.g., Ref. 8.4), we obtain

$$\frac{\partial p}{\partial \xi} + \frac{\rho U^2}{\sqrt{M^2 - 1}} \frac{\partial \theta}{\partial \xi} = 0 \tag{8.54a}$$

along the line having the direction

$$\frac{dn}{ds} = \tan \mu = \frac{1}{\sqrt{M^2 - 1}}$$

(i.e., the left-running characteristic ξ of Fig. 8.9). A *characteristic* is a line which exists only in supersonic flows. Characteristics should not be confused with finite-strength waves, such as shock waves. The ξ characteristic is inclined to the local streamline by the angle μ, which is the Mach angle,

$$\mu = \sin^{-1}\left(\frac{1}{M}\right)$$

Thus, it corresponds to the left-running Mach wave, which is so called because to an observer looking downstream, the Mach wave appears to be going downstream in a leftward direction. Equivalently,

$$\frac{\partial p}{\partial \eta} - \frac{\rho U^2}{\sqrt{M^2 - 1}} \frac{\partial \theta}{\partial \eta} = 0 \tag{8.54b}$$

along the line whose direction is

$$\frac{dn}{ds} = \tan(-\mu)$$

(i.e., the right-running characteristic η of Fig. 8.9). Equations (8.54a) and (8.54b) provide a relation between the local static pressure and the local flow inclination.

Euler's equation for a steady, inviscid flow, which can be derived by neglecting the viscous terms and the body forces in the momentum equation (3.1), states that

$$dp = -\rho U \, dU$$

Thus, for the left-running characteristic, equation (8.54a) becomes

$$\frac{dU}{U} = \frac{d\theta}{\sqrt{M^2 - 1}} \tag{8.55}$$

But from the adiabatic-flow relations for a perfect gas,

$$\left(\frac{U}{a_t}\right)^2 = M^2 \left(1 + \frac{\gamma - 1}{2} M^2\right)^{-1} \tag{8.56}$$

where a_t is the speed of sound at the stagnation conditions. Differentiating equation (8.56) and substituting the result into equation (8.55) yields

$$d\theta = \frac{\sqrt{M^2 - 1} \, dM^2}{2M^2 \{1 + [(\gamma - 1)/2]M^2\}} \tag{8.57}$$

Integration of equation (8.57) yields the relation, which is valid for a left-running characteristic:

$$\theta = \nu + \text{constant of integration}$$

where ν is a function of the Mach number as given by

$$\nu = \sqrt{\frac{\gamma + 1}{\gamma - 1}} \arctan \sqrt{\frac{\gamma - 1}{\gamma + 1}(M^2 - 1)} - \arctan \sqrt{M^2 - 1} \tag{8.58}$$

Tabulations of the *Prandtl–Meyer angle* (ν), the corresponding Mach number, and the corresponding Mach angle (μ) are presented in Table 8.2.

Thus, along a left-running characteristic,

$$\nu - \theta = R \tag{8.59a}$$

a constant. Similarly, along a right-running characteristic,

$$\nu + \theta = Q \tag{8.59b}$$

which is another constant. The use of equations (8.59a) and (8.59b) is simplified if the slope of the vehicle surface is such that one only needs to consider the waves of

TABLE 8.2 MACH NUMBER AND MACH ANGLE AS A FUNCTION OF PRANDTL-MEYER ANGLE

ν (deg)	M	μ (deg)	ν (deg)	M	μ (deg)
0.0	1.000	90.000	22.5	1.862	32.488
0.5	1.051	72.099	23.0	1.879	32.148
1.0	1.082	67.574	23.5	1.897	31.814
1.5	1.108	64.451	24.0	1.915	31.486
2.0	1.133	61.997	24.5	1.932	31.164
2.5	1.155	59.950	25.0	1.950	30.847
3.0	1.177	58.180	25.5	1.968	30.536
3.5	1.198	56.614	26.0	1.986	30.229
4.0	1.218	55.205	26.5	2.004	29.928
4.5	1.237	53.920	27.0	2.023	29.632
5.0	1.256	52.738	27.5	2.041	29.340
5.5	1.275	51.642	28.0	2.059	29.052
6.0	1.294	50.619	28.5	2.078	28.769
6.5	1.312	49.658	29.0	2.096	28.491
7.0	1.330	48.753	29.5	2.115	28.216
7.5	1.348	47.896	30.0	2.134	27.945
8.0	1.366	47.082	30.5	2.153	27.678
8.5	1.383	46.306	31.0	2.172	27.415
9.0	1.400	45.566	31.5	2.191	27.155
9.5	1.418	44.857	32.0	2.210	26.899
10.0	1.435	44.177	32.5	2.230	26.646
10.5	1.452	43.523	33.0	2.249	26.397
11.0	1.469	42.894	33.5	2.269	26.151
11.5	1.486	42.287	34.0	2.289	25.908
12.0	1.503	41.701	34.5	2.309	25.668
12.5	1.520	41.134	35.0	2.329	25.430
13.0	1.537	40.585	35.5	2.349	25.196
13.5	1.554	40.053	36.0	2.369	24.965
14.0	1.571	39.537	36.5	2.390	24.736
14.5	1.588	39.035	37.0	2.410	24.510
15.0	1.605	38.547	37.5	2.431	24.287
15.5	1.622	38.073	38.0	2.452	24.066
16.0	1.639	37.611	38.5	2.473	23.847
16.5	1.655	37.160	39.0	2.495	23.631
17.0	1.672	36.721	39.5	2.516	23.418
17.5	1.689	36.293	40.0	2.538	23.206
18.0	1.706	35.874	40.5	2.560	22.997
18.5	1.724	35.465	41.0	2.582	22.790
19.0	1.741	35.065	41.5	2.604	22.585
19.5	1.758	34.673	42.0	2.626	22.382
20.0	1.775	34.290	42.5	2.649	22.182
20.5	1.792	33.915	43.0	2.671	21.983
21.0	1.810	33.548	43.5	2.694	21.786
21.5	1.827	33.188	44.0	2.718	21.591
22.0	1.844	32.834	44.5	2.741	21.398

TABLE 8.2 (*Continued*)

ν (deg)	M	μ (deg)	ν (deg)	M	μ (deg)
45.0	2.764	21.207	67.5	4.133	14.002
45.5	2.788	21.017	68.0	4.173	13.865
46.0	2.812	20.830	68.5	4.214	13.729
46.5	2.836	20.644	69.0	4.255	13.593
47.0	2.861	20.459	69.5	4.297	13.459
47.5	2.886	20.277	70.0	4.339	13.325
48.0	2.910	20.096	70.5	4.382	13.191
48.5	2.936	19.916	71.0	4.426	13.059
49.0	2.961	19.738	71.5	4.470	12.927
49.5	2.987	15.561	72.0	4.515	12.795
50.0	3.013	19.386	72.5	4.561	12.665
50.5	3.039	19.213	73.0	4.608	12.535
51.0	3.065	19.041	73.5	4.655	12.406
51.5	3.092	18.870	74.0	4.703	12.277
52.0	3.119	18.701	74.5	4.752	12.149
52.5	3.146	18.532	75.0	4.801	12.021
53.0	3.174	18.366	75.5	4.852	11.894
53.5	3.202	18.200	76.0	4.903	11.768
54.0	3.230	18.036	76.5	4.955	11.642
54.5	3.258	17.873	77.0	5.009	11.517
55.0	3.287	17.711	77.5	5.063	11.392
55.5	3.316	17.551	78.0	5.118	11.268
56.0	3.346	17.391	78.5	5.175	11.145
56.5	3.375	17.233	79.0	5.231	11.022
57.0	3.406	17.076	79.5	5.289	10.899
57.5	3.436	16.920	80.0	5.348	10.777
58.0	3.467	16.765	80.5	5.408	10.656
58.5	3.498	16.611	81.0	5.470	10.535
59.0	3.530	16.458	81.5	5.532	10.414
59.5	3.562	16.306	82.0	5.596	10.294
60.0	3.594	16.155	82.5	5.661	10.175
60.5	3.627	16.006	83.0	5.727	10.056
61.0	3.660	15.856	83.5	5.795	9.937
61.5	3.694	15.708	84.0	5.864	9.819
62.0	3.728	15.561	84.5	5.864	9.819
62.5	3.762	15.415	85.0	6.006	9.584
63.0	3.797	15.270	85.5	6.080	9.467
63.5	3.832	15.126	86.0	6.155	9.350
64.0	3.868	14.983	86.5	6.232	9.234
64.5	3.904	14.840	87.0	6.310	9.119
65.0	3.941	14.698	87.5	6.390	9.003
65.5	3.979	14.557	88.0	6.472	8.888
66.0	4.016	14.417	88.5	6.556	8.774
66.5	4.055	14.278	89.0	6.642	8.660
67.0	4.094	14.140	89.5	6.729	8.546

TABLE 8.2 (Continued)

ν (deg)	M	μ (deg)	ν (deg)	M	μ (deg)
90.0	6.819	8.433	97.5	8.480	6.772
90.5	6.911	8.320	98.0	8.618	6.664
91.0	7.005	8.207	98.5	8.759	6.556
91.5	7.102	8.095	99.0	8.905	6.448
92.0	7.201	7.983	99.5	9.055	6.340
92.5	7.302	7.871	100.0	9.210	6.233
93.0	7.406	7.760	100.5	9.371	6.126
93.5	7.513	7.649	101.0	9.536	6.019
94.0	7.623	7.538	101.5	9.708	5.913
94.5	7.735	7.428	102.0	9.885	5.806
95.0	7.851	7.318			
95.5	7.970	7.208			
96.0	8.092	7.099			
96.5	8.218	6.989			
97.0	8.347	6.881			

a single family, i.e., all the waves are either left-running waves or right-running waves. To make the application clear, let us work through a sample problem.

Example 8.4

Consider the infinitesimally thin airfoil which has the shape of a parabola:

$$x^2 = -\frac{c^2}{z_{max}} (z - z_{max})$$

where $z_{max} = 0.10c$, moving through the air at $M_\infty = 2.059$. The leading edge of the airfoil is parallel to the free stream. The thin airfoil will be represented by five linear segments, as shown in Fig. 8.10. For each segment Δx will be $0.2c$. Thus, the slopes of these segments are:

Segment	a	b	c	d	e
θ	$-1.145°$	$-3.607°$	$-5.740°$	$-8.048°$	$-10.370°$

For the free-stream flow,

$$\nu_\infty = 28.000° \qquad \frac{p_\infty}{p_{t1}} = 0.11653 \qquad \theta_\infty = 0°$$

Since the turning angles are small, it will be assumed that both the acceleration of the flow over the upper surface and the deceleration of the flow over the lower surface are isentropic processes. Note that the expansion waves on the upper surface diverge as the flow accelerates, but the compression waves of the lower surface coalesce. Since the flow is isentropic, we can use equations (8.59a) and (8.59b). Furthermore, the stagnation pressure is constant throughout the flow field and equal to p_{t1} (which is the value for the free-stream flow).

In going from the free-stream (region ∞ in Fig. 8.10) to the first segment on the

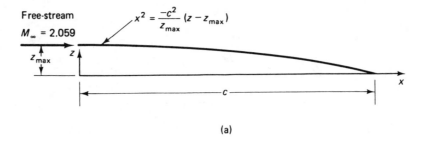

(a)

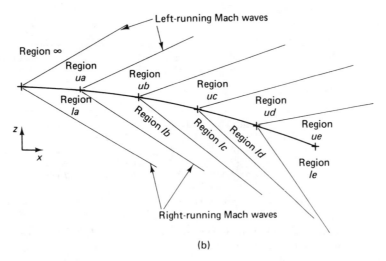

(b)

Figure 8.10 Mach waves for supersonic flow past a thin airfoil: (a) airfoil section; (b) wave pattern.

upper surface (region ua), we move along a right-running characteristic to cross the left-running Mach wave shown in the figure. Thus,

$$v + \theta = Q$$

or

$$dv = -d\theta$$

so

$$v_{ua} = v_\infty - (\theta_{ua} - \theta_\infty)$$

$$= 28.000° - (-1.145°) = 29.145°$$

and

$$M_{ua} = 2.1018$$

Using Table 8.1, $p_{ua}/p_{t1} = 0.1091$.

Similarly, in going from the free stream to the first segment on the lower surface (region la) we move along a left-running characteristic to cross the right-running Mach wave shown in the figure. Thus,

$$\nu - \theta = R$$

or

$$d\nu = d\theta$$

so

$$\nu_{la} = \nu_\infty + (\theta_{la} - \theta_\infty)$$

$$= 28.000° + (-1.145°) = 26.855°$$

Summarizing,

	Upper surface			Lower surface		
Segment	ν_u	M_u	$\dfrac{p_u}{p_{t1}}$	ν_l	M_l	$\dfrac{p_l}{p_{t1}}$
a	29.145°	2.1018	0.1091	26.855°	2.0173	0.1244
b	31.607°	2.1952	0.0942	24.393°	1.9286	0.1428
c	33.740°	2.2784	0.0827	22.260°	1.8534	0.1604
d	36.048°	2.3713	0.0715	19.952°	1.7733	0.1813
e	38.370°	2.4679	0.0615	17.630°	1.6940	0.2045

Let us now calculate the lift coefficient and the drag coefficient for the airfoil. Note that we have not been given the free-stream pressure (or, equivalently, the altitude at which the airfoil is flying) or the chord length of the airfoil. But that is not critical, since we seek the force coefficients.

$$C_l = \frac{l}{\frac{1}{2}\rho_\infty U_\infty^2 c}$$

Referring to equation (8.46), it is evident that, for a perfect gas,

$$q_\infty = \frac{1}{2}\rho_\infty U_\infty^2 = \frac{\gamma}{2}p_\infty M_\infty^2 \tag{8.60}$$

Thus,

$$C_l = \frac{l}{(\gamma/2)p_\infty M_\infty^2 c} \tag{8.61}$$

Referring again to Fig. 8.10, the incremental lift force acting on any segment (i.e., the ith segment) is

$$dl_i = (p_{li} - p_{ui})\, ds_i \cos\theta_i = (p_{li} - p_{ui})\, dx_i$$

Similarly, the incremental drag force for any segment is

$$dd_i = (p_{li} - p_{ui})\, ds_i \sin\theta_i = (p_{li} - p_{ui})\, dx_i \tan\theta_i$$

Segment	$\dfrac{p_{li}}{p_\infty}$	$\dfrac{p_{ui}}{p_\infty}$	$\dfrac{dl_i}{p_\infty}$	$\dfrac{dd_i}{p_\infty}$
a	1.070	0.939	0.0262c	0.000524c
b	1.226	0.810	0.0832c	0.004992c
c	1.380	0.710	0.1340c	0.01340c
d	1.559	0.614	0.1890c	0.0264c
e	1.759	0.529	0.2460c	0.0443c
Sum			0.6784c	0.0896c

Finally,

$$C_l = \frac{\Sigma\, dl_i}{(\gamma/2)p_\infty M_\infty^2 c} = \frac{0.6784c}{0.7(4.24)c} = 0.2286$$

$$C_d = \frac{\Sigma\, dd_i}{(\gamma/2)p_\infty M_\infty^2 c} = \frac{0.0896c}{0.7(4.24)c} = 0.0302$$

$$\frac{l}{d} = \frac{C_l}{C_d} = 7.57$$

SHOCK WAVES

The formation of a shock wave occurs when a supersonic flow decelerates in response to a sharp increase in pressure or when a supersonic flow encounters a sudden, compressive change in direction. For flow conditions where the gas is a continuum, the shock wave is a narrow region (on the order of several molecular mean free paths thick, $\sim 6 \times 10^{-6}$ cm) across which there is an almost instantaneous change in the values of the flow parameters. Because of the large streamwise variations in velocity, pressure, and temperature, viscous and heat-conduction effects are important within the shock wave. The difference between a shock wave and a Mach wave should be kept in mind. A *Mach wave* represents a surface across which some derivatives of the flow variables (such as the thermodynamic properties of the fluid and the flow velocity) may be discontinuous while the variables themselves are continuous. A *shock wave* represents a surface across which the thermodynamic properties and the flow velocity are essentially discontinuous. Thus, the characteristic curves, or Mach lines, are patching lines for continuous flows, whereas shock waves are patching lines for discontinuous flows.

Consider the curved shock wave illustrated in Fig. 8.11. The flow upstream of the shock wave, which is stationary in the body-fixed-coordinate system, is supersonic. At the plane of symmetry, the shock wave is normal (or perpendicular) to the freestream flow, and the flow downstream of the shock wave is subsonic. Away from the plane of symmetry, the shock wave is oblique and the downstream flow is often supersonic. The velocity and the thermodynamic properties upstream of the shock wave are designated by the subscript 1. Note that whereas the subscript 1 designates

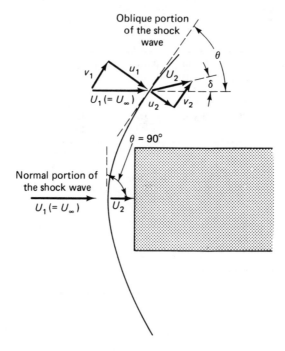

Figure 8.11 Curved shock wave illustrating nomenclature for normal shock wave and for oblique shock wave.

the free-stream (∞) properties for flows such as those in Fig. 8.11, it designates the local flow properties just upstream of the shock wave when it occurs in the midchord region of a transonic airfoil (see Chapter 9). The downstream values are designated by the subscript 2. We will analyze oblique shock waves by writing the continuity, the momentum, and the energy equations for the flow through the control volume, shown in Fig. 8.12. For a steady flow, the integral equations of motion yield the following relations for the flow across an oblique segment of the shock wave:

1. Continuity:

$$\rho_1 u_1 = \rho_2 u_2 \qquad (8.62)$$

2. Normal component of momentum:

$$p_1 + \rho_1 u_1^2 = p_2 + \rho_2 u_2^2 \qquad (8.63)$$

3. Tangential component of momentum:

$$\rho_1 u_1 v_1 = \rho_2 u_2 v_2 \qquad (8.64)$$

4. Energy:

$$h_1 + \tfrac{1}{2}(u_1^2 + v_1^2) = h_2 + \tfrac{1}{2}(u_2^2 + v_2^2) \qquad (8.65)$$

In addition to describing the flow across an oblique shock wave such as shown in Fig. 8.12, these relations can be used to describe the flow across a normal shock wave, or that portion of a curved shock wave which is perpendicular to the free stream, by letting $v_1 = v_2 = 0$.

Equation (8.63) can be used to calculate the maximum value of the pressure coefficient in the hypersonic limit as $M_1 \to \infty$. In this case, the flow is essentially stagnated ($U_2 \approx 0$) behind a normal shock wave. Thus, equation (8.63) becomes

$$p_2 - p_1 \approx \rho_1 u_1^2 \tag{8.66}$$

As a result,

$$C_{p_{max}} = \frac{p_2 - p_1}{\frac{1}{2}\rho_1 U_1^2} \approx 2 \tag{8.67}$$

Note that, at the stagnation point of a vehicle in a hypersonic stream, C_p approaches 2.0. The value of C_p at the stagnation point of a vehicle in a supersonic stream is a function of the free-stream Mach number and is greater than 1.0. Recall that it is 1.0 for a low-speed stream independent of the velocity, provided that the flow is incompressible.

Comparing equation (8.62) with (8.64), one finds that for the oblique shock wave,

$$v_1 = v_2 \tag{8.68}$$

That is, the tangential component of the velocity is constant across the shock wave and we need not consider equation (8.64) further. Thus, the energy equation becomes

$$h_1 + \frac{1}{2}u_1^2 = h_2 + \frac{1}{2}u_2^2 \tag{8.69}$$

There are four unknowns (p_2, ρ_2, u_2, h_2) in the three equations (8.62), (8.63), and (8.69). Thus, we need to introduce an equation of state as the fourth equation. For hypervelocity flows where the shock waves are strong enough to cause dissociation or ionization, one can solve these equations numerically using the equation of state in tabular or in graphical form (e.g., Ref. 8.5). However, for a perfect-gas flow,

$$p = \rho R T$$

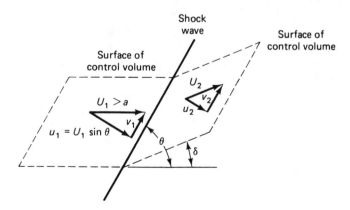

Figure 8.12   Control volume for analysis of flow through an oblique shock wave.

and

$$h = c_p T$$

A comparison of the properties downstream of a normal shock wave as computed using the charts for air in thermodynamic equilibrium with those computed using the perfect-gas model for air is presented in Chapter 13.

Note that equations (8.62), (8.63), and (8.69) involve only the component of velocity normal to the shock wave:

$$u_1 = U_1 \sin \theta \tag{8.70}$$

Hence, the property changes across an oblique shock wave are the same as those across a normal shock wave when they are written in terms of the upstream Mach number component perpendicular to the shock. The tangential component of the velocity is unchanged. This is the *sweepback principle*, that the oblique flow is reduced to the normal flow by a uniform translation of the axes (i.e., a Galilean transformation). Note that the tangential component of the Mach number does change, since the temperature (and therefore the speed of sound) changes across the shock wave.

Since the flow through the shock wave is adiabatic, the entropy must increase as the flow passes through the shock wave. Thus, the flow must decelerate (i.e., the pressure must increase) as it passes through the shock wave. One obtains the relation between the shock-wave (θ) and the deflection angle (δ),

$$\cot \delta = \tan \theta \left[\frac{(\gamma + 1)M_1^2}{2(M_1^2 \sin^2 \theta - 1)} - 1 \right] \tag{8.71}$$

From equation (8.71) it can be seen that the deflection angle is zero for two "shock"-wave angles. (1) The flow is not deflected when $\theta = \mu$, since the Mach wave results from an infinitesimal disturbance (i.e., a zero-strength shock wave). (2) The flow is not deflected when it passes through a normal shock wave (i.e., when $\theta = 90°$).

Solutions to equation (8.71) are presented in graphical form in Fig. 8.13a. Note that for a given deflection angle δ, there are two possible values for the shock-wave angle θ. The larger of the two values of θ is called the *strong* shock wave, while the smaller value is called the *weak* shock wave. In practice, the weak shock wave typically occurs in external aerodynamic flows. However, the strong shock wave occurs if the downstream pressure is sufficiently high. The high downstream pressure may occur in flows in wind tunnels, in engine inlets, or in other ducts.

If the deflection angle exceeds the maximum value for which it is possible that a weak shock can be generated, a strong, detached shock wave will occur. For instance, a flat plate airfoil can be inclined 34° to a Mach 3.0 stream and still generate a weak shock wave. This is the maximum deflection angle for a weak shock wave to occur. If the airfoil were to be inclined at 35° to the Mach 3.0 stream, a strong curved shock wave would occur with a complex subsonic/supersonic flow downstream of the shock wave.

Once the shock-wave angle θ has been found for the given values of M_1 and δ, the other downstream properties can be found using the following relations:

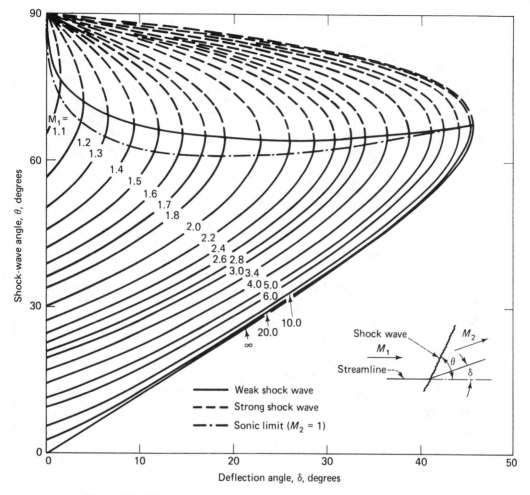

Figure 8.13 Variation of shock-wave parameters with wedge flow-deflection angle for various upstream Mach numbers, $\gamma = 1.4$: (a) shock-wave angle.

$$\frac{p_2}{p_1} = \frac{2\gamma M_1^2 \sin^2 \theta - (\gamma - 1)}{\gamma + 1} \tag{8.72}$$

$$\frac{\rho_2}{\rho_1} = \frac{(\gamma + 1)M_1^2 \sin^2 \theta}{(\gamma - 1)M_1^2 \sin^2 \theta + 2} \tag{8.73}$$

$$\frac{T_2}{T_1} = \frac{[2\gamma M_1^2 \sin^2 \theta - (\gamma - 1)][(\gamma - 1)M_1^2 \sin^2 \theta + 2]}{(\gamma + 1)^2 M_1^2 \sin^2 \theta} \tag{8.74}$$

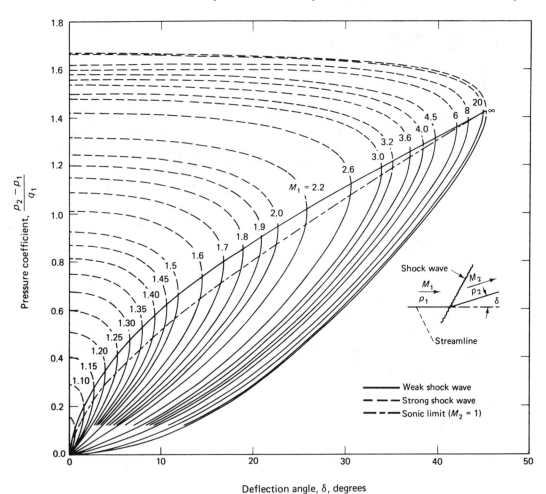

Figure 8.13 (*Continued*) (b) pressure coefficient.

$$M_2^2 = \frac{(\gamma - 1)M_1^2 \sin^2 \theta + 2}{[2\gamma M_1^2 \sin^2 \theta - (\gamma - 1)] \sin^2 (\theta - \delta)} \tag{8.75}$$

$$\frac{p_{t2}}{p_{t1}} = e^{-\Delta s/R}$$

$$= \left[\frac{(\gamma + 1)M_1^2 \sin^2 \theta}{(\gamma - 1)M_1^2 \sin^2 \theta + 2}\right]^{\gamma/(\gamma - 1)} \left[\frac{\gamma + 1}{2\gamma M_1^2 \sin^2 \theta - (\gamma - 1)}\right]^{1/(\gamma - 1)} \tag{8.76}$$

and

$$C_p = \frac{p_2 - p_1}{q_1} = \frac{4(M_1^2 \sin^2 \theta - 1)}{(\gamma + 1)M_1^2} \tag{8.77}$$

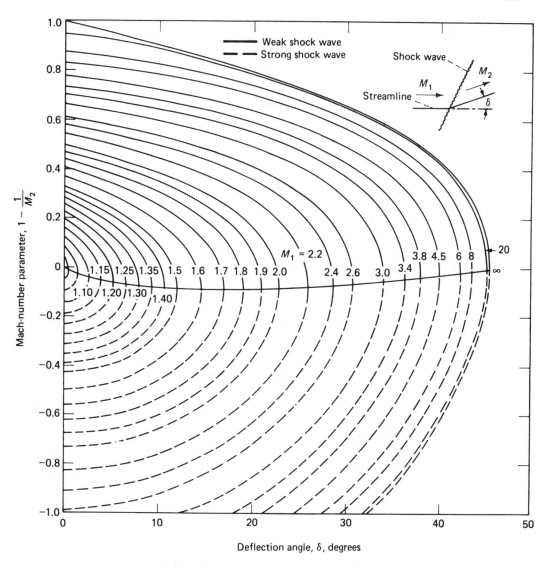

Figure 8.13 (*Continued*) (c) downstream Mach number.

The pressure coefficient is presented in Fig. 8.13b as a function of δ and M_1. Equation (8.77) is consistent with equation (8.67) for a normal shock since $\gamma \rightarrow 1$ as $M_1 \rightarrow \infty$ due to the dissociation of molecules in the air at high Mach numbers. The values for many of these ratios are presented for a normal shock wave in Table 8.3 and in Fig. 8.14. The values for the pressure ratios, the density ratios, and the temperature ratios for an oblique shock wave can be read from Table 8.3 provided that $M_1 \sin \theta$ is used instead of M_1 in the first column. Note that since it is the tangential component of the velocity which is unchanged and not the tangential component of the Mach number, we cannot use Table 8.3 to calculate the downstream Mach number. The downstream

TABLE 8.3 CORRELATION OF FLOW PROPERTIES ACROSS A NORMAL SHOCK WAVE AS A FUNCTION OF THE UPSTREAM MACH NUMBER FOR AIR, $\gamma = 1.4$

M_1	M_2	$\dfrac{p_2}{p_1}$	$\dfrac{\rho_2}{\rho_1}$	$\dfrac{T_2}{T_1}$	$\dfrac{p_{t2}}{p_{t1}}$
1.00	1.00000	1.00000	1.00000	1.00000	1.00000
1.05	0.95312	1.1196	1.08398	1.03284	0.99987
1.10	0.91177	1.2450	1.1691	1.06494	0.99892
1.15	0.87502	1.3762	1.2550	1.09657	0.99669
1.20	0.84217	1.5133	1.3416	1.1280	0.99280
1.25	0.81264	1.6562	1.4286	1.1594	0.98706
1.30	0.78596	1.8050	1.5157	1.1909	0.97935
1.35	0.76175	1.9596	1.6027	1.2226	0.96972
1.40	0.73971	2.1200	1.6896	1.2547	0.95819
1.45	0.71956	2.2862	1.7761	1.2872	0.94483
1.50	0.70109	2.4583	1.8621	1.3202	0.92978
1.55	0.68410	2.6363	1.9473	1.3538	0.91319
1.60	0.66844	2.8201	2.0317	1.3880	0.89520
1.65	0.65396	3.0096	2.1152	1.4228	0.87598
1.70	0.64055	3.2050	2.1977	1.4583	0.85573
1.75	0.62809	3.4062	2.2781	1.4946	0.83456
1.80	0.61650	3.6133	2.3592	1.5316	0.81268
1.85	0.60570	3.8262	2.4381	1.5694	0.79021
1.90	0.59562	4.0450	2.5157	1.6079	0.76735
1.95	0.58618	4.2696	2.5919	1.6473	0.74418
2.00	0.57735	4.5000	2.6666	1.6875	0.72088
2.05	0.56907	4.7363	2.7400	1.7286	0.69752
2.10	0.56128	4.9784	2.8119	1.7704	0.67422
2.15	0.55395	5.2262	2.8823	1.8132	0.65105
2.20	0.54706	5.4800	2.9512	1.8569	0.62812
2.25	0.54055	5.7396	3.0186	1.9014	0.60554
2.30	0.53441	6.0050	3.0846	1.9468	0.58331
2.35	0.52861	6.2762	3.1490	1.9931	0.56148
2.40	0.52312	6.5533	3.2119	2.0403	0.54015
2.45	0.51792	6.8362	3.2733	2.0885	0.51932
2.50	0.51299	7.1250	3.3333	2.1375	0.49902
2.55	0.50831	7.4196	3.3918	2.1875	0.47927
2.60	0.50387	7.7200	3.4489	2.2383	0.46012
2.65	0.49965	8.0262	3.5047	2.2901	0.44155
2.70	0.49563	8.3383	3.5590	2.3429	0.42359
2.75	0.49181	8.6562	3.6119	2.3966	0.40622
2.80	0.48817	8.9800	3.6635	2.4512	0.38946
2.85	0.48470	9.3096	3.7139	2.5067	0.37330
2.90	0.48138	9.6450	3.7629	2.5632	0.35773
2.95	0.47821	9.986	3.8106	2.6206	0.34275
3.00	0.47519	10.333	3.8571	2.6790	0.32834
3.50	0.45115	14.125	4.2608	3.3150	0.21295
4.00	0.43496	18.500	4.5714	4.0469	0.13876
4.50	0.42355	23.458	4.8119	4.8761	0.09170
5.00	0.41523	29.000	5.0000	5.8000	0.06172

TABLE 8.3 (Continued)

M_1	M_2	$\dfrac{p_2}{p_1}$	$\dfrac{\rho_2}{\rho_1}$	$\dfrac{T_2}{T_1}$	$\dfrac{p_{t2}}{p_{t1}}$
6.00	0.40416	41.833	5.2683	7.941	0.02965
7.00	0.39736	57.000	5.4444	10.469	0.01535
8.00	0.39289	74.500	5.5652	13.387	0.00849
9.00	0.38980	94.333	5.6512	16.693	0.00496
10.00	0.38757	116.50	5.1743	20.388	0.00304
∞	0.37796	∞	6.000	∞	0

Mach number is presented in Fig. 8.13c as a function of the deflection angle and of the upstream Mach number. An alternative procedure to calculate the Mach number behind the shock wave would be to convert the value of M_2 in Table 8.3 (which is the normal component of the Mach number) to the normal component of velocity using T_2 to calculate the local speed of sound. Then we can calculate the total velocity downstream of the shock wave:

$$U_2 = \sqrt{u_2^2 + v_2^2}$$

from which we can calculate the downstream Mach number.

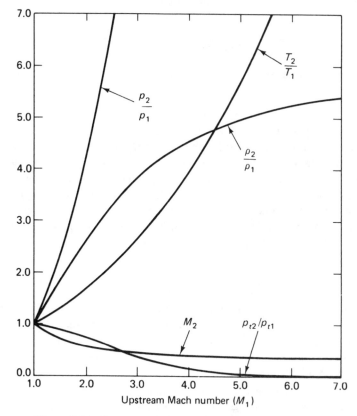

Figure 8.14 Property variations across a normal shock wave.

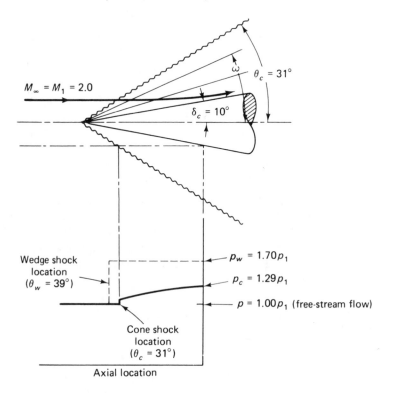

$M_\infty = M_1 = 2.0$

$\theta_c = 31°$

ω

$\delta_c = 10°$

Wedge shock
location
$(\theta_w = 39°)$

$p_w = 1.70 p_1$

$p_c = 1.29 p_1$

$p = 1.00 p_1$ (free-stream flow)

Cone shock
location
$(\theta_c = 31°)$

Axial location

Figure 8.15   Supersonic flow past a sharp cone at zero angle of attack.

For supersonic flow past a cone at zero angle of attack, the shock wave angle θ_c depends on the upstream Mach number M_1 and the cone half-angle δ_c. Whereas all properties are constant downstream of the weak, oblique shock wave generated when supersonic flow encounters a wedge, this is not the case for the conical shock wave. In this case, properties are constant along rays (identified by angle ω) emanating from the vertex of the cone, as shown in the sketch of Fig. 8.15. Thus, the static pressure varies with distance back from the shock along a line parallel to the cone axis. The shock-wave angle, the pressure coefficient

$$C_p = \frac{p_c - p_1}{q_1}$$

(where p_c is the static pressure along the surface of the cone), and the Mach number of the inviscid flow at the surface of the cone M_c are presented in Fig. 8.16 as a function of the cone semivertex angle δ_c and the free-stream Mach number M_1.

Example 8.5

Consider the cone whose semivertex angle is $10°$ exposed to a Mach 2 stream, as shown in Fig. 8.15. Using Fig. 8.16a, the shock wave angle is $31°$. The pressure just downstream of the shock wave is given by equation (8.72) as

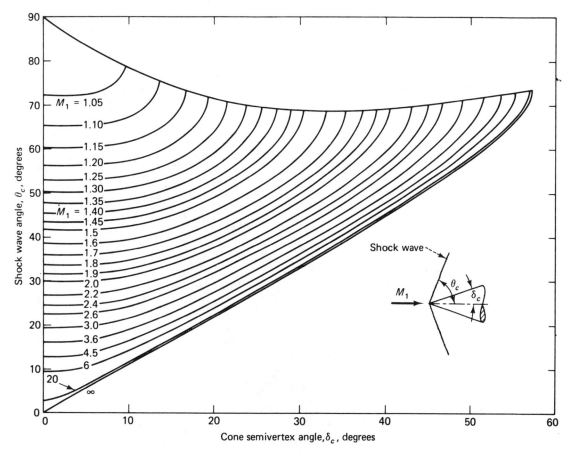

Figure 8.16 Variations of shock-wave parameters with cone semivertex angle for various upstream Mach numbers, $\gamma = 1.4$: (a) shock-wave angle.

$$\frac{p_2}{p_1} = \frac{2(1.4)(4)(0.5150)^2 - 0.4}{2.4}$$

$$= 1.07$$

The pressure increases across the shock layer (i.e., moving parallel to the cone axis), reaching a value of $1.29p_1$ (or $1.29p_\infty$) at the surface of the cone, as can be calculated using Fig. 8.16b. Included for comparison is the pressure downstream of the weak, oblique shock for a wedge with the same turning angle. Note that the shock wave angle θ_w and the pressure in the shock layer p_w are greater for the wedge. The difference is due to axisymmetric effects, which allow the flow to spread around the cone and which do not occur in the case of the wedge.

VISCOUS BOUNDARY LAYER

In our analysis of boundary layers in Chapter 4, we considered only flows for which the density is constant. The correlations for the skin-friction coefficient which were

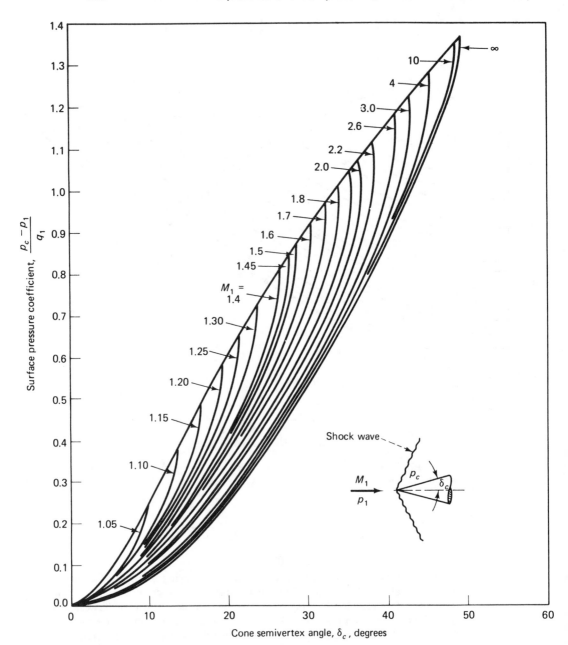

Figure 8.16 (*Continued*) (b) pressure coefficient.

developed for these low-speed flows were a function of the Reynolds number only. However, as the free-stream Mach number approaches the transonic regime, shock waves occur at various positions on the configuration, as discussed in Chapter 9. The presence of the boundary layer and the resultant shock-wave/boundary-layer interaction can radically alter the flow field. Furthermore, when the free-stream Mach number

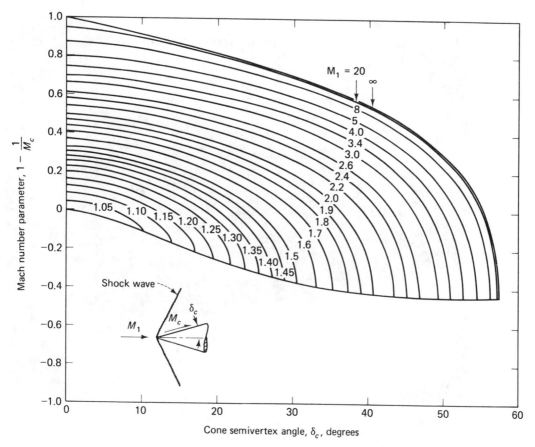

Figure 8.16 (*Continued*) (c) Mach number on the surface of the cone.

exceeds two, the work of compression and of viscous energy dissipation produces considerable increases in the static temperature in the boundary layer. Because the temperature-dependent properties, such as the density and the viscosity, are no longer constant, our solution technique must include the energy equation as well as the continuity equation and the momentum equation.

Because the density gradients in the compressible boundary layer affect a parallel beam of light passing through a wind-tunnel test section, we can photographically record the boundary layer. A shadowgraph of the flow field for a supersonic cone ($\delta_c = 12°$) in a Mach 11.5 stream is presented in Fig. 8.17. Boundary-layer transition occurs approximately one-quarter of the way along that portion of the conical generator which appears in the photograph. Upstream (nearer the apex), the laminar boundary layer is thin and "smooth." Downstream (toward the base of the cone), the vortical character of the turbulent boundary layer is evident. The reader might also note that the shock wave angle is approximately 14.6°. The theoretical value, as calculated using Fig. 8.16a, is 14.3°. That a slightly larger angle is observed experimentally is due in part to the displacement effect of the boundary layer.

In addition to the calculations of the aerodynamic forces and moments, the fluid

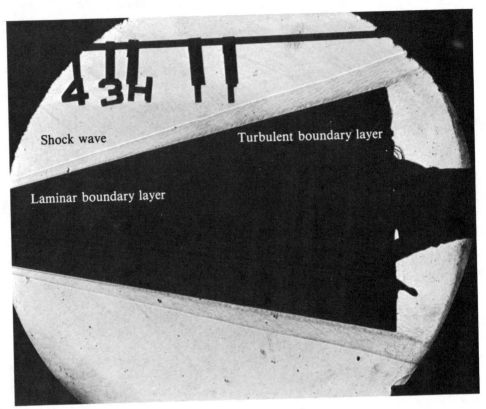

Figure 8.17 Hypersonic flow past a slender cone: $M_\infty = 11.5$; $Re_{\infty,L} = 4.28 \times 10^6$; $\delta_c = 12°$. (Courtesy of Vought Corp.)

dynamicist must address the problem of heat transfer. In this section we will discuss briefly:

1. The effects of compressibility
2. Shock-wave/boundary-layer interactions

Effects of Compressibility

As noted above, considerable variations in the static temperature occur in the supersonic flow field around a body. We can calculate the maximum temperature that occurs in the flow of a perfect gas by using the energy equation for an adiabatic flow [i.e., equation (8.34)]. This maximum temperature, which is the *stagnation temperature*, is

$$T_t = T_\infty \left(1 + \frac{\gamma - 1}{2} M_\infty^2 \right)$$

For example, let us calculate the stagnation temperature for a flow past a vehicle flying at a Mach number of 4.84 and at an altitude of 20 km. Referring to Table 1.2, T_∞ is 216.65 K. Also, since a_∞ is 295.069 m/s, U_∞ is 1428 m/s (or 5141 km/h).

$$T_t = 216.65[1 + 0.2(4.84)^2] = 1231.7 \text{ K} = T_{te}$$

which is the stagnation temperature of the air outside the boundary layer (where the effects of heat transfer are negligible). Thus, for this flow, we see that the temperature of the air at the stagnation point is sufficiently high that we could not use an aluminum structure. We have calculated the stagnation temperature, which exists only where the flow is at rest relative to the vehicle and where there is no heat transferred from the fluid. However, because heat is transferred from the boundary layer of these flows, the static temperature does not reach this value. A numerical solution for the static temperature distribution across a laminar boundary layer on a flat-plate wing exposed to this flow is presented in Fig. 8.18. Although the maximum value of the static temperature is well below the stagnation temperature, it is greater than either the temperature at the wall or at the edge of the boundary layer. Thus, the designer of vehicles that fly at supersonic speeds must consider problems related to convective heat transfer (i.e., the heat transfer due to fluid motion).

The correlation between convective heat transfer and the shear forces acting at the wall, which is known as *Reynolds analogy* and was discussed in Chapter 4, is clearly illustrated in Fig. 8.19. The streaks in the oil-flow pattern obtained in a wind tunnel show that the regions of high shear correspond to the regions of high heating, which is indicated by the char patterns on the recovered spacecraft. For further information about convective heat transfer, the reader is referred to Chapter 4 and to texts such as Refs. 8.6 and 8.7.

For high-speed flow past a flat plate, the skin-friction coefficient depends on the local Reynolds number, the Mach number of the inviscid flow, and the temperature ratio, T_w/T_e. Thus,

$$C_f = C_f\left(\text{Re}_x, M_e, \frac{T_w}{T_e}\right) \tag{8.78}$$

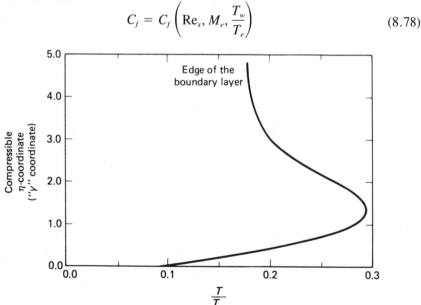

Figure 8.18 Static temperature distribution across a compressible, laminar boundary layer: $M_e = 4.84$; $T_w = 0.095T_{te}$.

Figure 8.19 Comparison between the oil flow pattern (indicating skin friction) obtained in the wind tunnel and the char patterns on a recovered Apollo spacecraft. (Courtesy of NASA.)

Spalding and Chi (Ref. 8.8) developed a calculation procedure based on the assumption that there is a unique relation between $C_f F_c$ and $Re_x F_{Re}$, where C_f is the skin-friction coefficient, Re_x is the local Reynolds number, and F_c and F_{Re} are correlation parameters which depend only on the Mach number and on the temperature ratio. F_c and F_{Re} are presented as functions of M_e and of T_w/T_e in Tables 8.4 and 8.5, respectively. Thus, given the flow conditions and the surface temperature, we can calculate F_c and F_{Re}. Then we can calculate the product $F_{Re} Re_x$ and find the corresponding value of $F_c C_f$ in Table 8.6. Since F_c is known, we can solve for C_f.

The ratio of the experimental skin-friction coefficient to the incompressible value at the same Reynolds number (as taken from Ref. 8.9) is presented in Fig. 8.20 as a function of the Mach number. The experimental skin-friction coefficients are for adiabatic flows; that is, the surface temperature was such that there was no heat transferred from the fluid to the wall. The experimental values are compared with those given by the Spalding-Chi correlation.

TABLE 8.4 VALUES OF F_c AS A FUNCTION OF M_e AND T_w/T_e

$\dfrac{T_w}{T_e}$	M_e						
	0.0	1.0	2.0	3.0	4.0	5.0	6.0
0.05	0.3743	0.4036	0.4884	0.6222	0.7999	1.0184	1.2759
0.10	0.4331	0.4625	0.5477	0.6829	0.8628	1.0842	1.3451
0.20	0.5236	0.5530	0.6388	0.7756	0.9584	1.1836	1.4491
0.30	0.5989	0.6283	0.7145	0.8523	1.0370	1.2649	1.5337
0.40	0.6662	0.6957	0.7821	0.9208	1.1069	1.3370	1.6083
0.50	0.7286	0.7580	0.8446	0.9839	1.1713	1.4031	1.6767
0.60	0.7873	0.8168	0.9036	1.0434	1.2318	1.4651	1.7405
0.80	0.8972	0.9267	1.0137	1.1544	1.3445	1.5802	1.8589
1.00	1.0000	1.0295	1.1167	1.2581	1.4494	1.6871	1.9684
2.00	1.4571	1.4867	1.5744	1.7176	1.9130	2.1572	2.4472
3.00	1.8660	1.8956	1.9836	2.1278	2.3254	2.5733	2.8687
4.00	2.2500	2.2796	2.3678	2.5126	2.7117	2.9621	3.2611
5.00	2.6180	2.6477	2.7359	2.8812	3.0813	3.3336	3.6355
6.00	2.9747	3.0044	3.0927	3.2384	3.4393	3.6930	3.9971
8.00	3.6642	3.6938	3.7823	3.9284	4.1305	4.3863	4.6937
10.00	4.3311	4.3608	4.4493	4.5958	4.7986	5.0559	5.3657

Source: Ref. 8.8.

TABLE 8.5 VALUES OF F_{Re} AS A FUNCTION OF M_e AND T_w/T_e

$\dfrac{T_w}{T_e}$	M_e						
	0.0	1.0	2.0	3.0	4.0	5.0	6.0
0.05	221.0540	232.6437	256.5708	278.2309	292.7413	300.8139	304.3061
0.10	68.9263	73.0824	82.3611	91.2557	97.6992	101.7158	103.9093
0.20	20.4777	22.0029	25.4203	28.9242	31.6618	33.5409	34.7210
0.30	9.8486	10.6532	12.5022	14.4793	16.0970	17.2649	18.0465
0.40	5.7938	6.2960	7.4742	8.7703	9.8686	10.6889	11.2618
0.50	3.8127	4.1588	4.9812	5.9072	6.7120	7.3304	7.7745
0.60	2.6969	2.9499	3.5588	4.2576	4.8783	5.3658	5.7246
0.80	1.5487	1.7015	2.0759	2.5183	2.9247	3.2556	3.5075
1.00	1.0000	1.1023	1.3562	1.6631	1.9526	2.1946	2.3840
2.00	0.2471	0.2748	0.3463	0.4385	0.5326	0.6178	0.6903
3.00	0.1061	0.1185	0.1512	0.1947	0.2410	0.2849	0.3239
4.00	0.0576	0.0645	0.0829	0.1079	0.1352	0.1620	0.1865
5.00	0.0356	0.0400	0.0516	0.0677	0.0856	0.1036	0.1204
6.00	0.0240	0.0269	0.0349	0.0460	0.0586	0.0715	0.0834
8.00	0.0127	0.0143	0.0187	0.0248	0.0320	0.0394	0.0466
10.00	0.0078	0.0087	0.0114	0.0153	0.0198	0.0246	0.0294

TABLE 8.6   VALUES OF F_cC_f AS A FUNCTION OF $F_{Re}Re_x$

F_cC_f	$F_{Re}Re_x$	F_cC_f	$F_{Re}Re_x$
0.0010	5.758×10^{10}	0.0055	8.697×10^4
0.0015	4.610×10^8	0.0060	5.679×10^4
0.0020	4.651×10^7	0.0065	3.901×10^4
0.0025	9.340×10^6	0.0070	2.796×10^4
0.0030	2.778×10^6	0.0075	2.078×10^4
0.0035	1.062×10^6	0.0080	1.592×10^4
0.0040	4.828×10^5	0.0085	1.251×10^4
0.0045	2.492×10^5	0.0090	1.006×10^4
0.0050	1.417×10^5		

Source: Ref. 8.8.

Example 8.6

What is the skin-friction coefficient for a turbulent boundary layer on a flat plate, when $M_e = 2.5$, $Re_x = 6.142 \times 10^6$, and $T_w = 3.0T_e$?

Solution   For these calculations,

$$F_c = 2.056 \text{ (see Table 8.4)}$$

and

$$F_{Re} = 0.1729 \text{ (see Table 8.5)}$$

Thus,

$$Re_x F_{Re} = 1.062 \times 10^6$$

Using Table 8.6, we obtain

$$F_c C_f = 0.0035$$

so that

$$C_f = 1.70 \times 10^{-3}$$

Shock-Wave/Boundary-Layer Interactions

Severe problems of locally high heating or premature boundary-layer separation may result due to viscous/inviscid interactions which occur during flight at supersonic Mach numbers. The shock wave generated by a deflected flap will interact with the upstream boundary layer. The interaction will generally cause the upstream boundary layer to separate with locally high heating rates occurring when the flow reattaches. The extent of the separation, which can cause a loss of control effectiveness, depends on the character of the upstream boundary layer. Other viscous interaction problems can occur when the shock waves generated by the forebody and other external components impinge on downstream surfaces of the vehicle. Again, locally severe heating rates or boundary-layer separation may occur.

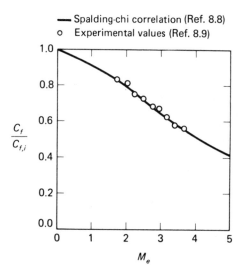

Figure 8.20 Ratio of the compressible, turbulent experimental skin friction coefficient to the incompressible value at the same Reynolds number as a function of Mach number.

The basic features of the interaction between a shock wave and a laminar boundary layer for a two-dimensional flow are shown in Fig. 8.21a. The pressure rise induced by the shock wave is propagated upstream through the subsonic portion of the boundary layer. Recall that pressure disturbances can affect the upstream flow only if the flow is subsonic. As a result, the boundary-layer thickness increases and the momentum decreases. The thickening boundary layer deflects the external stream and creates a series of compression waves to form a λ-like shock structure. If the shock-induced adverse pressure gradient is great enough, the skin friction will be reduced to zero and the boundary layer will separate. The subsequent behavior of the flow is a strong function of the geometry. For a flat plate, the flow reattaches at some distance downstream. In the case of a convex body, such as an airfoil, the flow may or may not reattach, depending upon the body geometry, the characteristics of the boundary layer, and the strength of the shock wave.

If the flow reattaches, a Prandtl–Meyer expansion fan results as the flow turns back toward the surface. As the flow reattaches and turns parallel to the plate, a second shock wave (termed the *reattachment shock*) is formed. Immediately downstream of reattachment, the boundary-layer thickness reaches a minimum. It is in this region that the maximum heating rates occur.

In the case of a shock interaction with a turbulent boundary layer (Fig. 8.21b), the length of the interaction is considerably shorter than the interaction length for a laminar boundary layer. This results because a turbulent boundary layer has greater momentum than does a laminar boundary layer and can therefore overcome a greater adverse pressure gradient. Furthermore, since the subsonic portion of a turbulent boundary layer is relatively thin, the region through which the shock-induced pressure rise can propagate upstream is limited. As a result, a much greater pressure rise is required to cause a turbulent boundary layer to separate.

Pressure distributions typical of the shock-wave/boundary-layer interactions are presented in Fig. 8.21 for a laminar boundary layer and for a turbulent one. As one

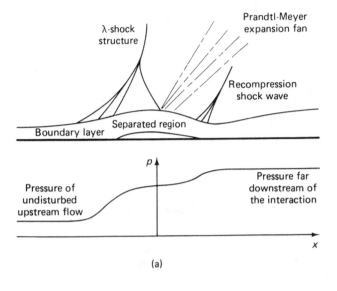

(a)

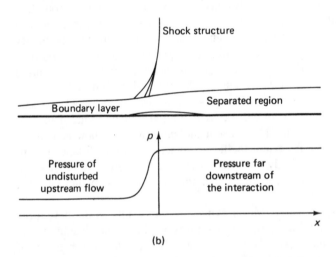

(b)

Figure 8.21   Flow field for shock-wave boundary-layer interaction: (a) laminar boundary layer; (b) turbulent boundary layer.

would expect from the preceding description of the shock interaction, the pressure rise is spread over a much longer distance when the boundary layer is laminar.

PROBLEMS

8.1. The test section of the wind tunnel of Example 8.3 has a square cross section that is 1.22 m × 1.22 m.

   (a) What is the mass-flow rate of air through the test section for the flow conditions of the example (i.e., $M_\infty = 3.5$, $p_\infty = 54.23$ mm Hg, and $T_\infty = 90.18$ K)?

   (b) What is the volume flow rate of air through the test section?

8.2. A practical limit for the stagnation temperature of a continuously operating supersonic wind tunnel is 1000 K. With this value for the stagnation (or total) temperature, prepare a graph of the static free-stream temperature as a function of the test-section Mach number. The fluid in the free stream undergoes an isentropic expansion. The static temperature should not be allowed to drop below 50 K, since at low pressure oxygen begins to liquefy at this temperature. With this as a lower bound for temperature, what is the maximum Mach number for the facility? Assume that the air behaves as a perfect gas with $\gamma = 1.4$.

8.3. Tunnel B at the Arnold Engineering Development Center (AEDC) in Tennessee is often used for determining the flow field and/or the heating rate distributions. The Mach number in the test section is 8. If the stagnation temperature is 725 K and the stagnation pressure can be varied from 6.90×10^5 N/m^2 to 5.90×10^6 N/m^2, what is the range of Reynolds number that can be obtained in this facility? Assume that the characteristic dimension is 0.75 m, which could be the length of a National Aerospace Plane model.

8.4. Given the flow conditions discussed in Problem 8.3.
   (a) What is the (range of the) static temperature in the test section? If there is only one value of the static temperature, state why. To what range of altitude (as given in Table 1.2), if any, do these temperatures correspond?
   (b) What is the (range of the) velocity in the test section? Does it depend on the pressure?
   (c) What is the range of static pressure in the test section? To what range of altitude (as given in Table 1.2) do these pressures correspond?

8.5. A convergent-only nozzle, which exhausts into a large tank, is used as a transonic wind tunnel (Fig. P8.5). Assuming that the air behaves as a perfect gas, answer the following.
   (a) If the pressure in the tank is atmospheric (i.e., 1.01325×10^5 N/m^2), what should the stagnation pressure in the nozzle reservoir be so that the Mach number of the exhaust flow is 0.80?

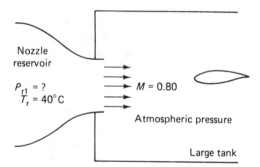

Nozzle reservoir

$P_{t1} = ?$
$T_t = 40°C$

$M = 0.80$

Atmospheric pressure

Large tank   **Figure P8.5**

   (b) If the stagnation temperature is 40°C, what is the static temperature in the test stream?
   (c) A transonic airfoil with a 30-cm chord is located in the test stream. What is Re$_c$ for the airfoil?

$$Re_c = \frac{\rho_\infty U_\infty c}{\mu_\infty}$$

   (d) What is the pressure coefficient C_p at the stagnation point of the airfoil?

8.6. Air flows through the insulated variable-area streamtube such that it may be considered one-dimensional and steady. At one end of the streamtube: $A_1 = 5.0$ ft^2 and $M_1 = 3.0$. At the other end of the streamtube: $p_{t2} = 2116$ psfa, $p_2 = 2101$ psfa, $T_{t2} = 500°R$, and

$A_2 = 5.0$ ft². What is the flow direction; that is, is the flow from (1) to (2) or from (2) to (1)?

8.7. A pitot tube in a supersonic stream produces a curved shock wave standing in front of the nose part, as shown in Fig. P8.7. Assume that the probe is at zero angle of attack and that the shock wave is normal in the vicinity of the nose. The probe is designed to sense the stagnation pressure behind a normal shock (p_{t2}) and the static pressure behind a normal shock (p_2). Derive an expression for M_∞ in terms of the stagnation (p_{t2}) and static (p_2) pressures sensed by the probe.

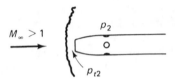

Figure P8.7

8.8. Consider the flow in a streamtube as it crosses a normal shock wave (Fig. P8.8).
(a) Determine the ratio A_2^*/A_1^*.
(b) What are the limits of A_2^*/A_1^* as $M_1 \to 1$ and as $M_1 \to \infty$?
(c) What is the significance of A_1^*? Of A_2^*?

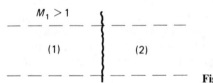

Figure P8.8

8.9. You are to measure the surface pressure on simple models in a supersonic wind tunnel. To evaluate the experimental accuracy, it is necessary to obtain theoretical pressures for comparison with the data. If a 30° wedge is to be placed in a Mach 3.5 stream (Fig. P8.9), calculate:
(a) The surface pressure in N/m².

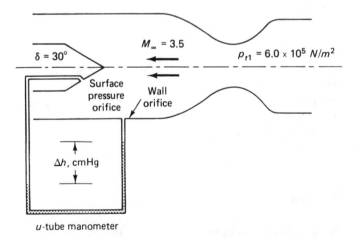

Figure P8.9

(b) The pressure difference (in cm Hg) between the columns of mercury in a U-tube manometer between the pressure experienced by the surface orifice and the wall orifice (which is used to measure the static pressure in the test section).

(c) The dynamic pressure of the free-stream flow.

Other measurements are: the pressure in the reservoir is 6.0×10^5 N/m^2 and the barometric pressure is 75.2 cm Hg.

8.10. It is desired to turn a uniform stream of air compressively by 10°. The upstream Mach number is 3.

(a) Determine the final Mach number and net change in entropy, if the turning is accomplished by (1) a single 10° sharp turn, (2) two successive 5° sharp turns, or (3) an infinite number of infinitesimal turns.

(b) What do you conclude from the results in part (a)?

(c) In light of the results in part (b), can we make any conclusions as to whether it is better to make *expansive* turns gradually or abruptly, when the flow is supersonic?

8.11. A flat-plate airfoil, whose length is c, is in a Mach 2.0 stream at an angle of attack of 10° (Fig. P8.11).

(a) Use the oblique shock-wave relations to calculate the static pressure in region (2) in terms of the free-stream value p_1.

(b) Use the Prandtl–Meyer relations to calculate the static pressure in region (3) in terms of p_1.

(c) Calculate C_l, C_d, and $C_{m0.5c}$ (the pitching moment about the midchord). Do these coefficients depend on the free-stream pressure (i.e., the altitude)?

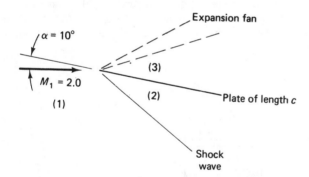

Figure P8.11

8.12. A conical spike whose half-angle (δ_c) is 8° is located in the inlet of a turbojet engine (Fig. P8.12). The engine is operating in a $M_\infty = 2.80$ stream such that the angle of attack of the spike is zero. If the radius of the engine inlet is 1.0 m, determine l, the length of the spike extension, such that the conical shock just grazes the lip of the nacelle.

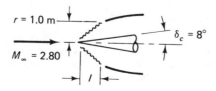

Figure P8.12

8.13. Consider the two-dimensional inlet for a turbojet engine. The upper surface deflects the flow 10°, the lower surface deflects the flow 5° (Fig. P8.13). The free-stream Mach number is 2.5 (i.e., $M_1 = 2.5$) and the pressure p_1 is 5.0×10^3 N/m². Calculate the static pressure, the Mach number, and the flow direction in regions (2), (3), (4), and (5). Note that since regions (4) and (5) are divided by a fluid/fluid interface,

$$p_4 = p_5 \quad \text{and} \quad \theta_4 = \theta_5$$

That is, the static pressure in region (4) is equal to that in region (5), and the flow direction in region (4) is equal to that in region (5).

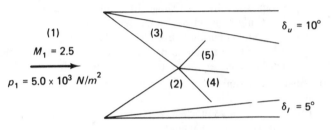

Figure P8.13

8.14. A single wedge airfoil is located on the centerline of the test section of a Mach 2.0 wind tunnel (Fig. P8.14). The airfoil, which has a half-angle δ of 5°, is at zero angle of attack. When the weak, oblique shock wave generated at the leading edge of the airfoil encounters the wall, it is reflected so that the flow in region (3) is parallel to the tunnel wall. If the test section is 30.0 cm high, what is the maximum chord length (c) of the airfoil so that it is not struck by the reflected shock wave? Neglect the effects of shock-wave/boundary-layer interactions at the wall.

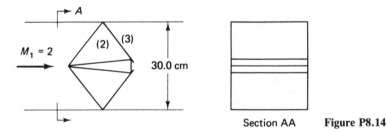

Section AA   **Figure P8.14**

8.15. An airplane flies 600 mi/h at an altitude of 35,000 ft where the temperature is $-75°F$ and the ambient pressure is 474 psfa. What is the temperature and the pressure of the air (outside the boundary layer) at the nose (stagnation point) of the airplane? What is the Mach number for the airplane?

8.16. A three-dimensional bow shock wave is generated by the Shuttle Orbiter during entry. However, there is a region where the shock wave is essentially normal to the free-stream flow, as illustrated in Fig. 8.11. The velocity of the Shuttle is 7.62 km/s at an altitude of 75.0 km. The free-stream temperature and the static pressure at this altitude are 200.15 K and 2.50 N/m², respectively. Use the normal shock relations to calculate the values of the following parameters downstream of the normal shock wave:

p_2:   static pressure

p_{t2}:   stagnation pressure

T_2:   static temperature

T_{t2}:   stagnation temperature

M_2:   Mach number

What is the pressure coefficient at the stagnation point; that is,

$$C_{p,t} = \frac{p_{t2} - p_\infty}{q_\infty} = ?$$

Use the perfect-gas relations and assume that $\gamma = 1.4$. The perfect-gas assumption is valid for the free-stream flow. However, it is not a good assumption for calculating the flow across the shock wave. Therefore, many of the perfect-gas theoretical values for the shocked-flow properties will not even be close to the actual values. See Chapter 13 for further discussions of this problem.

8.17. Consider the hypersonic flow past the cone shown in Fig. 8.17. The cone semivertex angle (δ_c) is $12°$. The free-stream flow is defined by

$$M_\infty = 11.5$$

$$T_t = 1970 \text{ K}$$

$$p_\infty = 1070 \text{ N/m}^2$$

Assume that the flow is inviscid, except for the shock wave itself (i.e., neglect the boundary layer on the cone). Assume further that the gas obeys the perfect-gas laws with $\gamma = 1.4$. Using Fig. 8.16b, calculate the static pressure at the surface of the cone (p_c). Using Fig. 8.16c, calculate the Mach number at the surface of the cone (M_c), which is, in practice, the Mach number at the edge of the boundary layer. Calculate the stagnation pressure of the flow downstream of the shock wave. Calculate the Reynolds number at a point 10.0 cm from the apex of the cone.

8.18. Repeat Problem 8.17 for a planar symmetric wedge that deflects the flow $12°$ (i.e., $\delta = 12°$ in Fig. 8.13).

REFERENCES

8.1. Staff, "Equations, Tables, and Charts for Compressible Flow," *Report 1135*, NACA, 1953.

8.2. Arnold, J. W., "High Speed Wind Tunnel Handbook," *AER-EIR-13552-B*, Vought Aeronautics Division, June 1968.

8.3. Hayes, W. D., and R. F. Probstein, *Hypersonic Flow Theory*, Vol. I, *Inviscid Flows*, Academic Press, New York, 1966.

8.4. Wayland, H., *Differential Equations Applied in Science and Engineering*, Van Nostrand Co., Princeton, N.J., 1957.

8.5. Moeckel, W. E., and K. C. Weston, "Composition and Thermodynamic Properties of Air in Chemical Equilibrium," *TN 4265*, NACA, Apr. 1958.

8.6. Kays, W. M., *Convective Heat and Mass Transfer*, McGraw-Hill Book Company, New York, 1966.

8.7. Chapman, A. J., *Heat Transfer*, Macmillan Publishing Co., New York, 1974.

8.8. Spalding, D. B., and S. W. Chi, ''The Drag of a Compressible Turbulent Boundary Layer on a Smooth Flat Plate with and Without Heat Transfer,'' *Journal of Fluid Mechanics*, Jan. 1964, Vol. 18, Part 7, pp. 117–143.

8.9. Stalmach, C. J., Jr., ''Experimental Investigation of the Surface Impact Pressure Probe Method of Measuring Local Skin Friction at Supersonic Speeds,'' *DRL-410*, Defense Research Laboratory, The University of Texas at Austin, Jan. 1958.

9

COMPRESSIBLE, SUBSONIC FLOWS AND TRANSONIC FLOWS

In Chapters 3 through 7 flow-field solutions were generated for a variety of configurations using the assumption that the density was constant throughout the flow field. As noted when discussing Fig. 8.6, an error of less than 1% results when the incompressible-flow Bernoulli equation is used to calculate the local pressure provided that the local Mach number is less than or equal to 0.5 in air. Thus, if the flight speed is small compared with the speed of sound, the changes in pressure which are generated by the vehicle motion are small relative to the free-stream static pressure, and the influence of compressibility can be neglected. As shown in Fig. 9.1, the streamlines converge as the incompressible flow accelerates past the midsection of the airfoil. The widening of the streamtubes near the nose and the contraction of the streamtubes in the regions of increased velocity lead to a progressive reduction in the curvature of the streamlines. As a result, there is a rapid attenuation of the flow disturbance with distance from the airfoil.

As the flight speed is increased, the flow may no longer be considered as incompressible. Even though the flow is everywhere subsonic, the density decreases as the pressure decreases (or, equivalently, as the velocity increases). The variable-density flow requires a relatively high velocity and diverging streamlines in order to get the mass flow past the midsection of the airfoil. The expansion of the minimum cross section of the streamtubes forces the streamlines outward so that they conform more nearly to the curvature of the airfoil surface, as shown in Fig. 9.1. Thus, the disturbance caused by the airfoil extends vertically to a greater distance.

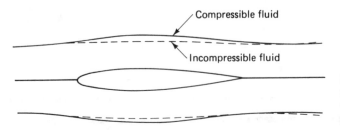

Figure 9.1 Comparison of streamlines for an incompressible flow past an airfoil with those for a subsonic, compressible flow.

Increasing the flight speed further, we reach the *critical Mach number*, the name given to the lowest (subsonic) free-stream Mach number for which the maximum value of the local velocity first becomes sonic. Above the critical Mach number, the flow field contains regions of locally subsonic and locally supersonic velocities in juxtaposition. Such mixed subsonic/supersonic flow fields are termed *transonic flows*.

COMPRESSIBLE, SUBSONIC FLOW

For completely subsonic flows, a compressible flow retains a basic similarity to an incompressible flow. In particular, the compressible flow can be considered to be an irrotational potential motion in many cases. In addition to the absence of significant viscous forces, the existence of potential motion for a compressible flow depends on the existence of a unique relation between the pressure and the density. In an inviscid flow, the fluid elements are accelerated entirely by the action of the pressure gradient. Thus, if the density is a function of the pressure only, the direction of the pressure gradient will coincide with that of the density gradient at all points. The force on each element will then be aligned with the center of gravity of the fluid element and the pressure forces will introduce no rotation.

Linearized Theory for Compressible Subsonic Flow About a Thin Wing at Relatively Small Angles of Attack

The continuity equation for steady, three-dimensional flow is

$$\frac{\partial(\rho u)}{\partial x} + \frac{\partial(\rho v)}{\partial y} + \frac{\partial(\rho w)}{\partial z} = 0 \tag{9.1}$$

In the inviscid region of the flow field (i.e., outside the thin boundary layer), the components of the momentum equation may be written as

$$u\frac{\partial u}{\partial x} + v\frac{\partial u}{\partial y} + w\frac{\partial u}{\partial z} = -\frac{1}{\rho}\frac{\partial p}{\partial x} \tag{9.2a}$$

$$u\frac{\partial v}{\partial x} + v\frac{\partial v}{\partial y} + w\frac{\partial v}{\partial z} = -\frac{1}{\rho}\frac{\partial p}{\partial y} \tag{9.2b}$$

$$u\frac{\partial w}{\partial x} + v\frac{\partial w}{\partial y} + w\frac{\partial w}{\partial z} = -\frac{1}{\rho}\frac{\partial p}{\partial z} \tag{9.2c}$$

The speed of sound is defined as the change in pressure with respect to the change in density for an isentropic process. Thus,

$$a^2 = \left(\frac{\partial p}{\partial \rho}\right)_s$$

However, since the flow we are studying actually is isentropic, this may be written in terms of the actual pressure and density changes which result due to the fluid motion. Thus,

$$\left(\frac{\partial p}{\partial \rho}\right) = a^2 \tag{9.3}$$

Combining equations (9.1) through (9.3) and noting that the flow is irrotational, one obtains

$$\left(1 - \frac{u^2}{a^2}\right)\frac{\partial u}{\partial x} + \left(1 - \frac{v^2}{a^2}\right)\frac{\partial v}{\partial y} + \left(1 - \frac{w^2}{a^2}\right)\frac{\partial w}{\partial z} - 2\frac{uv}{a^2}\frac{\partial u}{\partial y}$$

$$- 2\frac{vw}{a^2}\frac{\partial v}{\partial z} - 2\frac{wu}{a^2}\frac{\partial w}{\partial x} = 0 \tag{9.4}$$

A useful simplification of equation (9.4) can be made for the case of a slender body moving in the x direction at the velocity U_∞. As shown in Fig. 9.2, the magnitude and the direction of the local velocity are changed only slightly from the free-stream velocity. Thus, the resultant velocity at any point can be represented as the vector sum of the free-stream velocity (a constant) together with the perturbation velocities, u', v', and w'. Thus,

$$u = U_\infty + u'$$

$$v = v'$$

$$w = w'$$

Since the perturbation velocities are considered to be small in magnitude when compared with the free-stream velocity, equation (9.4) becomes

$$\left(1 - \frac{u^2}{a^2}\right)\frac{\partial u}{\partial x} + \frac{\partial v}{\partial y} + \frac{\partial w}{\partial z} = 0 \tag{9.5}$$

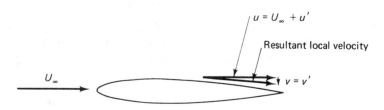

Figure 9.2 Velocity components for subsonic, compressible flow past a thin airfoil at a small angle of attack.

where u/a is essentially equal to the local Mach number. Equation (9.5) can be simplified further if one recalls that the local speed of sound can be determined using the energy equation for adiabatic flow:

$$\frac{a^2}{a_\infty^2} = 1 - \frac{\gamma - 1}{2}\left(\frac{u^2 + v^2 + w^2}{U_\infty^2} - 1\right)M_\infty^2 \tag{9.6}$$

Since only small perturbations are considered, the binomial theorem can be used to generate the relation

$$\frac{a_\infty^2}{a^2} = 1 + \frac{\gamma - 1}{2}M_\infty^2\left(2\frac{u'}{U_\infty} + \frac{u'^2 + v'^2 + w'^2}{U_\infty^2}\right) \tag{9.7}$$

To simplify equation (9.5), note that

$$1 - \frac{u^2}{a^2} = 1 - \frac{(U_\infty + u')^2}{a^2}\frac{U_\infty^2}{U_\infty^2}\frac{a_\infty^2}{a_\infty^2}$$
$$= 1 - \frac{U_\infty^2 + 2u'U_\infty + u'^2}{U_\infty^2}M_\infty^2\frac{a_\infty^2}{a^2} \tag{9.8}$$

Substituting equation (9.7) into equation (9.8) and neglecting the higher-order terms yields the expression

$$1 - \frac{u^2}{a^2} = 1 - M_\infty^2\left[1 + \frac{2u'}{U_\infty}\left(1 + \frac{\gamma - 1}{2}M_\infty^2\right)\right] \tag{9.9}$$

Using equation (9.9), equation (9.5) can be rewritten as

$$(1 - M_\infty^2)\frac{\partial u}{\partial x} + \frac{\partial v}{\partial y} + \frac{\partial w}{\partial z} = M_\infty^2\left(1 + \frac{\gamma - 1}{2}M_\infty^2\right)\frac{2u'}{U_\infty}\frac{\partial u}{\partial x}$$

This equation can be rewritten in terms of the perturbation velocities as

$$(1 - M_\infty^2)\frac{\partial u'}{\partial x} + \frac{\partial v'}{\partial y} + \frac{\partial w'}{\partial z} = \frac{2}{U_\infty}\left(1 + \frac{\gamma - 1}{2}M_\infty^2\right)M_\infty^2 u'\frac{\partial u'}{\partial x} \tag{9.10}$$

Furthermore, the term on the right-hand side often can be neglected, as it is of second order in the perturbation velocity components. As a result, one obtains the linearized equation

$$(1 - M_\infty^2)\frac{\partial u'}{\partial x} + \frac{\partial v'}{\partial y} + \frac{\partial w'}{\partial z} = 0 \tag{9.11}$$

Since the flow is everywhere isentropic, it is irrotational. The condition of irrotationality allows us to introduce a velocity potential ϕ, which is a point function with continuous derivatives. Let ϕ be the potential function for the perturbation velocity:

$$u' = \frac{\partial \phi}{\partial x} \qquad v' = \frac{\partial \phi}{\partial y} \qquad w' = \frac{\partial \phi}{\partial z} \tag{9.12}$$

The resultant expression, which applies to a completely subsonic, compressible flow, is the linearized potential equation

$$(1 - M_\infty^2)\, \phi_{xx} + \phi_{yy} + \phi_{zz} = 0 \tag{9.13}$$

By using a simple coordinate transformation, equation (9.13) can be reduced to Laplace's equation, which we used to describe incompressible, irrotational flows. If the "affine" transformation

$$x' = \frac{x}{\sqrt{1 - M_\infty^2}} \tag{9.14a}$$

$$y' = y \tag{9.14b}$$

$$z' = z \tag{9.14c}$$

is introduced, equation (9.13) becomes

$$\phi_{x'x'} + \phi_{y'y'} + \phi_{z'z'} = 0 \tag{9.15}$$

Thus, if the potential field for incompressible flow past a given configuration is known, a corresponding solution for the linearized compressible flow can be readily obtained. The potential distribution for an incompressible flow and the corresponding "foreshortened" distribution (which satisfies the compressible flow equation) are compared in Fig. 9.3 at points having the same value of ϕ. Although the calculation of a compressible flow field from a known incompressible flow is relatively straightforward, care must be taken in the determination of the boundary conditions satisfied by the compressible flow field.

Referring to equation (9.14) we see that the transformation in effect changes the ratio of the x dimension to the y and z dimensions. Although the spanwise dimensions are unaltered, the transformed chordwise dimension is. Thus, although the airfoil section of the corresponding wings remain geometrically similar, the aspect ratios for the wings differ. The compressible flow over a wing of aspect ratio AR at the Mach number M_∞ is related to the incompressible flow over a wing of aspect ratio AR $\sqrt{1 - M_\infty^2}$. This is illustrated in the sketch of Fig. 9.4. A study of the changes

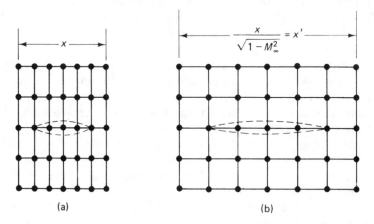

Figure 9.3 Distribution of points having equal values of ϕ in the linearized transformation for subsonic, compressible flow: (a) compressible flow; (b) corresponding incompressible flow.

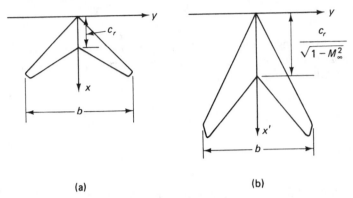

Figure 9.4   Wings for flows related by the linearized transformation: (a) wing for compressible flow; (b) corresponding wing for incompressible flow.

in a completely subsonic flow field around a given wing as the Mach number is increased corresponds to an investigation of the incompressible flow around a series of wings of progressively reduced aspect ratio.

Using the linearized approximation, the pressure coefficient for the compressible flow is given by

$$C_p = -\frac{2u'}{U_\infty} = -\frac{2}{U_\infty}\frac{\partial \phi}{\partial x} \tag{9.16a}$$

which is related to the pressure coefficient for the corresponding incompressible flow (C_p') through the correlation

$$C_p = \frac{C_p'}{\sqrt{1 - M_\infty^2}} \tag{9.16b}$$

The effect of compressibility on the flow past an airfoil system is to increase the horizontal perturbation velocities over the airfoil surface by the factor $1/\sqrt{1 - M_\infty^2}$. The correlation is known as the *Prandtl–Glauert formula*. The resultant variation for the lift-curve slope with Mach number is presented for a two-dimensional unswept airfoil in Fig. 9.5.

Although the Prandtl–Glauert relation provides a simple method for calculating the flow around an airfoil, Jones and Cohen (Ref. 9.1) warn that the method generally underestimates the effect of compressibility on the magnitude of disturbances for airfoils of finite thickness. As the free-stream Mach number approaches a value of unity, the quantity $1/\sqrt{1 - M_\infty^2}$ approaches infinity, causing various perturbation parameters to approach infinity. Hence, the Prandtl–Glauert formula begins to show increasing departures from reality as the Mach number approaches a value of unity. The relative inaccuracy at a particular Mach number depends on parameters such as the section thickness and the angle of attack.

Foss and Blay (Ref. 9.2) described a test flight of the YP-38 in September 1940. Major Signa Gilkey peeled off into a steep dive starting at 35,000 ft. "As he headed down and the airspeed built up, the airplane began to shake, mildly at first and then more violently. Even worse, the airplane wanted to nose down further and increase the

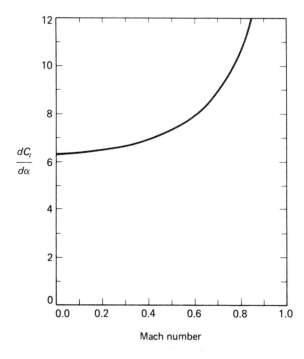

Figure 9.5 Variation of lift-curve slope with Mach number using Prandtl–Glauert formula.

Mach number

severity of the dive. The control column was shaking and the control forces had become heavy, and it was impossible to pull the column back to counteract the nose-down tendency. The airplane was out of control!'' As noted by Foss and Blay, Major Gilkey elected to stay with the plane and recovered at 7000 ft. ''Major Gilkey had become the first pilot to encounter a new, uncharted high-speed flight regime. He had attained air speeds where new aerodynamic phenomena known as compressibility effects were created.''

Using a P-38 model in the high-speed wind tunnel of the Ames Research Center, model drag-rise characteristics were measured for the first time. The wind-tunnel correlations, as taken from Ref. 9.2, are reproduced in Fig. 9.6. Pressure data and flow visualization clearly revealed the shock-stall conditions and helped confirm the flight-test findings.

TRANSONIC FLOW PAST UNSWEPT AIRFOILS

The lift coefficient measurements presented in Ref. 9.3 as a function of Mach number are reproduced in Fig. 9.7. The data indicate that the flow is essentially unchanged up to approximately one-half the speed of sound. The variations in the lift coefficient with Mach number indicate complex changes in the flow field through the transonic speed range. Attention is called to the section-lift coefficient at five particular Mach numbers (identified by the letters a through e). Significant differences exist between the flow fields at these five Mach numbers. To illustrate the essential changes in the flow, line drawings made from schlieren photographs are reproduced in Fig. 9.8.

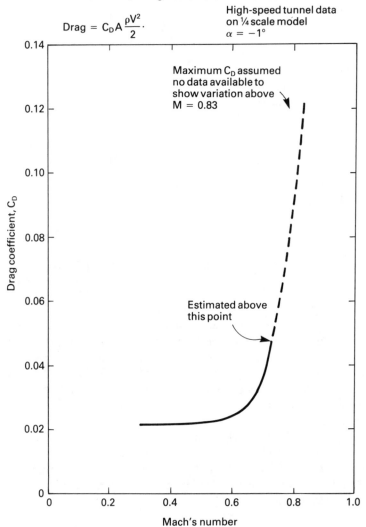

Figure 9.6 High-speed drag variation derived from wind-tunnel tests. (From Ref. 9.2.)

(a) When the free-stream Mach number is 0.75, the flow past the upper surface decelerates from local flow velocities which are supersonic without a shock wave. The lift coefficient is approximately 60% greater than the low-speed values at the same angles of attack.

(b) At $M_\infty = 0.81$, the lift coefficient reaches its maximum value, which is approximately twice the low-speed value. As indicated in Figs. 9.8b and 9.9a, flow is supersonic over the first 70% of the surface, terminating in a shock wave. The flow on the lower surface is subsonic everywhere. Because the viscous flow

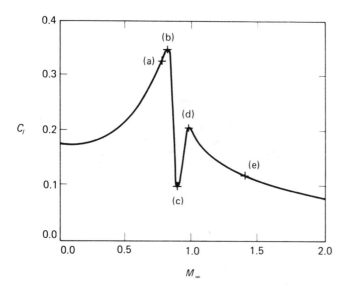

Figure 9.7 Lift coefficient as a function of Mach number to illustrate the effect of compressibility. (From Ref. 9.3.) Refer to Fig. 9.8 for the flow fields corresponding to the lettered points on this graph.

separates at the foot of the shock wave, the wake is appreciably wider than for (a).

(c) At $M_\infty = 0.89$, flow is supersonic over nearly the entire lower surface and deceleration to subsonic speed occurs through a shock wave at the trailing edge. As a result, the lower surface pressures are lower at $M_\infty = 0.89$ than at $M_\infty = 0.81$. Flow on the upper surface is not greatly different than that for Fig. 9.8b. As a result, the lift is drastically reduced. Separation at the foot of the upper surface shock wave is more conspicuous and the turbulent wake is wide. The shock wave at the trailing edge of the lower surface effectively isolates the upper surface from the lower surface. As a result, the pressure on the upper surface near the trailing edge is greater than that on the lower surface. The corresponding pressure and local Mach number distributions are presented in Fig. 9.9b.

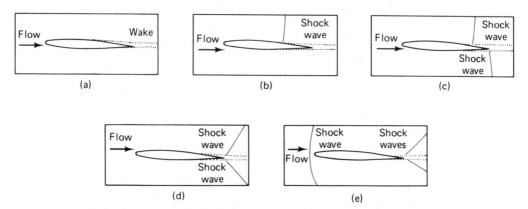

Figure 9.8 Flow field around an airfoil in transonic streams based on schlieren photographs: (a) Mach number $M_\infty = 0.75$; (b) Mach number $M_\infty = 0.81$; (c) Mach number $M_\infty = 0.89$; (d) Mach number $M_\infty = 0.98$; (e) Mach number $M_\infty = 1.4$. (From Ref. 9.3.)

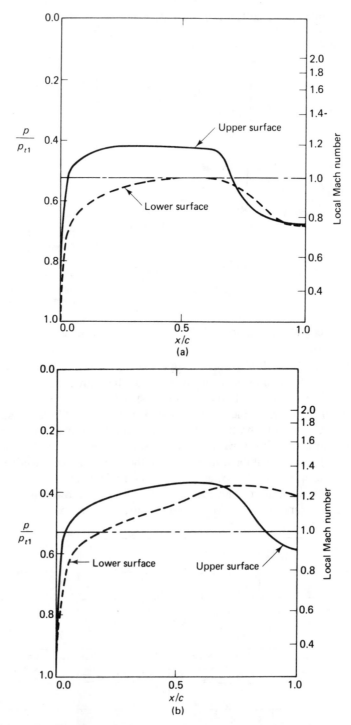

Figure 9.9 Pressure distribution and local Mach number distribution for transonic flows around an airfoil: (a) flow at the trailing edge is subsonic, $M_\infty = 0.81$; (b) flow at the lower surface trailing edge is supersonic, $M_\infty = 0.89$. (From Ref. 9.3.)

(d) When the free-stream Mach number is 0.98, the shock waves both for the upper surface and for the lower surface have reached the trailing edge. The local Mach number is supersonic for most of the airfoil (both upper and lower surfaces).

(e) When the free-stream flow is supersonic, a bow shock wave (i.e., the detached shock wave in front of the leading edge) is generated. The flow around the airfoil is supersonic everywhere except very near the rounded nose. The shock waves at the trailing edge remain, but they have become weaker.

The data presented in Figs. 9.7 through 9.9 illustrate the effect of Mach number for a given airfoil section at a particular angle of attack. Parameters such as thickness ratio, camber, and nose radius also influence the magnitude of the compressibility effects. Transonic flows are very sensitive to the contour of the body surface, since changes in the surface slope affect the location of the shock wave and, therefore, the inviscid flow field as well as the downstream boundary layer.

Furthermore, as discussed in Chapter 8, the shock-wave/boundary-layer interaction and the possible development of separation downstream of the shock wave are sensitive to the character of the boundary layer, to its thickness, and to its velocity profile at the interaction location. Since a turbulent boundary layer can negotiate higher adverse pressure gradients than can a laminar one, the shock-wave/boundary-layer

Figure 9.10 F-111 flying at transonic speeds with shock waves clearly visible. (Courtesy of R. C. Maydew and S. McAlees of the Sandia National Laboratories.)

interaction is smaller for a turbulent boundary layer. Thus, it is important to consider the Reynolds number when simulating a flow in the wind tunnel. A large Reynolds number difference between the desired flow and its simulation may produce significant differences in shock-wave location and the resultant flow field. The use of artificial trips to force transition to occur at a specified point (as discussed in Chapter 5) may be unsatisfactory for transonic flows, since the shock-wave location as well as the extent of flow separation can become a function of the artificial tripping.

The schlieren technique, which allows someone to see shock waves in a wind-tunnel test, such as depicted on the cover and in Fig. 9.8, is based on density gradients in the flow field that bend light rays passing through the flow. Under certain circumstances, these density gradients allow us to "see" the shock waves for an airplane in flight, as shown for the transonic flight of an F-111 in Fig. 9.10.

The experimentally determined lift coefficients for an untwisted rectangular wing whose aspect ratio is 2.75 and whose airfoil section is a NACA 65A005 (a symmetric profile for which $t = 0.05c$) are presented in Fig. 9.11 as a function of angle of attack. The corresponding drag polars are presented in Fig. 9.12. These data from Ref. 9.4 were obtained at Reynolds numbers between 1.0×10^6 and 1.8×10^6. The lift-curve slope is seen to be a function of the free-stream Mach number. Furthermore, the linear

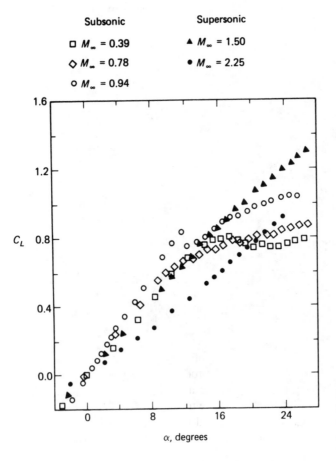

Figure 9.11 Effect of Mach number on the lift-coefficient/angle-of-attack correlation for a rectangular wing, AR = 2.75. (Data from Ref. 9.4.)

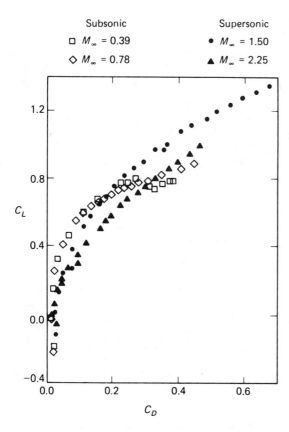

Subsonic

□ $M_\infty = 0.39$

◇ $M_\infty = 0.78$

Supersonic

● $M_\infty = 1.50$

▲ $M_\infty = 2.25$

Figure 9.12 Effect of the Mach number on drag polars for a rectangular wing, AR = 2.75. (Data from Ref. 9.4.)

relation between the lift coefficient and the angle of attack remains valid to higher angles of attack for supersonic flows.

Three features which contribute to the location of the shock wave and, therefore, to the pressure distribution (Ref. 9.5) are:

1. The flow in the supersonic region ahead of the shock wave
2. The pressure rise across the shock wave itself (which involves considerations of the interaction with the boundary layer)
3. The subsonic flow downstream of the shock wave (which involves considerations of the boundary-layer development between the shock wave and the trailing edge and of the flow in the near wake)

If the trailing-edge pressure changes as a result of flow separation from the upper surface, the lower surface flow must adjust itself to produce a similar change in pressure (since the pressure near the trailing edge must be approximately equal for the two surfaces, unless the flow is locally supersonic at the trailing edge). The conditions for divergence of the trailing-edge pressure correspond to those for a rapid drop in the lift coefficient and to the onset of certain unsteady flow phenomena, such as buffeting.

Supercritical Airfoil Sections

The Mach-number/lift-coefficient flight envelopes of modern jet aircraft operating at transonic speeds are limited by the compressibility drag rise and by the buffeting phenomenon. Airfoil section designs which alleviate or delay the onset of the drag rise and buffeting can contribute to higher maximum speeds (transport applications) or better lift performance (fighter applications). Using intuitive reasoning and substantiating experiment (see Ref. 9.6), R. T. Whitcomb and coworkers have developed a "supercritical" airfoil shape which delays the subsonic drag rise. The principal differences between the transonic flow field for a conventional airfoil and that for a supercritical airfoil are illustrated by the data presented in Fig. 9.13. At supercritical Mach numbers, a broad region of locally supersonic flow extends vertically from both airfoils, as indicated by the pressure coefficients above the sonic value and by the shaded areas of the flow fields in Fig. 9.13. The region of locally supersonic flow usually terminates in a shock wave, which results in increased drag. The much flatter shape of the upper surface of the supercritical airfoil causes the shock wave to occur downstream and therefore reduces its strength. Thus, the pressure rise across the shock wave is reduced with a corresponding reduction in drag. However, the diminished curvature of the upper surface also results in a reduction of the lift carried by the midchord region of the airfoil section. To compensate for the loss in lift from the midchord region, additional lift must be generated from the region of the airfoil behind the shock wave, particularly on the lower surface. The increase of lift in this area is achieved by substantial positive camber and incidence of the aft region of the airfoil, especially of the lower surface. The increased lift generated by the concave surface

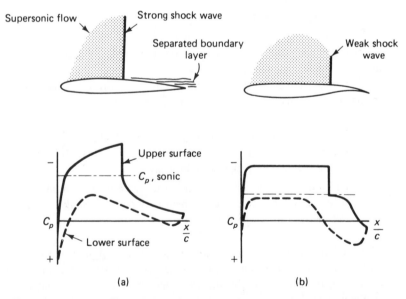

Figure 9.13 Comparison of transonic flow over a NACA 64A series airfoil with that over a "supercritical" airfoil section: (a) NACA 64A series, $M = 0.72$; (b) supercritical airfoil, $M = 0.80$. (From Ref. 9.7.)

near the trailing edge of the lower surface is evident in the experimental pressure distributions presented in Fig. 9.14. The midchord region of the lower surface should be designed to maintain subcritical flow over the range of operating conditions. If not, when the pressure rise associated with a shock wave is superimposed on the pressure rise caused by the cusp, separation of the lower-surface boundary layer would occur. To minimize the surface curvatures (and, therefore, the induced velocities) in the midchord regions both for the upper and the lower surfaces, the leading-edge radius is relatively large.

SWEPT WINGS AT TRANSONIC SPEEDS

In the late 1930s, two aerodynamicists who had been taught by Prandtl, Adolf Busemann and Albert Betz, discovered that drag at transonic and supersonic speeds could be reduced by sweeping back the wings. (The interested reader is referred to Ref. 9.9.) In his paper to the Volta conference on high-speed flight in Rome in 1935, Busemann showed that sweepback would reduce drag at supersonic speeds. Betz, in 1939, was the first person to draw attention to the significant reduction in transonic drag which

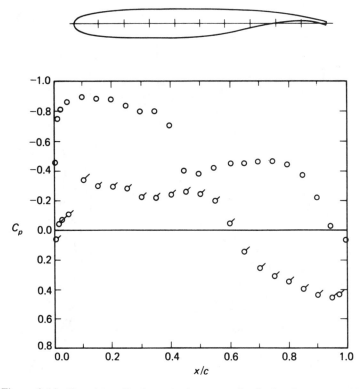

Figure 9.14 Experimentally determined pressure distribution for a supercritical airfoil section, $M_\infty = 0.80$, $C_l = 0.54$, $Re_c = 3.0 \times 10^6$ (flagged symbols are for the lower surface). (Data from Ref. 9.8.)

comes when the wing is swept back enough to avoid the formation of shock waves which occur when the flow over the wing is locally supersonic. The basic principle is that the component of the main flow parallel to the wing leading edge is not perturbed by the wing, so the critical conditions are reached only when the component of the free-stream velocity normal to the leading edge has been locally accelerated at some point on the wing to the local sonic speed. This simple principle is obviously only true (if at all) on an infinite span wing of constant section. Nevertheless, the initial suggestion of Betz led to wind-tunnel tests which substantiated the essence of the theory. The results of the wind-tunnel measurements performed at Göttingen in 1939 by H. Ludwieg (as presented in Ref. 9.10) are reproduced in Fig. 9.15. These data show that the effect of the shock waves that occur on the wing at high subsonic speeds are delayed to higher Mach numbers by sweepback.

The experimentally determined lift coefficients and drag polars for a delta wing whose aspect ratio is 2.31 and whose airfoil section is NACA 65A005 (a symmetric profile for which $t = 0.05c$) are presented in Fig. 9.16. These data from Ref. 9.4 were obtained at Reynolds numbers between 1.0×10^6 and 1.8×10^6. At subsonic speeds, the lift-coefficient/angle-of-attack correlation for the delta wing is markedly different than that for a rectangular wing. (In addition to the data presented in Figs. 9.11 and 9.16, the reader is referred to Chapter 7.) Even in subsonic streams, the lift

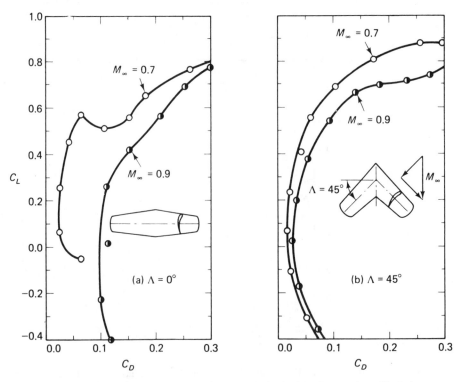

Figure 9.15 Comparison of the transonic drag polar for an unswept wing with that for a swept wing. (Data from Ref. 9.10.)

is a linear function of the angle of attack up to relatively high inclinations. However, over the angle-of-attack range for which the lift coefficient is a linear function of the angle of attack, the lift-curve slope $(dC_L/d\alpha)$ is greater for the rectangular wing for all the free-stream Mach numbers considered.

One can use a variable-geometry (or swing-wing) design to obtain a suitable combination of low-speed and high-speed characteristics. In the highly swept, low-aspect-ratio configuration, the variable-geometry wing provides low wave drag and eliminates the need for wing fold in the case of naval aircraft. At the opposite end of the sweep range, one obtains efficient subsonic cruise and loiter and good maneuverability at the lower speeds. Negative factors in a swing-wing design are complexity, a loss of internal fuel capacity, and the considerable weight of the hinge/pivot structure. A variable-geometry design, the Rockwell International B1, is presented in Fig. 9.17.

Wing–Body Interactions and the "Area Rule"

The chordwise pressure distribution on a plane, finite-span, swept-back wing varies across the span such that the maximum local velocity is reached much farther aft at the root and farther forward at the tip compared to the basic infinite-wing distribution

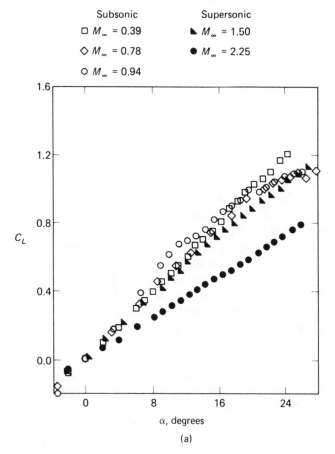

Figure 9.16   Effect of Mach number on the aerodynamic characteristics of a delta wing; AR = 2.31, Λ_{LE} = 60°: (a) lift-coefficient/angle-of-attack correlation. (Data from Ref. 9.4.)

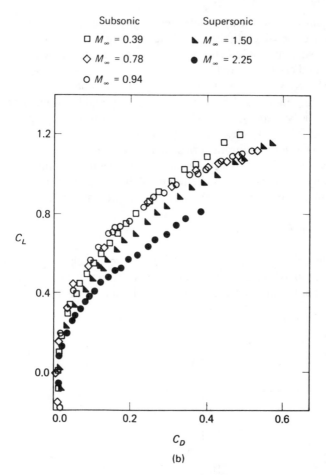

(b)

Figure 9.16 (*Continued*) (b) drag polar.

(Ref. 9.11). These distortions at the root and at the tip determine the flow pattern. On a swept wing with a square-cut tip, a shock wave first occurs near the tip. This "initial tip shock" is relatively short in extent. As the Mach number is increased, a rear shock is developed which dominates the flow. Because of the variation of the component of velocity normal to the wing leading edge, streamlines over the wing surface will tend to be curved. However, the flow at the wing root is constrained to follow the fuselage. Therefore, for a straight fuselage, a set of compression waves originate in the wing-root region to turn the flow parallel to the fuselage. As shown in Fig. 9.18, the compression waves propagate across the wing span and ultimately may coalesce near the tip to form the rear shock.

Thus, it is clear that the interaction between the central fuselage and the swept wing has a significant effect on the transonic flow over the wing. In the early 1950s, Küchemann (Ref. 9.12) recognized that a properly shaped waistlike indentation on the fuselage could be designed to produce a velocity distribution on the wing near the root similar to that on the corresponding infinite-span wing. Küchemann noted that "the critical Mach number can be raised and the drag reduced in the transonic and the

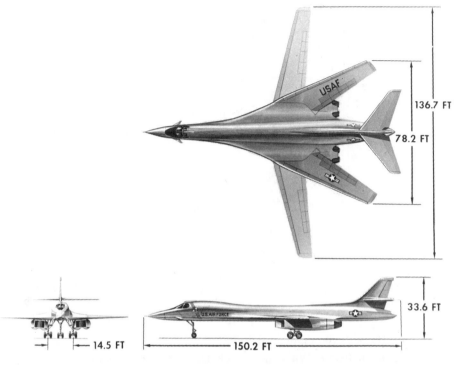

Figure 9.17 Variable geometry (swing-wing) aircraft, the Rockwell International B-1: (a) three-view sketches illustrating the variable-geometry wing; (b) low-speed configuration. (Courtesy of Rockwell International.)

supersonic type of flow, the wing behaving in some respects like a thinner wing.'' Whitcomb (Ref. 9.13) found that the zero-lift drag rise is due primarily to shock waves. Furthermore, the shock-wave formations about relatively complex swept-wing/body combinations at zero lift near the speed of sound are similar to those which occur for

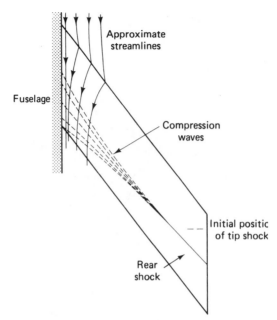

Figure 9.18 Formation of rear shock from compression waves associated with flow near the root. (From Ref. 9.11.)

a body of revolution with the same axial development of cross-sectional area normal to the airstream. Whitcomb concluded that "near the speed of sound, the zero-lift drag rise of a low-aspect-ratio thin-wing/body combination is primarily dependent on the axial development of the cross-sectional areas normal to the air stream." Therefore, the drag-rise increments near the speed of sound are less for fuselage/wing configurations which have a more gradual change in the cross-sectional area (including the fuselage and the wing) with axial position (as well as a reduction in the relative magnitude of the maximum area). Whitcomb noted that it would be expected that indenting the body of a wing–body combination, so that the combination has nearly the same axial distribution of cross-sectional area as the original body alone, would result in a large reduction in the transonic drag rise. Design applications of this "theory," which is often known as *Whitcomb's area rule* for transonic configurations, are illustrated in Fig. 9.19. Early flight tests revealed that the prototypes of the YF-102 had serious deficiencies in performance. A major redesign effort included the first application of Whitcomb's area rule to reduce the transonic drag. The modified YF-102A achieved supersonic flight satisfactorily, providing the U.S. Air Force with its first operational delta-wing design. For additional information, the reader is referred to Ref. 9.15.

The area rule is essentially a linear-theory concept for zero lift. In Ref. 9.16, Whitcomb noted that "to achieve the most satisfactory drag characteristics at lifting conditions the fuselage shape had to be modified from that defined by the simple application of the area rule as previously described to account for the nonlinearity of the flow at such conditions. For lifting conditions at near sonic speeds there is a substantial local region of supercritical flow above the wing surface which results in local expansions of the streamtube areas. In the basic considerations of the area rule

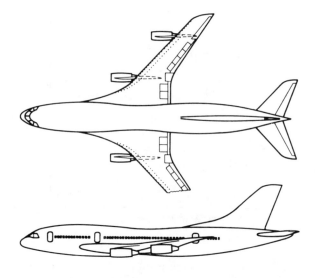

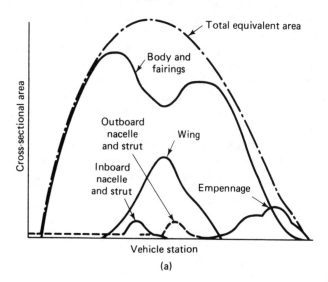

Cross-sectional area

Total equivalent area

Body and fairings

Outboard nacelle and strut

Wing

Inboard nacelle and strut

Empennage

Vehicle station

(a)

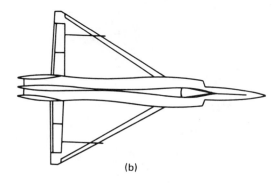

(b)

Figure 9.19 Application of the area rule to the axial distribution of the cross-sectional area: (a) area rule applied to the design of a near-sonic transport (From Ref. 9.14); (b) sketch of the Convair F 102 A.

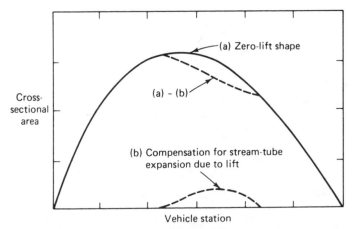

Figure 9.20 Second-order area rule considerations. (From Ref. 9.16.)

concept, this expansion is equivalent to an increase in the physical thickness of the wing. To compensate for this effect the fuselage indentation required to eliminate the far-field effects of the wing must be increased." The additional correction to the cross-sectional areas required for a transonic transport, as taken from Ref. 9.16, is illustrated in Fig. 9.20. "The fuselage indentation based on this corrected cross-sectional area distribution resulted in a significant (0.02) delay in the drag rise Mach number compared with that for the indentation based on the zero lift distribution." However, as noted by Ayers (Ref. 9.7), the fuselage cross-sectional area needed for storing landing gear and aircraft subsystems and for accommodating passenger seating and headroom, conflicts with the area-rule requirements in some cases.

Applications of the area rule in the design of the Northrop F-5 are illustrated in Figs. 9.21 through 9.23. The benefit due to area ruling, considering a straight-sided fuselage as a base, is a large decrease in drag at transonic speeds, as shown in Fig.

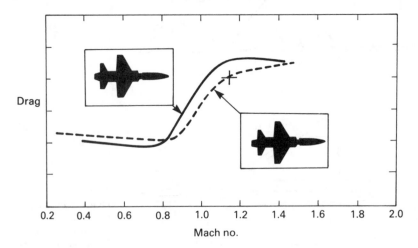

Figure 9.21 Reduction in the transonic drag rise due to the application of area rule to the Northrop F-5 fuselage. (From Ref. 9.17.)

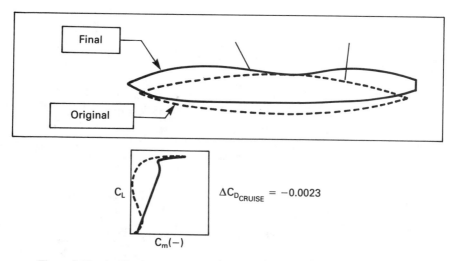

$$\Delta C_{D_{CRUISE}} = -0.0023$$

Figure 9.22 Application of area rule to the design of the tip tank for the F-5A/B aircraft. (From Ref. 9.17.)

9.21. As a result, there is a corresponding increase in transonic acceleration, which is very important to this fighter's design, since (according to Refs. 9.17 and 9.18) the major and decisive portions of air-to-air combat take place at altitudes below 30,000 ft at speeds from Mach 0.5 to 1.0.

Area ruling was also applied to the tip-tank design for the F-5A/B aircraft. Initial prototype flight and wind-tunnel tests indicated that transonic buffet and longitudinal instabilities existed with the originally proposed wingtip tank. Continued development

Figure 9.23 Illustration of the application of the area rule to the tip tanks of the F-5. (From the author's collection.)

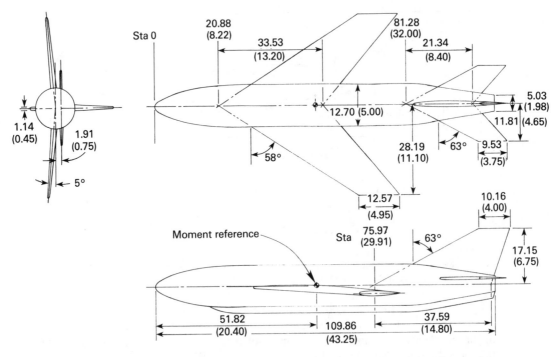

Figure 9.24 Details of a model of a swept-wing cruise missile. Linear dimensions are in centimeters (inches). (From Ref. 9.19.)

tests in the transonic wind tunnel proved that an area ruling concept (as shown in Fig. 9.22) could be used to essentially eliminate the pitch instabilities in the Mach 0.90 to 0.95 region. A cruise drag benefit also existed because of the improvement in wingtip airflow characteristics. A slight instability which remained in the wind-tunnel data and was thought to be Reynolds number dependent was never found during the flight tests. The photograph of an F5 (Fig. 9.23) illustrates these applications of the area rule.

By carefully designing the wing and the tail configurations, one can obtain a smooth variation of the cross-section area even without adjusting the fuselage cross section. Such is the case for the swept-wing cruise missile which is presented in Fig. 9.24. As shown in Fig. 9.25, the cross-section area distribution is relatively smooth and symmetric about the midbody station. This result is due, in part, to the high sweep angles for the wing and for the tail. As might be expected, this shape is conducive to low transonic drag and the drag-rise Mach number is relatively high (i.e., approximately 0.95). Furthermore, the configuration could be trimmed with zero control deflection near the lift coefficient required for the maximum lift-to-drag ratio at a Mach number of 0.95.

FORWARD SWEPT WING

The aerodynamic effects (e.g., increased critical Mach number, decreased lift curve slope, etc.) resulting from wing sweepback are also present for forward swept wings

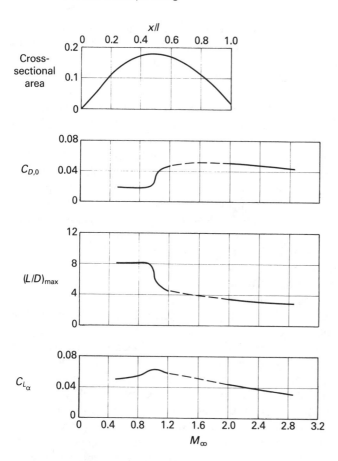

Figure 9.25 Area distribution and various aerodynamic parameters for a swept-wing cruise missile. (From Ref. 9.19.)

(FSW) as well. However, there are some comparative advantages to a FSW in relation to an aft swept wing (ASW), which will be discussed in this section.

Historically, the use of FSW in aircraft designs was, with minor exceptions, precluded by the relatively low (in comparison with ASW) Mach numbers at which they experience aeroelastic flutter and divergence when they are constructed of conventional aircraft materials such as metals. The reasons for this will be discussed below, but, as an example, Diederich and Budiansky (Ref. 9.20) show that uniform FSWs with sweep angles of from − 20 to − 60° exhibit decreases as high as 78% in aeroelastic divergence dynamic pressures in comparison to straight wings. To eliminate this problem using conventional materials requires a prohibitive structural weight penalty. Thus, until recently, the FSW, along with some of its desirable characteristics, has been excluded from the aircraft designer's stable of design options. However, recent advances in the development of composite materials for aircraft structures have permitted the design and fabrication of aeroelastically tailored wings which substantially eliminate the divergence problem.

A swept wing constructed of conventional materials twists under load. An ASW twists down at the tips as the lift increases while a FSW twists up. In the former case, the tips tend to unload, while in the latter case they tend to load the structure further. Thus, the FSW constructed of conventional materials has a much lower speed for structural divergence which occurs when the elastic restoring forces can no longer overcome the aerodynamic forces causing the deformation. The twisting behavior occurs because the axis of section centers of pressure along the span of an ASW is behind the structural axis of the wing in typical designs, while just the opposite is true for the FSW (Ref. 9.21). The structural axis can be thought of as the locus of points along the wing where a load produces only wing bending with no twist.

The solution to this problem for the FSW is to design an aeroelastically tailored wing using composite materials arranged in unidirectional layers or plies which lie at selected angles relative to one another. By careful design and fabrication, the wing can be made to deform under loading so that the divergence problem described above can be delayed to a much higher Mach number. This permits the designer to capitalize on some of the advantages of FSWs compared to ASWs without paying an exhorbitant weight penalty. Incidentally, aeroelastic tailoring is not limited to FSWs, but can be applied to straight and aft swept wings as well (Ref. 9.22).

The relative advantages of the FSW over the ASW are, of course, a function of the aircraft mission and fall into aerodynamic and other categories. Excellent overall discussions of these FSW characteristics are given in Ref. 9.23 and 9.24. We restrict our discussion here to aerodynamic advantages of the FSW. Key aerodynamic advantages of the FSW fall into two categories—reduced drag and enhanced maneuverability at transonic Mach numbers and high angles of attack.

Modern aircraft designs rely on wing sweep and supercritical airfoil sections to delay adverse compressibility effects and reduce their severity when they do occur. Good design practices force the shock on the upper surface of the wing to occur as close to the trailing edge as possible and to be as weak as possible. Typical shock locations for a supercritical wing occur in the vicinity of the 70% chord line along the span. As noted by Whitford (Ref. 9.23), this results in a more highly swept shock for an FSW compared to an ASW with the same leading-edge sweep angle, wing area, and taper ratio. The more highly swept shock, of course, results in a lower wave drag penalty. Further, for designs with the same shock sweep angle, the FSW requires less leading-edge sweep, which results in a higher lift curve slope than the comparable ASW and a reduction in induced drag at subsonic conditions. And although the lower leading-edge sweep will result in higher wave drag at supersonic conditions, drag due to shock-induced separation has been found to be less. Thus, as noted earlier, careful attention to the aircraft mission will determine the design solution to be adopted in each case. Whitford (Ref. 9.23) also points out that an FSW with the same planform area, shock sweep, and span as an ASW has its aerodynamic center closer to the wing root and hence experiences a lower wing root bending moment. Conversely, an FSW can be designed with a higher aspect ratio than a comparable ASW and produce the same bending moment at the wing root. The higher aspect ratio in turn produces lower induced drag.

Flow over an ASW has a component along the span and toward the tips which results in a thicker boundary layer in the tip region than would otherwise exist. This, coupled with the more highly loaded tips of the ASW, can lead to tip stall, with the stall region moving inward toward the root as the stall progresses. This phenomenon can result in loss of aileron control effectiveness at high angles of attack with a consequent loss of maneuverability. An additional and undesirable offshoot of this phenomenon is an overall pitch-up tendency since the loss of lift typically occurs aft of the aircraft center of gravity. For the FSW the spanwise flow component is toward the wing root leading to root stall first. Thus, aileron effectiveness is preserved up to higher angles of attack with a consequent improvement in maneuverability. However, the FSW with root stall still suffers the pitch-up tendency since the loss of lift is once again typically aft of the aircraft center of gravity.

To investigate further and validate by flight test these and other potential advantages of the FSW relative to the ASW, the Defense Advanced Research Projects Agency (DARPA), NASA, and the Air Force have funded the design and fabrication of the X-29. This aircraft, produced by Grumman, is an experimental flight test vehicle designed to explore a variety of advanced technologies, including FSW and close-coupled canards. The primary objective of the X-29 program is to give designers confidence in the FSW approach and to validate the advantages of FSW and other technologies for use in future aircraft designs (Ref. 9.25).

TRANSONIC AIRCRAFT

Ayers (Ref. 9.7) also notes that "wind-tunnel studies have indicated that combining the supercritical airfoil, the area rule, and wing sweep can push the cruising speed of subsonic aircraft very near Mach 1.0." Using computer-oriented numerical techniques, numerous investigators have solved the nonlinear equation for the disturbance velocity potential in transonic flow, which can be written (refer to Ref. 9.26)

$$(1 - M_\infty^2)\phi_{xx} + \phi_{yy} + \phi_{zz} = K\phi_x\phi_{xx} \qquad (9.17)$$

Comparing equation (9.17) with equation (9.10), one would find that, for the particular assumptions made during the development of equation (9.10),

$$K = \frac{2}{U_\infty}\left(1 + \frac{\gamma - 1}{2}M_\infty^2\right)M_\infty^2$$

One numerical procedure commonly used to solve the transonic flow potential equation is the *relaxation process*. The relaxation process is an iterative technique which starts with an initial approximation to the flow field and calculates successive aproximations to the flow until it reaches a state that is both invariant with further iteration and independent of the initial guess. To translate the governing differential equation [e.g., equation (9.17)] into finite-difference form, a grid is formed within the space bounded by the boundary conditions. At each grid intersection, the derivatives may be replaced by a difference approximation. The difference scheme used may differ from one investigator to another (and from one type of flow to another). It may be a central difference, a forward difference, or a backward difference, and it can have various orders of accuracy. For example, we can use the notation of Fig. 9.26 to write the following finite-difference expressions for $\phi_x = \partial\phi/\partial x$. A central difference:

$$\frac{\partial\phi}{\partial x} = \frac{\phi(i + 1,j,k) - \phi(i - 1,j,k)}{2\Delta x} = \phi_x(i,j,k) \qquad (9.18a)$$

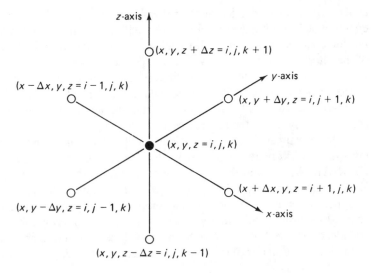

Figure 9.26 Computational notation.

A forward difference:

$$\frac{\partial \phi}{\partial x} = \frac{\phi(i+1,j,k) - \phi(i,j,k)}{\Delta x} = \phi_x(i+\tfrac{1}{2},j,k) \qquad (9.18b)$$

A backward difference:

$$\frac{\partial \phi}{\partial x} = \frac{\phi(i,j,k) - \phi(i-1,j,k)}{\Delta x} = \phi_x(i-\tfrac{1}{2},j,k) \qquad (9.18c)$$

Combining the expressions for a forward difference and for a backward difference, we can obtain an expression for the second derivative:

$$\begin{aligned}
\frac{\partial^2 \phi}{\partial x^2} &= \frac{\partial}{\partial x}\left(\frac{\partial \phi}{\partial x}\right) \\
&= \frac{1}{\Delta x}\left[\frac{\phi(i+1,j,k) - \phi(i,j,k)}{\Delta x} - \frac{\phi(i,j,k) - \phi(i-1,j,k)}{\Delta x}\right] \qquad (9.19) \\
&= \frac{\phi(i+1,j,k) - 2\phi(i,j,k) + \phi(i-1,j,k)}{\Delta x^2}
\end{aligned}$$

In this way, a set of difference equations for the unknown, dependent variables can be formulated. The number of equations is equal to the product of the number of grid points times the number of dependent variables. For additional information about relaxation techniques, the reader is referred to any of the excellent survey articles on the subject, such as Ref. 9.27.

As an example of one of these numerical techniques, consider the work of Bailey and Ballhaus (Ref. 9.28), who used a relaxation procedure to solve the transonic small-disturbance equation:

$$(1 - M_\infty^2)\phi_{xx} + \phi_{yy} + \phi_{zz} = \left(\frac{\gamma + 1}{2} M_\infty^n \phi_x^2\right)_x \qquad (9.20)$$

The parameter n reflects the nonuniqueness of the equation and, quoting Ref. 9.28, "can be adjusted to better approximate the exact sonic pressure coefficient." A finite-difference equation is derived by applying the divergence theorem to the integral of equation (9.20) over an elemental, rectangular, computation volume (or cell). The boundary conditions for the wing include the Kutta condition, which requires that the pressure, which is proportional to ϕ_x, be continuous at the trailing edge. This fixes the section circulation, which is equal to the difference in potential at the section trailing edge linearly extrapolated from points above and below. The solution of the difference equation is obtained by a relaxation scheme with the iterations viewed as steps in pseudo-time. The combination of new and old values in the difference operators is chosen so that the related time-dependent equation represents a properly posed problem whose steady-state solution approaches that of the steady-state equation. The calculated and the experimental pressure distribution for a swept-wing/fuselage configuration at $M_\infty = 0.93$ and $\alpha = 0°$ are compared in Fig. 9.27. The wing has an aspect ratio of 4 and a taper ratio of 0.6. The quarter-chord of the wing is swept 45° and the streamwise airfoil section is a NACA 65A006. The computed results were obtained using a Cartesian grid (x, y, z) of $91 \times 59 \times 27$ for the wing as well as for the fuselage. The experimental data were obtained at $Re_{\bar{c}} = 2.0 \times 10^6$. As was noted earlier in this chapter, the maximum velocity is reached farther aft for the stations near the root and moves toward the leading edge at stations nearer the tip. The agreement with experiment on the fuselage centerline and the two inboard panels is good. In the computed results, the wing-root shock propagates laterally to $2y/b = 0.60$, but the experimental shock dissipates before reaching that point. Thus, there is a discrepancy

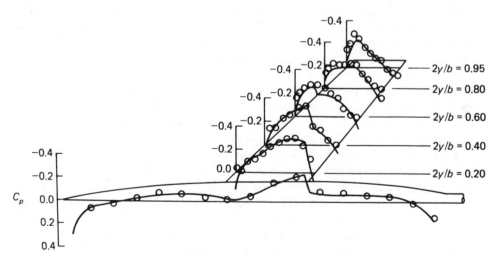

Figure 9.27 Comparison of computed and experimental pressure coefficients C_p for swept-wing/fuselage configuration, $M_\infty = 0.93$; $\alpha = 0°$, $\Lambda_{c/4} = 45°$, AR = 4, $\lambda = 0.6$, NACA 65A006 streamwise section. (From Ref. 9.28.)

between the experimental values and the theoretical predictions. This deficiency results for wings with moderate-to-large sweep angles. Ballhaus, Bailey, and Frick (Ref. 9.29) note that a modified small-disturbance equation, containing cross-flow terms which have been previously neglected, would be a suitable approximation to the full potential equation over a wide range of sweep angles. Thus, a small-disturbance differential equation which could be used to describe the resultant three-dimensional flow is

$$(1 - M_\infty^2)\phi_{xx} + \phi_{yy} + \phi_{zz} - (\gamma + 1)M_\infty^n\phi_x\phi_{xx}$$

$$= 2M_\infty^2\phi_y\phi_{xy} + (\gamma - 1)M_\infty^2\phi_x\phi_{yy} + \frac{\gamma + 1}{2}M_\infty^2\phi_x^2\phi_{xx}$$

(9.21)

Note that the terms on the left-hand side of this equation are those of equation (9.20) and those on the right-hand side are the additional terms of the modified small disturbance formulation. Boppe (Ref. 9.30) notes that the cross-flow terms, $\phi_y\phi_{xy}$ and $\phi_x\phi_{yy}$, provide the ability to define shock waves which are swept considerably relative to the free-stream flow. Boppe recommends that the higher-order term, $\phi_x^2\phi_{xx}$, be included to provide an improved approximation to the full potential equation at the critical velocity. Ballhaus et al. (Ref. 9.29) note that the use of an improved form of the governing equation alone does not guarantee that the shock waves will be properly represented. The finite-difference scheme must also adequately describe the physics of the problem.

Flow-field solutions for complex airplane configurations flying at transonic speeds can be computed using numerical programs that solve the Euler equations. The Euler equations for three-dimensional flows are presented in Appendix A. Using a structured, surface-fitted grid, Kusunose et al. (Ref. 9.31) generated solutions for the flow over advanced turboprop (with an ultra-high-bypass propulsion system) airplane configurations. A structured surface-fitted grid on the configuration surface, as shown in Fig. 9.28, was generated first and then the flow-field grids were obtained using an elliptic grid generation scheme. Kusunose et al. (Ref. 9.31) chose this scheme because it offered "two major advantages: (1) the structural grid requires minimum storage for the flow solution since the grid points are well ordered and (2) the surface-fitted grid allows easier and more accurate implementation of the boundary conditions." The presence of trailing vortex wakes produced by the wing, the strut, and the horizontal tail as well as the slipstream and the rotational flow (with different total pressure and total temperature) produced by the propellor led Kusunose et al. to use the Euler equations in formulating the solution algorithm. "When these are solved, trailing vortex wakes come out automatically as part of the solution, and the rotational flows are captured naturally through the Euler equation formulation." To account for the viscous effects on the wing, the Euler solver is coupled with a three-dimensional boundary-layer method (e.g., Ref 9.32). For the approach of Kusunose et al., the viscous effects are simulated by using the boundary-layer displacement thickness to create an effective "displacement body." Miranda (Ref. 9.33) indicates that modifying the boundary condition (i.e., the surface transpiration concept) is preferable to the displacement body approach because the surface geometry and the computational grid are not affected by the boundary layer.

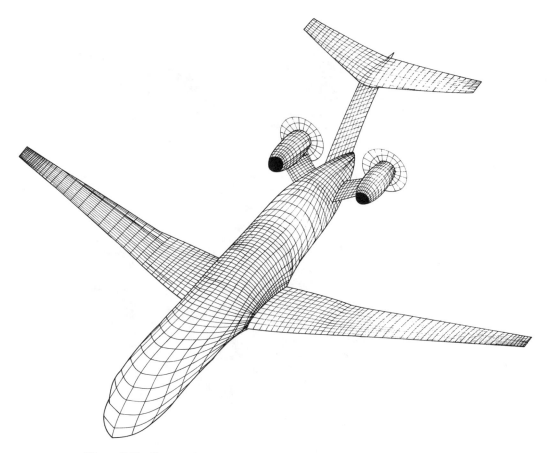

Figure 9.28   Structured surface-fitted grid for an advanced propfan airplane. (From Ref. 9.31.)

Miranda (Ref. 9.33) notes that the problem of grid generation is a crucial one for the practical application of numerical flow simulation. Both methods of solution, singularity and field, require the construction of complex computational grids: mesh-surface grids for singularity methods and spatial grids for field methods. Takanashi et al. (Ref. 9.34) have developed an analytic method for generating boundary fitted grids about aircraft configurations, such as shown in Fig. 9.29. The procedure is based on the successive use of analytical transformations which include classical conformal mapping and three-dimensional sharing/twisting/stretching/clustering transformations. By using this procedure the whole flow field can be transformed into a single computational domain and applied to Navier–Stokes simulations of complicated transonic viscous flows.

Although finite-difference formulations are by far the most commonly used approach in the United States at this writing, the finite element approach has been applied to the numerical solution of nonlinear problems in fluid dynamics. For instance, finite-element methods have been used to generate numerical solutions for potential (inviscid)

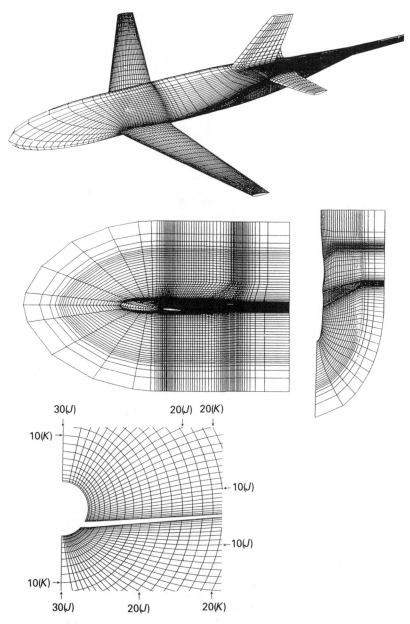

Figure 9.29 Grid system developed by Takanashi et al. (Ref. 9.34) for compressible viscous flows.

transonic flows. The simulation of transonic potential flow for the tri-jet engine Falcon 50 of Avions Marcel Dassault/Breguet Aviation (AMD/BA) at $M_\infty = 0.85$ and an angle of attack of $1°$ is presented in Fig. 9.30. Shown are the trace of the three-dimensional finite-element mesh on the aircraft and the Mach number distribution at the surface (dark: low Mach numbers; light: high Mach numbers).

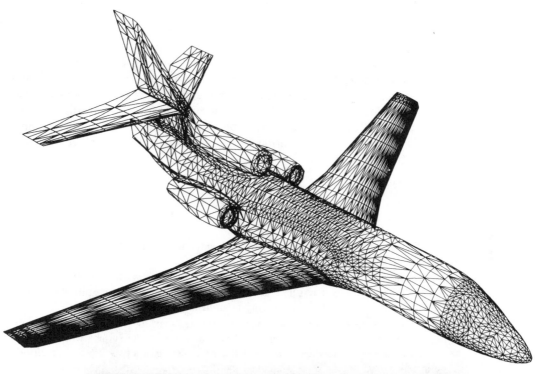

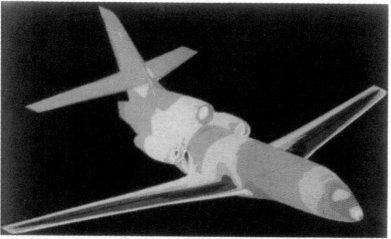

Figure 9.30 Transonic flow simulation by finite elements for a Falcon 50 ($M_\infty = 0.85$, $\alpha = 1°$) (Avions Marcel Dassault/Bréguet Aviation) (a) Finite-element mesh; (b) Mach distribution (from Ref. 9.35).

SUMMARY

Transonic flows are inherently nonlinear and highly three-dimensional for wing–body combination. The nonlinear inviscid flow and the resultant shock waves present a considerable challenge to the analyst. Strong shock-wave/boundary-layer interactions cause rapid increases in drag and can lead to buffeting. Shock-induced boundary-layer separation can also limit the lift coefficient. However, experimental investigations have provided considerable insights into desirable design practices for transonic configurations. Furthermore, three-dimensional transonic flow computational techniques are developing at a rapid pace.

PROBLEMS

9.1. Starting with the energy equation for steady, one-dimensional, adiabatic flow of a perfect gas [i.e., equation (8.29)], derive equation (9.7).

9.2. A rectangular wing having an aspect ratio of 3.5 is flying at $M_\infty = 0.85$ at 12 km. A NACA 0006 airfoil section is used at all spanwise stations. What is the airfoil section and the aspect ratio for the equivalent wing in an incompressible flow?

9.3. When discussing Whitcomb's area rule, it was noted that, for lifting conditions, there is a substantial local region of supercritical flow above the wing surface which results in local expansions of the streamtube areas. Consider flow past a two-dimensional airfoil in a stream where $M_\infty = 0.85$. Calculate the distance between two streamlines, dy_e at a point where the local inviscid Mach number, M_e, is 1.20 in terms of the distance between these two streamlines in the undisturbed flow, dy_∞. Use the integral form of the continuity equation [equation (2.5)],

$$dy_e = \frac{\rho_\infty U_\infty}{\rho_e U_e} dy_\infty$$

Assume an isentropic expansion of a perfect gas.

9.4. Write equation (9.20) in finite-difference form using the nomenclature of Fig. 9.26 and the difference expressions given in equations (9.18) and (9.19).

REFERENCES

9.1. Jones, R. T., and D. Cohen, *High Speed Wing Theory*, Princeton Aeronautical Paperbacks, Princeton University Press, Princeton, N.J., 1960.

9.2. Foss, R. L., and R. Blay, "From Propellors to Jets in Fighter Aircraft Design," *Lockheed Horizons*, Apr. 1987, Issue 23, pp. 2–17.

9.3. Farren, W. S., "The Aerodynamic Art," *Journal of the Royal Aeronautical Society*, July 1956, Vol. 60, No. 547, pp. 431–449.

9.4. Stahl, W., and P. A. Mackrodt, "Dreikomponentenmessungen bis zu grossen Anstellwinkeln an fünf Tragflügeln mit verschieden Umrissformen in Unterschall und Überschallströmung," *Zeitschrift fuer Flugwissenschaften*, Dec. 1965, Vol. 13, No. 12, pp. 447–453.

9.5. Holder, D. W., "The Transonic Flow Past Two-Dimensional Aerofoils," *Journal of the Royal Aeronautical Society*, Aug. 1964, Vol. 68, No. 664, pp. 501–516.

9.6. Whitcomb, R. T., and L. R. Clark, "An Airfoil Shape for Efficient Flight at Supercritical Mach Numbers," *TMX-1109*, NASA, July 1965.

9.7. Ayers, T. G., "Supercritical Aerodynamics: Worthwhile over a Range of Speeds," *Astronautics and Aeronautics*, Aug. 1972, Vol. 10, No. 8, pp. 32–36.

9.8. Hurley, F. X., F. W. Spaid, F. W. Roos, L. S. Stivers, Jr., and A. Bandettini, "Detailed Transonic Flow Field Measurements About a Supercritical Airfoil Section," *TMX-3244*, NASA, July 1975.

9.9. Miller, R., and D. Sawers, *The Technical Development of Modern Aviation*, Praeger Publishers, New York, 1970.

9.10. Schlichting, H., "Some Developments in Boundary Layer Research in the Past Thirty Years," *Journal of the Royal Aeronautical Society*, Feb. 1960, Vol. 64, No. 590, pp. 64–79.

9.11. Rogers, E. W. E., and I. M. Hall, "An Introduction to the Flow About Plane Swept-Back Wings at Transonic Speeds," *Journal of the Royal Aeronautical Society*, Aug. 1960, Vol. 64, No. 596, pp. 449–464.

9.12. Küchemann, D., "Methods of Reducing the Transonic Drag of Swept-Back Wings at Zero Lift," *Journal of the Royal Aeronautical Society*, Jan. 1957, Vol. 61, No. 553, pp. 37–42.

9.13. Whitcomb, R. T., "A Study of Zero-Lift Drag-Rise Characteristics of Wing-Body Combinations near the Speed of Sound," *Report 1273*, NACA, 1954.

9.14. Goodmanson, L. T., and L. B. Gratzer, "Recent Advances in Aerodynamics for Transport Aircraft," *Aeronautics and Astronautics*, Dec. 1973, Vol. 11, No. 12, pp. 30–45.

9.15. Taylor, J. W. R. (Ed.), *Combat Aircraft of the World*, G. P. Putnam's Sons, New York, 1969.

9.16. Whitcomb, R. T., "Advanced Transonic Aerodynamic Technology," presented in *CP 2001*, *Advances in Engineering Science*, Vol. 4, NASA, Nov. 1976.

9.17. Stuart, W. G., *"Northrop F-5 Case Study in Aircraft Design,"* AIAA Professional Study Series, Sept. 1978.

9.18. Bradley, R. G., "Practical Aerodynamic Problems—Military Aircraft," *Transonic Aerodynamics*, Chap. III, edited by David Nixon, Progress in Astronautics and Aeronautics, Vol. 81, AIAA, 1982.

9.19. Spearman, M. L., and I. K. Collins, "Aerodynamic Characteristics of a Swept-Wing Cruise Missile at Mach Numbers from 0.50 to 2.86," *TN D-7069*, NASA, Nov. 1972.

9.20. Diederich, F. W., and B. Budiansky, "Divergence of Swept Wings," *TN 1680*, NACA, 1948.

9.21. Uhuad, G. C., T. M. Weeks, and R. Large, "Wind Tunnel Investigation of the Transonic Aerodynamic Characteristics of Forward Swept Wings," *Journal of Aircraft*, Mar. 1983, Vol. 20, No. 3, pp. 195–202.

9.22. Brown, L. E., Jr., M. A. Price, and P. B. Gingrich, "Aeroelastically Tailored Wing Design," *AIAA paper 80-3046*, 1980.

9.23. Whitford, R., *Design for Air Combat*, Jane's Publishing, London, 1987.

9.24. Krone, N. J., "Forward Swept Wing Design," *AIAA paper 80-3047*, 1980.

9.25. Putnam, T. W., "The X-29 Flight Research Program," *AIAA Student Journal*, Fall 1984, Vol. 22, No. 3, pp. 2–12, 39.

9.26. Newman, P. A., and D. O. Allison, "An Annotated Bibliography on Transonic Flow Theory," *TMX-2353*, NASA, Sept. 1971.

9.27. Lomax, H., and J. L. Steger, "Relaxation Methods in Fluid Mechanics," from the *Annual Review of Fluid Mechanics*, Vol. 7, Annual Reviews, Palo Alto, Calif., 1975.

9.28. Bailey, F. R., and W. F. Ballhaus, "Comparisons of Computed and Experimental Pressures for Transonic Flows About Isolated Wings and Wing-Fuselage Configurations," presented in *SP-347*, *Aerodynamic Analyses Requiring Advanced Computers*, Part II, NASA, Mar. 1975.

9.29. Ballhaus, W. F., F. R. Bailey, and J. Frick, "Improved Computational Treatment of Transonic Flow About Swept Wings," presented in *CP 2001*, *Advances in Engineering Science*, Vol. 4, NASA, Nov. 1976.

9.30. Boppe, C. W., "Computational Transonic Flow About Realistic Aircraft Configurations," *AIAA Paper 78–104*, presented at the AIAA 16th Aerospace Sciences Meeting, Huntsville, Ala., Jan. 1978.

9.31. Kusunose, K., D. L. Mascum, H. C. Chen, and N. J. Yu, "Transonic Analysis for Complex Airplane Configurations," *AIAA Paper 87–1196*, presented at the AIAA 19th Fluid Dynamics, Plasma Dynamics and Laser Conference, Honolulu, Hawaii, June 1987.

9.32. McLean, J. D., and J. L. Randall, "Computer Program to Calculate Three-Dimensional Boundary-Layer Flows over Wing with Wall Mass Transfer," *CR–3123*, NASA, 1978.

9.33. Miranda, L. R., "Application of Computational Aerodynamics to Airplane Design," *Journal of Aircraft*, June 1984, Vol. 21, No. 6, pp. 355–370.

9.34. Takanashi, S., S. Obayashi, K. Matsushima, and K. Fujii, "Numerical Simulation of Compressible Viscous Flows Around Practical Aircraft Configurations," *AIAA Paper 87–2410*, presented at the AIAA 5th Applied Aerodynamics Conference, Monterey, Calif., Aug. 1987.

9.35. Bristeau, M. O., O. Pironneau, R. Glowinski, J. Periaux, P. Perrier, and G. Poirier, "On the Numerical Solution of Nonlinear Problems in Fluid Dynamics by Least Squares and Finite Element Methods, Part II; Applications to Transonic Flow Simulations," *Computer Methods in Applied Mechanics and Engineering*, Vol. 51, 1985, pp. 363–394.

10

TWO-DIMENSIONAL, SUPERSONIC FLOWS AROUND THIN AIRFOILS

The equations that describe inviscid supersonic flows around thin airfoils at low angles of attack will be developed in this chapter. The airfoil is assumed to extend to infinity in both directions from the plane of symmetry (i.e., it is a wing of infinite aspect ratio). Thus, the flow field is the same for any cross section perpendicular to the wing and the flow is two-dimensional. However, even though the relations developed in this chapter neglect the effects of viscosity, there will be a significant drag force on a two-dimensional airfoil in a supersonic stream. This drag component is known as *wave drag*.

In Chapter 8 we derived the Prandtl–Meyer relations to describe the isentropic flow which results when a supersonic flow undergoes an expansive or a compressive change in direction which is sufficiently small that shock waves do not occur. In Example 8.4, the Prandtl–Meyer relations were used to calculate the aerodynamic coefficients for supersonic flow past a thin airfoil. When relatively large compressive changes in the flow direction occur, it is necessary to use the relations describing the nonisentropic flow through an oblique shock wave. (As will be shown, when the compressive changes in direction are only a few degrees, the pressure increase calculated using the Prandtl–Meyer relations is essentially equal to that calculated using the oblique shock-wave relations.) Provided that the assumptions made in the derivations of these techniques are valid, they can be combined to solve for the two-dimensional flow about an airfoil if the shock wave(s) at the leading edge is (are) attached and

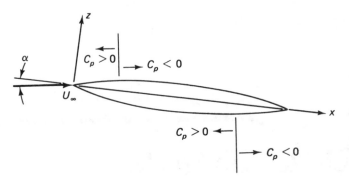

Figure 10.1 General features for linearized supersonic flow past a thin airfoil.

planar. When the leading-edge shock wave(s) is (are) planar, the flow downstream of the shock wave(s) is isentropic. Thus, the isentropic relations developed in Chapter 8 can be used to describe the subsequent acceleration of the flow around the airfoil.

Experience has shown that the leading edge and the trailing edge of supersonic airfoils should be sharp and the section relatively thin. If the leading edge is not sharp (or only slightly rounded), the leading-edge shock wave will be detached and relatively strong, causing relatively large wave drag. We shall consider, therefore, profiles of the general cross section shown in Fig. 10.1. For these thin airfoils at relatively small angles of attack, we can apply the method of small perturbations to obtain theoretical approximations to the aerodynamic characteristics of the two-dimensional airfoils.

By "thin" airfoil, we mean that the thickness, camber, and angle of attack of the section are such that the local flow direction at the airfoil surface deviates only slightly from the free-stream direction. First we treat the Ackeret, or linearized, theory for thin airfoils and then the Busemann, or second-order, theory. The coefficients calculated using the linearized and Busemann theories will be compared with the values calculated using the techniques of Chapter 8.

LINEAR THEORY

The basic assumption of linear theory is that pressure waves generated by thin sections are sufficiently weak that they can be treated as Mach waves. Under this assumption, the flow is isentropic everywhere. The pressure and the velocity changes for a small expansive change in flow direction (an acceleration) have already been derived in Chapter 8 [i.e., equation (8.55)]. Let us define the free-stream flow direction to be given by $\theta_\infty = 0$. For small changes in θ, we can use equation (8.55) to calculate the change in pressure:

$$p - p_\infty = -\rho_\infty U_\infty (U - U_\infty) \tag{10.1a}$$

$$\frac{U_\infty - U}{U_\infty} = \frac{\theta}{\sqrt{M_\infty^2 - 1}} \tag{10.1b}$$

We will define the angle θ so that we obtain the correct sign for the pressure coefficient both for left-running characteristics and for right-running characteristics. Combining these relations yields

$$C_p = + \frac{2\theta}{\sqrt{M_\infty^2 - 1}} \qquad (10.1c)$$

A positive pressure coefficient is associated with a compressive change in flow direction relative to the free-stream flow. If the flow is turned toward the upstream Mach waves, the local pressure coefficient is positive and is greatest where the local inclination is greatest. Thus, for the double-convex-arc airfoil section shown in Fig. 10.1, the pressure is greatest at the leading edge, being greater on the lower surface when the airfoil is at a positive angle of attack. Flow accelerates continuously from the leading edge to the trailing edge for both the lower surface and the upper surface. The pressure coefficient is zero (i.e., the local static pressure is equal to the free-stream value) at those points where the local surface is parallel to the free stream. Downstream, the pressure coefficient is negative, which corresponds to an expansive change in flow direction.

The pressure coefficients calculated using the linearized approximation and Busemann's second-order approximation (to be discussed in the next section) are compared in Fig. 10.2 with the exact values of Prandtl–Meyer theory for expansive

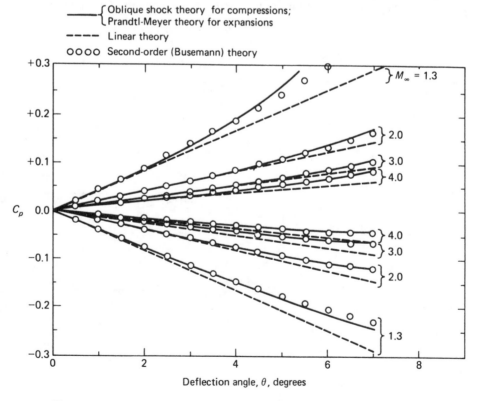

Figure 10.2   Theoretical pressure coefficients as a function of the deflection angle (relative to the stream) for various two-dimensional theories.

turns and of oblique shock-wave theory for compressive turns. For small deflections, linear theory provides suitable values for engineering calculations.

Since the slope of the surface of the airfoil section measured with respect to the free-stream direction is small, we can set it equal to its tangent. Referring to Fig. 10.3, we can write

$$\theta_u = \frac{dz_u}{dx} - \alpha \qquad (10.2a)$$

$$\theta_l = -\frac{dz_l}{dx} + \alpha \qquad (10.2b)$$

The lift, the drag, and the moment coefficients for the section can be determined using equations (10.1c) and (10.2).

Lift

Referring to Fig. 10.4, we see that the incremental lift force (per unit span) acting on the chordwise segment $ABCD$ of the airfoil section is

$$dl = p_l \, ds_l \cos \theta_l - p_u \, ds_u \cos \theta_u \qquad (10.3)$$

Employing the usual thin-airfoil assumptions, equation (10.3) can be written as

$$dl \approx (p_l - p_u) \, dx \qquad (10.4)$$

In coefficient form, we have

$$dC_l \approx (C_{pl} - C_{pu}) \, d\left(\frac{x}{c}\right) \qquad (10.5)$$

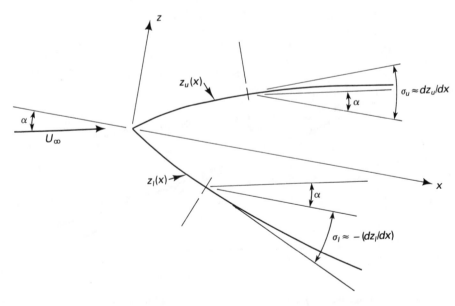

Figure 10.3 Detailed sketch of an airfoil section.

Using equations (10.1c) and (10.2), equation (10.5) becomes

$$dC_l = \frac{2}{\sqrt{M_\infty^2 - 1}} \left(2\alpha - \frac{dz_l}{dx} - \frac{dz_u}{dx} \right) d\left(\frac{x}{c} \right) \tag{10.6}$$

where, without loss of generality, we have assumed that positive values both for θ_u and θ_l represent compressive changes in the flow direction from the free-stream flow.

We can calculate the total lift of the section by integrating equation (10.6) from $x/c = 0$ to $x/c = 1$. Note that since the $z_u = z_l = 0$ at both the leading edge and the trailing edge,

$$\int_0^1 \frac{dz_l}{dx} d\left(\frac{x}{c} \right) = 0 \tag{10.7a}$$

and

$$\int_0^1 \frac{dz_u}{dx} d\left(\frac{x}{c} \right) = 0 \tag{10.7b}$$

Thus,

$$C_l = \frac{4\alpha}{\sqrt{M_\infty^2 - 1}} \tag{10.8}$$

We see that, in the linear approximation for supersonic flow past a thin airfoil, the lift coefficient is independent of the camber and of the thickness distribution. Furthermore, the angle of attack for zero lift is zero. The lift-curve slope is seen to be only a function of the free-stream Mach number, since

$$\frac{dC_l}{d\alpha} = \frac{4}{\sqrt{M_\infty^2 - 1}} \tag{10.9}$$

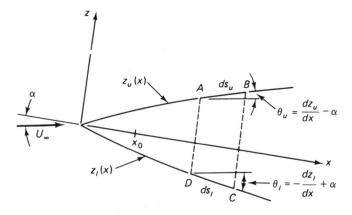

Figure 10.4 Thin-airfoil geometry for determining C_l, C_d, and C_{m0}.

Examining equation (10.9), we see that for $M_\infty \gtrsim 1.185$, the lift-curve slope is less than the theoretical value for incompressible flow past a thin airfoil, which is 2π per radian.

Drag

The incremental drag force due to the inviscid flow acting on the arbitrary chordwise element $ABCD$ of Fig. 10.4 is

$$dd = p_l \, ds_l \sin \theta_l + p_u \, ds_u \sin \theta_u \qquad (10.10)$$

Again, using the assumptions common to small deflection angles, equation (10.10) becomes

$$dd = p_l \theta_l \, dx + p_u \theta_u \, dx \qquad (10.11)$$

In coefficient form, we have

$$dC_d = (C_{p_l} \theta_l + C_{p_u} \theta_u) \, d\left(\frac{x}{c}\right) + \frac{2}{\gamma M_\infty^2} (\theta_l + \theta_u) \, d\left(\frac{x}{c}\right) \qquad (10.12)$$

Using equation (10.1c) for compressive turns and equation (10.2) to approximate the angles, equation (10.12) yields

$$dC_d = \frac{2}{\sqrt{M_\infty^2 - 1}} \left[2\alpha^2 + \left(\frac{dz_u}{dx}\right)^2 + \left(\frac{dz_l}{dx}\right)^2 \right] d\left(\frac{x}{c}\right)$$
$$+ \left[\frac{-4\alpha}{\sqrt{M_\infty^2 - 1}} \left(\frac{dz_l}{dx} + \frac{dz_u}{dx}\right) + \frac{2}{\gamma M_\infty^2} \left(\frac{dz_u}{dx} - \frac{dz_l}{dx}\right) \right] d\left(\frac{x}{c}\right) \qquad (10.13)$$

Using equation (10.7), we find that the integration of equation (10.13) yields

$$C_d = \frac{4\alpha^2}{\sqrt{M_\infty^2 - 1}} + \frac{2}{\sqrt{M_\infty^2 - 1}} \int_0^1 \left[\left(\frac{dz_u}{dx}\right)^2 + \left(\frac{dz_l}{dx}\right)^2 \right] d\left(\frac{x}{c}\right) \qquad (10.14)$$

Note that for the small-angle assumptions commonly used in analyzing flow past a thin airfoil,

$$\frac{dz_u}{dx} = \tan \sigma_u \approx \sigma_u$$

Thus,

$$\frac{1}{c} \int_0^c \left(\frac{dz_u}{dx}\right)^2 dx = \overline{\sigma_u^2} \qquad (10.15a)$$

Similarly, we can write

$$\frac{1}{c} \int_0^c \left(\frac{dz_l}{dx}\right)^2 dx = \overline{\sigma_l^2} \qquad (10.15b)$$

We can use these relations to replace the integrals of equation (10.14) by the average values which they represent. Thus, the section-drag coefficient for this frictionless flow model is

$$C_d = \frac{d}{q_\infty c} = \frac{4\alpha^2}{\sqrt{M_\infty^2 - 1}} + \frac{2}{\sqrt{M_\infty^2 - 1}}(\overline{\sigma_u^2} + \overline{\sigma_l^2}) \tag{10.16}$$

Note that the drag is not zero even though the airfoil is of infinite span and the viscous forces have been neglected. This drag component, which is not present in subsonic flows, is known as *wave drag*. Note also that, as this small perturbation solution shows, it is not necessary that shock waves be present for wave drag to exist. Such was also the case in Example 8.4, which examined the shock-free flow past an infinitesimally thin, parabolic arc airfoil.

Let us examine the character of the terms in equation (10.16). Since the lift is directly proportional to the angle of attack and is independent of the section thickness, the first term is called the *wave drag due to lift* or the *induced drag* and is independent of the shape of the airfoil section. The second term is often referred to as the *wave drag due to thickness* and depends only on the shape of the section. Equation (10.16) also indicates that, for a given configuration, the wave-drag coefficient decreases with increasing Mach number. If we were to account for the effects of viscosity, we could write

$$C_d = C_{d, \text{ due to lift}} + C_{d, \text{ thickness}} + C_{d, \text{ friction}} \tag{10.17a}$$

where

$$C_{d, \text{ due to lift}} = \frac{4\alpha^2}{\sqrt{M_\infty^2 - 1}} = \alpha C_l \tag{10.17b}$$

and

$$C_{d, \text{ thickness}} = \frac{2}{\sqrt{M_\infty^2 - 1}}(\overline{\sigma_u^2} + \overline{\sigma_l^2}) \tag{10.17c}$$

Note also that $C_{d, \text{ friction}}$ is the C_{d0} of previous chapters.

Pitching Moment

Let us now use linear theory to obtain an expression for the pitching moment coefficient. Referring to Fig. 10.4, the incremental moment (taken as positive for nose up relative to the free stream) about the arbitrary point x_0 on the chord is

$$dm_{x_0} = (p_u - p_l)(x - x_0)\, dx \tag{10.18}$$

where we have incorporated the usual small-angle assumptions and have neglected the contributions of the chordwise components of p_u and p_l to the pitching moment.

In coefficient form, we have

$$dC_{m_{x_0}} = (C_{p_u} - C_{p_l})\frac{x - x_0}{c}\, d\left(\frac{x}{c}\right) \tag{10.19}$$

Substituting equation (10.1c) into equation (10.19) yields

$$dC_{m_{x_0}} = \frac{2}{\sqrt{M_\infty^2 - 1}} (\theta_u - \theta_l) \frac{x - x_0}{c} d\left(\frac{x}{c}\right) \qquad (10.20)$$

Substituting equation (10.2) into equation (10.20) and integrating along the chord gives

$$C_{m_{x_0}} = \frac{-4\alpha}{\sqrt{M_\infty^2 - 1}} \left(\frac{1}{2} - \frac{x_0}{c}\right) + \frac{2}{\sqrt{M_\infty^2 - 1}} \int_0^1 \left(\frac{dz_u}{dx} + \frac{dz_l}{dx}\right) \frac{x - x_0}{c} d\left(\frac{x}{c}\right) \qquad (10.21)$$

Note that the average of the upper surface coordinate z_u and the lower surface coordinate z_l defines the mean camber coordinate z_c,

$$\tfrac{1}{2}(z_u + z_l) = z_c$$

We can then write equation (10.21) as

$$C_{m_{x_0}} = \frac{-4\alpha}{\sqrt{M_\infty^2 - 1}} \left(\frac{1}{2} - \frac{x_0}{c}\right) + \frac{4}{\sqrt{M_\infty^2 - 1}} \int_0^1 \frac{dz_c}{dx} \frac{x - x_0}{c} d\left(\frac{x}{c}\right) \qquad (10.22)$$

where we have assumed that $z_u = z_l$ both at the leading edge and at the trailing edge.

As discussed in Chapter 5, the aerodynamic center is that point about which the pitching moment coefficient is independent of the angle of attack. It may also be considered to be that point along the chord at which all changes in lift effectively take place. Thus, equation (10.22) shows that the aerodynamic center is at midchord for a thin airfoil in a supersonic flow. This is in contrast to the thin airfoil in an incompressible flow where the aerodynamic center is at the quarter-chord.

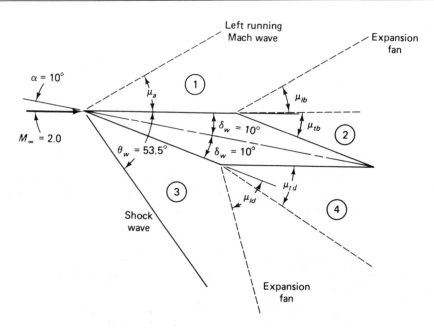

Figure 10.5 Wave pattern for a double-wedge airfoil in a Mach 2 stream.

Example 10.1

Let us use the linear theory to calculate the lift coefficient, the wave-drag coefficient, and the pitching moment coefficient for the airfoil section whose geometry is illustrated in Fig. 10.5. For purposes of discussion, the flow field has been divided into numbered regions, which correspond to each of the facets of the double-wedge airfoil, as shown. In each region the flow properties are such that the static pressure and the Mach number are constant, although they differ from region to region. We seek the lift coefficient, the drag coefficient, and the pitching moment coefficient per unit span of the airfoil given the free-stream flow conditions, the angle of attack, and the geometry of the airfoil neglecting the effect of the viscous boundary layer. The only forces acting on the airfoil are the pressure forces. Therefore, once we have determined the static pressure in each region, we can then integrate to find the resultant forces and moments.

Solution   Let us now evaluate the various geometric parameters required for the linearized theory.

$$z_u(x) = \begin{cases} x \tan 10° & \text{for } 0 \le x \le \dfrac{c}{2} \\[2em] (c - x) \tan 10° & \text{for } \dfrac{c}{2} \le x \le c \end{cases}$$

$$z_l(x) = \begin{cases} -x \tan 10° & \text{for } 0 \le x \le \dfrac{c}{2} \\[2em] -(c - x) \tan 10° & \text{for } \dfrac{c}{2} \le x \le c \end{cases}$$

Furthermore,

$$\overline{\sigma_l^2} = \int_0^1 \sigma_l^2 \, d\left(\frac{x}{c}\right) = \delta_w^2$$

and

$$\overline{\sigma_u^2} = \int_0^1 \sigma_u^2 \, d\left(\frac{x}{c}\right) = \delta_w^2$$

We can use equation (10.8) to calculate the section lift coefficient for a 10° angle of attack at $M_\infty = 2.0$:

$$C_l = \frac{4(10\pi/180)}{\sqrt{2^2 - 1}} = 0.4031$$

Similarly, the drag coefficient can be calculated using equation (10.16):

$$C_d = \frac{4(10\pi/180)^2}{\sqrt{2^2 - 1}} + \frac{2}{\sqrt{2^2 - 1}} \left[\left(\frac{10}{57.296}\right)^2 + \left(\frac{10}{57.296}\right)^2 \right]$$

Therefore,

$$C_d = 0.1407$$

The lift/drag ratio is

$$\frac{l}{d} = \frac{C_l}{C_d} = \frac{0.4031}{0.1407} = 2.865$$

Note that in order to clearly illustrate the calculation procedures, this airfoil section is much thicker (i.e., $t = 0.176c$) than typical supersonic airfoil sections for which $t \approx 0.05c$ (see Table 5.1). The result is a relatively low lift/drag ratio.

Similarly, equation (10.22) for the pitching moment coefficient gives

$$C_{m x_0} = \frac{-4\alpha}{\sqrt{M_\infty^2 - 1}} \left(\frac{1}{2} - \frac{x_0}{c} \right)$$

since the mean camber coordinate x_c is everywhere zero. At midchord, we have

$$C_{m 0.5c} = 0$$

This is not a surprising result, since equation (10.22) indicates that the moment about the aerodynamic center of a symmetric (zero camber) thin airfoil in supersonic flow vanishes.

BUSEMANN'S THEORY

Equation (10.1c) is actually the first term in a Taylor series expansion of Δp in powers of θ. Busemann showed that a more accurate expression for the pressure change resulted if the θ^2 term were retained in the expansion. His result (given in Ref. 10.1) in terms of the pressure coefficient is

$$C_p = \frac{2\theta}{\sqrt{M_\infty^2 - 1}} + \left[\frac{(\gamma + 1)M_\infty^4 - 4M_\infty^2 + 4}{2(M_\infty^2 - 1)^2} \right] \theta^2 \qquad (10.23a)$$

or

$$C_p = C_1\theta + C_2\theta^2 \qquad (10.23b)$$

Again, θ is positive for a compression turn and negative for an expansion turn. We note that the θ^2 term in equation (10.23) is always a positive contribution. Table 10.1 gives C_1 and C_2 for various Mach numbers in air.

It is important to note that since the pressure waves are treated as Mach waves, the turning angles must be small. These assumptions imply that the flow is isentropic everywhere. Thus, equations (10.5), (10.12), and (10.19), along with equation (10.23), can still be used to find C_l, C_d, and $C_{m x_0}$. Note that Fig. 10.2 shows that Busemann's theory agrees even more closely with the results obtained from oblique shock and Prandtl–Meyer expansion theory than does linear theory.

Example 10.2

Let us calculate the pressure coefficient on each panel of the airfoil in Example 10.1 using Busemann's theory. We will use equation (10.23). Since panel 1 is parallel to the free stream, $C_{p1} = 0$ as before. For panel 2:

TABLE 10.1 COEFFICIENTS C_1 AND C_2
FOR THE BUSEMANN THEORY FOR
PERFECT AIR, $\gamma = 1.4$

M_∞	C_1	C_2
1.10	4.364	30.316
1.12	3.965	21.313
1.14	3.654	15.904
1.16	3.402	12.404
1.18	3.193	10.013
1.20	3.015	8.307
1.22	2.862	7.050
1.24	2.728	6.096
1.26	2.609	5.356
1.28	2.503	4.771
1.30	2.408	4.300
1.32	2.321	3.916
1.34	2.242	3.599
1.36	2.170	3.333
1.38	2.103	3.109
1.40	2.041	2.919
1.42	1.984	2.755
1.44	1.930	2.614
1.46	1.880	2.491
1.48	1.833	2.383
1.50	1.789	2.288
1.52	1.747	2.204
1.54	1.708	2.129
1.56	1.670	2.063
1.58	1.635	2.003
1.60	1.601	1.949
1.70	1.455	1.748
1.80	1.336	1.618
1.90	1.238	1.529
2.00	1.155	1.467
2.50	0.873	1.320
3.00	0.707	1.269
3.50	0.596	1.248
4.00	0.516	1.232
5.00	0.408	1.219
10.0	0.201	1.204
∞	0	1.200

$$C_{p2} = \frac{2(-20\pi/180)}{\sqrt{2^2 - 1}} + \frac{(1.4 + 1)(2)^4 - 4(2)^2 + 4}{2(2^2 - 1)^2}\left(\frac{20\pi}{180}\right)^2$$

$$= -0.4031 + 0.1787$$

$$= -0.2244$$

As noted, the airfoil in this sample problem is relatively thick, and therefore the turning angles are quite large. As a result, the differences between linear theory and higher-order approximations are significant but not unexpected. For panel 3:

$$C_{p_3} = 0.4031 + 0.1787$$

$$= 0.5818$$

For panel 4:

$$C_{p_4} = 0$$

since the flow along surface 4 is parallel to the free stream.

Having determined the pressures acting on the individual facets of the double-wedge airfoil, let us now determine the section lift coefficient.

$$C_l = \frac{\Sigma p \cos \theta (0.5c/\cos \delta_w)}{(\gamma/2)p_\infty M_\infty^2 c} \qquad (10.24a)$$

where δ_w is the half-angle of the double-wedge configuration. We can use the fact that the net force in any direction due to a constant pressure acting on a closed surface is zero to get

$$C_l = \frac{1}{2 \cos \delta_w} \Sigma C_p \cos \theta \qquad (10.24b)$$

where the signs assigned to the C_p terms account for the direction of the force. Thus,

$$C_l = \frac{1}{2 \cos 10°} (-C_{p_2} \cos 20° + C_{p_3} \cos 20°)$$

$$= 0.3846$$

Similarly, we can calculate the section wave-drag coefficient:

$$C_d = \frac{\Sigma p \sin \theta (0.5c/\cos \delta_w)}{(\gamma/2)p_\infty M_\infty^2 c} \qquad (10.25a)$$

or

$$C_d = \frac{1}{2 \cos \delta_w} \Sigma C_p \sin \theta \qquad (10.25b)$$

Applying this relation to the airfoil section of Fig. 10.4, the section wave-drag coefficient for $\alpha = 10°$ is

$$C_d = \frac{1}{2 \cos 10°} (C_{p_3} \sin 20° - C_{p_2} \sin 20°)$$

$$= 0.1400$$

Let us now calculate the moment coefficient with respect to the midchord of the airfoil section (i.e., relative to $x = 0.5c$). As we have seen, the theoretical solutions for linearized flow show that the midchord point is the aerodynamic center for a thin airfoil in a supersonic flow. Since the pressure is constant on each of the facets of the double-wedge airfoil of Fig. 10.5 (i.e., in each numbered region), the force acting on a given facet will be normal to the surface and will act at the midpoint of the panel. Thus,

$$C_{m_{0.5c}} = \frac{m_{0.5c}}{\frac{1}{2}\rho_\infty U_\infty^2 c^2} = (-p_1 + p_2 + p_3 - p_4) \frac{c^2/8}{\frac{1}{2}\rho_\infty U_\infty^2 c^2}$$

$$+ (p_1 - p_2 - p_3 + p_4) \frac{(c^2/8) \tan^2 \delta_w}{\frac{1}{2}\rho_\infty U_\infty^2 c^2} \qquad (10.26)$$

Note that, as usual, a nose-up pitching moment is considered positive. Also note that we have accounted for terms proportional to $\tan^2 \delta_w$. Since the pitching moment due to a uniform pressure acting on any closed surface is zero, equation (10.26) can be written as

$$C_{m0.5c} = (-C_{p_1} + C_{p_2} + C_{p_3} - C_{p_4})\frac{1}{8}$$
$$+ (C_{p_1} - C_{p_2} - C_{p_3} + C_{p_4})\frac{\tan^2 \delta_w}{8} \tag{10.27}$$

The reader is referred to equations (5.10) through (5.14) for a review of the technique. Thus,

$$C_{m0.5c} = 0.04329$$

SHOCK-EXPANSION TECHNIQUE

The techniques discussed thus far assume that compressive changes in flow direction are sufficiently small that the inviscid flow is everywhere isentropic. In reality, a shock wave is formed as the supersonic flow encounters the two-dimensional double-wedge airfoil of the previous example problems. Since the shock wave is attached to the leading edge and is planar, the downstream flow is isentropic. Thus, the isentropic Prandtl–Meyer relations developed in Chapter 8 can be used to describe the acceleration of the flow around the airfoil. Let us use this shock-expansion technique to calculate the flow field around the airfoil shown in Fig. 10.5.

Example 10.3

For purposes of discussion, the flow field has been divided into numbered regions that correspond to each of the facets of the double-wedge airfoil, as shown. As was true for the approximate theories, the flow properties in each region, such as the static pressure and the Mach number, are constant, although they differ from region to region. We will calculate the section lift coefficient, the section drag coefficient, and the section pitching moment coefficient for the inviscid flow.

Solution Since the surface of region 1 is parallel to the free stream, the flow does not turn in going from the free-stream conditions (∞) to region 1. Thus, the properties in region 1 are the same as in the free stream. The pressure coefficient on the airfoil surface bounding region 1 is zero. Thus,

$$M_1 = 2.0 \qquad \nu_1 = 26.380° \qquad \theta_1 = 0° \qquad C_{p_1} = 0.0$$

Furthermore, since the flow is not decelerated in going to region 1, a Mach wave (and not a shock wave) is shown as generated at the leading edge of the upper surface. Since the Mach wave is of infinitesimal strength, it has no effect on the flow. However, for completeness, let us calculate the angle between the Mach wave and the free-stream direction. The Mach angle is

$$\mu_a = \sin^{-1}\frac{1}{M_\infty} = 30°$$

Since the surface of the airfoil in region 2 "turns away" from the flow in region 1, the flow accelerates isentropically in going from region 1 to region 2. To cross the left-running Mach waves dividing region 1 from region 2, we move along right-running characteristics. Therefore,

$$dv = -d\theta$$

Since the flow direction in region 2 is

$$\theta_2 = -20°$$

v_2 is

$$v_2 = v_1 - \Delta\theta = 46.380°$$

Therefore,

$$M_2 = 2.83$$

To calculate the pressure coefficient for region 2,

$$C_{p2} = \frac{p_2 - p_\infty}{\frac{1}{2}\rho_\infty U_\infty^2} = \frac{p_2 - p_\infty}{(\gamma/2)p_\infty M_\infty^2}$$

Thus,

$$C_{p2} = \left(\frac{p_2}{p_\infty} - 1\right)\frac{2}{\gamma M_\infty^2}$$

Since the flow over the upper surface of the airfoil is isentropic,

$$p_{t\infty} = p_{t1} = p_{t2}$$

Therefore,

$$C_{p2} = \left(\frac{p_2}{p_{t2}}\frac{p_{t\infty}}{p_\infty} - 1\right)\frac{2}{\gamma M_\infty^2}$$

Using Table 8.1, or equation (8.36), to calculate the pressure ratios given the values for M_∞ and for M_2,

$$C_{p2} = \left(\frac{0.0352}{0.1278} - 1\right)\frac{2}{1.4(4)} = -0.2588$$

The fluid particles passing from the free stream to region 3 are turned by the shock wave through an angle of 20°. The shock wave decelerates the flow and the pressure in region 3 is relatively high. To calculate the pressure coefficient for region 3, we must determine the pressure increase across the shock wave. Since we know that $M_\infty = 2.0$ and $\theta = 20°$, we can use Fig. 8.13b to find the value of C_{p3} directly: it is 0.66. As an alternative procedure, we can use Fig. 8.13a to find the shock-wave angle θ_w; $\theta_w = 53.5°$. Therefore, as discussed on page 330, we can use $M_\infty \sin\theta$ (instead of M_∞), which is 1.608, and the correlations of Table 8.3 to calculate the pressure increase across the oblique shock wave:

$$\frac{p_3}{p_\infty} = 2.848$$

Thus,

$$C_{p3} = \left(\frac{p_3}{p_\infty} = 1\right) \frac{2}{\gamma M_\infty^2} = 0.66$$

Using Fig. 8.13c, we find that $M_3 = 1.20$.
Having determined the flow in region 3,

$$M_3 = 1.20 \qquad \nu_3 = 3.558° \qquad \theta_3 = -20°$$

one can determine the flow in region 4 using the Prandtl–Meyer relations. One crosses the right-running Mach waves dividing regions 3 and 4 on left-running characteristics. Thus,

$$d\nu = d\theta$$

Since $\theta_4 = 0°$, $d\theta = +20°$ and

$$\nu_4 = 23.558°$$

so

$$M_4 = 1.90$$

Note that because of the dissipative effect of the shock wave, the Mach number in region 4 (whose surface is parallel to the free stream) is less than the free-stream Mach number.
Whereas the flow from region 3 to region 4 is isentropic, and

$$p_{t3} = p_{t4}$$

the presence of the shock wave causes

$$p_{t3} < p_{t\infty}$$

To calculate

$$C_{p4} = \left(\frac{p_4}{p_\infty} - 1\right) \frac{2}{\gamma M_\infty^2}$$

$$= \left(\frac{p_4}{p_3}\frac{p_3}{p_\infty} - 1\right) \frac{2}{\gamma M_\infty^2}$$

the ratio

$$\frac{p_4}{p_3} = \frac{p_4}{p_{t4}}\frac{p_{t3}}{p_3}$$

can be determined since both M_3 and M_4 are known. The ratio p_3/p_∞ has already been found to be 2.848. Thus,

$$C_{p4} = \left[\frac{0.1492}{0.4124}(2.848) - 1.0\right] \frac{2}{1.4(4.0)} = 0.0108$$

We can calculate the section lift coefficient using equation (10.24).

$$C_l = \frac{1}{2 \cos 10°}(-C_{p1} - C_{p2} \cos 20° + C_{p3} \cos 20° + C_{p4})$$

$$= 0.4438$$

Similarly, using equation (10.25) to calculate the section wave-drag coefficient,

$$C_d = \frac{1}{2 \cos 10°} (C_{p_3} \sin 20° - C_{p_2} \sin 20°)$$

Thus,

$$C_d = 0.1595$$

The lift/drag ratio in our example is

$$\frac{l}{d} = 2.782$$

for this airfoil section.

In some cases, it is of interest to locate the leading and trailing Mach waves of the Prandtl–Meyer expansion fans at b and d. Thus, using the subscripts l and t to indicate leading and trailing Mach waves, respectively, we have

$$\mu_{lb} = \sin^{-1} \frac{1}{M_1} = 30°$$

$$\mu_{tb} = \sin^{-1} \frac{1}{M_2} = 20.7°$$

$$\mu_{ld} = \sin^{-1} \frac{1}{M_3} = 56.4°$$

$$\mu_{td} = \sin^{-1} \frac{1}{M_4} = 31.8°$$

Each Mach angle is shown in Fig. 10.5.

To calculate the pitching moment about the midchord point, we substitute the values we have found for the pressure coefficients into equation (10.27) and get

$$C_{m_{0.5c}} = 0.04728$$

Example 10.3 illustrates how to calculate the aerodynamic coefficients using the *shock-expansion technique*. This approach is exact provided the relevant assumptions are satisfied. A disadvantage of the technique is that it is essentially a numerical method which does not give a closed-form solution for evaluating airfoil performance parameters, such as the section lift and drag coefficients. However, if the results obtained by the method applied to a variety of airfoils are studied, one observes that the most efficient airfoil sections for supersonic flow are thin with little camber and have sharp leading edges. (The reader is referred to Table 5.1 to see that these features are used on the high-speed aircraft.) Otherwise, wave drag becomes prohibitive.

Thus, we have used a variety of techniques to calculate the inviscid flow field and the section aerodynamic coefficients for the double-wedge airfoil at an angle of attack of $10°$ in an airstream where $M_\infty = 2.0$. The theoretical values are compared in Table 10.2. Although the airfoil section considered in these sample problems is much thicker than those actually used on supersonic airplanes, there is reasonable agreement between the aerodynamic coefficients calculated using the various tech-

TABLE 10.2 COMPARISON OF THE AERODYNAMIC PARAMETERS FOR THE TWO-DIMENSIONAL AIRFOIL SECTION OF FIG. 10.5, $M_\infty = 2.0$, $\alpha = 10°$

	Linearized (Ackeret) theory	Second-order (Busemann) theory	Shock-expansion technique
C_{p_1}	0.0000	0.0000	0.0000
C_{p_2}	−0.4031	−0.2244	−0.2588
C_{p_3}	+0.4031	+0.5818	+0.660
C_{p_4}	0.0000	0.0000	+0.0108
C_l	0.4031	0.3846	0.4438
C_d	0.1407	0.1400	0.1595
$C_{m_{0.5c}}$	0.0000	0.04329	0.04728

niques. Thus, the errors in the local pressure coefficients tend to compensate for each other when the aerodynamic coefficients are calculated.

The theoretical values of the section aerodynamic coefficients as calculated using these three techniques are compared in Fig. 10.6, with experimental values taken from

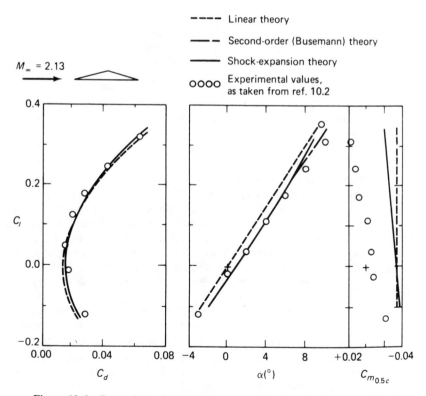

Figure 10.6 Comparison of the experimental and the theoretical values of C_l, C_d, and $C_{m_{0.5c}}$ for supersonic flow past an airfoil.

Ref. 10.2. The airfoil is reasonably thin and the theoretical values for the section lift coefficient and for the section wave-drag coefficient are in reasonable agreement with the data. The experimental values of the section moment coefficient exhibit the angle-of-attack dependence of the shock-expansion theory, but they differ in magnitude. Note that, for the airfoil shown in Fig. 10.6, C_l is negative at zero angle of attack. This is markedly different from the subsonic result, where the section lift coefficient is positive for a cambered airfoil at zero angle of attack. This is another example illustrating that the student should not apply intuitive ideas for subsonic flow to supersonic flows.

PROBLEMS

10.1. Consider supersonic flow past the thin airfoil shown in Fig. P10.1. The airfoil is symmetric about the chord line. Use linearized theory to develop expressions for the lift coefficient, the drag coefficient, and the pitching moment about the midchord. The resultant expressions should include the free-stream Mach number, the constants a_1 and a_2, and the thickness ratio t/c ($\equiv \tau$). Show that, for a fixed thickness ratio, the wave drag due to thickness is a minimum when $a_1 = a_2 = 0.5$.

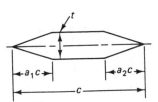

Figure P10.1

10.2. Consider the infinitesimally thin airfoil which has the shape of a parabola:

$$x^2 = -\frac{c^2}{z_{max}}(z - z_{max})$$

The leading edge of the profile is tangent to the direction of the oncoming airstream. This is the airfoil of Example 8.4. Use linearized theory for the following.
  (a) Find the expressions for the lift coefficient, the drag coefficient, the lift/drag ratio, and the moment coefficient about the leading edge. The resultant expressions should include the free-stream Mach number and the parameter, z_{max}/c.
  (b) Graph the pressure coefficient distribution as a function of x/c for the upper surface and for the lower surface. The calculations are for $M_\infty = 2.059$ and $z_{max} = 0.1c$. Compare the pressure distributions for linearized theory with those of Example 8.4.
  (c) Compare the lift coefficient and the drag coefficient calculated using linearized theory for $M_\infty = 2.059$ and for $z_{max} = 0.1c$ with those calculated in Example 8.4.

10.3. Consider the double-wedge profile airfoil shown in Fig. P10.3. If the thickness ratio is 0.04 and the free-stream Mach number is 2.0 at an altitude of 12 km, use linearized theory to compute the lift coefficient, the drag coefficient, the lift/drag ratio, and the moment coefficient about the leading edge. Also, compute the static pressure and the Mach number in each region of the sketch. Make these calculations for angles of attack of 2.29° and 5.0°.

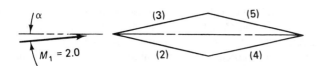

Figure P10.3

10.4. Repeat Problem 10.3 using second-order (Busemann) theory.

10.5. Repeat Problem 10.3 using the shock-expansion technique. What is the maximum angle of attack at which this airfoil can be placed and still generate a weak shock wave?

10.6. Calculate the lift coefficient, the drag coefficient, and the coefficient for the moment about the leading edge for the airfoil of Problem 10.3 and for the same angles of attack, if the flow were incompressible subsonic.

10.7. Verify the theoretical correlations presented in Fig. 10.6. Note that for this airfoil section,

$$\tau = \frac{t}{c} = 0.063$$

Furthermore, the free-stream Mach number is 2.13.

10.8. For linearized theory, it was shown [i.e., equation (10.17c)] that the section drag coefficient due to thickness is

$$C_{d,\text{ thickness}} = \frac{2}{\sqrt{M_i^2 - 1}}\,(\overline{\sigma_u^2} + \overline{\sigma_l^2})$$

If τ is the thickness ratio, show that

$$C_{d,\text{ thickness}} = \frac{4\tau^2}{\sqrt{M_i^2 - 1}}$$

for a symmetric, double-wedge airfoil section and that

$$C_{d,\text{ thickness}} = \frac{5.33\tau^2}{\sqrt{M_i^2 - 1}}$$

for a biconvex airfoil section. In doing this problem, we are verifying the values for K_1 given in Table 11.1.

10.9. Consider the symmetric, diamond-shaped airfoil section (Fig. P10.9; as shown in the sketch all four facets have the same value of δ_w) exposed to a Mach 2.20 stream in a wind tunnel. For the wind tunnel, $p_{t1} = 125$ psia, $T_t = 600°R$. The airfoil is such that the maximum thickness t equals to $0.07c$. The angle of attack is 6°.

(a) Using the shock-wave relations where applicable and the isentropic expansion relations (Prandtl–Meyer) where applicable, calculate the pressures in regions 2 through 5.

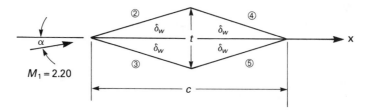

Figure P10.9

(b) Using the linear-theory relations, calculate the pressures in regions 2 through 5.

(c) Calculate C_A, C_N, C_l, C_d, and $C_{M0.5}$ for this configuration. C_A is the axial force coefficient for the force along the axis (i.e., parallel to the chordline of the airfoil) and C_N is the normal force coefficient (i.e., normal to the chordline of the airfoil).

10.10. Consider the "cambered," diamond-shaped airfoil exposed to a Mach 2.00 stream in a wind tunnel (Fig. P10.10). For the wind tunnel, $p_{t1} = 125$ psia, $T_t = 650°R$. The airfoil is such that $\delta_2 = 8°$, $\delta_3 = 2°$, the maximum thickness t equals to $0.07c$ and is located at $x = 0.40c$. The angle of attack is $10°$.

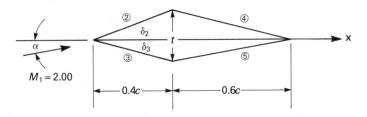

Figure P10.10

(a) Using the shock-wave relations where applicable and the isentropic expansion relations (Prandtl–Meyer) where applicable, calculate the pressures in regions 2 through 5.

(b) Using the linear-theory relations, calculate the pressures in regions 2 through 5.

(c) Calculate C_A, C_N, C_l, C_d, and $C_{M0.5}$ for this configuration. C_A is the axial force coefficient for the force along the axis (i.e., parallel to the chordline of the airfoil) and C_N is the normal force coefficient (i.e., normal to the chordline of the airfoil).

REFERENCES

10.1. Edmondson, N., F. D. Murnaghan, and R. M. Snow, "The Theory and Practice of Two-Dimensional Supersonic Pressure Calculations," *Bumblebee Report 26*, JHO/APL, Dec. 1945.

10.2. Pope, A., *Aerodynamics of Supersonic Flight*, 2nd ed., Pitman Publishing Corp., New York, 1958.

11

SUPERSONIC FLOWS OVER WINGS AND AIRPLANE CONFIGURATIONS

The density variations associated with the flow over an aircraft in supersonic flight significantly affect the aerodynamic design considerations relative to those for subsonic flight. As noted in Chapter 10, the inviscid pressure distribution results in a drag component, known as wave drag, even if one assumes the flow to be isentropic (i.e., neglects the effects of shock waves). Wave drag represents a significant fraction of the total drag in supersonic flows and is related to the bluntness of the configuration. Thus, as illustrated in Table 5.1, the wing sections for supersonic aircraft tend to have a relatively small thickness ratio. Furthermore, the wing planform is such that the aspect ratio is relatively small.

Techniques by which the aerodynamic forces and moments can be computed can be classed either as panel (or singularity) methods or field methods. For panel methods, the configuration is modeled by a large number of quadrilateral panels, approximating the aircraft surface. One or more types of singularity distributions (e.g., sources) are assigned to each elementary panel to simulate the effect that panel has on the flow field. A variety of panel methods have been developed (e.g., that described in Ref.11.1) which can be used to generate solutions for compressible flow. Panel methods yield good results for slender bodies at small angles of attack.

However, as the design lift coefficient is increased, methods based on linear-theory approximations no longer provide an adequate simulation of the complex flow fields. At high-lift conditions, the flow becomes extremely nonlinear, requiring the use of higher-order prediction techniques which take into account the various nonlinear flow phenomenon in determining the configuration aerodynamics. The high-lift flow regime can be modeled using a fully three-dimensional, inviscid, attached-flow computational approach which employs the full-potential equation.

Field methods include a wide variety of assumptions in the flow models. Walkley and Smith (Ref. 11.2) describe a technique which uses a finite-difference formulation based on the characteristics theory of signal propagation to solve the conservative form of the full potential equation, including flows with embedded shock waves and subsonic flows. Additional rigor is obtained (at the cost of additional computational time and expense) if one solves the Euler equations (e.g., Ref. 11.3). Solution techniques employing the Euler equations allow one to incorporate entropy terms that are neglected in the full potential model. For relatively slender vehicles at small angles of attack (so that the flow is attached and the boundary layer is thin), one should expect reasonable accuracy for the pressure distributions and force coefficients if one solves the flow field using the Euler equations. To obtain suitable accuracy when computing the flow fields for higher angles of attack or for configurations where viscous/inviscid interactions are important, it is necessary to model the viscous effects and their interdependence on the inviscid flow. For such applications, Navier–Stokes formulations are needed (e.g., Ref. 11.4).

It is beyond the scope of this text to treat the various methods available by which one can calculate the forces and moments acting on supersonic vehicles. Furthermore, even computer codes that are based on the Navier–Stokes equations employ simplifying assumptions in the solution algorithm. Thus, in reviewing techniques for predicting wing leading-edge vortex flows at supersonic speeds, Wood and Miller (Ref. 11.5) note that, based on their comparisons between computed flows and experimental data, (at least) one Euler code was not well suited for the analysis of wings with separated flow at high lift and low supersonic speeds. Instead, a code based on a linearized-theory method that was modified to account for both nonlinear attached-flow effects (lower surface) and nonlinear separated-flow effects (upper surface) (Ref. 11.6), provided the best correlation with the experimentally measured vortex strength, vortex position, and total lifting characteristics. Thus, it is important to develop an understanding of the general features of supersonic flows and their analysis.

In this chapter we consider steady, supersonic, irrotational flow about wings of finite aspect ratio. The objective is to determine the influence of geometric parameters on the lift, the drag, and the pitching moment for supersonic flows past finite wings. In Chapter 10 we evaluated the effect of the section geometry for flows that could be treated as two-dimensional (i.e., wings of infinite aspect ratio). In this chapter three-dimensional effects will be included since the wings are of finite span. However, we will continue to concentrate on configurations that can be handled by small disturbance (linear) theories.

After a discussion of the general characteristics of flow about supersonic wings, we proceed to a development of the governing equation and boundary conditions for the supersonic wing problem. We then outline the consequences (particularly as they pertain to determining drag) of linearity on the part of the equation and the boundary conditions, and proceed to discuss two solution methods: the conical-flow and the singularity-distribution methods. Example problems are worked using the latter method.

We close the chapter with discussions of aerodynamic interaction effects among aircraft components in supersonic flight and of some design considerations for supersonic aircraft.

GENERAL REMARKS ABOUT SUPERSONIC WINGS

The unique characteristics of supersonic flow lead to some interesting conclusions about wings in a supersonic stream. Consider the rectangular wing of Fig. 11.1. The pressure at a given point $P(x, y)$ on the wing is influenced only by pressure disturbances generated at points within the upstream Mach cone (determined by $\mu = \sin^{-1} 1/M_\infty$) emanating from P. As a result, the wingtips affect the flow only in the regions BAC and DFE. The remainder of the wing ($ACDF$) is not influenced by the tips and can be treated using the two-dimensional theory developed in Chapter 10.

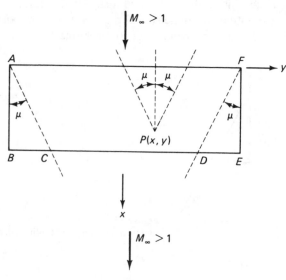

Figure 11.1 Rectangular wing in a supersonic stream.

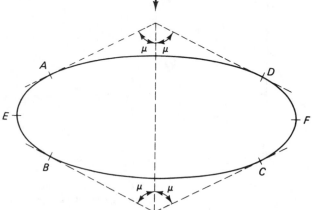

Figure 11.2 Wing of arbitrary planform in a supersonic stream.

In the case of an arbitrary planform (see Fig. 11.2) we have the following definitions:

1. A supersonic (subsonic) leading edge is that portion of the wing leading edge where the component of the free-stream velocity normal to the edge is supersonic (subsonic).
2. A supersonic (subsonic) trailing edge is that portion of the wing trailing edge where the component of the free-stream velocity normal to the edge is supersonic (subsonic).

In Fig. 11.2, AD and BC are supersonic leading and trailing edges, respectively. AE and DF are subsonic leading edges, and EB and FC are subsonic trailing edges. Note that the points A, D, B, and C are the points of tangency of the free-stream Mach cone with the leading and trailing edges.

The delta wing of Fig. 11.3 has supersonic leading and trailing edges, while the arrow wing of Fig. 11.4 has a subsonic leading edge and a supersonic trailing edge.

Points on the upper surface within two-dimensional regions which are bounded by supersonic edges have flows that are independent of lower surface flow and vice

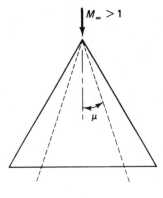

Figure 11.3 Delta wing with supersonic leading and trailing edges.

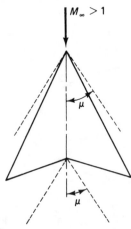

Figure 11.4 Arrow wing with a subsonic leading edge and a supersonic trailing edge.

versa, for points on the lower surface. Thus, in many cases, portions of supersonic wings can be treated by the two-dimensional theory of Chapter 10.

The conclusion drawn in Chapter 10—that good aerodynamic efficiency in supersonic flight depends on thin-airfoil sections with sharp leading and trailing edges— carries over to finite aspect ratio wings. We also find that the benefits of sweepback (discussed in Chapter 9) in decreasing wave drag are also present in the supersonic regime.

GOVERNING EQUATION AND BOUNDARY CONDITIONS

In Chapter 9 we derived the small perturbation (or linear potential) equation (9.13). Although the derivation was for the subsonic case, the assumptions made in that derivation are satisfied by thin wings in supersonic flow as well.

In mathematical form, the assumptions made in deriving equation (9.13) are

$$M_\infty^2 \left(\frac{u'}{U_\infty}\right)^2 \ll 1 \qquad M_\infty^2 \left(\frac{v'}{U_\infty}\right)^2 \ll 1 \qquad M_\infty^2 \left(\frac{w'}{U_\infty}\right)^2 \ll 1$$

$$M_\infty^2 \left(\frac{u'v'}{U_\infty^2}\right) \ll 1 \qquad M_\infty^2 \left(\frac{u'w'}{U_\infty^2}\right) \ll 1 \qquad M_\infty^2 \left(\frac{v'w'}{U_\infty^2}\right) \ll 1 \qquad (11.1)$$

$$\frac{M_\infty^2}{|1 - M_\infty^2|} \left(\frac{u'}{U_\infty}\right) \ll 1 \qquad M_\infty^2 \left(\frac{v'}{U_\infty}\right) \ll 1 \qquad M_\infty^2 \left(\frac{w'}{U_\infty}\right) \ll 1$$

One observes from equation (11.1) that the assumptions are satisfied for thin wings in supersonic flow provided the free-stream Mach number (M_∞) is not too close to one (transonic regime), nor too great (hypersonic regime). Practically speaking, this restricts supersonic linear theory to the range $1.2 \le M_\infty \le 5$.

Rewriting equation (9.13) in standard form (to have a positive factor for the ϕ_{xx} term) yields

$$(M_\infty^2 - 1)\phi_{xx} - \phi_{yy} - \phi_{zz} = 0 \qquad (11.2)$$

where ϕ is the perturbation potential. This is a linear, second-order partial differential equation of the hyperbolic type, whereas equation (9.13) is of the elliptic type (when $M_\infty < 1$). This fundamental mathematical difference between the equations governing subsonic and supersonic small perturbation flow has already been discussed in Chapter 8. There, we saw that a small disturbance in a subsonic stream affects the flow upstream and downstream of the disturbance, whereas in a supersonic flow the influence of the disturbance is present only in the "zone of action" defined by the Mach cone emanating in the downstream direction from the disturbance. These behaviors are characteristic of the solutions to elliptic and hyperbolic equations, respectively.

A boundary condition imposed on the flow is that it must be tangent to the surface at all points on the wing. Mathematically, we have

$$\left(\frac{w'}{U_\infty + u'}\right)_{\text{surface}} = \frac{dz_s}{dx} \qquad (11.3)$$

which is the same as the condition imposed on the flow in the subsonic case. See the equation immediately following equation (7.40). Note that the equation of the surface is given by $z_s = z_s(x, y)$.

Consistent with the definition of the perturbation potential, equation (11.3) becomes

$$\left(\frac{\phi_z}{U_\infty + \phi_x}\right)_{\text{surface}} = \frac{dz_s}{dx} \tag{11.4}$$

Applying the assumption that the flow perturbations are small, we have

$$(\phi_z)_{z=0} = U_\infty \frac{dz_s}{dx} \tag{11.5}$$

as the flow tangency boundary condition since $U_\infty + u' \approx U_\infty$ and the surface corresponds to $z_s \approx 0$. An additional condition that must be applied at a subsonic trailing edge is the *Kutta condition*. This condition is

$$C_{p_{u_{te}}} = C_{p_{l_{te}}} \tag{11.6}$$

where the subscripts u_{te} and l_{te} stand for upper and lower wing surface at the trailing edge, respectively. Physically, the condition represented by equation (11.6) means that the local lift at a subsonic trailing edge is zero.

CONSEQUENCES OF LINEARITY

In Chapter 10 it was shown that the effects of angle of attack, camber, and thickness distribution were additive; refer to equations (10.8), (10.11), and (10.22). In general, a wing can, for the purpose of analysis, be replaced by three components (see Fig. 11.5): (a) a flat plate of the same planform at the same angle of attack, (b) a thin plate with the same planform and camber at zero angle of attack, and (c) a wing of the same planform and thickness distribution but with zero camber and angle of attack. The perturbation potential for each of these components can be determined separately and added together to get the combined potential describing the flow about the actual wing.

The contributions to the wing drag from angle of attack and camber are known as the *drag due to lift*. The total drag coefficient is given by the sum of the drag due to friction, drag due to thickness, and drag due to lift coefficients. Thus,

$$C_D = C_{D,\text{friction}} + C_{D,\text{thickness}} + C_{D,\text{due to lift}}$$

Experimental evidence indicates that this can be written as

$$C_D = C_{D0} + kC_L^2$$

where C_{D0} is the zero-lift-drag coefficient ($C_{D0} = C_{D,\text{friction}} + C_{D,\text{thickess}}$) and k is the drag-due-to-lift factor. Note that $C_{D,\text{friction}}$ can be determined using boundary-layer theory (see Chapters 4 and 8).

As noted in Ref 11.7, wave drag can be calculated using either far-field methods

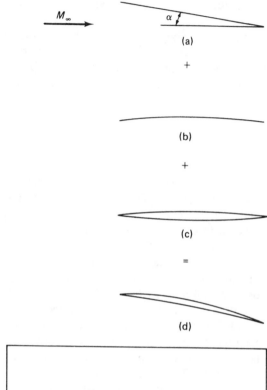

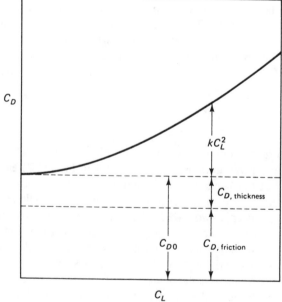

Figure 11.5 Effects of angle of attack, camber, and thickness are additive in linear theory: (a) angle of attack; (b) camber distribution; (c) thickness distribution; (d) resultant wing.

Figure 11.6 Drag components at supersonic speed.

(supersonic area rule) or near-field methods (surface-pressure integration). The far-field method offers advantages for fuselage optimization according to area-rule concepts. The near-field method is used as an analysis tool when detailed pressure distribution are of interest. For supersonic flows, k is a strong function of the Mach number. Figure 11.6 shows, schematically, how each component of the drag contributes to the total drag.

Thus, the linear nature of the governing equation and the boundary conditions allows us to break the general wing problem into parts, solve each part by methods appropriate to it, and linearly combine the results to arrive at the final flow description. The ability to treat thin wings in this manner greatly simplifies what would otherwise be very difficult problems.

SOLUTION METHODS

We learned in Chapter 9 that subsonic, compressible flows could be treated by applying an affine transformation [equation (9.14)] to equation (9.13) and the boundary conditions. This resulted in an equivalent incompressible flow problem [equation (9.15)], which could be handled by the methods of Chapters 6 and 7 for two- and three-dimensional flows, respectively. However, no affine transformation exists which can be used to transform equation (11.2) into equation (9.15).

We will discuss two solution methods for equation (11.2). The first, the conical flow method, was first proposed by Busemann (Ref. 11.8) and was used extensively before the advent of high-speed digital computers. The second, the singularity-distribution method, has known for some time (Refs. 11.9, 11.10, and 11.11) but was not generally exploited until the high-speed digital computer was commonly available. The latter method is particularly suited to such computers and is more easily applied to general configurations than the former. Thus, it is more widely used today. However, the solutions generated using the conical-flow method (where applicable) serve as comparison checks for solutions using the computerized singularity-distribution method.

CONICAL-FLOW METHOD

A conical flow exists when flow properties, such as the velocity, the static pressure, and the static temperature, are invariant along rays (e.g., *PA* in Fig. 11.7) emanating from a point (e.g., point *P* at the wing tip). If equation (11.2) is transformed from the *x*, *y*, *z* coordinate system to a conical coordinate system (as was done in Ref. 11.12), the resulting equation has only two independent variables, since properties are invariant along rays from the apex of the cone. A further transformation (Ref. 11.10) results in Laplace's equation in two independent variables, which is amenable to solution by well-known methods (complex variable theory, Fourier series, etc.).

Since the conical-flow technique is not generally applicable and is not as adaptable to computers, we will not go through the mathematical details of its development. However, we will present some results that are applicable to simple wing shapes. The interested reader is referred to Refs. 11.10, 11.11, 11.13, and 11.14 for in-depth presentations of conical flow theory and its applications.

Regions of various wings that can be treated using conical-flow theory are illustrated in Fig. 11.8.

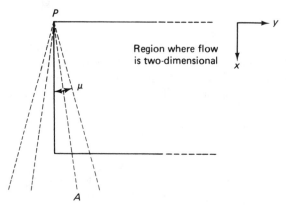

Region
influenced by
wingtip

Figure 11.7 In a conical flow,
properties are invariant along rays
emanating from a point.

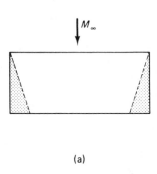

(a)

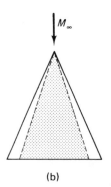

(b)

Shaded regions: can be analyzed with conical theory
Unshaded regions: can be analyzed with two-dimensional theory

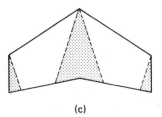

(c)

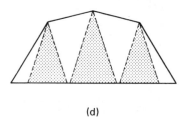

(d)

Figure 11.8 Examples of regions that can be treated with conical theory:
(a) rectangular wing; (b) delta wing; (c) swept swing; (d) "double" delta wing.

Rectangular Wings

Bonney (Ref. 11.15) has shown that the lift inside the Mach cone at the tip of a rectangular wing is one-half the lift of a two-dimensional flow region of equal area. This is illustrated in the pressure distribution for an isolated rectangular wing tip, which is presented in Fig. 11.9. The curve of Fig. 11.9 represents the equation

$$\frac{\Delta p}{\Delta p_{2-d}} = \frac{2}{\pi} \sin^{-1} \sqrt{\frac{\tan \mu'}{\tan \mu}}$$

The analysis can be extended to the interaction of the two tip flows when their respective Mach cones intersect (or overlap) on the wing surface. The case where the entire trailing edge of the wing is in the overlap region of the tip Mach cones is illustrated in Fig. 11.10. The pressure distribution in the region of overlap (when $1 \leq \beta \cdot AR \leq 2$) is determined by adding the pressures due to each tip and subtracting from them the two-dimensional pressure field determined by Busemann's second-order equation (10.23).

The extent of the overlap region is determined by the parameter $\beta \cdot AR$. Three

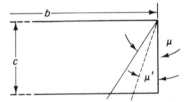

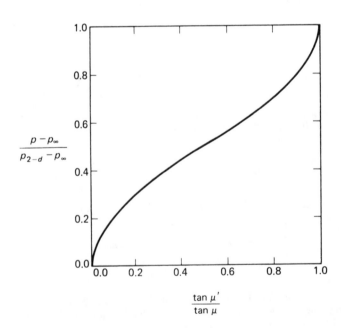

Figure 11.9 Pressure distribution at the tip of a rectangular wing (p = actual pressure due to tip loss, p_{2-d} = corresponding two-dimensional pressure). (From Ref. 11.15.)

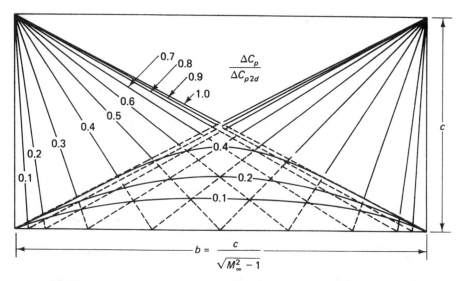

Figure 11.10 Effect of a subsonic wing tip on the pressure distribution for a rectangular wing for which $\beta \cdot AR = 1$.

cases are shown in Fig. 11.11. Conical-flow theory is not applicable in the regions indicated in the figure. Such cases will occur when $\beta \cdot AR < 1$.

The key assumptions used in this development are:

1. Secondary tip effects originating at the point of maximum thickness of a double-wedge airfoil are neglected.
2. Tip effects extend to the limits of the Mach cone defined by the free-stream Mach number M_∞ and not to the Mach cone defined by the local Mach number.
3. Flow separation does not occur.
4. Linear theory applies.

A summary of the results taken from Ref. 11.15 for the case of nonoverlapping tip effects is given in Table 11.1. Conclusions to be drawn from this analysis are:

1. A decrease in aspect ratio for a given supersonic Mach number and airfoil section results in decreases in the coefficients for the drag due to lift, the lift, and the pitching moment. Note that the behavior of the drag due to lift here is in direct contrast to its behavior in subsonic flow. The center of pressure will move forward with a decrease in aspect ratio.
2. When the thickness ratio is increased, the coefficients of lift and of drag due to lift for finite-span wings increase slightly, but the moment coefficient about the leading edge decreases. Further, the center of pressure moves forward both for airfoils and for wings as the thickness ratio is increased.
3. Airfoils having the same cross-sectional area will have the same center of pressure location.

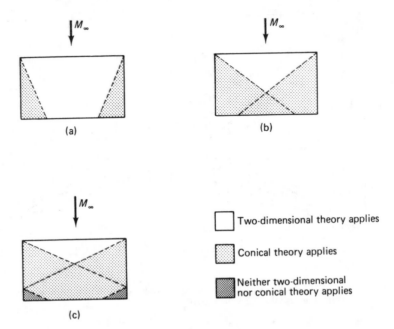

Figure 11.11   Regions where conical flow applies for rectangular wings: (a) $\beta \cdot AR$ > 2: no overlap, (b) $1 \le \beta \cdot AR \le 2$: overlap along the trailing edge; (c) $\beta \cdot AR$ < 1: overlap extends beyond the wing tips.

4. Thickness drag will vary with the square of the thickness ratio for a given cross-sectional shape.

5. Airfoils of symmetrical cross section with a maximum thickness at the mid-chord point will have the least drag for a given thickness ratio.

A comparison of conical flow predictions with data obtained from Ref. 11.16 is given in Fig. 11.12 for a double-wedge airfoil for the conditions shown in the figure.

Swept Wings

If the wing leading edge is swept aft of the Mach cone originating at the apex of the wing, as shown in Fig. 11.13, the disturbances propagated along the Mach lines "warn" the approaching flow of the presence of the wing. As noted earlier, a leading edge that is swept within the Mach cone is referred to as subsonic leading edge and the flow approaching the wing is similar to subsonic flow, even though the flight speed is supersonic. Thus, when the leading edge is subsonic, the wing can be treated by the methods of Chapters 7 and 9. Furthermore, for sufficiently swept leading edges, they can be rounded similar to those used for subsonic speeds.

There are penalties associated with sweepback (some of which show up at subsonic speeds) including reduction of the lift-curve slope, increased drag due to lift, tip stalling, and reduced effectiveness of high-lift devices. There are structural considerations also. A swept wing has a greater structural span than a straight wing having

TABLE 11.1 CONICAL-FLOW RESULTS FOR RECTANGULAR WINGS USING BUSEMANN SECOND-ORDER APPROXIMATION THEORY

Type	Flat-plate airfoil	Flat-plate wing	Finite-thickness airfoil	Finite-thickness wing
C_L	$\dfrac{4\alpha}{\beta}$	$\dfrac{4\alpha}{\beta}\left(1 - \dfrac{1}{2AR\cdot\beta}\right)$	$\dfrac{4\alpha}{\beta}$	$\dfrac{4\alpha}{\beta}\left[1 - \dfrac{1}{2AR\cdot\beta}(1 - C_3A')\right]$
C_D	$\dfrac{4\alpha^2}{\beta} + C_{D,\text{friction}}$	$\dfrac{4\alpha^2}{\beta}\left(1 - \dfrac{1}{2AR\cdot\beta}\right) + C_{D,\text{friction}}$	$\dfrac{K_1\tau^2}{\beta} + \dfrac{4\alpha^2}{\beta} + C_{D,\text{friction}}$	$\dfrac{K_1\tau^2}{\beta} + C_{D,\text{friction}} + \dfrac{4\alpha^2}{\beta}\left[1 - \dfrac{1}{2AR\cdot\beta}(1 - C_3A')\right]$
C_{M_0}	$\dfrac{2\alpha}{\beta}$	$\dfrac{2\alpha}{AR\cdot\beta^2}\left(AR\cdot\beta - \dfrac{2}{3}\right)$	$\dfrac{2\alpha}{\beta}(1 - C_3A')$	$\dfrac{2\alpha}{AR\cdot\beta^2}\left[AR\cdot\beta - \dfrac{2}{3} - C_3A'(AR\cdot\beta - 1)\right]$
x_{cp}	$\dfrac{c}{2}$	$\left[\dfrac{AR\cdot\beta - \frac{2}{3}}{2AR\cdot\beta - 1}\right]c$	$(1 - C_3A')\dfrac{c}{2}$	$\left[\dfrac{AR\cdot\beta - \frac{2}{3} - C_3A'(AR\cdot\beta - 1)}{2AR\cdot\beta - 1 + C_3A'}\right]c$

$^a\; C_3 = \dfrac{\gamma M_\infty^4 + (M_\infty^2 - 2)^2}{2(M_\infty^2 - 1)^{3/2}}$; $A' = \dfrac{\text{airfoil cross-sectional area}}{\text{chord squared}}$; α in radians.

Values of A', K_1:

Type of airfoil	A'	K_1
Double wedge	$\frac{1}{2}\tau$	4
Modified double wedge	$\frac{2}{3}\tau$	6
Biconvex	$\frac{2}{3}\tau$	5.33

Source: Ref. 11.15.

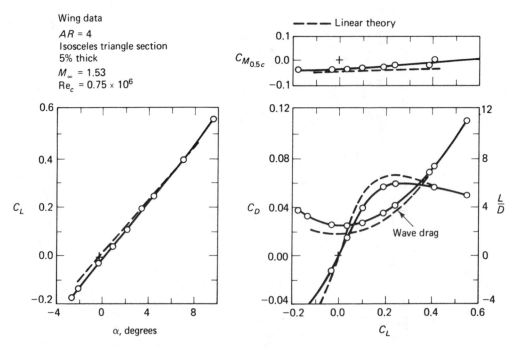

Wing data

$AR = 4$
Isosceles triangle section
5% thick

$M_\infty = 1.53$
$Re_c = 0.75 \times 10^6$

- - - Linear theory

Figure 11.12 Comparison of linear theory results with experimental data for a rectangular wing. (Data from Ref. 11.16.)

the same area and aspect ratio. Sweepback introduces additional (possibly severe) torsion because the applied loads on the wing act aft of the wing root.

Consider the flow depicted in Fig. 11.14, where the leading and trailing edges of the wing are supersonic. The tips and center portion of a swept wing can be treated with conical-flow theory while the remaining portion of the wing can be analyzed by the two-dimensinal techniques of Chapter 10, if an appropriate coordinate transformation is made. Refer to Fig. 11.15, in which a segment of an infinitely long swept-back wing with sweepback angle Λ and angle of attack α are presented. The free-

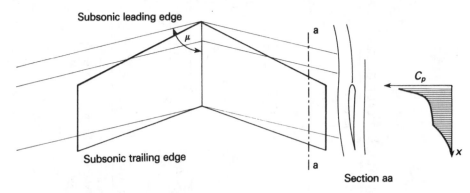

Figure 11.13 Pressure distribution over the wing chord for a section of a swept wing with subsonic leading and trailing edges.

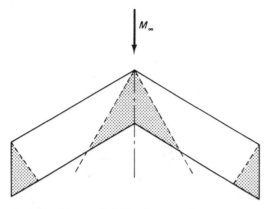

M_∞

Shaded regions: conical flow theory applies
Unshaded regions: two-dimensional flow theory applies

Figure 11.14 Regions of conical flow and of two-dimensional flow for a swept wing.

stream Mach number can be broken into three components, as shown in the figure. The component tangent to the leading edge is unaffected by the presence of the wing (if we neglect viscous effects). Thus, we may consider the equivalent free-stream Mach number M_{∞_e} normal to the leading edge. This will be the flow as seen by an observer moving spanwise at the tangential Mach number $M_\infty \sin \Lambda \cos \alpha$. Note that the airfoil section exposed to M_{∞_e} will be that taken in a plane normal to the leading edge. The flow at M_{∞_e} about this section can be treated with the two-dimensional theory of Chapter 10.

Referring to the geometry presented in Fig. 11.15, we can see that

$$M_{\infty_e} = [(M_\infty \sin \alpha)^2 + (M_\infty \cos \alpha \cos \Lambda)^2]^{0.5}$$

or

$$M_{\infty_e} = M_\infty (1 - \sin^2 \Lambda \cos^2 \alpha)^{0.5} \tag{11.7}$$

Also,

$$\alpha_e = \tan^{-1} \frac{M_\infty \sin \alpha}{M_\infty \cos \alpha \cos \Lambda} = \tan^{-1} \frac{\tan \alpha}{\cos \Lambda} \tag{11.8}$$

$$\tau_e = \frac{t}{c \cos \Lambda} = \frac{\tau}{\cos \Lambda} \tag{11.9}$$

where $\tau \equiv t/c$ is the thickness ratio.

Now, the total lift per unit span is not changed by the spanwise motion of the observer; and only M_{∞_e} generates pressure forces necessary to create lift. Thus,

$$C_l = \frac{l}{(\gamma/2)p_\infty M_\infty^2 c}$$

and

$$C_{l_e} = \frac{l}{(\gamma/2)p_\infty M_{\infty_e}^2 c \cos \Lambda(1/\cos \Lambda)} \tag{11.10}$$

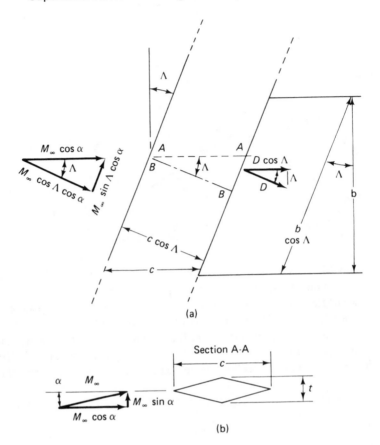

(a)

(b)

(c)

Figure 11.15 Nomenclature for flow around a sweptback wing of infinite aspect ratio: (a) view in plane of wing; (b) view in plane parallel to direction of flight; (c) view in plane normal to the leading edge. (From Ref. 11.10.)

Similarly, ignoring viscous effects and noting that the wave drag is normal to the leading edge:

$$C_{de} = \frac{d}{(\gamma/2)p_\infty M_{\infty e}^2 c \cos \Lambda \, (1/\cos \Lambda)} \tag{11.11}$$

while

$$C_d = \frac{d \cos \Lambda}{(\gamma/2)p_\infty M_\infty^2 c} \tag{11.12}$$

where $d \cos \Lambda$ is the drag component in the free-stream direction.

Combining equations (11.7) and (11.10), we get

$$C_l = C_{le} \left(\frac{M_{\infty e}}{M_\infty}\right)^2 = C_{le}(1 - \sin^2 \Lambda \cos^2 \alpha) \tag{11.13}$$

For the drag, we combine equations (11.7), (11.11), and (11.12) to get

$$C_d = C_{de} \cos \Lambda \left(\frac{M_{\infty e}}{M_\infty}\right)^2 = C_{de} \cos \Lambda (1 - \sin^2 \Lambda \cos^2 \alpha) \tag{11.14}$$

The results derived above are true in general. If we restrict ourselves to the assumptions of the linear (Ackeret) theory, which was discussed in Chapter 10, then

$$C_{le} = \frac{4\alpha_e}{\sqrt{M_{\infty e}^2 - 1}} \tag{11.15}$$

$$C_{de} = \frac{4}{\sqrt{M_{\infty e}^2 - 1}} \left(\alpha_e^2 + \frac{\overline{\sigma_{u e}^2} + \overline{\sigma_{l e}^2}}{2}\right) \tag{11.16}$$

These results are identical to those obtained in Chapter 10 for infinite aspect ratio wings with leading edges normal to the free-stream flow direction. See equations (10.8) and (10.16)

The results of Ivey and Bowen (Ref. 11.17) for flow about swept-back airfoils with double-wedge profiles are presented in Fig. 11.16. Note that significant improvement in performance can be realized with sweep back. The results in Fig. 11.16 are based on the exact relations [i.e., equations (11.13) and (11.14)], the shock-expansion theory (not linear theory), and the assumption that the skin-friction drag coefficient per unit span is 0.006.

Delta and Arrow Wings

In Ref. 11.18, Puckett and Stewart used a combination source-distribution and conical-flow theory to investigate the flow about delta- and arrow-shaped planforms (see Fig. 11.17). Cases studied included subsonic and/or supersonic leading and trailing edges with double-wedge airfoil sections. Stewart (Ref. 11.19) and Puckett (Ref. 11.20) used conical flow theory to investigate the flow about simple delta planforms.

Two significant conclusions about delta and arrow planforms that can be drawn from these studies are:

1. For wings where the sweepback of both leading and trailing edges is relatively small, the strong drag peak at Mach 1 (characteristic of two-dimensional wings) is replaced by a weaker peak at a higher Mach number, corresponding to co-incidence of the Mach wave with the leading edge.

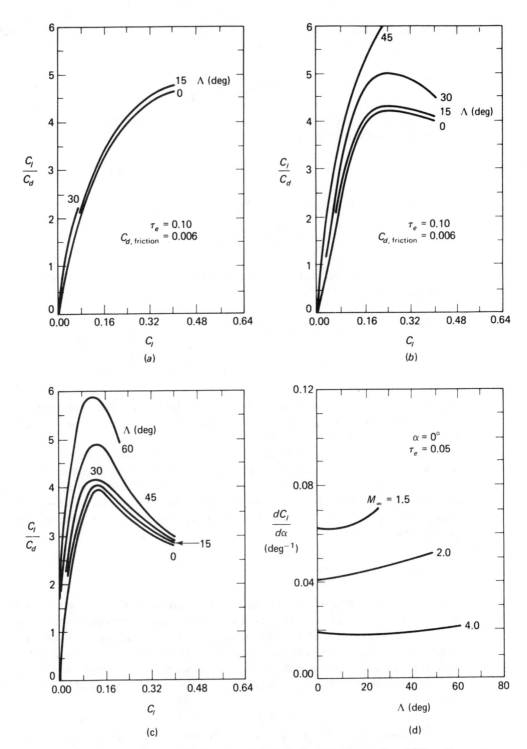

Figure 11.16 Theoretical effect of sweepback for double-wedge section airfoils with supersonic leading edges: (a) lift-to-drag ratio for $M_\infty = 1.5$; (b) lift-to-drag ratio for $M_\infty = 2.0$; (c) lift-to-drag ratio for $M_\infty = 4.0$; (d) lift-curve slope. (From Ref. 11.17.)

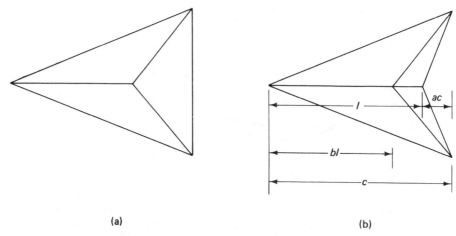

(a) (b)

Figure 11.17 Delta- and arrow-wing planforms with double-wedge sections: (a) delta planform; (b) arrow planform.

2. Delta and arrow wings with subsonic leading edges can have lift curve slopes $(dC_L/d\alpha)$ approaching the two-dimensional value $(4/\beta)$ with much lower values of C_D/τ^2 than those characteristic of two-dimensional wings of the same thickness.

A theoretical comparison of a rectangular, delta, and arrow wing is given in Table 11.2. As noted in Ref. 11.21: "One of the prominent advantages of the arrow wing is in the area of induced drag. This is illustrated qualitatively in Fig. 11.18,

TABLE 11.2 COMPARISON OF AERODYNAMIC COEFFICIENTS FOR RECTANGULAR, DELTA, AND ARROW WING PLANFORMS FOR $M_\infty = 1.50$

Wing planform	Rectangular	Delta[a]	Arrow[a]
	$\Lambda = 70°$	$\Lambda = 70°$	$\Lambda = 70°$
		$b = 0.2$	$b = 0.2$
		$\beta \cot \Lambda = 0.4$	$\beta \cot \Lambda = 0.4$
	$AR = 1$	$a = 0$	$a = 0.25$
$\dfrac{\beta}{4}\dfrac{dC_L}{d\alpha}$	0.554	0.544	0.591
$\dfrac{dC_L}{d\alpha}$	1.98	1.94	2.11
Relative area, S	1.00	1.018	0.938
Relative root chord, l	1.00	1.69	1.26
Root thickness ratio, τ	0.10	0.059	0.080
$C_{D,\text{thickness}}$	0.0119	0.0048	0.0070
$C_{D,\text{friction}}$	0.0060	0.0060	0.0060
C_{D0}	0.0179	0.0108	0.0130
$\left(\dfrac{C_L}{C_D}\right)_{\text{max}}$	5.25	8.6	9.3

[a] See Fig. 11.17 for a definition of a and b.

Source: Ref. 11.18.

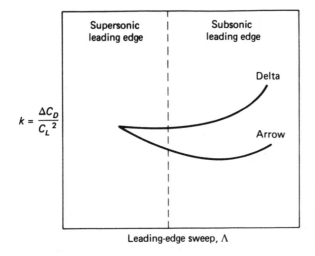

$$k = \frac{\Delta C_D}{C_L{}^2}$$

Figure 11.18 Comparison of induced drag for delta- and arrow-wing planforms. *Note*: $\Delta C_D - C_{D0}$. (From Ref. 11.21.)

where the planform with a trailing edge cut out or notch ratio is shown to have lower induced drag. The second advantage of the arrow wing is its ability to retain a subsonic round leading edge at an aspect ratio that is of the same level as that of the lesser swept supersonic leading edge delta. The advantages of the subsonic leading edge are a lower wave drag at cruise and a high L/D for subsonic flight operations due to increased leading edge suction.''

SINGULARITY-DISTRIBUTION METHOD

The second method that can be used to solve equation (11.2) is the singularity-distribution method. Detailed treatment of the mathematical aspects of the theory as well as applications of it to various wing planforms are presented in Refs. 11.9, 11.10, 11.14, and 11.22. For simple planforms, the singularity-distribution method can provide exact, analytical closed-form solutions to three-dimensional wing problems (see Refs. 11.10, 11.14, and 11.18). However, the method is quite adaptable for use with computers to solve for flow about complex shapes, and this is where it is most extensively applied today.

In Chapter 3 we learned that the governing equation [i.e., equation (3.26)] for incompressible, irrotational flow is linear even without the assumption of small disturbances. This allowed us to combine elementary solutions (i.e., source, sink, doublet, vortex, etc.) of the governing equation to generate solutions for incompressible flows about shapes of aerodynamic interest. In supersonic flow, where the small disturbance assumption is necessary to linearize the governing equation [e.g., equation (11.2)], there are analogs to the simple solutions for the incompressible case. Owing to a mathematical similarity to their incompressible counterparts, the supersonic solutions are quite naturally referred to as supersonic sources, sinks, doublets, and vortices. However, the physical relationship to their subsonic counterparts is not quite so direct and will not be pursued here.

The supersonic source (recall that a sink is simply a negative source), the doublet, and the horseshoe vortex potentials given by Ref. 11.9 are as follows:

$$\text{Source: } \phi_s = -\frac{Q}{r_c} \tag{11.17a}$$

$$\text{Doublet: } \phi_d = +\frac{Qz\beta^2}{r_c^3} \tag{11.17b}$$

where the axis of the doublet is in the positive z direction.

$$\text{Vortex: } \phi_v = -\frac{Qzv_c}{r_c} \tag{11.17c}$$

In equation (11.17), Q is the strength of the singularity and

$$r_c = \{(x - x_1)^2 - \beta^2[(y - y_1)^2 + z^2]\}^{0.5}$$

$$\beta^2 = M_\infty^2 - 1$$

$$v_c = \frac{x - x_1}{(y - y_1)^2 + z^2}$$

One can verify by direct substitution that equations (11.17a) through (11.17c) satisfy equation (11.2). Note that the point (x_1, y_1, z_1) is the location of the singularity. Since the wing is in the $z = 0$ plane, $z_1 = 0$ for every singularity for this approximation. This is because, in the singularity distribution method, the wing is replaced by a distribution of singularities in the plane of the wing. The hyperbolic radius, r_c, is seen to be imaginary outside the Mach cone extending downstream from the location of the singularity in each instance. Thus, the influence of the singularity is only present in the zone of action downstream of the point $(x_1, y_1, 0)$.

Using high-speed digital computers, modern numerical techniques consist of modeling a wing or body by replacing it with singularities at discrete points. These singularities are combined linearly to create a flow pattern similar to that about the actual body. The strengths of the singularities are determined so that the boundary condition requiring that the flow be tangent to the body's surface is satisfied at selected points. For configurations with sharp trailing edges it is also necessary to satisfy the Kutta condition at those sharp trailing edges which are subsonic.

Once the singularity distribution is determined, the potential at a given point is obtained by summing the contributions of all the singularities to the potential at that point. Velocity and pressure distributions follow from equations (9.12) and (9.16a).

Four types of problems that can be treated by the singularity distribution method (see Ref. 11.9) are as follows:

Two nonlifting cases:

1. Given the thickness distribution and the planform shape, find the pressure distribution on the wing.

2. Given the pressure distribution on a wing of symmetrical section, find the wing shape (i.e., find the thickness distribution and the planform).

Two lifting cases:

3. Given the pressure distribution on a lifting surface (zero thickness) find the slope at each point of the surface.
4. Given a lifting surface, find the pressure distribution on it. Here, it is necessary to impose the Kutta condition for subsonic trailing edges when they are present.

Cases 1 and 3 are called "direct" problems bcause they involve integrations with known integrands. Cases 2 and 4 are "indirect" or "inverse" problems, since the unknown to be found appears inside the integral sign. Thus, the solution of inverse problems involves the inversion of an integral equation.

One might expect that cases 1 and 4 would be the only ones of practical interest. However, this is not the case. Many times, a designer wishes to specify a given loading distribution (e.g., either for structural or for stability analyses) and solve for the wing shape which will give that prescribed loading distribution. Thus, the engineer may encounter any one of the four cases in aircraft design work. A variation of cases 2 and 3 is to specify the potential on a surface instead of the pressure distribution.

Cases 1 and 2 are most conveniently solved using source or doublet distributions, while cases 3 and 4 are most often treated using vortex distributions.

Consider a distribution of supersonic sources in the xy plane. The contribution to the potential at any point $P(x, y, z)$ due to an infinitesimal source at $P'(x_1, y_1, 0)$ in the plane is, from equation (11.17a),

$$d\phi(x, y, z) = -\frac{C(x_1, y_1)\, dx_1\, dy_1}{\sqrt{(x - x_1)^2 - \beta^2[(y - y_1)^2 + z^2]}} \qquad (11.18)$$

where $C(x_1, y_1)$ is the source strength per unit area. Consistent with the linearity assumption, the flow tangency condition [equation (11.5)] gives the vertical (z direction) velocity component in the xy plane as

$$w'(x, y, 0) = \left[\frac{\partial\phi(x, y, z)}{\partial z}\right]_{z=0} = U_\infty \frac{dz_s(x, y)}{dx} \qquad (11.19)$$

Taking the derivative with respect to z of equation (11.18) gives

$$\frac{\partial[d\phi(x, y, z)]}{\partial z} = dw'(x, y, z) = -\frac{C(x_1, y_1)\beta^2 z\, dx_1\, dy_1}{\{(x - x_1)^2 - \beta^2[(y - y_1)^2 + z^2]\}^{1.5}} \qquad (11.20)$$

Note that x_1 and y_1 are treated as constants. Taking the limit of equation (11.20) as $z \to 0$, we get $dw'(x, y, 0) = 0$, except very near the point (x_1, y_1) where the limit is indeterminant (i.e., of the form 0/0). We conclude that the vertical velocity at a point in the xy plane is due only to the source at that point and to no other sources. In other words, a source induces a vertical velocity only at its location and nowhere else. A source does, however, contribute to u' and v' (i.e., the x and y components of the perturbation velocity) at other locations. We still must determine the contribution of the source at the point (x_1, y_1) to the vertical velocity at $P'(x_1, y_1, 0)$. Puckett (Ref. 11.20) shows that the latter contribution is

$$dw'(x_1, y_1, 0) = \pi C(x_1, y_1) \tag{11.21a}$$

However, since this is the entire contribution,

$$w'(x_1, y_1, 0) = dw'(x_1, y_1, 0) = \pi C(x_1, y_1) \tag{11.21b}$$

Using equation (11.19), we see that

$$C(x_1, y_1) = \frac{U_\infty}{\pi} \left[\frac{dz_s(x_1, y_1)}{dx_1} \right] \equiv \frac{U_\infty \lambda(x_1, y_1)}{\pi} \tag{11.21c}$$

where λ is the local slope of the wing section. Thus, we obtain the important result that the source distribution strength at a point is proportional to the local surface slope at the point.

Substituting equation (11.21c) into equation (11.18) and integrating gives

$$\phi(x, y, 0) = -\frac{U_\infty}{\pi} \iint_S \frac{\lambda(x_1, y_1) \, dx_1 \, dy_1}{[(x - x_1)^2 - \beta^2(y - y_1)^2]^{0.5}} \tag{11.22}$$

where S is the region in the xy plane within the upstream Mach cone with apex at $P(x, y)$. Once the potential is known, the pressure distribution follows from equation (9.16); that is,

$$C_p = -\frac{2(\partial\phi/\partial x)}{U_\infty} \tag{9.16a}$$

Since source distributions are used where the airfoil section is symmetric (see cases 1 and 2 above), the pressure distribution determined by equations (11.22) and (9.16) is the same on the upper and lower surfaces. Because linear theory requires that the deflection angles are small, the wave-drag coefficient at zero lift is

$$C_{D_w} = 2 \iint_{S_u} C_{p_u}(x, y)\lambda_u(x, y) \, dx \, dy \tag{11.23}$$

where S is the surface of the wing and the subscript u indicates the upper surface. The formula contains the factor 2 to include the contribution of the lower surface since the section is symmetric.

Example 11.1

Let us determine the pressure distribution for the simple wing shape shown in Fig. 11.19, which is at zero angle of attack. This is a single-wedge delta with subsonic leading edges. The leading edge is swept by the angle Λ. Granted, this wing is not practical because of its blunt trailing edge. However, neglecting the presence of the boundary layer, the effects of the trailing edge are not propagated upstream in the supersonic flow. Thus, to obtain the flow about a wing with a sharp trailing edge we can add (actually subtract) the solutions for the flow around two delta wings of constant slope, such that the desired airfoil section can be obtained. This additive process is illustrated in Fig. 11.20. Thus, the simple case considered here can be used as a building block to construct more complex flow fields about wings of more practical shape (see Refs. 11.10 and 11.19).

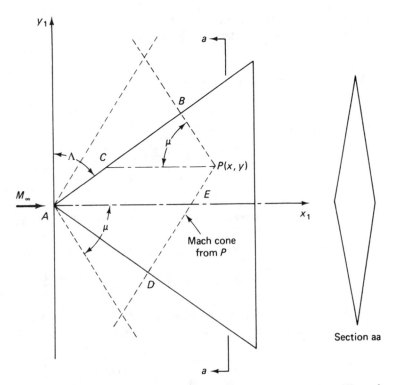

Figure 11.19   Nomenclature and geometry for single-wedge delta wing of Example 11.1.

Solution   Consider the point $P(x, y)$ in Fig. 11.19. The flow conditions at P are a result of the combined influences of all the sources within the upstream Mach cone from P. From symmetry, the vertical velocity perturbation vanishes ahead of the wing, and the source distribution simulating the wing extends only to the leading edges. Thus, the source distribution which affects P is contained entirely in the region $ABPD$.

The potential at P is given by equation (11.22):

$$\phi(x, y, 0) = -\frac{\lambda}{\pi} U_\infty \iint\limits_{ABPD} \frac{dx_1\, dy_1}{[(x - x_1)^2 - \beta^2(y - y_1)^2]^{0.5}}$$

where we have moved λ outside the integral sign, since it is a constant in this example.

To carry out the integration, it is convenient to break $ABPD$ into three separate areas, thus:

$$\phi(x, y, 0) = -\frac{\lambda}{\pi} U_\infty \left\{ \iint\limits_{ADE} \frac{dx_1\, dy_1}{[(x - x_1)^2 - \beta^2(y - y_1)^2]^{0.5}} \right.$$

$$+ \iint\limits_{AEPC} \frac{dx_1\, dy_1}{[(x - x_1)^2 - \beta^2(y - y_1)^2]^{0.5}}$$

$$+ \left. \iint\limits_{CPB} \frac{dx_1\, dy_1}{[(x - x_1)^2 - \beta^2(y - y_1)^2]^{0.5}} \right\}$$

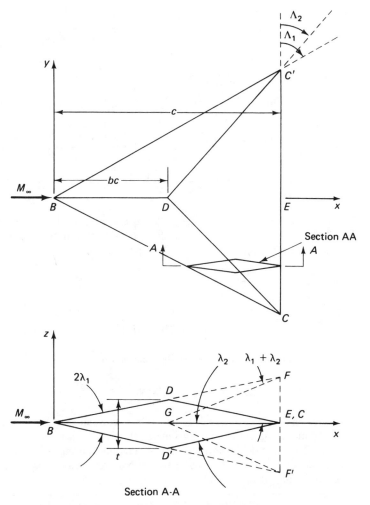

Figure 11.20 Geometry for a delta-wing planform with a double-wedge section. (From Ref. 11.10.)

To define the limits of integration, we note the following relationships from geometry:

$$\text{Along } AD: \quad x_1 = -y_1 \tan \Lambda$$

$$\text{Along } ACB: \quad x_1 = +y_1 \tan \Lambda$$

$$\text{Along } BP: \quad x_1 = x - \beta(y_1 - y)$$

$$\text{Along } DEP: \quad x_1 = x + \beta(y_1 - y)$$

Finally, the coordinates of points B and D are

$$B\left[\frac{(x + \beta y) \tan \Lambda}{\tan \Lambda + \beta}, \frac{x + \beta y}{\tan \Lambda + \beta}\right]$$

$$D\left[\frac{(x - \beta y)\tan \Lambda}{\tan \Lambda + \beta,} \quad \frac{-(x - \beta y)}{\tan \Lambda + \beta}\right]$$

Thus, we have

$$\phi(x, y, 0) = -\frac{\lambda}{\pi} U_\infty \left[\int_{\frac{-(x-\beta y)}{\tan \Lambda + \beta}}^0 dy_1 \int_{-y_1 \tan \Lambda}^{x + \beta(y_1 - y)} \frac{dx_1}{[(x - x_1)^2 - \beta^2(y - y_1)^2]^{0.5}} \right.$$

$$+ \int_0^y dy_1 \int_{y_1 \tan \Lambda}^{x + \beta(y_1 - y)} \frac{dx_1}{[(x - x_1)^2 - \beta^2(y - y_1)^2]^{0.5}}$$

$$\left. + \int_y^{\frac{(x+\beta y)}{\tan \Lambda + \beta}} dy_1 \int_{y_1 \tan \Lambda}^{x - \beta(y_1 - y)} \frac{dx_1}{[(x - x_1)^2 - \beta^2(y - y_1)^2]^{0.5}} \right]$$

One can use standard integral tables and relationships involving inverse hyperbolic functions (e.g., see Ref. 11.23) to show that the result of this integration and subsequent differentiation with respect to x is (see Ref. 11.20)

$$u'(x, y, 0) = -\frac{\partial \phi}{\partial x}$$

$$= -\frac{\lambda}{\pi} U_\infty \frac{2}{\beta[(\tan^2 \Lambda/\beta^2) - 1]^{0.5}} \cosh^{-1}\left\{\left(\frac{\tan \Lambda}{\beta}\right)\left[\frac{1 - (\beta y/x)^2}{1 - (y^2 \tan^2 \Lambda)/x^2}\right]^{0.5}\right\}$$

By equation (9.16a) the pressure distribution on the wing is

$$C_p(x, y, 0) = -\frac{2u'}{U_\infty}$$

Notice that C_p is invariant along rays ($y/x = $ constant) from the apex of the wing. Thus, as we might suspect from the geometry, this is a conical flow.

The wave-drag coefficient can be determined using equation (11.23). Figure 11.21 presents pressure distributions and wave drag for various configurations of single- and double-wedge delta wings (see Figs. 11.19 and 11.20).

Carlson and Miller (Ref. 11.22) present a numerical application of the vortex distribution method which is applicable to thin wings of arbitrary profile and planform. We will discuss the application of their method to determine the lifting pressure distribution on a wing of arbitrary planform.

The equation governing the differential pressure coefficient ($\Delta C_p = C_{pl} - C_{pu}$) is

$$\Delta C_p(x, y) = -\frac{4}{\beta}\frac{\partial z_c(x, y)}{\partial x} + \frac{1}{\pi}\oiint_S R(x - x_1, y - y_1)\Delta C_p(x_1, y_1)\, d\beta y_1\, dx_1 \quad (11.24)$$

where

$$R(x - x_1, y - y_1) \equiv \frac{x - x_1}{\beta^2(y - y_1)^2[(x - x_1)^2 - \beta^2(y - y_1)^2]^{0.5}}$$

and $z_c(x, y)$ is the z coordinate of the camber line. The function R may be thought of as an influence function relating the local loading at the point (x_1, y_1) to its influence on the flow field.

The integral in equation (11.24) represents the influence of a continuous distribution of horseshoe vortices originating from wing elements with vanishingly small chords and spans. The region of integration, S, originating at the field point (x, y), is shown in Fig. 11.22. The integral gives the appearance of being improper and divergent because of the singularity at $y_1 = y$ within the region of integration. However, the integral can be treated according to the concept of the generalization of the Cauchy

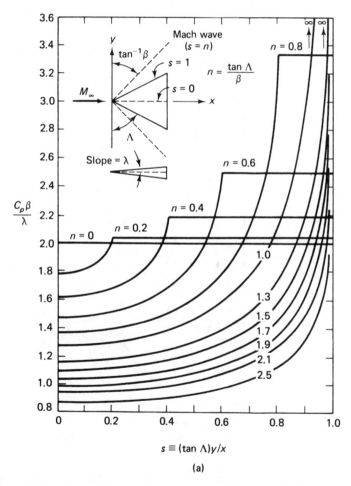

Figure 11.21 Theoretical solutions for a delta wing (from Ref. 11.20.): (a) pressure distribution for a single-wedge delta wing at $\alpha = 0$.

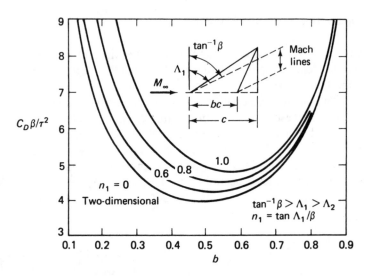

Figure 11.21 (*Continued*) (b) Thickness drag of a double-wedge delta wing with a supersonic leading edge and a supersonic line of maximum thickness.

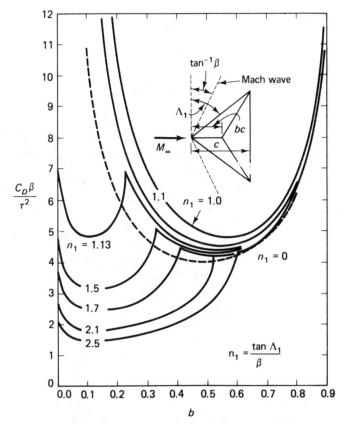

Figure 11.21 (*Concluded*) (c) Thickness drag of a double-wedge delta wing with a subsonic line of maximum thickness.

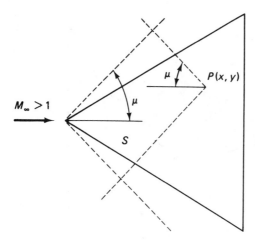

Figure 11.22 S is the region of integration for the supersonic vortex-lattice method.

principal value (see Ref. 11.9), as indicated by the double-dash marks on the integral signs.

Equation (11.24) can be rearranged

$$\frac{\partial z_c(x, y)}{\partial x} = -\frac{\beta}{4} \Delta C_p(x, y)$$

$$+ \frac{\beta}{4\pi} \oiint_S R(x - x_1, y - y_1) \, \Delta C_p(x_1, y_1) \, d\beta y_1 \, dx_1 \qquad (11.25)$$

Equation (11.25) can be used to determine the wing surface shape necessary to support a specified lift distribution.

In order to replace the indicated integration in equations (11.24) and (11.25) with an algebraic summation, it is first necessary to replace the Cartesian coordinate system shown in Fig. 11.23a with the grid system shown in Fig. 11.23b. The region of integration, originally bound by the wing leading edge and the Mach lines, now consists of a set of grid elements approximating that region shown by the shaded area of Fig. 11.23b. Inclusion of partial as well as full grid elements provides a better definition of the wing leading edge and tends to reduce any irregularities that may arise in local surface slopes for elements in the vicinity of the leading edge.

Thus, we will determine the lifting pressure distribution numerically using a system of grid elements similar to those shown in Fig. 11.23b. Note that, in practice, many more elements would be used. The numbers L and N identify the spaces in the grid which replace the integration element $dx_1 \, d\beta y_1$. L^* and N^* identify the element associated with, and immediately in front of, the field point $(x, \beta y)$. Note that $L^* = x$ and $N^* = \beta y$, and x and βy take on only integer values. The region of integration, originally bounded by the leading edge and the forecone Mach lines from $(x, \beta y)$ is now approximated by the grid elements within the Mach forecone emanating from $(x, \beta y)$. Note that in the $(x, \beta y)$ coordinate system, the Mach cone half-angle is always $45°$.

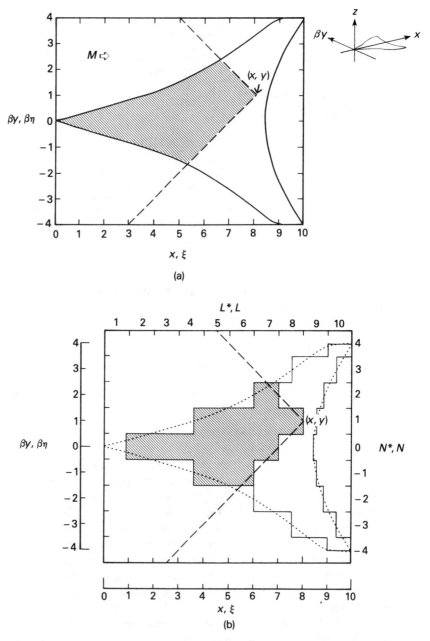

Figure 11.23 Coordinate systems for the linearized, supersonic vortex lattice method: (a) Cartesian coordinate system; (b) grid system used in numerical solution.

The summation approximation to equation (11.24) then becomes

$$\Delta C_p(L^*, N^*) = -\frac{4}{\beta}\frac{\partial z_c(L^*, N^*)}{\partial x} +$$

$$\frac{1}{\pi}\sum_{N_{min}}^{N_{max}}\sum_{L_{le}}^{L^* - |N^* - N|}\overline{R}(L^* - L, N^* - N)A(L, N)\,B(L, N)\,C(L, N)\,\Delta C_p(L, N)$$

(11.26)

where $\overline{R}$ is the average value of $\overline{R}$ within an element and is given by

$$\overline{R}(L^* - L, N^* - N) = \frac{[(L^* - L + 0.5)^2 - (N^* - N - 0.5)^2]^{0.5}}{(L^* - L + 0.5)(N^* - N - 0.5)}$$

$$- \frac{[(L^* - L + 0.5)^2 - (N^* - N + 0.5)^2]^{0.5}}{(L^* - L + 0.5)(N^* - N + 0.5)}$$

(11.27)

A graphical representation of this factor is presented in Fig. 11.24. Note the relatively small variations of the factor in the x (or L)-direction contrasted with the larger variations in the y (or N)-direction. For a given L^*–L set of elements, the spanwise summation of the R values is found to be zero, due to the single negative value at $N^* - N = 0$ balancing all the others. At $L^* - L = 0$, where there is only one element in the spanwise summation, the $\overline{R}$ value of that element is zero. This fact ensures that an element will have no influence on itself.

The limits on L in the summation of equation (11.26) are those of the wing leading edge (i.e., $L_{le} = 1 + [x_{le}]$, where $[x_{le}]$ designates the whole-number part of the quantity) and of the Mach forecone at the selected N value. The vertical lines are used in $|N^* - N|$ to designate the absolute value of the enclosed quantity.

The $A(L, N)$, $B(L, N)$, and $C(L, N)$ terms are weighting factors which account for the leading edge, trailing edge, and wing tips, respectively. Their values are given by

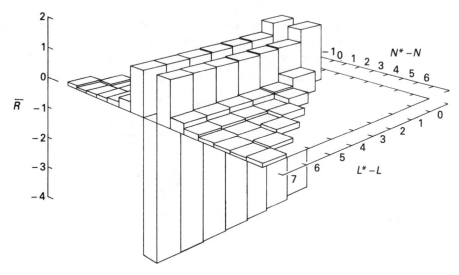

Figure 11.24 Numerical representation of the influence factor $\overline{R}$ (the $\overline{R}$ function).

$$A(L, N) = 0 \qquad\qquad L - x_{le} \le 0$$

$$A(L, N) = L - x_{le} \qquad\qquad 0 < L - x_{le} < 1$$

$$A(L, N) = 1 \qquad\qquad L - x_{le} \ge 1$$

$$B(L, N) = 0 \qquad\qquad L - x_{te} \ge 1$$

$$B(L, N) = 1 - (L - x_{te}) \qquad\qquad 0 < L - x_{te} < 1$$

$$B(L, N) = 1 \qquad\qquad L - x_{te} \le 0$$

$$C(L, N) = 0.5 \qquad\qquad N = N_{max}$$

$$C(L, N) = 1 \qquad\qquad N \ne N_{max}$$

Note that an element has no influence on itself since $\bar{R}(0, 0) = 0$ from equation (11.27). Thus, $\Delta C_p(L = L^*, N = N^*)$ is not required in the summation term of equation (11.26). Also, note that equation (11.26) is the summation to approximate what was originally an integral equation (see Ref. 11.9) and thus already accounts for the flow tangency boundary condition, equation (11.19). This condition is satisfied exactly only at control points located at the midspan of the trailing edge of each grid element (see Fig. 11.25b). If calculations are performed in the proper sequence (i.e., from the wing apex rearward), at no time will there be an unknown $\Delta C_p(L, N)$ in the summation equation (11.26).

The $\Delta C_p(L^*, N^*)$ given by equation (11.26) is defined at the trailing edge of the (L^*, N^*) element. To eliminate large oscillations in the pressure coefficient which can occur with this numerical technique, a smoothing operation is required. The procedure, taken directly from Ref. 11.22, is as follows:

1. Calculate and retain, temporarily, the preliminary ΔC_p values for a given row, with $L^* =$ constant. Designate this as $\Delta C_{p,a}(L^*, N^*)$.
2. Calculate and retain, temporarily, ΔC_p values for the following row with $L^* =$ constant $+ 1$, by using the $\Delta C_{p,a}$ values obtained in the previous step for contributions from the row with $L^* =$ constant. Designate this as $\Delta C_{p,b}(L^*, N^*)$.
3. Calculate a final ΔC_p value from a fairing of integrated preliminary ΔC_p results.

For leading-edge elements, defined as $L^* - x_{le}(N^*) \le 1$,

$$\Delta C_p(L^*, N^*) = \frac{1}{2}\left[1 + \frac{A(L^*, N^*)}{1 + A(L^*, N^*)}\right]\Delta C_{p,a}(L^*, N^*)$$
$$+ \frac{1}{2}\left[\frac{1}{1 + A(L^*, N^*)}\right]\Delta C_{p,b}(L^*, N^*) \tag{11.28}$$

For all other elements, defined as $L^* - x_{le}(N^*) > 1$,

$$\Delta C_p(L^*, N^*) = \tfrac{3}{4}\Delta C_{p,a}(L^*, N^*) + \tfrac{1}{4}\Delta C_{p,b}(L^*, N^*) \tag{11.29}$$

where

$$A(L^*, N^*) = A(L, N)$$

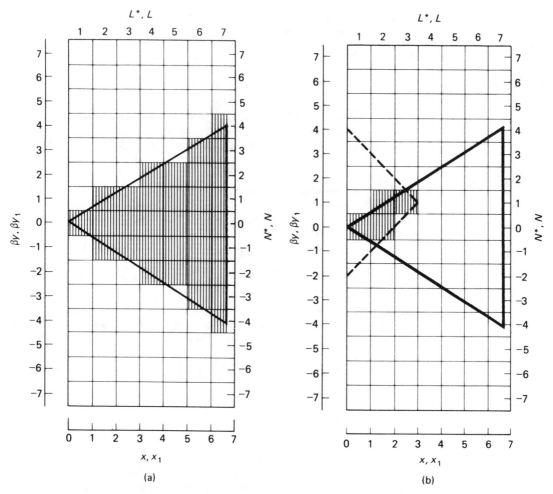

Figure 11.25   Grid element geometry for the supersonic vortex lattice method: (a) general pattern; (b) region of integration for element (3, 1) of Example 11.2.

The nomenclature $A(L^*, N^*)$ is used here to be consistent with that of Ref. 11.22. Note that there is an error in Ref. 11.22 for equation (11.28). The formulation given here is correct (see Ref. 11.24).

Example 11.2

To illustrate the method, let us show how to manually calculate the pressure distribution at $M_\infty = 1.5$ for the flat-plate (zero-camber) delta planform of Fig. 11.25a for the subsonic leading edge case where $\beta \cot \Lambda_{LE} = 0.6$. We will use the grid element set shown in Fig. 11.25a in order to keep the number of manual calculations within bounds. In an actual application, of course, one would use a much larger number of elements. Note that in Fig. 11.25a we use some partial elements along the leading and the trailing edges. Since ΔC_p is treated as a constant over a given element, partial elements affect only the reference area used when integrating the pressures to calculate the lift coefficient, the drag due-to-lift coefficient, etc.

Solution   For this case,

$$\frac{\partial z_c(x, y)}{\partial x} = \frac{dz_c}{dx} = -\tan \alpha \approx -\alpha$$

since, even here, we are still restricting ourselves to linear theory, which implies that all changes in flow direction about the free-stream direction are small.

For this case, where the free-stream Mach number is 1.5:

$$\beta \equiv \sqrt{M_\infty^2 - 1} = 1.118$$

Therefore, the leading-edge sweepback angle corresponding to $\beta \cot \Lambda_{LE} = 0.6$ is 61.78°. Thus, when the transformation is made from the (x, y) to the $(x, \beta y)$ plane, the angle that the leading edge makes with the βy axis is 59.04°. This is shown in Fig. 11.25a along with the grid system.

The equation of the wing leading and trailing edges in the $(x, \beta y)$ system is

$$x_{le} = \frac{\beta|y|}{0.6} \qquad x_{te} = \frac{N_{max}}{\beta \cot \Lambda_{LE}} = 6.6667$$

Note that the value of x_{te} is defined by selection of a maximum N value, N_{max}. For the purposes of this example, we arbitrarily select $N_{max} = 4$. The wing is then scaled to give a semispan of N_{max}. This ensures that the weighting factor $C(L, N) = 0.5$ for $N = N_{max}$ is appropriate. This is important for streamwise tips of non-zero chord. Using this and the relations defining $A(L, N)$, $B(L, N)$, and $C(L, N)$, we have the following.

$A(L, N)$

N \ L	1	2	3	4	5	6	7
0	1.0000	1.0000	1.0000	1.0000	1.0000	1.0000	1.0000
±1	0.0000	0.3333	1.0000	1.0000	1.0000	1.0000	1.0000
±2	0.0000	0.0000	0.0000	0.6667	1.0000	1.0000	1.0000
±3	0.0000	0.0000	0.0000	0.0000	0.0000	1.0000	1.0000
±4	0.0000	0.0000	0.0000	0.0000	0.0000	0.0000	0.3333

$B(L, N)$:

$$B(7, N) = 0.6667$$

$$B(L, N) = 1 \qquad \text{for all other grid elements}$$

$C(L, N)$:

$$C(7, \pm 4) = 0.5$$

$$C(L, N) = 1 \qquad \text{for all other grid elements}$$

We now proceed to calculate values of ΔC_p for several grid elements to show the technique. We begin at the apex with element (1,0).

Element $(1, 0)$. For this element, $L^* = 1$, $N^* = 0$, and there are no other elements which contribute to the differential pressure coefficient at this element. Thus,

$$\Delta C_{p,a}(1, 0) = \frac{4\alpha}{\beta}$$

To determine $\Delta C_p(1, 0)$, we must calculate a preliminary value $\Delta C_{p,b}(1, 0)$. This involves consideration of the element at $(2, 0)$. The only element contributing to the pressure differential at $(2, 0)$ is the one at $(1, 0)$. Thus,

$$\Delta C_{p,b}(1, 0) = \frac{4\alpha}{\beta} + \frac{1}{\pi} \left\{ \frac{[(2 - 1 + 0.5)^2 - (0 - 0 - 0.5)^2]^{0.5}}{(2 - 1 + 0.5)(0 - 0 - 0.5)} \right.$$
$$\left. - \frac{[(2 - 1 + 0.5)^2 - (0 - 0 + 0.5)^2]^{0.5}}{(2 - 1 + 0.5)(0 - 0 + 0.5)} \right\} \frac{4\alpha}{\beta}$$

or

$$\Delta C_{p,b}(1, 0) = \frac{4\alpha}{\beta} (1 - 1.2004) = -0.8016 \frac{\alpha}{\beta}$$

Therefore,

$$\Delta C_p(1, 0) = \frac{1}{2} \left(1 + \frac{1}{1 + 1} \right) \frac{4\alpha}{\beta} + \frac{1}{2} \left(\frac{1}{1 + 1} \right) (-0.8016) \frac{\alpha}{\beta}$$

or

$$\Delta C_p(1, 0) = 2.7996 \frac{\alpha}{\beta}$$

Element $(2, 0)$. For this element, $L^* = 2$, $N^* = 0$, and the only element that contributes to the differential pressure coefficient at this element is the one at $(1, 0)$. Thus,

$$\Delta C_{p,a}(2, 0) = 4 \frac{\alpha}{\beta} + \frac{1}{\pi} \left\{ \frac{[(2 - 1 + 0.5)^2 - (0 - 0 - 0.5)^2]^{0.5}}{(2 - 1 + 0.5)(0 - 0 - 0.5)} \right.$$
$$\left. - \frac{[(2 - 1 + 0.5)^2 - (0 - 0 + 0.5)^2]^{0.5}}{(2 - 1 + 0.5)(0 - 0 + 0.5)} \right\} 2.7996 \frac{\alpha}{\beta}$$

or

$$\Delta C_{p,a}(2, 0) = 0.6393 \frac{\alpha}{\beta}$$

To determine $\Delta C_p(2, 0)$, we must calculate a preliminary value $\Delta C_{p,b}(2, 0)$. This involves consideration of the element at $(3, 0)$. But, we note that elements at $(1, 0)$, $(2, 0)$, $(2, 1)$, and $(2, -1)$ contribute to the differential pressure at $(3, 0)$. Thus, we must go ahead and find $\Delta C_{p,a}(2, 1)$ and $\Delta C_{p,a}(2, -1)$. These, of course, are equal from symmetry.

The only element influencing the element at $(2, 1)$ is the one at $(1, 0)$. Thus,

$$\Delta C_{p,a}(2, 1) = \frac{4\alpha}{\beta} + \frac{1}{\pi} \left\{ \frac{[(2 - 1 + 0.5)^2 - (1 - 0 - 0.5)^2]^{0.5}}{(2 - 1 + 0.5)(1 - 0 - 0.5)} \right.$$
$$\left. - \frac{[(2 - 1 + 0.5)^2 - (1 - 0 + 0.5)^2]^{0.5}}{(2 - 1 + 0.5)(1 - 0 + 0.5)} \right\} 2.7996 \frac{\alpha}{\beta}$$

or

$$\Delta C_{p,a}(2, 1) = 5.6804 \frac{\alpha}{\beta} = \Delta C_{p,a}(2, -1)$$

Therefore,

$$
\begin{aligned}
\Delta C_{p,b}(2, 0) = \frac{4\alpha}{\beta} + \frac{1}{\pi} &\left\{ \frac{[(3 - 1 + 0.5)^2 - (0 - 0 - 0.5)^2]^{0.5}}{(3 - 1 + 0.5)(0 - 0 - 0.5)} \right. \\
&\left. - \frac{[(3 - 1 + 0.5)^2 - (0 - 0 + 0.5)^2]^{0.5}}{(3 - 1 + 0.5)(0 - 0 + 0.5)} \right\} 2.7996 \frac{\alpha}{\beta} \\
+ \frac{1}{\pi} &\left\{ \frac{[(3 - 2 + 0.5)^2 - (0 - 0 - 0.5)^2]^{0.5}}{(3 - 2 + 0.5)(0 - 0 - 0.5)} \right. \\
&\left. - \frac{[(3 - 2 + 0.5)^2 - (0 - 0 + 0.5)^2]^{0.5}}{(3 - 2 + 0.5)(0 - 0 + 0.5)} \right\} 0.6393 \frac{\alpha}{\beta} \\
+ \frac{2}{\pi} &\left\{ \frac{[(3 - 2 + 0.5)^2 - (0 - 1 - 0.5)^2]^{0.5}}{(3 - 2 + 0.5)(0 - 1 - 0.5)} \right. \\
&\left. - \frac{[(3 - 2 + 0.5)^2 - (0 - 1 + 0.5)^2]^{0.5}}{(3 - 2 + 0.5)(0 - 1 + 0.5)} \right\} (0.3333)(5.6804) \frac{\alpha}{\beta}
\end{aligned}
$$

where the factor of 2 in front of the last bracketed term accounts for the fact that elements $(2, 1)$ and $(2, -1)$ have an equal influence on element $(3, 0)$.

Thus,

$$\Delta C_{p,b}(2, 0) = 2.0128 \frac{\alpha}{\beta}$$

Finally,

$$\Delta C_p(2, 0) = \frac{3}{4}(0.6393) \frac{\alpha}{\beta} + \frac{1}{4}(2.0128) \frac{\alpha}{\beta}$$

or

$$\Delta C_p(2, 0) = 0.9827 \frac{\alpha}{\beta}$$

Element $(2, 1)$. For this element, $L^* = 2$, $N^* = 1$, and the only element that contributes to the differential pressure coefficient at this element is the one at $(1, 0)$. Thus, as shown above,

$$
\begin{aligned}
\Delta C_{p,a}(2, 1) = \frac{4\alpha}{\beta} + \frac{1}{\pi} &\left\{ \frac{[(2 - 1 + 0.5)^2 - (1 - 0 - 0.5)^2]^{0.5}}{(2 - 1 + 0.5)(1 - 0 - 0.5)} \right. \\
&\left. - \frac{[(2 - 1 + 0.5)^2 - (1 - 0 + 0.5)^2]^{0.5}}{(2 - 1 + 0.5)(1 - 0 + 0.5)} \right\} 2.7996 \frac{\alpha}{\beta}
\end{aligned}
$$

or

$$\Delta C_{p,a}(2, 1) = 5.6804 \frac{\alpha}{\beta}$$

To determine $\Delta C_p(2, 1)$, we must calculate a preliminary value $\Delta C_{p,b}(2, 1)$. This involves consideration of the element at $(3, 1)$. We note, however, that elements at $(1, 0)$, $(2, 0)$, and $(2, 1)$ contribute to the differential pressure at $(3, 1)$. Thus,

$$\Delta C_{p,b}(2, 1) = \frac{4\alpha}{\beta} + \frac{1}{\pi} \left\{ \frac{[(3 - 1 + 0.5)^2 - (1 - 0 - 0.5)^2]^{0.5}}{(3 - 1 + 0.5)(1 - 0 - 0.5)} \right.$$

$$\left. - \frac{[(3 - 1 + 0.5)^2 - (1 - 0 + 0.5)^2]^{0.5}}{(3 - 1 + 0.5)(1 - 0 + 0.5)} \right\} 2.7996 \frac{\alpha}{\beta}$$

$$+ \frac{1}{\pi} \left\{ \frac{[(3 - 2 + 0.5)^2 - (1 - 0 - 0.5)^2]^{0.5}}{(3 - 2 + 0.5)(1 - 0 - 0.5)} \right.$$

$$\left. - \frac{[(3 - 2 + 0.5)^2 - (1 - 0 + 0.5)^2]^{0.5}}{(3 - 2 + 0.5)(1 - 0 + 0.5)} \right\} 0.6393 \frac{\alpha}{\beta}$$

$$+ \frac{1}{\pi} \left\{ \frac{[(3 - 2 + 0.5)^2 - (1 - 1 - 0.5)^2]^{0.5}}{(3 - 2 + 0.5)(1 - 1 - 0.5)} \right.$$

$$\left. - \frac{[(3 - 2 + 0.5)^2 - (1 - 1 + 0.5)^2]^{0.5}}{(3 - 2 + 0.5)(1 - 1 + 0.5)} \right\} (0.3333)(5.6804) \frac{\alpha}{\beta}$$

or

$$\Delta C_{p,b}(2, 1) = 3.3820 \frac{\alpha}{\beta}$$

Note that we use preliminary (i.e., $\Delta C_{p,a}$) values for the ΔC_p's of influencing elements in the same "L" row as the field-point element under consideration. Thus, we have used $\Delta C_{p,a}(2, 0) = 0.6393$ instead of $\Delta C_p(2, 0) = 0.9827$ as the factor of the second major bracketed term in the equation above. As an alternative procedure we could have used ΔC_p instead of $\Delta C_{p,a}$, but we have chosen to be consistent with the method of Ref. 11.22. For influencing elements not in the same "L" row as the field-point element, we use the final (i.e., averaged ΔC_p values) as given by equations (11.28) and (11.29).

Finally,

$$\Delta C_p(2, 1) = \frac{1}{2} \left(1 + \frac{0.3333}{1 + 0.3333} \right) 5.6804 \frac{\alpha}{\beta}$$

$$+ \frac{1}{2} \left(\frac{1.00}{1 + 0.3333} \right) 3.3820 \frac{\alpha}{\beta}$$

or

$$\Delta C_p(2, 1) = 4.8185 \frac{\alpha}{\beta} = \Delta C_p(2, -1)$$

from symmetry.

We have now calculated the differential pressure coefficient for elements $(1, 0)$, $(2, 0)$, and $(2, \pm 1)$. One can continue in this manner and determine ΔC_p for the remaining elements. A summary of the computed values is given in Table 11.3.

TABLE 11.3 DIFFERENTIAL PRESSURE COEFFICIENT RESULTS FOR EXAMPLE 11.2, β COT $\Lambda_{LE} = 0.6$, $M_\infty = 1.5$

N^* \ L^*	$\Delta C_p \dfrac{\beta}{\alpha}$						
	1	2	3	4	5	6	7
0	2.80	0.98	1.92	1.56	1.81	1.99	1.34
1		4.82	2.94	2.06	2.09	1.58	1.81
2			5.54	2.84	2.37	3.51	
3				8.11	6.20	1.92	
4						9.80	

Typical numerical results for this case at three chordwise stations are presented in Fig. 11.26. The number of grid elements used in Fig. 11.26 is approximately 2000.

Lifting-pressure distributions for a flat-plate delta wing (β cot $\Lambda_{LE} = 1.20$) as computed by Middleton and Carlson (Ref. 11.25) are reproduced in Fig. 11.27. Agreement between the numerical and analytical solutions is quite good, although the numerical solution does not exhibit the sharp break at the Mach line that is characteristic of the analytical solution. The unevenness of the pressure coefficients observed in the subsonic leading-edge case does not occur because of the reduced effect of the averaging techniques with the less severe variations in the pressure near the leading edge. Figure 11.28 shows a comparison of the numerical results with those of exact linear theory for flat delta planforms such as the one used in Example 11.2.

Note that the numerical method is also applicable to wings of arbitrary camber and planform (e.g., supersonic or subsonic leading and/or trailing edges, etc.). To accommodate nonzero camber, one only need specify $\partial z_c(x, y)/\partial x$ in equation (11.26). A companion numerical method, also described in Ref. 11.22, provides a means for the design of a camber surface corresponding to a specified loading distribution or to an optimum combination of loading distributions.

Once the lifting pressure coefficients, ΔC_p, have been determined for all the elements, it is possible to compute the aerodynamic coefficients for the wing. The lift coefficient for symmetric loads may be obtained from the following summation over all the elements:

$$C_L = \frac{2}{\beta S} \sum_{N^*=0}^{N^*=N_{max}} \sum_{L^*=L_{le}}^{L^*=L_{te}} \left[\frac{3}{4} \Delta C_p(L^*, N^*) \right.$$

$$\left. + \frac{1}{4} \Delta C_p(L^* + 1, N^*) \right] A(L^*, N^*) B(L^*, N^*) C(L^*, N^*) \tag{11.30}$$

The pitching-moment coefficient about $x = 0$ is

$$C_M = \frac{2}{\beta S \bar{c}} \sum_{N^*=0}^{N^*=N_{max}} \sum_{L^*=L_{le}}^{L^*=L_{te}} (L^*) \left[\frac{3}{4} \Delta C_p(L^*, N^*) \right.$$

$$\left. + \frac{1}{4} \Delta C_p(L^* + 1, N^*) \right] A(L^*, N^*) B(L^*, N^*) C(L^*, N^*) \tag{11.31}$$

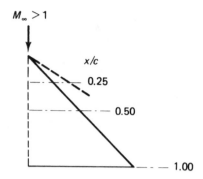

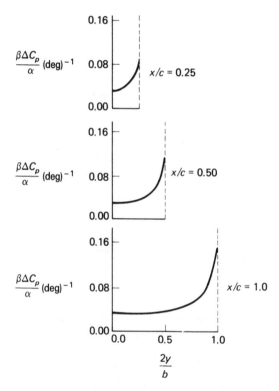

Figure 11.26 Numerical results for a flat-plate delta planform ($\beta \cot \Lambda_{LE} = 0.6$). (From Ref. 11.22.)

The drag coefficient may be expressed as

$$
C_D = -\frac{2}{\beta S} \sum_{N^*=0}^{N^*=N_{\max}} \sum_{L^*=L_{le}}^{L^*=L_{te}} \left[\frac{3}{4} \Delta C_p(L^*, N^*) + \frac{1}{4} \Delta C_p(L^* + 1, N^*) \right]
$$

$$
\left[\frac{3}{4} \frac{\partial z_c(L^*, N^*)}{\partial x} + \frac{1}{4} \frac{\partial z_c(L^* - 1, N^*)}{\partial x} \right] A(L^*, N^*) B(L^*, N^*) C(L^*, N^*)
$$

(11.32)

This relationship does not consider any contribution of the theoretical leading-edge-suction force or of any separated flow effects associated with its exclusion and accounts only for the inclination of the normal force to the relative wind.

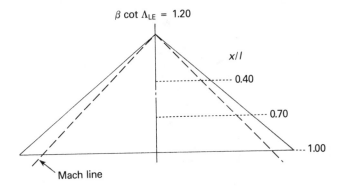

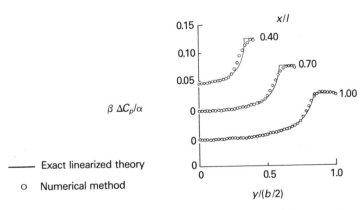

Figure 11.27 Lifting-pressure distributions for a flat-plate delta wing with a supersonic leading edge, $N_{max} = 50$.

The element weighting factors are defined as follows. The leading-edge field-point-element weighting factor takes on values from 0 to 1.5.

$$A(L^*, N^*) = 0 \qquad\qquad (L^* - x_{le} \leq 0)$$

$$A(L^*, N^*) = L^* - x_{le} + 0.5 \qquad (0 < L^* - x_{le} < 1)$$

$$A(L^*, N^*) = 1 \qquad\qquad (L^* - x_{le} \geq 1)$$

The trailing-edge field-point-element weighting factor also takes on values from 0 to 1.5.

$$B(L^*, N^*) = 0 \qquad\qquad (L^* - x_{te} \geq 0)$$

$$B(L^*, N^*) = 0.5 - (L^* - x_{te}) \qquad (0 > L^* - x_{te} > -1)$$

$$B(L^*, N^*) = 1 \qquad\qquad (L^* - x_{te} \leq -1)$$

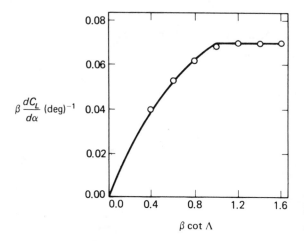

Figure 11.28 Comparison of linear theory with vortex lattice results for a delta wing. (From Ref. 11.22.)

The centerline or wingtip grid element weighting factor is defined as

$$C(L^*, N^*) = 0.5 \qquad (N^* = 0)$$

$$C(L^*, N^*) = 1 \qquad (0 < N^* < N_{max})$$

$$C(L^*, N^*) = 0.5 \qquad (N^* = N_{max})$$

The wing area used in the expressions for the aerodynamic coefficients may be computed using the summation

$$S = \frac{2}{\beta} \sum_{N^*=0}^{N^* = N_{max}} \sum_{L^* = 1 + [x_{le}]}^{L^* = 1 + [x_{te}]} A(L^*, N^*) B(L^*, N^*) C(L^*, N^*) \tag{11.33}$$

DESIGN CONSIDERATIONS FOR SUPERSONIC AIRCRAFT

We have seen that theory predicts that highly swept wings (such that the leading edge is subsonic) have the potential for very favorable drag characteristics at supersonic flight speeds. However, experiment has shown that many of the theoretical aerodynamic benefits of leading-edge sweepback are not attained in practice, because of separation of the flow over the upper surface of the wing. Elimination of separated flow can only be achieved in design by a careful blending of the effects of leading- and trailing-edge sweepback angles, leading-edge nose radius, camber, twist, body shape, wing/body junction aerodynamics, and wing planform shape and thickness distribution. Theory does not give all the answers here, and various empirical design criteria have been developed to aid in accounting for all of these design variables.

Stanbrook and Squire (Ref. 11.26) and Kulfan and Sigalla (Ref. 11.27) have studied the types of flow around highly swept edges. Sketches of the main types of flow on highly swept wings, as taken from their work, are presented in Fig. 11.29. Based on these investigations, we could classify the types of separation as:

1. Leading-edge separation due to high suction pressures
2. Separation due to spanwise flow

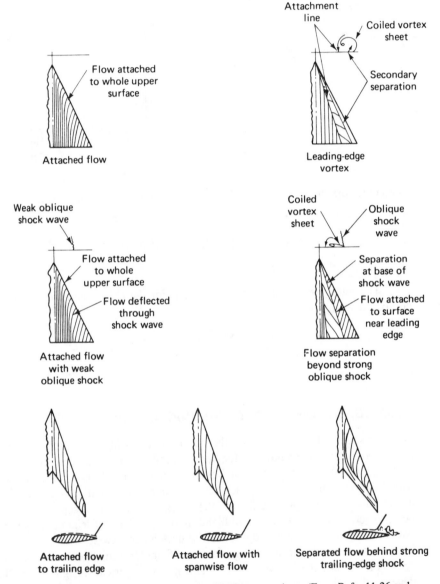

Figure 11.29 Main types of flow on highly swept wings. (From Refs. 11.26 and 11.27.)

3. Inboard shock separation

4. Trailing-edge shock separation

In our analysis of swept wings, we found that the effective angle of attack of the wing was given by equation (11.8) as

$$\alpha_e = \tan^{-1} \frac{\tan \alpha}{\cos \Lambda}$$

for small angles. Thus, for highly swept wings with subsonic leading edges, separation can occur quite readily even for small values of wing angle of attack. This is particularly true if the leading edge is sharp. The observed flow from the leading edge is very similar to that for the delta wing in subsonic flow (see Chapter 7). The coiled vortices that form at the leading edges affect the flow below the wing as well, since the leading edge is subsonic, and therefore the top and bottom surfaces can "communicate."

The phenomenon of spanwise flow on a swept wing has also been discussed in Chapter 7. Spanwise flow results in a thickening of the boundary layer near the wing tips, and a thick boundary layer will separate more readily than a thin one exposed to the same pressure gradient.

Inboard shock separation is primarily a function of the geometry of the wing/body junction near the leading edge of the wing. The upper surface flow in this region is toward the body, and therefore a shock is formed to turn the stream tangent to the body. The strength of the shock is, of course, dependent on the turning angle. If the shock is strong enough, the resulting pressure rise will separate the boundary layer.

Trailing-edge shock separation can occur if the wing trailing edge is supersonic (as is the case for delta wings, for example). A shock wave near the trailing edge is required to adjust the upper surface pressure back to the free-stream value. Again, if the required shock strength is too great, it will induce separation of the boundary layer.

The design criteria that have been developed to eliminate separation are given in Ref. 11.27, and can be summarized as follows:

1. *Leading-edge separation.* Reject designs where the theoretical suction pressure exceeds 70% of vacuum pressure.

2. *Separation due to spanwise flow.* Use thin wingtips. (*Note:* An additional technique is "washout" or a lower section incidence at the wing tips relative to the incidence angles of inboard sections.)

3. *Inboard shock separation.* Use body contouring to keep the pressure rise across the inboard shock wave to less than 50%.

4. *Trailing-edge shock separation.* Keep the pressure ratio across the trailing-edge shock below $1 + 0.3M_1^2$, where M_1 is the local Mach number ahead of the shock. For swept trailing edges, use the local normal Mach number ($M_{N1} = M_1 \cos \Lambda_{te}$) in the equation above.

Application of these criteria will not guarantee that the flow will not separate, however, and wind-tunnel tests of proposed designs must be undertaken. This is particularly true for military or sport aircraft designed for maneuvering at high load factors.

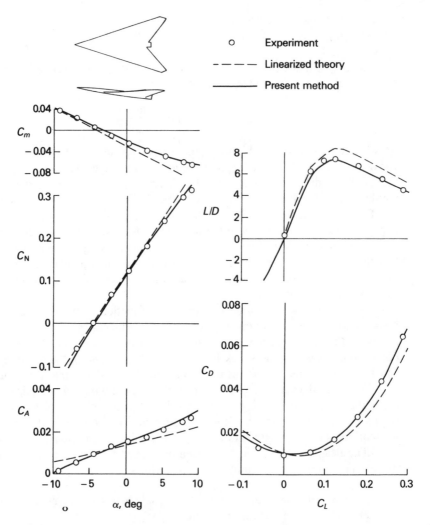

Figure 11.30 Comparison of the predicted and the measured forces and moments for 72.65° swept leading-edge arrow wing; $M_\infty = 2.6$; design value of $C_L = 0.12$.

Carlson and Mack (Ref. 11.6) note that there are compensatory errors in linearized theory and the failure to account for nonlinearities may introduce little error in prediction of lift and drag. However, significant errors in the prediction of the pitching moment are common, especially for wings that depart from a delta planform. Additionally, for wings with twist and camber, appreciable errors in the prediction of drag due to the surface distortion (camber drag) often occur. In particular, linearized-theory methods do not indicate the proper camber surface for drag minimization (a function of the design lift coefficient). The method of Carlson and Mack (Ref. 11.6), which includes the estimation of nonlinearities associated with leading-edge thrust and the detached leading-edge vortex flow, provides improved prediction over linearized theory as illustrated by the correlations reproduced in Fig. 11.30.

AERODYNAMIC INTERACTION

Owing to the complexity of flow fields about most flight vehicles, aerodynamicists often develop theories for the flows about components of such vehicles. However, the designer is faced with the fact that aerodynamic loads on an entire aircraft are not simply the sum of the loads on the individual components; the difference is commonly referred to as *aerodynamic interaction*. Wind-tunnel experiments, flight tests, and physical reasoning have shown that interaction loads can be a significant contribution to the total loading on an aircraft. It was not until the high-speed digital computer was generally available that systematic treatment of the loads on rather arbitrary shapes was even feasible. Theories are still being developed to account for the effects of three-dimensional separated flows about arbitrary shapes. Thus, this is a rapidly changing area of aerodynamic theory. Significant progress has been made, however, and our purpose here is to give a brief account of the physical reasoning associated with the effects of aerodynamic interaction in supersonic flight, and to describe methods of determining these effects.

Consider Fig. 11.31, where a simple rectangular wing is mounted on an ogive cylinder with circular cross section. Two interaction effects are present: the effect of the wing on the body and the effect of the body on the wing. In supersonic flight, the suction pressure on the upper surface of the wing and the relatively higher pressure on the lower surface will be confined to the regions bounded by the Mach wedges from the leading and trailing edges of the wing, as shown. This pressure differential will carry over onto the body and generate a net lift and wave-drag force on it. Thus, we have an example of the effect of the wing on the body.

Conversely, the effect of the body at angle of attack will be to produce an upwash about the sides of the body (see Fig. 11.32) which will increase the effective angle of attack of the wing. Provided that the resulting angle of attack will not cause the flow to separate on the upper wing surface (i.e., provided that the wing does not stall), the result will be greater wing lift in accordance with equation (10.8).

Hilton (Ref. 11.28) notes that the combined effects of the wing and body for the

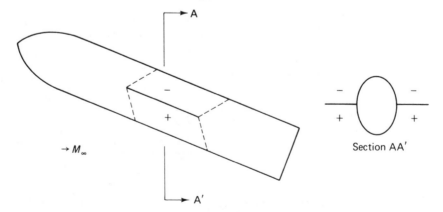

Figure 11.31   Wing-on-body interaction showing regions of positive and negative C_p. (From Ref. 11.28.)

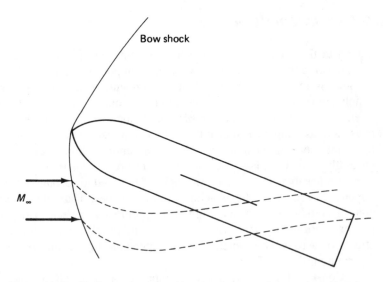

Figure 11.32 Body-on-wing interaction showing upwash-over-wing effect due to the body at angle of attack. (From Ref. 11.28.)

case shown in Fig. 11.33 are such that the wings produce the full lift to be expected from two-dimensional theory without tip effects. Taking into account tip effects and ignoring interaction, one would expect a wing lift of only three-fourths of this value. Thus, by introducing the body interaction effect, a full 25% increase in wing lift is experienced.

Other interaction effects can also be present (e.g., wing/tail, body/tail, etc.). Some simplifications arise in supersonic flow due to the fact that disturbances cannot propagate upstream. Thus, while one may want to consider the effects of the wings

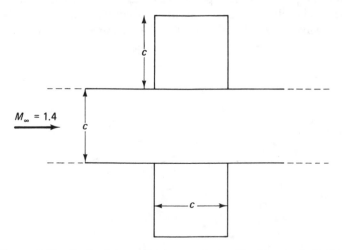

Figure 11.33 Configuration giving increased wing lift. (From Ref. 11.28.)

on the tail, the effects of the tail on the wings can usually be neglected unless they are very close to one another.

Many treatments of interacting flows rely on the small disturbance theory governed by equation (11.2). The flow tangency boundary condition at surfaces and the Kutta condition at sharp trailing edges are involved in determining solutions. Problems are solved by distributing a series of singularities (e.g., sources, sinks, doublets, vortices) to simulate the vehicle in a uniform supersonic stream. The strength of these singularities are determined so as to satisfy the boundary conditions at discrete selected locations on the vehicle. In the method of Ref. 11.29, wing camber and incidence are simulated by vortex distributions, while thickness is simulated by source and sink distributions. Thickness, camber, and incidence of the body are represented by line sources and doublets along the axis of the body. The effect of the interaction of the wing on the body is represented by a distribution of vortices on a cylinder whose radius is related to the average radius of the body, while the body-on-wing interaction is represented by a distribution of vortices on the wing camber surface (see Ref. 11.30).

An integrated supersonic design and analysis system is presented by Middleton and Lundry (Ref. 11.7). The analysis program can be used to predict the lifting pressure calculations for a wing camber surface at a selected angle of attack. It can also be used to calculate lifting pressure distributions and force coefficients for complete configurations over a range of angles of attack. The program described by Middleton and Lundry carries two solutions along: one for the configuration at its input angle of attack, the other an incremental solution per degree angle of attack (called the flat-wing solution). The interference terms associated with the two solutions acting on the other surface shape are also calculated. The summation of these effects into the drag polar and the other force coefficients is performed by superposition. Calculation of the complete configuration lifting pressure solution can involve any or all of the following tasks:

1. Determine the isolated fuselage upwash field.
2. Determine the nacelle pressure field acting on the wing.
3. Compute the pressure field due to asymmetrical fuselage volume.
4. Generate a wing/canard solution in the presence of the fuselage upwash field.
5. Calculate the effects of the wing pressure field acting on the nacelles.
6. Determine the fuselage lift distribution in the presence of the wing downwash field.
7. Obtain the horizontal tail solution in the presence of the fuselage and wing flow fields.
8. Generate a solution for the complete configuration by superposition of the elemental solutions.

For the reader interested in pursuing the subject of aerodynamic interactions, there often appear surveys of the relevant phenomena and solution techniques (e.g., Ref. 11.31).

AERODYNAMIC ANALYSIS FOR COMPLETE CONFIGURATIONS IN A SUPERSONIC STREAM

There is a heirarchy of prediction/computation techniques that can be used to estimate the aerodynamics forces and moments for suitably complex configurations in a supersonic stream. Techniques based on linear theory have been described at length in this chapter. Techniques based on impact methods, on solutions of the Euler equations, and on solutions of the Navier–Stokes equations are discussed in Chapter 13.

Numerous techniques exist using flow models whose complexity lies between linear-theory methods and Euler/Navier–Stokes solvers. One such technique uses the conservative form of the full potential equation (e.g., Refs. 11.32 and 11.33). Shankar (Ref. 11.32) notes that the conservation form of the full potential equation in cartesian coordinates can be written as

$$\frac{\partial(\rho u)}{\partial x} + \frac{\partial(\rho v)}{\partial y} + \frac{\partial(\rho w)}{\partial z} = 0 \tag{11.34}$$

Note that this is merely the differential form of the continuity equation for a steady, three-dimensional, compressible flow, which as noted in Appendix A, is already written in conservative form. The velocity components can be calculated as the gradient of the potential:

$$\vec{V} = \nabla\phi$$

Combining the energy equation for an adiabatic flow [equation (8.29)] with the isentropic-flow relation [equation (8.20c)], the density can be computed using the isentropic relation

$$\frac{\rho}{\rho_\infty} = \left[1 + M_\infty^2 \frac{\gamma - 1}{2} \left(1 - \frac{u^2 + v^2 + w^2}{U_\infty^2} \right) \right]^{1/(\gamma - 1)} \tag{11.35}$$

An implicit factorization algorithm together with an upwind-biased representation for the density is used to generate flow-field solutions. The reader is referred to Ref. 11.32 for a detailed description of the technique. As usual, the reader is advised to review the literature regularly, since a technique that is appropriate at the time this manuscript is written may have been replaced by a technique employing a more correct flow-model approximation or a more rapid solution algorithm at the time the manuscript is read.

A development of this aerodynamic prediction technique which is based on the steady form of the full-potential equation was applied to a variety of three-dimensional supersonic flow problems containing embedded subsonic regions. One such solution is for the isolated Space Shuttle Orbiter at $M_\infty = 1.4$ at $\alpha = 0°$. The circumferential pressure distribution for $x = 1200$ in, as taken from Ref. 11.33, is reproduced in Figure 11.34. The authors of Ref. 11.33 note "that the pressures along the vertical tail and the orbital maneuvering subsystem (OMS) pods are well predicted." It is the authors' experience that it is much easier to generate CFD solutions for the pressure than for heat transfer or skin friction.

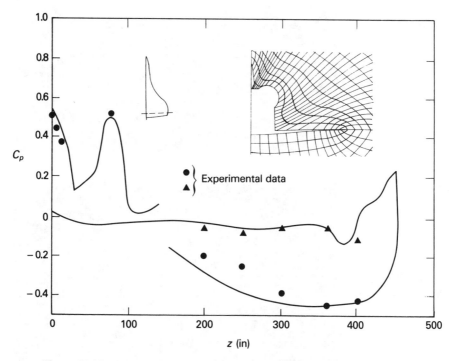

Figure 11.34   Circumferential pressure distribution for the Space Shuttle Orbiter at $x = 1200$ in, $M_\infty = 1.4$, $\alpha = 0°$. (From Ref. 11.33.)

Three-view oil-flow photographs of the Space Shuttle Orbiter are presented in Fig. 11.35 for

$M_\infty = 1.25$

$\alpha = 10°$

$\delta_{E,L} = -18.8°$ (the left elevon is deflected upward and the oil-flow pattern indicates a shock wave at the lower right of Fig. 11.35c, just below the OMS pod)

$\delta_{E,R} = +14.4°$ [the right elevon is deflected downward and the surface oil flows through the space between the OMS pod and the right elevon (Fig. 11.35a)]

$\delta_{SB} = 87.2°$ and $\delta_R = -25°$ [the speed brake is open 87.2° and deflected 25° to serve as a rudder, creating a relatively strong shock wave with an attendant viscous/inviscid interaction on one side of the vertical tail (Fig. 11.35a) but not on the other (Fig.11.35c)]

$\delta_{BF} = 23.7°$ (the body flap is deflected 23.7°)

Also evident in these flow visualization photographs are free-shear layer separations and feather patterns associated with reattaching vortical flows. Because the complex nature of the flow field makes it impossible to develop realistic flow models for numerical solutions, a total of 493 wind-tunnel tests using 52,993 hours were used in phase C/D of the Space Shuttle design development program in order to develop an aerodynamic data base (Ref. 11.34). However, because wind-tunnel flows are also simulations

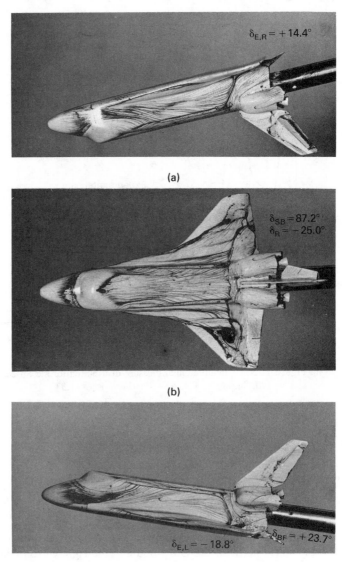

(a)

(b)

(c)

Figure 11.35   Oil flow patterns for Space Shuttle Orbiter at $M_\infty = 1.25$ and $\alpha = 10°$: (a) view of right side; (b) top view; (c) view of left side. (Courtesy of Johnson Space Center, NASA.)

(limited in model scale, high-temperature effects, etc.), wind-tunnel measurements should be correlated with the corresponding computed values based on solution techniques employing adequate flow models before extrapolating to flight.

PROBLEMS

11.1. Consider a flat-plate, rectangular wing. Derive the expressions for the lift coefficient, for the drag coefficient, for the coefficient of the moment about the leading edge, and for the location of the center of pressure given in Table 11.1. Assume that $\beta \cdot AR >$ 2 so that the Mach cones emanating from the tip do not overlap.

11.2. Verify the statement given in Table 11.1 that A', which is defined as airfoil cross-sectional area divided by the square of the chord, is equal to $\tau/2$ for a double-wedge airfoil section.

11.3. Consider a wing with a rectangular planform, whose aspect ratio is 4.0 and whose section is that shown in Fig. 10.5. Use Bonney's results in Table 11.1 to determine C_L and C_D for this wing for the flow conditions shown in Fig. 10.5.

11.4. (a) Using the relations given in Table 11.1, develop expressions for the lift coefficient as a function of α [i.e., $C_L(\alpha)$] and for the drag coefficient [i.e., $C_D(C_L)$] for the wing of Figs. 9.11 and 9.12. The wing has a rectangular planform with an aspect ratio of 2.75. Develop the relations for $M_\infty = 1.50$. Assume that the airfoil section is biconvex with a maximum thickness ratio of 0.05.

(b) Compare the theoretical values with the experimental values presented in Figs. 9.11 and 9.12. What value of $C_{D,\text{friction}}$ (at $M_\infty = 1.50$) will cause your theoretical results to agree most closely with the data in the figures?

11.5. Consider the wing of the Northrop F-5E (see Table 5.1). If the airplane is flying at a Mach number of 1.23, will the quarter-chord line of the wing be in a supersonic or a subsonic condition relative to the free-stream flow? What must M_∞ be for the quarter-chord line to be in a sonic condition?

11.6. Derive the equation of the leading-edge sweep angle Λ as a function of M_∞ for a sonic leading edge. Prepare a graph of the results. Assume small-angle approximations for α.

11.7. Show that the section-lift coefficient for a swept airfoil with a supersonic leading edge is given by

$$C_l = \frac{4 \cos \Lambda}{\sqrt{M_\infty^2 \cos^2 \Lambda - 1}} \alpha$$

The thickness ratio and the angle of attack of the airfoil are sufficiently small that the small-angle approximations may be used.

11.8. Discuss the limits of validity of the result derived in Problem 11.7.

11.9. Using a Taylor's series expansion about $z = 0$, derive equation (11.5) from equation (11.4).

11.10. Consider the flat-plate rectangular wing of Problem 11.1. Assume that there is a plain flap along the entire trailing edge with hinge line at $x_f = fc$, where $0 \le f \le 1$. Derive a formula for C_L as a function of the flap deflection angle δ_f (see Fig. P11.10). Assume that $\beta \cdot AR > 2$.

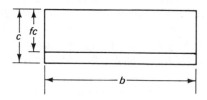

Figure P11.10

11.11. Derive the relation between the aspect ratio of a delta wing and the free-stream Mach number M_∞ if the leading edge is to be sonic.

11.12. Show that equation (11.27) follows from

$$R(x - x_1, y - y_1) \equiv \frac{x - x_1}{\beta^2(y - y_1)^2[(x - x_1)^2 - \beta^2(y - y_1)^2]^{0.5}}$$

The variation of R with x may be assumed to be small.

11.13. Show that subtracting the single wedge FGF' from the single wedge BFF' yields the relations in the double wedge $BDED'$ (see Fig. 11.20). Thus, show that the source strength in region DCD' will be

$$C(x, y) = -\lambda_2 \frac{U_\infty}{\pi}$$

where $C = C_1 + C_2$ and C_1 is the source strength due to BFF' and C_2 is the source (sink) strength due to FGF'.

11.14. Determine $\Delta C_p(3, 0)$ for the flow described in Example 11.2.

11.15. Determine $\Delta C_p(3, \pm 1)$ for the flow described in Example 11.2.

11.16. Consider a flat-plate (zero-camber) delta planform in a $M_\infty = 1.5$ stream. The leading-edge sweepback angle (Λ_{LE}) is 42.974°. Is the leading edge subsonic or supersonic? What is the value of $\beta \cot \Lambda_{LE}$ for this flow?

11.17. For the configuration described in Problem 11.16 and with $N_{max} = 8$, determine $\Delta C_p(3, 0)$.

11.18. For the configuration described in Problem 11.16 and with $N_{max} = 8$, determine $\Delta C_p(3 \pm 1)$.

11.19. Derive equation (11.35).

REFERENCES

11.1. Cenko, A., E. N. Tinoco, R. D. Dyer, and J. DeJongh, "PAN AIR Applications to Weapons Carriage and Separation," *Journal of Aircraft*, Feb. 1981, Vol. 18, No. 2, pp. 128–134.

11.2. Walkley, K. B., and G. E. Smith, "Wave Drag Analysis of Realistic Fighter Aircraft Using a Full-Potential Method," *Journal of Aircraft*, Sept. 1987, Vol. 24, No. 9, pp. 623–628.

11.3. Allen, J. M., and J. C. Townsend, "Application of a Supersonic Euler Code (SWINT) to Wing-Body-Tail Geometries," *Journal of Aircraft*, June 1986, Vol. 23, No. 6, pp. 513–519.

11.4. Schiff, L. B., and J. L. Steger, "Numerical Simulation of Steady Supersonic Viscous Flow," *AIAA Paper 79–0130*, presented at the 17th Aerospace Sciences Meeting, New Orleans, La., Jan. 1979.

11.5. Wood, R. M., and D. S. Miller, "Assessment of Preliminary Prediction Techniques for Wing Leading-Edge Vortex Flows at Supersonic Speeds," *Journal of Aircraft*, June 1985, Vol. 22, No. 6, pp. 473–478.

11.6. Carlson, H. W., and R. J. Mack, "Estimation of Wing Nonlinear Aerodynamic Characteristics at Supersonic Speeds," *TP-1718*, NASA, Nov. 1980.

11.7. Middleton, W. D., and J. L. Lundry, "A System for Aerodynamic Design and Analysis of Supersonic Aircraft, I: General Description and Theoretical Development," *CR-3351*, NASA, Dec. 1980.

11.8. Busemann, A., "Infinitesimal Conical Supersonic Flow," *Technical Memorandum 1100*, NACA, 1947.

11.9. Lomax, H., M. A. Heaslet, and F. B. Fuller, "Integrals and Integral Equations in Linearized Wing Theory," *Report 1054*, NACA, 1951.

11.10. Shapiro, A. H., *The Dynamics and Thermodynamics of Compressible Fluid Flow*, Vol. II, Chap. 18, The Ronald Press Co., New York, 1954.

11.11. Ferri, A., *Elements of Aerodynamics of Supersonic Flows*, Chap. 15, Macmillan Publishing Co., New York, 1949.

11.12. Snow, R. M., "Aerodynamics of Thin Quadrilateral Wings at Supersonic Speeds," *Quarterly of Applied Mathematics*, 1948, Vol. 5, No. 4, pp. 417–428.

11.13. Carafoli, E., *High Speed Aerodynamics*, Chap. 8, Pergamon Press, Elmsford, N.Y., 1956.

11.14. Jones, R. T., and D. Cohen, "Aerodynamics of Wings at High Speeds," in *Aerodynamic Components of Aircraft at High Speeds*, Vol. 7, Chap. 3, of *High Speed Aerodynamics and Jet Propulsion*, Princeton University Press, Princeton, N.J., 1957.

11.15. Bonney, E. A., "Aerodynamic Characteristics of Rectangular Wings at Supersonic Speeds," *Journal of the Aeronautical Sciences*, Feb. 1947, Vol. 14, No. 2, pp. 110–116.

11.16. Nielsen, J. N., F. H. Matteson, and W. G. Vincenti, "Investigation of Wing Characteristics at a Mach Number of 1.53, III: Unswept Wings of Differing Aspect Ratio and Taper Ratio," *Research Memorandum A8E06*, NACA, 1948.

11.17. Ivey, H. R., and E. H. Bowen, Jr., "Theoretical Supersonic Lift and Drag Characteristics of Symmetrical Wedge-Shape-Airfoil Sections as Affected by Sweepback Outside the Mach Cone," *TN 1226*, NACA, 1947.

11.18. Puckett, A. E., and H. J. Stewart, "Aerodynamic Performance of Delta Wings at Supersonic Speeds," *Journal of the Aeronautical Sciences*, Oct. 1947, Vol. 14, No. 10, pp. 567–578.

11.19. Stewart, H. J., "The Lift of a Delta Wing at Supersonic Speeds," *Quarterly of Applied Mathematics*, 1946, Vol. 4, No. 3, pp. 246–254.

11.20. Puckett, A. E., "Supersonic Wave Drag of Thin Airfoils," *Journal of the Aeronautical Sciences*, Sept. 1946, Vol. 13, No. 9, pp. 475–484.

11.21. Wright, B. R., F. Bruckman, and N. A. Radovich, "Arrow Wings for Supersonic Cruise Aircraft," *AIAA Paper 78–151*, presented at the AIAA 16th Aerospace Sciences Meeting, Huntsville, Ala., Jan. 1978.

11.22. Carlson, H. W., and D. S. Miller, "Numerical Analysis of Wings at Supersonic Speeds," *TN D-7713*, NASA, Dec. 1974.

11.23. Hodgman, C. D. (Ed), *C.R.C. Standard Mathematical Tables*, 12th ed., Chemical Rubber Company, Cleveland, Ohio, 1977.

11.24. Carlson, H. W., and R. J. Mack, "Estimation of Leading Edge Thrust for Supersonic Wings of Arbitrary Planform," *TP-1270*, NASA, 1978.

11.25. Middleton, W. D., and H. W. Carlson, "A Numerical Method for Calculating the Flat-Plate Pressure Distributions on Supersonic Wings of Arbitrary Planform," *TN D-2570*, NASA, Jan. 1965.

11.26. Stanbrook, A., and L. C. Squire, "Possible Types of Flow at Swept Leading Edges," *The Aeronautical Quarterly*, Feb. 1964, pp. 72–82.

11.27. Kulfan, R. M., and A. Sigalla, "Real Flow Limitations in Supersonic Airplane Design," *AIAA Paper 78-147*, presented at the AIAA 16th Aerospace Sciences Meeting, Huntsville, Ala., Jan. 1978.

11.28. Hilton, W.F., *High Speed Aerodynamics*, Chap. 12, Longmans, Green & Co., New York, 1951.

11.29. Carmichael, R. S., and F. A. Woodward, "Integrated Approach to the Analysis and Design of Wings and Wing-Body Combinations in Supersonic Flow," *TN D-3685*, NASA, 1966.

11.30. Woodward, F. A., "Analysis and Design of Wing-Body Combinations at Subsonic and Supersonic Speeds," *Journal of Aircraft*, Nov.–Dec. 1968, Vol. 5, No. 6, pp. 528–534.

11.31. Ashley, H., and W. P. Rodden, "Wing-Body Aerodynamic Interaction," *Annual Review of Fluid Mechanics*, Vol. 4, Annual Reviews, Palo Alto, Calif., 1972.

11.32. Shankar, V., "Conservative Full Potential, Implicit Marching Scheme for Supersonic Flows," *AIAA Journal*, Nov. 1982, Vol. 20, No. 11, pp. 1508–1514.

11.33. Szema, K. Y., W. L. Riba, V. Shankar, and J. J. Gorski, "Computation of Supersonic Flows over Three-Dimensional Configurations," *Journal of Aircraft*, Dec. 1985, Vol. 22, No. 12, pp. 1079–1084.

11.34. Whitnah, A. M., and E. R. Hillje, "Space Shuttle Wind Tunnel Testing Program Summary," *Reference Publication 1125*, NASA, 1984.

12

COMPUTATIONAL FLUID DYNAMICS

In the previous chapters, the equations defining fluid motion for a wide variety of applications were developed and discussed. For many applications, relatively simple equations can be used to obtain engineering values of the design parameters. For other applications, charts and tables incorporating empirical information can be used to generate the desired values. For still other applications, experiments are conducted to measure the desired parameters using small-scale models in simulated flow fields or using full-scale models in the actual design environment. All these techniques have a valid role in engineering technology. However, because of rapid advances in computer hardware capability and software procedures, evermore use is made of computer solutions to the flow fields of interest.

Computational fluid dynamics (CFD) is certainly not pure theoretical analysis. CFD changes the fundamental nature of the analysis from calculus to arithmetic and from the continuum domain to the discrete domain so that it can be solved using a computer. The existing mathematical theory for numerical solutions of nonlinear partial differential equations is continuously being improved. In computational fluid dynamics, it is necessary to rely heavily on the mathematical analysis of simpler, linearized, formulations, and on heuristic reasoning, physical intuition, wind-tunnel experience, and trial-and-error procedures.

COMPUTATIONAL FLUID DYNAMICS AS A DESIGN TOOL

Miranda (Ref. 12.1) states that the many activities that take place in an airplane design project can be considered as part of a three-step procedure that may be repeated many times and at varying levels of complexity and detail. The processes are: (1) configuration definition, (2) data acquisition, and (3) evaluation. The configuration geometry is defined during the first step, and the required engineering data for the range of expected environments are obtained during the second step. The third step involves the evaluation of the design using the acquired data to determine to what degree the design objectives and constraints have been met, and to specify what configuration changes, if any, are needed to satisfy the design criteria. Miranda also notes that computational aerodynamics has affected the process loop by providing (1) a new means of aerodynamic data acquisition, complementing the classical analytical techniques and wind-tunnel tests; and (2) the unique capability of executing the design loop in an inverse manner (i.e., specifying desired flow characteristics and calculating the configuration geometry that yields such characteristics).

Miranda (Ref. 12.1) divides the flow fields into zones: (1) the inviscid zone, where the effects of viscosity are negligible; and (2) the viscous zone, where the effects of viscosity have a significant role, such as in the boundary layer and in the near wake. Kutler (Ref. 12.2) states that "The use of zonal methods is advantageous because (1) the regions over which complex equation sets are required are minimized, (2) the simpler equation sets have faster solution algorithms which result in improved computational efficiency, and (3) the computer storage requirements for the overall procedure are not as great."

Chapman (Ref. 12.3) cities four major stages of successively refined approximations to the full Navier–Stokes equations in order of their evolution and complexity: (I) linearized inviscid, (II) nonlinear inviscid, (III) Reynolds averaged Navier–Stokes, and (IV) full Navier–Stokes. Numerical methods using the stage I approximation are termed "panel methods," since complex aircraft geometries are modeled by a large number of contiguous surface panels. The stage II approximations neglect only the viscous terms and have been used to predict transonic and hypersonic flow fields. For the stage III approximations, the basic equations are averaged over a time interval long compared to turbulent eddy fluctuations yet small compared to macroscopic flow changes. Since terms representing the time-averaged transport of turbulent momentum and energy must be modeled, the principal inaccuracy of stage III approximations is that of modeling turbulence. In its full complexity, the stage IV approximation involves the direct simulation of turbulent eddies from the complete time-dependent Navier–Stokes equations. Chapman states that "Progressive advances in computer power and in numerical methods have made possible the development of a variety of computational codes ranging from stage I codes for complex aircraft configurations to stage IV codes for very simple geometry." Although there have been significant advances both in computer hardware and in computer software since that statement first appeared, it remains qualitatively correct.

Panel methods have been discussed in Chapters 3, 7, and 11. Solutions to the equations for a steady, incompressible boundary layer were discussed in Chapter 4. Other computational techniques with specific applications are presented in this book (e.g., the method of characteristics, which is presented in Chapter 8). Although there are a variety of computational procedures and formulations, many of which are peculiar to characteristics of the flow field for the problem of interest, in this chapter we concentrate on the finite-difference formulation, which can be applied to a very wide range of fluid dynamic applications.

FINITE-DIFFERENCE FORMULATIONS

We seek to represent the derivatives that appear in the differential equations of motion using a finite-difference formulation. Consider the general function f, which represents a parameter that is a function of two spatial coordinates (x, y) and time (t), that is $f(x, y, t)$. The spatial coordinates are defined by a rectangular mesh, as shown in Fig. 12.1. We will use the subscript indexes i and j to represent x and y coordinates, respectively, and the superscript index n to designate time. The mesh spacings in the x and y directions are Δx and Δy, respectively. These incremental displacements will be assumed constant throughout the flow field for the present discussion. However, advanced techniques employ variable-sized meshes that are tailored to the problem. Thus, for the two points identified in Fig. 12.1,

$$f_{i,j}^n \quad \text{represents} \quad f(x, y, t)$$

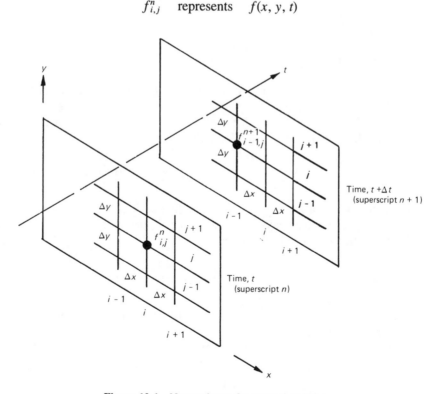

Figure 12.1 Nomenclature for space-time index.

and

$$f^{n+1}_{i-1,j} \quad \text{represents } f(x - \Delta x, y, t + \Delta t)$$

Let us assume the function f and its derivatives are continuous. The basic finite-difference forms for the partial derivatives can be derived from Taylor-series expansions. We will first develop the spatial derivatives. For simplicity, we will omit the superscript n for the temporal variation. The Taylor series for $f_{i+1,j}$ expanded about the point (i, j) is

$$
\begin{aligned}
f_{i+1,j} &= f_{i,j} + \left.\frac{\partial f}{\partial x}\right|_{i,j} (x_{i+1,j} - x_{i,j}) + \frac{1}{2}\left.\frac{\partial^2 f}{\partial x^2}\right|_{i,j} (x_{i+1,j} - x_{i,j})^2 + \cdots \\
&= f_{i,j} + \left.\frac{\partial f}{\partial x}\right|_{i,j} \Delta x + \frac{1}{2}\left.\frac{\partial^2 f}{\partial x^2}\right|_{i,j} \Delta x^2 + \text{HOT}
\end{aligned}
\tag{12.1}
$$

where HOT is an abbreviation for higher-order terms.

Solving for $\partial f/\partial x$ gives

$$
\begin{aligned}
\left.\frac{\partial f}{\partial x}\right|_{i,j} &= \frac{f_{i+1,j} - f_{i,j}}{\Delta x} - \frac{1}{2}\left.\frac{\partial^2 f}{\partial x^2}\right|_{i,j} \Delta x + \text{HOT} \\
&= \frac{f_{i+1,j} - f_{i,j}}{\Delta x} + O(\Delta x)
\end{aligned}
\tag{12.2}
$$

where $O(\Delta x)$ is read "terms of order Δx" and refers to additional terms with factors of Δx, Δx^2, Δx^3, and so on.

We denote any finite-difference analog of $\partial f/\partial x$ by the notation $\delta f/\delta x$. Then the forward difference approximation $\delta f/\delta x$ is

$$
\left.\frac{\delta f}{\delta x}\right|_{i,j} = \frac{f_{i+1,j} - f_{i,j}}{\Delta x}
\tag{12.3}
$$

and has a truncation error of order Δx, that is, is a first-order approximation.

By expanding backward to $(x - \Delta x, y)$, that is, to $(i - 1, j)$, we obtain another finite-difference analog for $\delta f/\delta x$:

$$
\left.\frac{\delta f}{\delta x}\right|_{i,j} = \frac{f_{i,j} - f_{i-1,j}}{\Delta x}
\tag{12.4}
$$

Equation (12.4) provides the backward difference approximation, which is also a first-order approximation.

The centered difference approximation for $\delta f/\delta x$ is obtained by subtracting the backward expansion from the forward expansion.

$$f_{i+1,j} = f_{i,j} + \frac{\partial f}{\partial x}\bigg|_{i,j} \Delta x + \frac{1}{2}\frac{\partial^2 f}{\partial x^2}\bigg|_{i,j} \Delta x^2 + \frac{1}{6}\frac{\partial^3 f}{\partial x^3}\bigg|_{i,j} \Delta x^3$$
$$+ \frac{1}{24}\frac{\partial^4 f}{\partial x^4}\bigg|_{i,j} \Delta x^4 + O(\Delta x^5) \tag{12.5}$$

$$f_{i-1,j} = f_{i,j} - \frac{\partial f}{\partial x}\bigg|_{i,j} \Delta x + \frac{1}{2}\frac{\partial^2 f}{\partial x^2}\bigg|_{i,j} \Delta x^2 - \frac{1}{6}\frac{\partial^3 f}{\partial x^3}\bigg|_{i,j} \Delta x^3$$
$$+ \frac{1}{24}\frac{\partial^4 f}{\partial x^4}\bigg|_{i,j} \Delta x^4 - O(\Delta x^5) \tag{12.6}$$

Subtracting equation (12.6) from (12.5) gives

$$f_{i+1,j} - f_{i-1,j} = 2\frac{\partial f}{\partial x}\bigg|_{i,j} \Delta x + \frac{1}{3}\frac{\partial^3 f}{\partial x^3}\bigg|_{i,j} \Delta x^3 + \text{HOT}$$

Solving for $\partial f/\partial x$ gives

$$\frac{\partial f}{\partial x}\bigg|_{i,j} = \frac{f_{i+1,j} - f_{i-1,j}}{2\,\Delta x} - \frac{1}{6}\frac{\partial^3 f}{\partial x^3}\bigg|_{i,j} \Delta x^2 + \text{HOT}$$
$$= \frac{f_{i+1,j} - f_{i-1,j}}{2\,\Delta x} + O(\Delta x^2) \tag{12.7}$$

Thus, the centered difference form of $\delta f/\delta x$ is

$$\frac{\delta f}{\delta x}\bigg|_{i,j} = \frac{f_{i+1,j} - f_{i-1,j}}{2\,\Delta x} \tag{12.8}$$

From equation (12.7), we see that the centered difference expression for the derivative is a second-order approximation, that is, has a truncation error of order $(\Delta x)^2$.

Analogous expressions follow immediately for y and t derivatives. For example, the centered difference analog of $\delta f/\delta t$ is

$$\frac{\delta f}{\delta t}\bigg|_{i,j}^{n} = \frac{f_{i,j}^{n+1} - f_{i,j}^{n-1}}{2\,\Delta t} \tag{12.9}$$

We derive a centered difference analog of $\partial^2 f/\partial x^2$ by adding the expansions and (12.6)

$$f_{i+1,j} + f_{i-1,j} = 2f_{i,j} + \frac{\partial^2 f}{\partial x^2}\bigg|_{i,j} \Delta x^2 + \frac{1}{12}\frac{\partial^4 f}{\partial x^4}\bigg|_{i,j} \Delta x^4 + \text{HOT} \tag{12.10}$$

Solving for $\partial^2 f/\partial x^2$ gives

$$\left.\frac{\partial^2 f}{\partial x^2}\right|_{i,j} = \frac{f_{i+1,j} - 2f_{i,j} + f_{i-1,j}}{\Delta x^2} + O(\Delta x^2) \tag{12.11}$$

The centered difference form of $\delta^2 f/\delta x^2$ is

$$\left.\frac{\delta^2 f}{\delta x^2}\right|_{i,j} = \frac{f_{i+1,j} - 2f_{i,j} + f_{i-1,j}}{\Delta x^2} \tag{12.12}$$

which is a second-order approximation.

EXPLICIT AND IMPLICIT FORMULATIONS OF UNSTEADY FLOWS

Let us apply these differencing techniques to Stokes's first problem, the suddenly accelerated plane wall. For this problem, the plate and entire fluid are at rest prior to a certain time, $t = 0$. The plate is then impulsively started from rest and moves in its own plane with a constant velocity. Selecting the x axis as coinciding with the wall and the direction of U_0, we obtain the Navier–Stokes equation:

$$\frac{\partial u}{\partial t} = \nu \frac{\partial^2 u}{\partial y^2} \tag{12.13}$$

as discussed in Ref. 12.4. The pressure is constant throughout the flow field. The boundary conditions are

$$t < 0: u = 0 \quad \text{for all } y \tag{12.14a}$$

$$t \geq 0: u = U_0 \quad \text{for } y = 0$$

$$u = 0 \quad \text{for } y \to \infty \tag{12.14b}$$

A variety of finite-difference approximations can be applied to replace the differentials in equation (12.13). The interested reader is referred to the discussion in Chapter 8 of Richtmyer and Morton (Ref. 12.5). Basically, two classes of methods are used in describing the time-dependent velocity distribution in the fluid: (1) explicit methods and (2) implicit methods.

Explicit Methods

Let us consider the explicit formulation where we write the time derivative of the velocity of the ith node in the forward-difference form:

$$\frac{\partial u}{\partial t} = \frac{u_i^{n+1} - u_i^n}{\Delta t} + O(\Delta t) \tag{12.15a}$$

and where we evaluate the spatial derivative using the values of the velocity at the nth time step:

$$\nu \frac{\partial^2 u}{\partial y^2} = \nu \frac{u_{i+1}^n - 2u_i^n + u_{i-1}^n}{(\Delta y)^2} + O(\Delta y^2) \tag{12.15b}$$

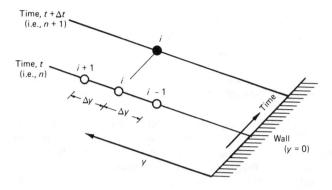

Figure 12.2 Nomenclature for explicit formulation.

Note that the spatial derivative is evaluated in terms of the velocities at the time when the flow field is known (i.e., n). Thus, equation (12.13) can be written

$$\frac{u_i^{n+1} - u_i^n}{\Delta t} = \nu \frac{u_{i+1}^n - 2u_i^n + u_{i-1}^n}{(\Delta y)^2} + O(\Delta t, \Delta y^2) \tag{12.16}$$

which is known as the Euler method. A sketch of this formulation is presented in Fig. 12.2. The use of the i index for the y coordinate should not confuse the reader, since there is no fixed convention between the index notation and the coordinates.

Using this explicit Euler method, we can solve directly for the velocity at the ith node at the time $(t + \Delta t)$, that is, $(n + 1)$, as represented by the solid symbol in Fig. 12.2, since that is the only term in equation (12.16) that is not known.

$$u_i^{n+1} = u_i^n + \frac{\nu \Delta t}{(\Delta y)^2} (u_{i+1}^n - 2u_i^n + u_{i-1}^n) \tag{12.17}$$

As noted in Ref. 12.5, the method is numerically stable provided that

$$\frac{\nu \Delta t}{(\Delta y)^2} \leq 0.5 \tag{12.18}$$

If we take too large a time step for the mesh size, the solution procedure is unstable and oscillatory errors will grow as the solution proceeds. Furthermore, as indicated in equation (12.16), the accuracy of the method is $O(\Delta t)$ and $O[(\Delta y)^2]$. Note, however, that improvement in the spatial resolution will cost dearly because the number of time steps to reach a given time will increase enormously.

Example 12.1

Let us use the explicit finite-difference representation to obtain solutions for Stokes's first problem. The plate and the entire fluid are at rest prior to a certain time, $t = 0$. The plate is then impulsively started from rest and moves in its own plane with a velocity of 10 m/s. For air at standard atmospheric conditions, the kinematic viscosity is 1.46×10^{-5} m²/s. Using the explicit formulation described by equation (12.17), compute the velocity as a function of y and t for the following step sizes: $\Delta y = 0.001$ m and $\Delta t = 0.10$ s.

Solution   Let us first calculate the value of the stability parameter, the diffusion number

$$D = \frac{\nu \, \Delta t}{(\Delta y)^2} = \frac{(1.46 \times 10^{-5})(0.10)}{(0.001)^2} = 1.46$$

Note that this value of the diffusion number exceeds the stability criterion stated in equation (12.18). Thus, we would expect the solution to be unstable with growing oscillatory errors.

A computer code has been written to generate a solution. We will use the nomenclature shown in Fig. 12.3. The $i = $ IMAX node is a point in the fluid "far" from the wall. The velocity at this point was assumed to be zero for all time. The wall is defined as the $i = 1$ node. The velocity $u_1^n = 10.0$ for all n zero or greater, that is, for all time t, since that is a given boundary condition. Let us calculate the velocity field for the first two time steps.

Initial conditions:

$$t = 0: \qquad u_1^0 = 10.0 \text{ m/s}$$

All other values of the velocity u are zero.

Flow after one time step:

$$n = 1; t = 0.10 \text{ s}; u_1^1 = 10.0 \text{ m/s}$$

Using Equation (12.17) to calculate the velocity at the $i = 2$ node:

$$u_2^1 = u_2^0 + D(u_3^0 - 2u_2^0 + u_1^0)$$

$$u_2^1 = 0.0 + 1.46 \, (0 - 0 + 10) = 14.6 \text{ m/s}$$

All other values of the velocity u are zero.

Flow after two time steps:

$$n = 2; t = 0.20 \text{ s}; u_1^2 = 10.0 \text{ m/s}$$

$$u_2^2 = u_2^1 + D(u_3^1 - 2u_2^1 + u_1^1)$$

$$u_2^2 = 14.6 + 1.46[0 - 2(14.6) + 10] = -13.432 \text{ m/s}$$

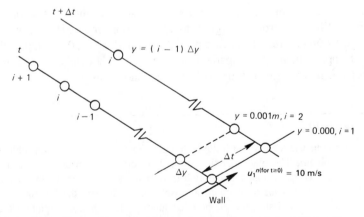

Figure 12.3   Nomenclature for the numerical solutions to Stokes' first problem of Examples 12.1 through 12.4.

$$u_3^2 = u_3^1 + D(u_4^1 - 2u_3^1 + u_2^1)$$

$$u_3^2 = 0.0 + 1.46(0 - 0 + 14.6) = 21.316 \text{ m/s}$$

All other values of the velocity u are zero.

These velocity distributions are presented graphically in Fig. 12.4. Note that except for the values at the wall, which are specified by the boundary condition, and those at points in the fluid not as yet affected by viscosity, the calculated velocities vary rapidly, changing sign with each time step. This oscillatory error growth characterizes a *dynamic instability*.

Additional values from the computer solution are presented in Table 12.1. It is obvious in this case that the numerical instability is entirely an analysis by-product and

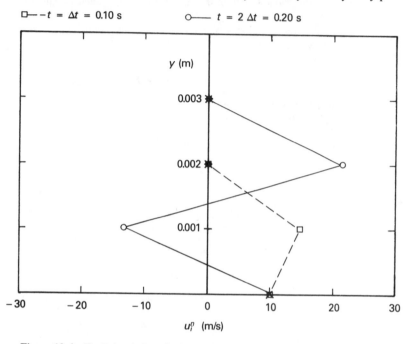

Figure 12.4 Explicit solutions for $D = 1.46$ (the boundary conditions are indicated by the symbols with the crosses through them).

TABLE 12.1 VALUES OF u (m/s) FOR EXAMPLE 12.1

y (m)	0.00	0.10	0.20	0.50	5.00
			t (s)		
0 (wall)	10.000	10.000	10.000	10.000	1.0000
0.001	0.000	14.600	−13.432	808.35	-2.0515×10^{32}
0.002	0.000	0.000	21.316	−1035.3	3.9072×10^{32}
0.003	0.000	0.000	0.000	739.23	-5.4018×10^{32}
0.004	0.000	0.000	0.000	−303.52	6.4246×10^{32}
0.005	0.000	0.000	0.000	66.338	-6.9321×10^{32}
0.010	0.000	0.000	0.000	0.000	4.0381×10^{32}
0.020	0.000	0.000	0.000	0.000	5.0250×10^{30}

not related to a physical phenomenon. In more complex physical problems, however, it is not always obvious if "erratic" results are physical or numerical.

Example 12.2

Repeat Example 12.1 for the following step sizes: $\Delta y = 0.001\ m$, $\Delta t = 0.025$ s.

Solution Let us first calculate the diffusion number

$$D = \frac{v\,\Delta t}{(\Delta y)^2} = \frac{(1.46 \times 10^{-5})(0.025)}{(0.001)^2} = 0.365$$

Note that this value is within the stability criterion stated in equation (12.18). Thus, we would expect the solution to be stable.

Using the established nomenclature, let us calculate the velocity distribution for the first two time steps.

Initial conditions:

$$t = 0;\ n = 0;\ u_1^0 = 10.0\ \text{m/s}$$

All other values of u are zero.

Flow after one time step:

$$n = 1 \qquad t = 0.025\ \text{s} \qquad u_1^1 = 10.0\ \text{m/s}$$

$$u_2^1 = u_2^0 + D(u_3^0 - 2u_2^0 + u_1^0)$$

$$u_2^1 = 0.0 + 0.365(0 - 0 + 10) = 3.650\ \text{m/s}$$

All other values of the velocity u are zero.

Flow after two time steps:

$$n = 2 \qquad t = 0.050\text{s} \qquad u_1^2 = 10.0\ \text{m/s}$$

$$u_2^2 = u_2^1 + D(u_3^1 - 2u_2^1 + u_1^1)$$

$$u_2^2 = 3.65 + 0.365(0 - 2(3.65) + 10) = 4.636\ \text{m/s}$$

$$u_3^2 = u_3^1 + D(u_4^1 - 2u_3^1 + u_2^1)$$

$$u_3^2 = 0.0 + 0.365(0 - 0 + 3.65) = 1.332\ \text{m/s}$$

Using these calculations and the output from the computer solution, we obtain Table 12.2. These results are presented graphically in Fig. 12.5.

Let us examine the finite-difference equation (FDE) that we have solved numerically, i.e., equation (12.16). Substituting the relevant equivalent of equation (12.2),

$$\frac{u_i^{n+1} - u_i^n}{\Delta t} = \frac{\partial u}{\partial t} + \frac{\Delta t}{2}\frac{\partial^2 u}{\partial t^2}$$

and of equation (12.10),

$$v\,\frac{u_{i+1}^n - 2u_i^n + u_{i-1}^n}{(\Delta y)^2} = v\,\frac{\partial^2 u}{\partial y^2} + \frac{v(\Delta y)^2}{12}\frac{\partial^4 u}{\partial y^4}$$

TABLE 12.2 VALUES OF u (m/s) FOR EXAMPLE 12.2

			t (s)			
y (m)	0.000	0.025	0.100	0.200	0.500	5.000
0 (wall)	10.000	10.000	10.000	10.000	10.000	10.000
0.001	0.000	3.650	5.854	6.902	7.967	9.341
0.002	0.000	0.000	2.698	4.236	6.061	8.687
0.003	0.000	0.000	0.880	2.270	4.387	8.042
0.004	0.000	0.000	0.177	1.038	3.011	7.410
0.005	0.000	0.000	0.000	0.392	1.952	6.795
0.010	0.000	0.000	0.000	0.000	0.083	4.085
0.020	0.000	0.000	0.000	0.000	0.000	0.972

into equation (12.16), we obtain

$$\frac{\partial u}{\partial t} + \frac{\Delta t}{2}\frac{\partial^2 u}{\partial t^2} = v\frac{\partial^2 u}{\partial y^2} + \frac{v(\Delta y)^2}{12}\frac{\partial^4 u}{\partial y^4} \tag{12.19}$$

Equation (12.19), which is known as the modified equation, is essentially the partial differential equation (PDE) that the numerical method is solving and depends both on the original PDE and on the method of differencing used. It should be noted that the

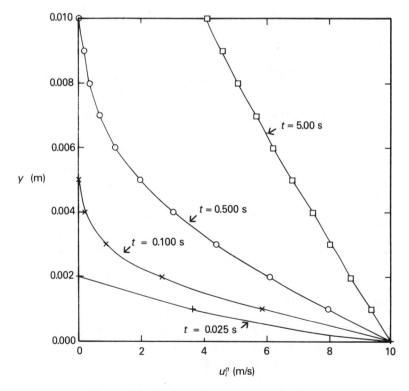

Figure 12.5 Explicit solutions for $D = 0.365$.

presence of a second derivative with respect to time (albeit with a small coefficient) makes the modified equation hyperbolic. The original PDE is parabolic. This has a significant effect on the relation between the computation mesh and the computed solution. For more information on this point, the reader is referred to Ref. 12.6.

As noted, the truncation error in the forward difference representation of the time derivative used in the Euler method is of order Δt. However, the truncation error of the centered difference form of equation (12.7) is of order $(\Delta t)^2$:

$$\frac{\partial u}{\partial t} = \frac{u_i^{n+1} - u_i^{n-1}}{2\Delta t} + O(\Delta t^2) \tag{12.20}$$

Substituting equation (12.20) into the governing equation (12.13), we obtain a finite-difference equation for which the truncation errors are second order both in time and in space $[O(\Delta t^2), O(\Delta y^2)]$.

$$\frac{u_i^{n+1} - u_i^{n-1}}{2\Delta t} = \nu \frac{u_{i+1}^n - 2u_i^n + u_{i-1}^n}{\Delta y^2} \tag{12.21}$$

However, this technique, known as Richardson's method, is unconditionally unstable and cannot be used to solve the diffusion equation.

The unstable Richardson method can be made stable by replacing u_i^n in the diffusion term by the expression

$$u_i^n = \frac{u_i^{n+1} + u_i^{n-1}}{2}$$

The resulting FDE is

$$\frac{u_i^{n+1} - u_i^{n-1}}{2\Delta t} = \nu \frac{u_{i+1}^n - u_i^{n+1} - u_i^{n-1} + u_{i-1}^n}{(\Delta y)^2} \tag{12.22}$$

We can solve equation (12.22) explicitly for u_i^{n+1}. Thus,

$$u_i^{n+1} = \frac{u_i^{n-1} + 2r(u_{i+1}^n - u_i^{n-1} + u_{i-1}^n)}{1 + 2r} \tag{12.23a}$$

where

$$2r = \frac{2\nu \Delta t}{(\Delta y)^2} \tag{12.23b}$$

The resultant three-time level, explicit scheme, which is known as the DuFort–Frankel method, is unconditionally stable.

Implicit Methods

For implicit methods, the gradient term is evaluated using the velocities at the end of the time step, for example, at time $(t + \Delta t)$, or at some intermediate time $t + \Delta t/2$. The resulting expression for the second derivative at an interior node (the ith node) contains three unknown velocities: that at the ith node and those at the two adjacent nodes.

Consider the Crank–Nicolson formulation, which is illustrated in Fig. 12.6a. The time derivative of the velocity of the ith node as a centered difference about the point $n + \frac{1}{2}$ with a time increment of $\Delta t/2$ can be written

$$\left(\frac{\partial u}{\partial t}\right)_i^{n+1/2} = \frac{u_i^{n+1} - u_i^n}{2(\Delta t/2)} + O(\Delta t^2) \tag{12.24a}$$

The centered spatial difference must also be written at time level $n + \frac{1}{2}$. This can be done by averaging the spatial difference expressions at time t and at time $t + \Delta t$. Thus,

$$\nu\left(\frac{\partial^2 u}{\partial y^2}\right)_i^{n+1/2} = \frac{\nu}{2}\left[\frac{u_{i+1}^{n+1} - 2u_i^{n+1} + u_{i-1}^{n+1}}{(\Delta y)^2} \right.$$
$$\left. + \frac{u_{i+1}^n - 2u_i^n + u_{i-1}^n}{(\Delta y)^2}\right] + O(\Delta y^2) \tag{12.24b}$$

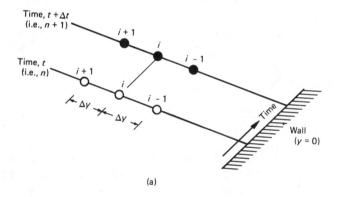

(a)

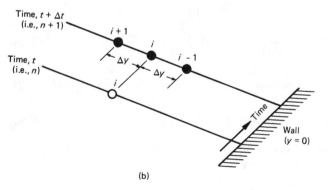

(b)

Figure 12.6 Nomenclature for implicit formulations: (a) Crank–Nicolson implicit formulation; (b) Laasonen implicit formulation.

Thus, equation (12.13) becomes:

$$\frac{u_i^{n+1} - u_i^n}{\Delta t} = \frac{\nu}{2}\left[\frac{u_{i+1}^{n+1} - 2u_i^{n+1} + u_{i-1}^{n+1}}{(\Delta y)^2}\right.$$
$$\left. + \frac{u_{i+1}^n - 2u_i^n + u_{i-1}^n}{(\Delta y)^2}\right] + O(\Delta t^2, \Delta y^2) \tag{12.25}$$

Note that in order to to advance the solution from time level n to time level $n + 1$, there are three unknowns in this equation: u_{i+1}^{n+1}, u_i^{n+1}, and u_{i-1}^{n+1}, as indicated by the filled symbols of Fig. 12.6a. As stated in Ref. 12.5, the accuracy of the method is $O[(\Delta t)^2]$ and $O[(\Delta y)^2]$. Furthermore, the method is unconditionally stable. That is, regardless of the time- or space-step size, the solution will not diverge.

Consider next the implicit formulation of Laasonen (in Ref. 12.5) which is illustrated in Fig. 12.6b. The time derivative of the velocity of the ith node is written as a backward time difference:

$$\left(\frac{\partial u}{\partial t}\right)_i^{n+1} = \frac{u_i^{n+1} - u_i^n}{\Delta t} + O(\Delta t) \tag{12.26a}$$

The spatial derivative is evaluated using the finite-difference expression at time $(t + \Delta t)$:

$$\nu\left(\frac{\partial^2 u}{\partial y^2}\right)_i^{n+1} = \nu\left[\frac{u_{i+1}^{n+1} - 2u_i^{n+1} + u_{i-1}^{n+1}}{(\Delta y)^2}\right] + O(\Delta y^2) \tag{12.26b}$$

Thus, equation (12.13) becomes

$$\frac{u_i^{n+1} - u_i^n}{\Delta t} = \nu\left[\frac{u_{i+1}^{n+1} - 2u_i^{n+1} + u_{i-1}^{n+1}}{(\Delta y)^2}\right] + O(\Delta t, \Delta y^2) \tag{12.27}$$

Note that there are again three unknowns in this equation: u_{i+1}^{n+1}, u_i^{n+1}, and u_{i-1}^{n+1}, as indicated by the filled symbols of Fig. 12.6b. As stated in Ref. 12.5, the accuracy of the method is $O[\Delta t]$ and $O[(\Delta y)^2]$. The method is always stable.

Rearranging the terms in equation (12.27), we obtain

$$u_{i-1}^{n+1} + \left[-2 - \frac{(\Delta y)^2}{\nu(\Delta t)}\right]u_i^{n+1} + u_{i+1}^{n+1} = \left[-\frac{(\Delta y)^2}{\nu(\Delta t)}u_i^n\right] \tag{12.28}$$

Thus, equation (12.28) can be written as a linear finite-difference equation of the general form

$$a_i^n u_{i-1}^{n+1} + b_i^n u_i^{n+1} + c_i^n u_{i+1}^{n+1} = d_i^n \tag{12.29}$$

The coefficients a, b, c, and d contain only constants or terms evaluated at the initial (or known) time step t. Comparing equations (12.28) and (12.29), we see that

$$a_i^n = 1; \quad b_i^n = \left[-2 - \frac{(\Delta y)^2}{\nu(\Delta t)}\right]; \quad c_i^n = 1; \quad d_i^n = \left[-\frac{(\Delta y)^2}{\nu(\Delta t)}u_i^n\right] \tag{12.30}$$

Equation (12.25) can also be written as the general, linear finite-difference equation (12.29). This equation leads to a tridiagonal system that can be easily solved by the technique of factorization (Gaussian elimination). Because of its implicit character, equation (12.29) is unconditionally stable.

A technique for inverting the tridiagonal system of equations represented by equation (12.29) is discussed in Ref. 12.5. Using the Gaussian elimination technique, we can calculate the velocity at the ith node if the velocity at the $(i + 1)$th node is known. Thus,

$$u_i^{n+1} = -h_i^n u_{i+1}^{n+1} + g_i^n \tag{12.31}$$

where

$$h_i^n = \frac{c_i^n}{b_i^n - a_i^n h_{i-1}^n} \tag{12.32}$$

and

$$g_i^n = \frac{d_i^n - a_i^n g_{i-1}^n}{b_i^n - a_i^n h_{i-1}^n} \tag{12.33}$$

Note that the solution procedure would involve first calculating g_i^n and h_i^n from node $i = $ IMAX-1 (the next-to-last node) to the node $i = 2$. The expressions for a_i^n, b_i^n, c_i^n, and d_i^n are presented for the Laasonen formulation in equation (12.30). Similar expressions for the Crank-Nicolson formulation are given in Example 12.3. The values of g_{IMAX}^n and h_{IMAX}^n required to calculate g_{IMAX-1}^n and h_{IMAX-1}^n are obtained by applying the boundary condition for the velocity at $i = $ IMAX, which for these numerical solutions represents a point far from the wall, and equation (12.31).

Having determined the values of g_i^n and of h_i^n from $i = $ IMAX $- 1$ to $i = 2$, we can calculate the values of u_i^{n+1} from $i = 2$ to $i = $ IMAX using equation (12.31). This can be done, since we know u_1^{n+1} from the wall boundary condition.

Example 12.3

Use the Crank-Nicolson implicit finite-difference representation to obtain solutions to Stokes's first problem. The plate is impulsively started from rest at time $t = 0$ and moves in its plane with a velocity of 10 m/s. For air at standard atmospheric conditions, the kinematic viscosity is 1.46×10^{-5} m²/s. Compute the velocity as a function of y and t for $\Delta y = 0.001$ m, $\Delta t = 0.10$ s.

Solution The Crank–Nicolson finite-difference formulation is given by equation (12.25). In order to obtain a solution using equations (12.31) through (12.33), we must first determine the relations for a_i^n, b_i^n, c_i^n, and d_i^n. Rearranging equation (12.25) into the general form of equation (12.29), it is clear that

$$a_i^n = 1.0 \tag{12.34a}$$

$$b_i^n = -2 - \frac{2}{D} \tag{12.34b}$$

$$c_i^n = 1.0 \tag{12.34c}$$

$$d_i^n = -u_{i+1}^n - u_i^n \left(\frac{2}{D} - 2\right) - u_{i-1}^n \tag{12.34d}$$

Note that for this problem the diffusion number is

$$D = \frac{\nu \, \Delta t}{(\Delta y)^2} = \frac{(1.46 \times 10^{-5})(0.10)}{(0.001)^2} = 1.46$$

This value exceeds the stability criterion for explicit solutions. Nevertheless, a stable solution can be obtained using the implicit formulation.

A computer code has been written to generate numerical solutions for this problem using the Crank–Nicolson formulation. The computed results are given in Table 12.3. Compare the values at the first node away from the wall (i.e., $y = 0.001$ m) as computed with the Crank–Nicolson implicit technique for $D = 1.46$ with the corresponding values computed using the explicit technique with $D = 0.365$ (Example 12.2) and those computed using the Crank–Nicolson technique with $D = 0.365$ (Example 12.4). It appears that there are errors at early times (i.e., $t = 0.1$ s and $t = 0.2$ s) which damp out. It has been shown (Ref. 12.7) that early-time solutions with this implicit method are indeed inaccurate for $D > 0.5$. This emphasizes the distinction between accuracy and stability. A necessary condition for accuracy is numerical stability. On the other hand, stability does not necessarily imply that a method is accurate.

TABLE 12.3   VALUES OF u (m/s) FOR EXAMPLE 12.3

	t (s)				
y (m)	0.000	0.100	0.200	0.500	5.000
0 (wall)	10.000	10.000	10.000	10.000	10.000
0.001	0.000	3.288	6.610	7.817	9.337
0.002	0.000	1.081	3.266	5.820	8.678
0.003	0.000	0.356	1.433	4.074	8.029
0.004	0.000	0.117	0.589	2.699	7.393
0.005	0.000	0.038	0.233	1.683	6.774
0.010	0.000	0.000	0.002	0.071	4.055
0.020	0.000	0.000	0.000	0.000	0.955

Example 12.4

Repeat Example 12.3 for the following step sizes: $\Delta y = 0.001$ m, $\Delta t = 0.025$ s.

Solution   The diffusion number for this problem is 0.365.

Thus, we have already obtained a solution to this problem using the explicit technique, Example 12.2. The computed results for the Crank–Nicolson implicit formulation are given in Table 12.4.

The velocity fields computed for Examples 12.2 through 12.4 at $t = 1.000$ s are compared in Fig. 12.7. The three solutions are in relatively good agreement with each other and with the exact (analytical) solution (Ref. 12.4). Thus, one could conclude that the explicit technique provides a valid solution if we stay within the stability criterion. These four sample problems are intended to demonstrate some of the considerations of accuracy and stability that we must consider when using finite-difference representations. We should not extrapolate the results from these solutions for a rel-

TABLE 12.4   VALUES OF u (m/s) FOR EXAMPLE 12.4

y (m)	t (s)				
	0.000	0.100	0.200	0.500	5.000
0 (wall)	10.000	10.000	10.000	10.000	10.000
0.001	0.000	5.287	6.674	7.904	9.339
0.002	0.000	2.154	3.920	5.953	8.683
0.003	0.000	0.700	2.021	4.261	8.036
0.004	0.000	0.192	0.918	2.893	7.402
0.005	0.000	0.046	0.371	1.861	6.786
0.010	0.000	0.000	0.001	0.091	4.072
0.020	0.000	0.000	0.000	0.000	0.967

◇ Explicit technique, D = 0.365

□ Crank-Nicolson technique, D = 1.46

○ Crank-Nicolson technique, D = 0.365

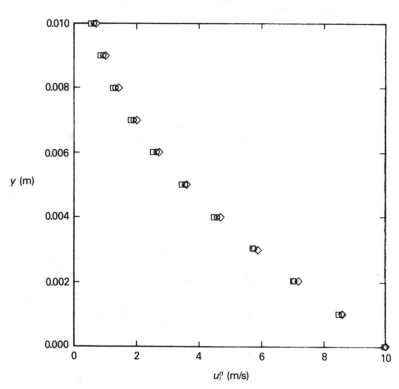

Figure 12.7 Comparison of the velocity distributions for Examples 12.2 through 12.4 at t = 1.000 s.

atively simple linear equation to significantly different problems. The choice of technique and of difference formulation may depend strongly on the problem to be solved.

LINEARIZATION OF THE NAVIER–STOKES EQUATIONS

Even for a steady, incompressible, two-dimensional laminar boundary layer, the Navier–Stokes equations are nonlinear, partial differential equations. For example, referring to equation (4.6), the x component of momentum is

$$\rho u \frac{\partial u}{\partial x} + \rho v \frac{\partial u}{\partial y} = \rho_e u_e \frac{du_e}{dx} + \mu \frac{\partial^2 u}{\partial y^2} \qquad (12.35)$$

Finite-difference approximations for the partial derivatives of the dependent variables are expressed as functions of their values at the streamwise coordinate $i + 1$, where the solutions are to be calculated, and at the upstream station i, where the flow field is already known. The nodal pattern for the finite-difference formulation is shown in Fig. 12.8.

The difference approximations for the partial derivatives involving products are defined so that the dependent variables at the $(i + 1)$ streamwise coordinate appear linearly in the resulting equation. A possible linearized finite-difference representation of equation (12.35) is

$$\rho u_{i,j} \frac{u_{i+1,j} - u_{i,j}}{\Delta x} + \rho v_{i,j} \frac{u_{i+1,j+1} - u_{i+1,j-1}}{2 \, \Delta y}$$

$$= \rho_e u_e \frac{du_e}{dx} + \mu \frac{u_{i+1,j+1} - 2u_{i+1,j} + u_{i+1,j-1}}{(\Delta y)^2}$$

Note that the term $\rho_e u_e \, (du_e/dx)$ involves only parameters evaluated at the edge of the boundary layer, where the flow is known. Rearranging this equation, we obtain

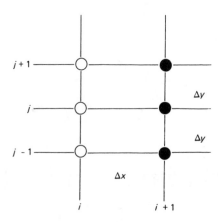

Figure 12.8 Nodal pattern for Navier–Stokes formulation.

$$
u_{i+1,j-1}\left[\frac{\mu}{(\Delta y)^2} + \frac{\rho v_{i,j}}{2(\Delta y)}\right] + u_{i+1,j}\left[-\frac{\rho u_{i,j}}{\Delta x} - \frac{2\mu}{(\Delta y)^2}\right]
$$

$$
+ u_{i+1,j+1}\left[\frac{\mu}{(\Delta y)^2} - \frac{\rho v_{i,j}}{2(\Delta y)}\right] = \left[-\frac{\rho(u_{i,j})^2}{\Delta x} - \rho_e u_e \frac{du_e}{dx}\right]
\tag{12.36}
$$

This may be rewritten as

$$
a_{i,j}u_{i+1,j-1} + b_{i,j}u_{i+1,j} + c_{i,j}u_{i+1,j+1} = d_{i,j}
\tag{12.37}
$$

Note the similarity between equations (12.37) and (12.29). Again, we have obtained a tridiagonal system that can be readily solved by the techniques of factorization (i.e., Gaussian elimination).

APPROACH TO DEVELOPING A CFD TOOL FOR DETERMINING THE AEROTHERMODYNAMICS OF A VEHICLE

According to Li (Ref. 12.8), the development and application of a computational fluid dynamics code involves seven steps:

1. Selecting the physical processes to be modeled.
2. Selecting the mathematical and topological models.
3. Defining the body geometry and constructing the spatial grid.
4. Developing a numerical solution algorithm.
5. Incorporating steps 1 through 4 in a computer code.
6. Calibrating and validating the code against benchmark experimental data or "classic," analytical solutions.
7. Using the CFD code to predict the desired aerodynamic coefficients, flow-field properties, and/or heating rate environment.

The development, calibration, and verification of a really useful CFD code may require several years.

Furthermore, the code developer should always be ready to incorporate improved differencing procedures, new solution algorithms, and improved flow models into his/her code. It has long been recognized that for a shock-capturing procedure to conserve mass across a shock wave, it is essential that the equations be cast in conservative form. But there are other considerations critical to this problem. Dwoyer (Ref. 12.8) notes: "Since the central difference approach does not contain sufficient dissipation to capture shocks, central difference methods are usually augmented with additional artificial viscosity to resolve the shock on the mesh and to control oscillations in the solution around the shocks. The second, and more recent, formulation is based on upwind differencing of the convective (i.e., Euler) part of the governing equations and central differencing of the viscous terms. Again, both explicit and implicit time-like integration is used. Unlike fully central difference methods, these hybrid upwind/ central methods are naturally dissipative and require no artificial viscosity to capture shocks.''

As noted above, a very important way of writing the equations of motion when using computational fluid dynamics techniques to obtain solutions for the flow field is the conservation form. The equations of motion are presented in conservation form in Appendix A.

SUMMARY

The purpose of this chapter is to introduce the reader to the nomenclature and principles of computational fluid dynamics. Examples were worked to demonstrate possible algorithms for representing the finite differences and different techniques for solving the equations. Solution procedures that work very well for one application may fail for another. Furthermore, subtle changes in the finite-difference analog may produce an FDE which is unconditionally stable, whereas the related FDE was unconditionally unstable. For the interested reader, there is a rapidly expanding library of journals and texts (e.g., Refs. 12.5 through 12.7 and 12.10) discussing such problems.

PROBLEMS

12.1 Consider Stokes' first problem with water as the bounding fluid medium. The kinematic viscosity of water is 1.078×10^{-5} ft²/s. If $\Delta y = 0.001$ in and $\Delta t = 0.01$ s, would you expect an explicit solution to be stable? The wall moves at a constant speed of 20 ft/s.

12.2. Using the explicit technique, calculate the velocity field for two time steps for the conditions of Problem 12.1.

12.3. Again, prior to time $t = 0$, the wall and the adjacent water are at rest. After time $t = 0$, the wall moves at 30 ft/s. The kinematic viscosity of water is 1.078×10^{-5} ft²/s. If $\Delta y = 0.001$ ft and $\Delta t = 0.01$ s, would you expect an explicit solution to be stable?

12.4. Using the explicit technique, calculate the velocity field for two time steps for the conditions of Problem 12.3.

12.5. Consider Stokes' first problem for an oil whose kinematic viscosity is 1×10^{-5} m²/ s. If you want information about the velocity field with a grid resolution (Δy) of 0.1 cm, what is the maximum time step you can take and still have a stable, explicit solution?

12.6. Repeat Problem 12.5 for oil whose kinematic viscosity is 2×10^{-4} m²/s.

12.7. Show that for the Crank-Nicolson implicit finite-difference representation, equation (12.29) are the appropriate expressions for a_i^n, b_i^n, c_i^n, and d_i^n.

12.8. Using the implicit Crank–Nicolson formulation, calculate the velocity field for two time steps for the condition of Problem 12.3.

12.9. Using a centered difference in time, so that one uses second-order expressions both for the space derivative and for the time derivative, we could write

$$\frac{u_i^{n+1} - u_i^{n-1}}{2\Delta t} = v \frac{u_{i+1}^n - 2u_i^n + u_{i-1}^n}{\Delta y^2}$$

Since the velocities on the left-hand side are evaluated at $t + \Delta t$ and at $t - \Delta t$, whereas they are evaluated at t on the right-hand side, this is known as the *leapfrog* technique. This method is numerically unstable for all $v > 0$ and for all time steps $\Delta t > 0$. Using the leapfrog technique, calculate the velocity field for the first three time steps for the conditions of Problem 12.3.

12.10. Let us replace the center node value of u_i^n in the diffusion term of the equation in Problem 12.9 by its average value at times $(n - 1)$ and $(n + 1)$, that is,

$$u_i^n = 0.5(u_i^{n-1} + u_i^{n+1})$$

Thus,

$$\frac{u_i^{n+1} - u_i^{n-1}}{2\Delta t} = v \frac{u_{i+1}^n - u_i^{n+1} - u_i^{n-1} + u_{i-1}^n}{\Delta y^2}$$

Solve this expression to obtain the explicit expression for u_i^{n+1}. The result, which is known as the *DuFort Frankel leapfrog* method, is stable. Use the expression to calculate the velocity field for the first three time steps for conditions of Problem 12.3.

REFERENCES

12.1. Miranda, L. R. "Application of Computational Aerodynamics in Airplane Design," *Journal of Aircraft*, June 1984, Vol. 21, No. 6, pp. 355–369.

12.2. Kutler, P., "A Perspective of Theoretical and Applied Computational Fluid Dynamics," *AIAA Paper 83–0037*, presented at the AIAA 21st Aerospace Sciences Meeting, Reno, Nev., Jan. 1983.

12.3. Chapman, D. R., "Computational Aerodynamics Development and Outlook," *AIAA Paper 79–0129*, presented at the AIAA 17th Aerospace Sciences Meeting, New Orleans, La., Jan. 1979.

12.4. H. Schlichting, *Boundary Layer Theory*, 7th ed., McGraw-Hill Book Company, New York, 1979.

12.5. Richtmyer, R. D., and K. W. Morton, *Difference Methods for Initial Value Problems*, Wiley-Interscience, New York, 1967.

12.6. Ferziger, J. H., *Numerical Methods for Engineering Applications*, Wiley-Interscience, New York, 1981.

12.7. Roache, P. J., *Computational Fluid Dynamics*, Hermosa Publishers, Albuquerque, N.M., 1976.

12.8. Li, C. P., "Computation of Hypersonic Flow Fields," *CP 87–208*, from a Short Course on Hypersonics conducted at the University of Texas at Austin, Oct. 1987.

12.9. Dwoyer, D. L., "CFD for Hypersonic, Airbreathing Airplanes," *CP 87–209*, from a Short Course on Hypersonics conducted at the University of Texas at Austin, Oct. 1987.

12.10. Anderson, D. A., J. C. Tannehill, and R. H. Pletcher, *Computational Fluid Mechanics and Heat Transfer*, McGraw-Hill Book Company, New York, 1984.

13

HYPERSONIC FLOWS

A vehicle flying through the atmosphere at hypersonic speeds generates a shock layer, that is, the region between the shock wave and the vehicle surface, in which the pressure, the temperature, and/or the density may change by two orders of magnitude, and more. Because the kinetic energy associated with hypersonic flight is converted into high temperatures within the shock layer, the flow-field trade studies conducted during the vehicle design include consideration both of the heat-transfer environment and of the aerodynamic forces and moments. Thus, one often speaks of the aerothermodynamic environment of a hypersonic vehicle.

By definition,

$$M_\infty = \frac{U_\infty}{a_\infty} \gg 1 \tag{13.1}$$

is the basic assumption for all hypersonic flow theories. Thus, the internal thermodynamic energy of the free-stream fluid particles is small compared with the kinetic energy of the free stream for hypersonic flows. In flight applications, this results because the velocity of the fluid particles is relatively large. The limiting case, where M_∞ approaches infinity because the free-stream velocity approaches infinity while the free-stream thermodynamic state remains fixed, produces extremely high temperatures in the shock layer. The high temperatures associated with hypersonic flight cannot be accommodated in ground test facilities. Therefore, in wind tunnel applications, hypersonic Mach numbers are achieved through relatively low speeds of sound; that is, for the wind tunnel, M_∞ approaches infinity because the speed of sound (the temperature) goes to "zero" while the free-stream velocity is held fixed. As a result, the fluid temperatures remain below the levels that would damage the wind tunnel and the model.

Another assumption common to hypersonic flow is that

$$\epsilon = \frac{\rho_\infty}{\rho_2} \ll 1 \tag{13.2}$$

which is known as the small-density-ratio assumption. Thus, this assumption relates primarily to the properties of the gas downstream of the shock wave. Recall that, for a perfect gas,

$$\epsilon = \frac{\gamma - 1}{\gamma + 1} \tag{13.3}$$

for a normal shock wave as $M_\infty \to \infty$. Thus, ϵ is $\frac{1}{6}$ for perfect air. Note that typical hypersonic wind tunnels operate at conditions where the test gas can be approximated by the perfect-gas relations. However, during the reentry of the Apollo Command Module, the density ratio approached $\frac{1}{20}$.

For slender configurations, such as sharp cones and wedges, the strong shock assumption is

$$M_\infty \sin \theta_b \gg 1 \tag{13.4}$$

which is mixed in nature, since it relates both to the flow and to the configuration. The concept termed the "Mach number independence principle" depends on this assumption.

NEWTONIAN FLOW MODEL

As the free-stream Mach number approaches infinity and the hypersonic independence principle applies to the flow, the shock layer becomes very thin. As a result, one can assume that the speed and direction of the gas particles in the free stream remain unchanged until they strike the solid surface exposed to the flow. For this flow model (termed Newtonian flow since it is similar in character to one described by Newton in the seventeenth century), the normal component of momentum of the impinging fluid particle is wiped out, while the tangential component of momentum is conserved. Thus, using the nomenclature of Fig. 13.1 and writing the integral form of the momentum equation for a constant-area streamtube normal to the surface,

$$p_\infty + \rho_\infty (U_{\infty,n})^2 = p_\infty + \rho_\infty (U_\infty \sin \theta_b)^2 = p_s \tag{13.5}$$

Rearranging so that the local pressure is written in terms of the pressure coefficient gives

$$C_p = \frac{p_s - p_\infty}{\frac{1}{2}\rho_\infty U_\infty^2} = 2 \sin^2 \theta_b = 2 \cos^2 \phi \tag{13.6}$$

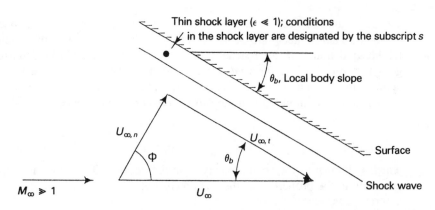

Figure 13.1 Nomenclature for Newtonian flow model.

The pressure coefficient, as defined in equation (13.6), is known as the Newtonian value, where the 2 represents the pressure coefficient at the stagnation point (which is designated $C_{p,t2}$), since $\theta_b = 90°$ at the Newtonian stagnation point.

The Newtonian flow model and the various theories for thin shock layers related to the Newtonian approximation are based on the small-density-ratio assumption. The small-density-ratio requirement for Newtonian theory also places implicit restrictions

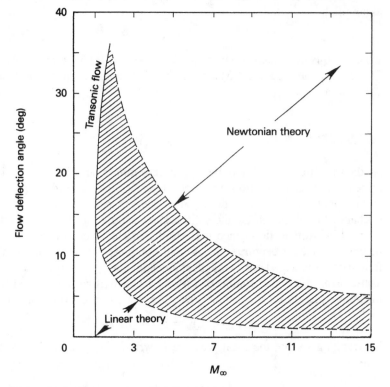

Figure 13.2 Regions of applicability of inviscid flow theories for the surface pressure on a sharp cone. (From Ref. 13.1.)

on the body shape in order that the shock layer be thin. The range of applicability for Newtonian theory, as defined by Marconi et al. (Ref. 13.1), is reproduced in Fig. 13.2. Small perturbation theory yields accurate results only for the flow over slender bodies at small angles of attack in a low supersonic Mach number stream. However, Newtonian theory provides useful results when the Mach number is large and/or the flow deflection angle is large. This is equivalent to the strong shock assumption

$$M_\infty \sin \theta_b \gg 1$$

Consider the pressure coefficients presented in Fig. 13.3. Theoretical values of C_p (taken from Chapter 8) are presented as a function of M_∞ for a cone whose semivertex angle is 15° and for a 15° wedge. Note that, as $M_\infty \to \infty$, the pressure coefficients become independent of the Mach number (the Mach number independence principle). The shock layer is thinner for the cone, and the limiting value of C_p is approached at a lower Mach number. Also presented is the Newtonian pressure coefficient (which is independent of the Mach number). As $M_\infty \to \infty$, all three techniques yield roughly the same value for C_p. Therefore, it can be seen that, at hypersonic speeds, the pressure coefficient for these simple shapes depends primarily on the flow deflection angle.

The Mach number independence principle was derived for inviscid flow by Oswatitsch. Since pressure forces are much larger than the viscous forces for blunt

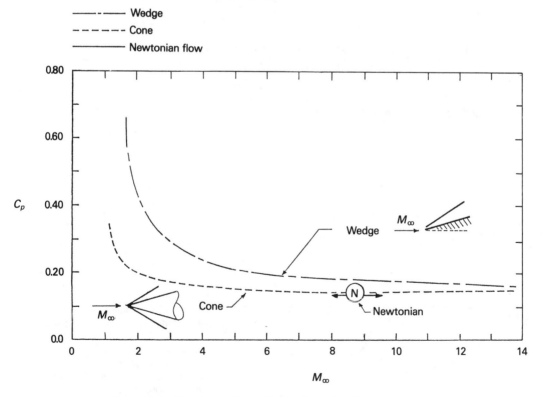

Figure 13.3 Pressure coefficient for the flow of perfect air past a wedge and a cone; deflection angle is 15°.

bodies or for slender bodies at relatively large angles of attack when the Reynolds number exceeds 10^5, one would expect the Mach number independence principle to hold at these conditions.

STAGNATION REGION FLOW-FIELD PROPERTIES

The nomenclature for the flow near the stagnation point of a vehicle in a hypersonic stream is illustrated in Fig. 13.4. The free-stream flow (designated by the subscript ∞ or 1) passes through the normal portion of the shock wave reaching state 2 and then decelerates isentropically to state $t\,2$, which constitutes the edge condition for the thermal boundary layer at the stagnation point. The streamline from the shock wave to the stagnation point may be curved for nonaxisymmetric flow fields. The pressure and the heating rate at the stagnation point are useful reference values for characterizing hypersonic flows.

The relations for steady, one-dimensional, inviscid, adiabatic flow in a constant-area streamtube were used to compute the conditions across a normal shock wave.

$$\rho_1 U_1 = \rho_2 U_2 \tag{13.7}$$

$$p_1 + \rho_1 U_1^2 = p_2 + \rho_2 U_2^2 \tag{13.8}$$

$$h_1 + \tfrac{1}{2}U_1^2 = h_2 + \tfrac{1}{2}U_2^2 = H_t \tag{13.9}$$

where H_t is the total (or stagnation) enthalpy of the flow.

If one assumes that the gas is thermally perfect,

$$p = \rho R T \tag{13.10}$$

and calorically perfect,

$$h = c_p T \tag{13.11}$$

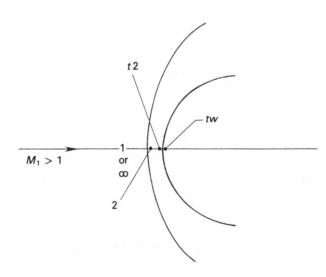

Figure 13.4 Nomenclature for the stagnation region. 1 or ∞, Free-stream conditions; 2, conditions immediately downstream of the shock wave; $t2$, conditions at the stagnation point (downstream of the normal portion of the shock wave) but outside the boundary layer; tw, conditions at the wall at the downstream stagnation point.

the ratio of the values of flow properties across the shock wave can be written as a unique function of M_1 (or M_∞, the free-stream Mach number) and γ (the ratio of specific heats). Referring to Chapter 8, the relations are

$$\frac{p_2}{p_1} = \frac{2\gamma M_1^2 - (\gamma - 1)}{\gamma + 1} \tag{13.12}$$

$$\frac{\rho_2}{\rho_1} = \frac{U_1}{U_2} = \frac{(\gamma + 1)M_1^2}{(\gamma - 1)M_1^2 + 2} \tag{13.13}$$

$$\frac{T_2}{T_1} = \frac{[2\gamma M_1^2 - (\gamma - 1)][(\gamma - 1)M_1^2 + 2]}{(\gamma + 1)^2 M_1^2} \tag{13.14}$$

If one assumes that the flow decelerates isentropically from the conditions at point 2 (immediately downstream of the normal portion of the shock wave) to the stagnation point outside of the thermal boundary layer (point $t2$),

$$\frac{p_{t2}}{p_1} = \left[\frac{(\gamma + 1)M_1^2}{2} \right]^{\gamma/(\gamma - 1)} \left[\frac{\gamma + 1}{2\gamma M_1^2 - (\gamma - 1)} \right]^{1/(\gamma - 1)} \tag{13.15}$$

$$\frac{T_{t2}}{T_1} = \frac{T_{t1}}{T_1} = 1 + \frac{\gamma - 1}{2} M_1^2 \tag{13.16}$$

Note that, whereas it is generally true that the stagnation enthalpy is constant across a normal shock wave for an adiabatic flow [see equation (13.9)], the stagnation temperature is constant across a normal shock wave only for the adiabatic flow of a perfect gas [see equation (13.16)]. Note also that for a perfect gas (i.e., one that is thermally perfect and calorically perfect), the ratio of the downstream value to the free-stream value for the flow properties (across a normal shock wave) can be written as a function of γ and $M_1 (= M_\infty)$ only. Thus, the perfect-gas values for the ratios defined by equations (13.12) through (13.16) do not depend specifically on the altitude.

In reality, for hypersonic flight, the gas molecules that pass through the bow shock wave are excited to higher vibrational and chemical energy modes. This lowers the specific-heat ratio of the gas below the free-stream value if it is assumed that equilibrium exists and that dissociation is not driven to completion. A large amount of the energy that would have gone into increasing the static temperature behind the bow shock wave for a perfect gas is used instead to excite the vibrational energy levels or to dissociate the gas molecules. As additional energy is absorbed by the gas molecules entering the shock layer, the conservation laws and the thermophysics dictate certain changes in the forebody flow. The static temperature, the speed of sound, and the velocity in the shock layer are less for the equilibrium, real-gas flow than for a perfect-gas flow. The real-gas value of the static pressure is slightly larger than the perfect-gas value. The density is increased considerably, and, as a result, the shock layer thickness is reduced.

Equations (13.7) through (13.9) are not restricted to the perfect-gas assumption and can be applied to a high-temperature, hypersonic flow. Let us use the U.S. Standard Atmosphere (Ref. 13.2) to define the free-stream properties (i.e., p_1, ρ_1, and h_1, at 150,000 ft). Since there are four unknowns in equations (13.7) through (13.9) (i.e.,

p_2, ρ_2, h_2, and U_2), but only three equations, additional relations are needed to obtain a solution. The graphs of Ref. 13.3 (see Fig. 13.5) were used in tabular form to define

$$\rho \quad (p, h)$$

$$s \quad (p, h)$$

$$T \quad (h, s)$$

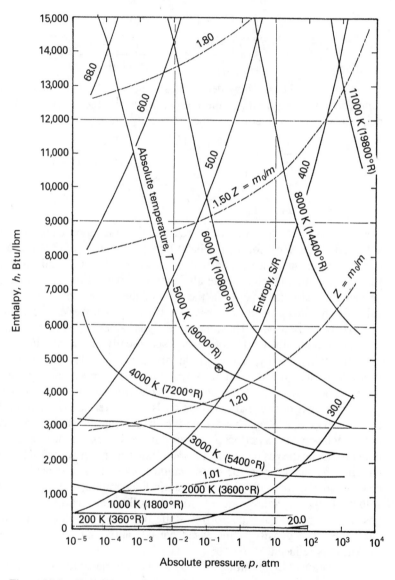

Figure 13.5 Thermodynamic properties of air in chemical equilibrium. (From Ref. 13.3.)

Let us calculate the flow downstream of a normal shock wave when $M_1 = 14$ at an altitude of 150,000 ft. Using the more generally applicable expression equation (13.9),

$$H_{t1} = h_1 + \tfrac{1}{2}U_1^2 = H_{t2} = H_t = 4636.77 \text{ Btu/lbm}$$

Assuming that the air remains in thermodynamic equilibrium as it crosses the shock wave, p_{t2} is 0.3386 atm (716.57 lb/ft²) and T_{t2} is 8969.6°R, as represented by the $\oplus$ in Fig. 13.5. Using the perfect-gas relations of equations (13.12) through (13.16), p_{t2} is 0.3256 atm (689.12 lb/ft²) and T_{t2} is 19,325°R.

The perfect-gas values and the equilibrium air values for T_{t2}/T_1, p_{t2}/p_1, and $C_{p,t2}$ (the pressure coefficient at the stagnation point) are presented as a function of the free-stream Mach number for an altitude of 150,000 ft in Figs. 13.6, 13.7, and 13.8, respectively. The comments made earlier regarding the qualitative differences between the perfect-gas values and those for equilibrium air (e.g., "The real-gas value of the static pressure is slightly larger than the perfect-gas value") are illustrated in Figs. 13.7 and 13.8.

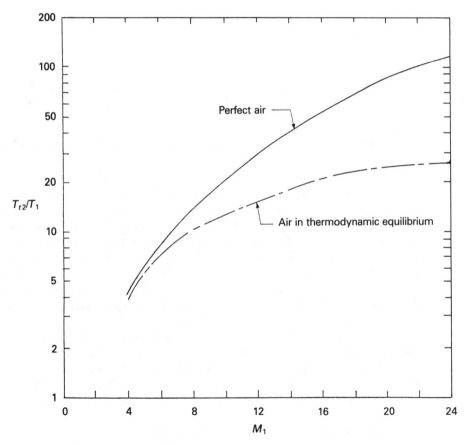

Figure 13.6 Temperature at the stagnation point (at the edge of the boundary layer) of a sphere ($R_N = 0.3048$ m) at an altitude of 45,721 m.

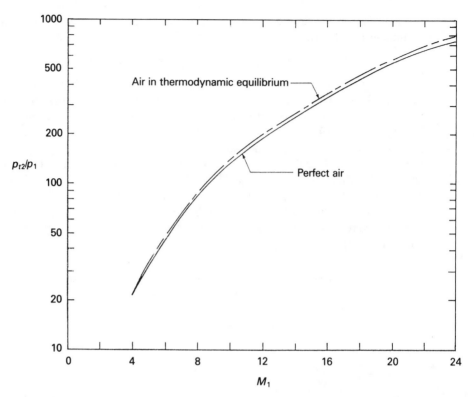

Figure 13.7 Pressure at the stagnation point of a sphere (R_N = 0.3048 m) at an altitude of 45,721 m.

The ratio of T_{t2}/T_1 is presented in Fig. 13.6. As noted earlier, the energy absorbed by the dissociation process causes the real-gas equilibrium temperature to be markedly lower than the perfect-gas value. The specific heat correlations that were presented by Hansen (Ref. 13.4) as a function of pressure and of temperature can be used to identify conditions where the dissociation of oxygen and of nitrogen affect the properties. Hansen notes that, at all pressures, the dissociation of oxygen is essentially complete before the dissociation of nitrogen begins. Based on Hansen's correlations, the oxygen dissociation reaction begins near Mach 7 and the nitrogen reaction begins near Mach 18 for the equilibrium air model at 150,000 ft.

The p_{t2}/p_1 ratio and the stagnation-point pressure coefficient are presented in Figs. 13.7 and 13.8, respectively. As noted earlier, the real-gas value of the static pressure is slightly greater than the perfect-gas value. Note that, at $M_1 = 4$, the stagnation pressure coefficient ($C_{p,t2}$) is approximately 1.8 for both perfect air and for air in thermodynamic equilibrium. At $M_1 = 24$, $C_{p,t2}$ is 1.932 for the equilibrium air model as compared with 1.838 for perfect air and 2 for Newtonian flow.

From the discussion above, it should be clear that the changes in a fluid property across a normal shock wave are a function of M_1 and γ only for a perfect gas [see equation (13.12)], that is,

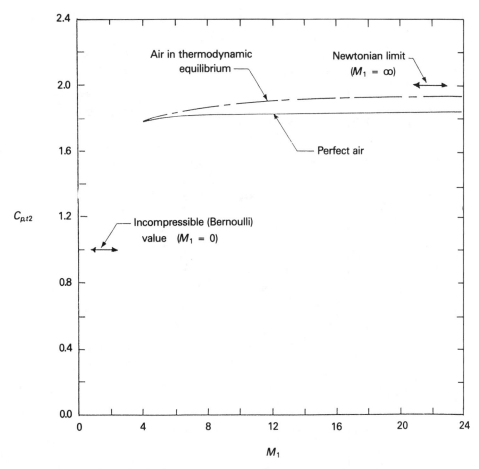

Figure 13.8 Pressure coefficient at the stagnation point of a sphere ($R_N = 0.3048$ m) at an altitude of 45,721 m.

$$\frac{p_2}{p_1} = f(M_1, \gamma)$$

However, for a reacting gas in chemical equilibrium, three free-stream parameters are necessary to obtain the ratios of properties across a normal shock wave (the free-stream velocity and two thermodynamic properties),

$$\frac{p_2}{p_1} = f(U_1, p_1, T_1)$$

MODIFIED NEWTONIAN FLOW

It is evident in the computed values for the pressure coefficient at the stagnation point presented in Fig. 13.8 that, even when $M_1 = 24$, $C_{p,t2}$ is 1.838 for perfect air and is 1.932 for air in thermodynamic equilibrium. Thus, as noted by Lees (Ref. 13.5), it

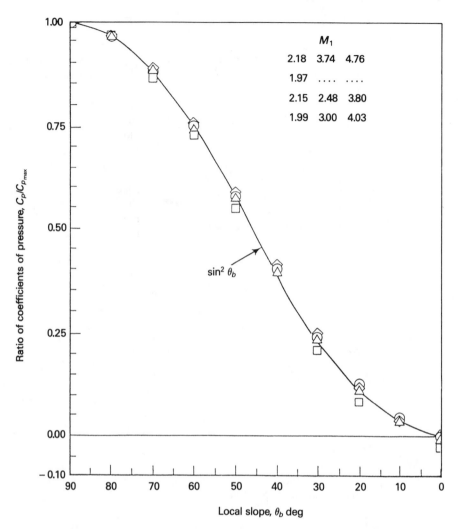

Figure 13.9   Correlation between $C_p/C_{p,\max}$ ratio and local body slope. (From Ref. 13.6.)

would be more appropriate to compare the ratio $C_p/C_{p,\max}$ with $\sin^2\theta_b$ (or, equivalently, $\cos^2\phi$). Such a comparison is presented in Fig. 13.9 using data for hemispherically capped cylinders. Even though the data of Fig. 13.9 are for free-stream Mach numbers from 1.97 to 4.76, the $\sin^2\theta_b$ relation represents the data quite adequately for this blunt configuration. Therefore, an alternative representation of the pressure coefficient for hypersonic flow is

$$C_p = C_{p,t2}\,\sin^2\theta_b = C_{p,t2}\,\cos^2\phi \qquad (13.17)$$

which will be termed modified Newtonian flow.

Example 13.1

Neglecting the effects of skin friction and using the modified Newtonian flow model to describe the pressure distribution, derive an expression for the drag coefficient for a sphere.

Solution   Because the sphere is a blunt configuration, the pressure forces are the principal component of the drag force at high Reynolds numbers. Thus, using the coordinate system shown in Fig. 13.10, the drag on the sphere due to the pressure is

$$D = \int p[2\pi y(ds)] \sin \theta_b = \int p2\pi y \, dy$$

where the incremental surface area on which the pressure acts is $[2\pi y(ds)]$, refer to the shaded region on the forebody in Fig. 13.10. As discussed in Chapter 5, the net resultant force in any direction due to a constant pressure acting over a closed surface is zero. Thus, the pressure coefficient can be used in our expression for the drag:

$$D = q_\infty \int C_p 2\pi y \, dy \qquad (13.18)$$

Using modified Newtonian theory to define the pressure, as given by equation (13.17), with the coordinates shown in Fig. 13.10,

$$C_p = C_{p,t2} \sin^2\theta_b = C_{p,t2} \left(\frac{dy}{ds}\right)^2 = C_{p,t2} \frac{(y')^2}{1 + (y')^2} \qquad (13.19)$$

where $y' = dy/dx$. Combining equations (13.18) and (13.19), the expression for the drag becomes

$$D = q_\infty C_{p,t2} \int_0^R \frac{(y')^2}{1 + (y')^2} 2\pi y \, y' \, dx \qquad (13.20)$$

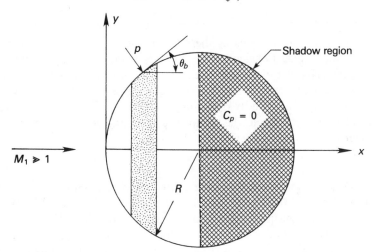

Figure 13.10   Sketch for calculating the modified Newtonian pressure drag on a sphere, Example 13.1.

The limits of the integration are $0 \leq x \leq R$, since, according to the Newtonian flow model, the pressure coefficient on the leeward side (i.e., the crosshatched region of the sphere in Fig. 13.10) is zero.

For the sphere,

$$y = (2xR - x^2)^{0.5}$$

and

$$\frac{dy}{dx} = \frac{R - x}{(2xR - x^2)^{0.5}}$$

Substituting these expressions into equation (13.20),

$$D = 2\pi q_\infty C_{p,t2} \int_0^R \frac{\left[\dfrac{R - x}{(2xR - x^2)^{0.5}}\right]^3}{1 + \dfrac{(R - x)^2}{2xR - x^2}} (2xR - x^2)^{0.5}\, dx$$

Simplifying and integrating yields

$$D = \frac{2\pi q_\infty C_{p,t2}}{R^2} \left[R^3 x - \frac{3}{2} x^2 R^2 + x^3 R - \frac{x^4}{4} \right]_0^R$$

$$= \frac{C_{p,t2}}{2} q_\infty \pi R^2$$

Thus,

$$C_D = \frac{D}{q_\infty \pi R^2} = \frac{C_{p,t2}}{2} \tag{13.21}$$

Let us consider the case where the modified Newtonian flow model is applied in a similar manner to calculate the pressure drag on a right circular cylinder whose axis is perpendicular to the free-stream flow. Although the pressure distribution is that given by Equation (13.19) and the cross section is a circle, as shown in Fig. 13.10, the flow around the cylinder is two-dimensional, whereas the flow around a sphere is axisymmetric. As a result, the section pressure drag coefficient (per unit span) for the cylinder is

$$C_{d,p} = \tfrac{2}{3} C_{p,t2} \tag{13.22}$$

The values of section pressure drag coefficient calculated using equation (13.22) are compared in Fig. 13.11 with data that were presented in Ref. 13.7. Despite the simplifications inherent in this technique to calculate the theoretical drag coefficient, the agreement between the data and the theoretical values is outstanding. Note also that the pressure drag for the blunt, right circular cylinder reaches its hypersonic limiting value by $M_\infty = 4$.

Data presented by Koppenwallner (Ref. 13.7) indicate a significant increase in the total drag coefficient for a right circular cylinder occurs due to the friction drag when the Knudsen number (which is the ratio of the length of the molecular mean

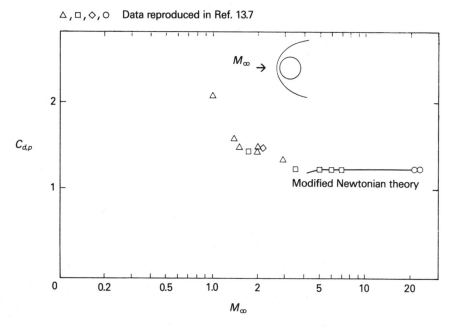

△ , □ , ◇ , ○   Data reproduced in Ref. 13.7

Figure 13.11 Pressure drag of a right circular cylinder as a function of Mach number. (From Ref. 13.7.)

free path to a characteristic dimension of the flow field) is greater than 0.01. Data presented by Koppenwallner are reproduced in Fig. 13.12. Using the Reynolds number based on the flow conditions behind a normal shock wave as the characteristic parameter,

$$\mathrm{Re}_2 = \frac{\rho_2 U_2 d}{\mu_2}$$

the friction drag for $\mathrm{Re}_2 > 10$ is given by

$$C_{d,f} = \frac{5.3}{(\mathrm{Re}_2)^{1.18}} \tag{13.23}$$

The data presented in Figs. 13.11 and 13.12 illustrate the significance of high-altitude effects on the aerodynamic coefficients.

The modified Newtonian flow model can be used to obtain a quick, engineering estimate of the pressure distribution. Consider the axisymmetric configuration shown in Fig. 13.13, where $\hat{n}$ is a unit vector that is normal to the surface element dA and is positive in the inward direction, θ is the local surface inclination, and β is the angular position of a point on the surface of the body. The angle η, the angle between the velocity vector $\vec{V}_\infty$ and the inward normal $\hat{n}$, is given by

$$\cos \eta = \frac{\vec{V}_\infty \cdot \hat{n}}{|\vec{V}_\infty||\hat{n}|} \tag{13.24}$$

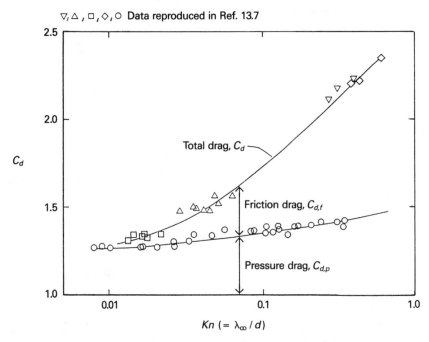

Figure 13.12 Contribution of the total drag due to friction drag and to pressure drag. (From Ref. 13.7.)

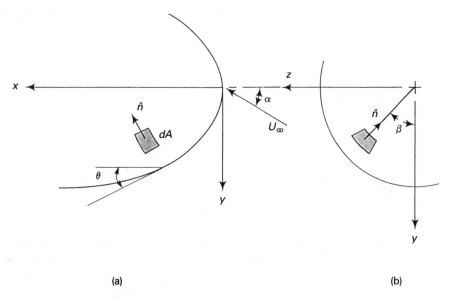

(a)   (b)

Figure 13.13 Coordinate system nomenclature for axisymmetric configurations; (a) side view; (b) front view.

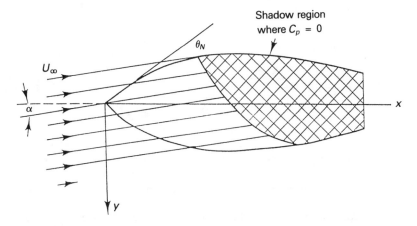

Figure 13.14 Region where $C_p = 0$, that is, that portion of the body which lies in the "shadow of the free stream."

where

$$\vec{V}_\infty = U_\infty \cos \alpha \hat{i} - U_\infty \sin \alpha \hat{j} \qquad (13.25)$$

and

$$\hat{n} = \hat{i} \sin \theta - \hat{j} \cos \theta \cos \beta - \hat{k} \cos \theta \sin \beta \qquad (13.26)$$

Thus,

$$\cos \eta = \cos \alpha \sin \theta + \sin \alpha \cos \theta \cos \beta \qquad (13.27)$$

so that the pressure coefficient is

$$C_p = C_{p,t2} \cos^2 \eta \qquad (13.28)$$

In the Newtonian (or the modified Newtonian) model, the free-stream flow does not impinge on those portions of the body surface which are inclined away from the free-stream direction and which may, therefore, be thought of as lying in the "shadow of the free stream." This is illustrated in Fig. 13.14. For the modified Newtonian flow model, $C_p = 0$ in the shaded region of Fig. 13.14.

Example 13.2

Consider hypersonic flow past a sharp cone where $-\theta_c \leq \alpha \leq \theta_c$, as shown in Fig. 13.15. Neglecting the effects of skin friction and using the modifed Newtonian flow model to describe the pressure distribution, derive expressions for the lift coefficient, the drag coefficient, and the pitching moment coefficient.

Solution For simplicity, let us first calculate the forces in a body-fixed coordinate system. As shown in Fig. 13.16, the forces in the body-fixed coordinate system are A, the force

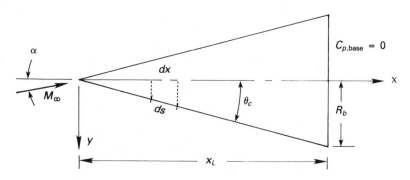

Figure 13.15 Nomenclature for hypersonic flow past a sharp cone.

along the axis of the body, and N, the force normal to the body axis. Once A and N have been calculated, the lift and the drag can be calculated since

$$L = N \cos \alpha - A \sin \alpha \tag{13.29}$$

and

$$D = N \sin \alpha + A \cos \alpha \tag{13.30}$$

Applying equations (13.24) through (13.28) specifically for the sharp cone flow depicted in Fig. 13.15,

$$C_p = C_{p,t2} (\cos \alpha \sin \theta_c + \sin \alpha \cos \theta_c \cos \beta)^2 \tag{13.31}$$

since the deflection angle is the cone semivertex angle, θ_c, which is a constant.

The axial force coefficient is found by integrating the pressure force over the entire (closed) surface of the cone:

$$C_A = \frac{1}{(\frac{1}{2}\rho_\infty U_\infty^2)(\pi R_b^2)} \oiint_S (p - p_\infty)(\hat{n}dS) \cdot \hat{\imath} \tag{13.32}$$

Recall that, since we are integrating over a closed surface, the net force in any direction due to a constant pressure acting on that closed surface is zero. Hence, subtracting p_∞, as is done in equation (13.32), does not affect the value of the axial force. A differential element for the surface of the cone (dS) is

$$dS = r \, d\beta \, ds = (x \tan \theta_c)d\beta \frac{dx}{\cos \theta_c} \tag{13.33}$$

Combining equations (13.31) through (13.33) with equation (13.26) and noting that, since $C_{p,base} = 0$, the limits of integration are $0 \leq \beta \leq \pi$ (if we multiply by 2) and $0 \leq x \leq x_L$, the axial force coefficient is

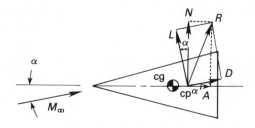

Figure 13.16 Resolving the forces acting on a vehicle.

$$C_A = \frac{2C_{p,t2}}{\pi R_b^2} \int_0^{x_L} \left[\int_0^\pi (\cos^2\alpha \, \sin^2\theta_c + 2 \sin\alpha \cos\alpha \sin\theta_c \cos\theta_c \cos\beta \right.$$

$$\left. + \sin^2\alpha \cos^2\theta_c \cos^2\beta)(x \tan\theta_c)d\beta \, \frac{dx}{\cos\theta_c} \sin\theta_c \right]$$

Integrating first over β yields

$$C_A = \frac{2C_{p,t2}}{\pi R_b^2} \int_0^{x_L} \left(\cos^2\alpha \, \sin^2\theta_c \pi + \sin^2\alpha \cos^2\theta_c \frac{\pi}{2} \right) \tan^2\theta_c x \, dx$$

Integrating with respect to x gives us

$$C_A = \frac{2C_{p,t2}}{R_b^2} \left(\cos^2\alpha \, \sin^2\theta_c + \frac{1}{2} \sin^2\alpha \cos^2\theta_c \right) \tan^2\theta_c \frac{x_L^2}{2}$$

Since $R_b = x_L \tan\theta_c$,

$$C_A = C_{p,t2} \left[\sin^2\theta_c + \tfrac{1}{2} \sin^2\alpha(1 - 3\sin^2\theta_c) \right] \tag{13.34}$$

Let us calculate the normal force coefficient due to the pressures acting over the closed surface of the cone. Since the normal force is positive in the negative y direction,

$$C_N = \frac{1}{(\frac{1}{2}\rho_\infty U_\infty^2)(\pi R_b^2)} \oiint_S (p - p_\infty)(\hat{n} \, dS) \cdot (-\hat{j}) \tag{13.35}$$

Combining equations (13.26), (13.31), (13.33), and (13.35) and using the limits of integration, $0 \le \beta \le \pi$ and $0 \le x \le x_L$ (as discussed previously), we have

$$C_N = \frac{2C_{p,t2}}{\pi R_b^2} \int_0^{x_L} \left[\int_0^\pi (\cos^2\alpha \, \sin^2\theta_c + 2 \sin\alpha \cos\alpha \sin\theta_c \cos\theta_c \cos\beta \right.$$

$$\left. + \sin^2\alpha \cos^2\theta_c \cos^2\beta)(x \tan\theta_c)d\beta \, \frac{dx}{\cos\theta_c} (\cos\theta_c \cos\beta) \right]$$

Integrating first with respect to β gives us

$$C_N = \frac{2C_{p,t2}}{\pi R_b^2} \int_0^{x_L} \left(2 \sin\alpha \cos\alpha \sin\theta_c \cos\theta_c \frac{\pi}{2} \right) \tan\theta_c x \, dx$$

Integrating with respect to x yields

$$C_N = \frac{2C_{p,t2}}{R_b^2} (\sin\alpha \cos\alpha \sin\theta_c \cos\theta_c) \tan\theta_c \frac{x_L^2}{2}$$

so that

$$C_N = \frac{C_{p,t2}}{2} \sin 2\alpha \cos^2\theta_c \tag{13.36}$$

Referring to equation (13.29), the lift coefficient is

$$C_L = C_N \cos\alpha - C_A \sin\alpha$$

Using the coefficients presented in equations (13.34) and (13.36), we have

$$C_L = C_{p,t2} \sin\alpha \left[\cos^2\alpha \cos^2\theta_c - \sin^2\theta_c - \tfrac{1}{2} \sin^2\alpha(1 - 3\sin^2\theta_c) \right] \tag{13.37}$$

Correspondingly,

$$C_D = C_{p,t2} \cos \alpha \, [\sin^2\alpha \, \cos^2\theta_c + \sin^2\theta_c + \tfrac{1}{2} \sin^2\alpha(1 - 3 \sin^2\theta_c)] \quad (13.38)$$

To calculate the pitching moment, note that the moment due to the incremental pressure force acting at a radial distance $\vec{r}$ from the origin is

$$\overrightarrow{dM} = \vec{r} \times \overrightarrow{dF} = \vec{r} \times p \, \hat{n} \, dS$$

Note that the incremental moment $\overrightarrow{dM}$ is a vector, which can be written in terms of its components:

$$\overrightarrow{dM} = dL \, \hat{\imath} + dN \, \hat{\jmath} + dM \, \hat{k} \quad (13.39)$$

In equation (13.39), L is the rolling moment (which is positive when causing the right wing to move down), N is the yawing moment (which is positive when causing the nose to move to the right), and M is the pitching moment (which is positive when causing a nose-up motion).

If we want to take the moments about the apex of the cone, the moment arm is

$$\vec{r} = x\hat{\imath} + x \tan \theta_c \cos \beta \hat{\jmath} + x \tan \theta_c \sin \beta \hat{k} \quad (13.40)$$

Let us examine the $\vec{r} \times \hat{n}$ term of the moment expression:

$$\vec{r} \times \hat{n} = \begin{vmatrix} \hat{\imath} & \hat{\jmath} & \hat{k} \\ x & x \tan \theta_c \cos \beta & x \tan \theta_c \sin \beta \\ \sin \theta_c & -\cos \theta_c \cos \beta & -\cos \theta_c \sin \beta \end{vmatrix}$$

Thus,

$$\begin{aligned} \vec{r} \times \hat{n} = {} & \hat{\imath}[0] + \hat{\jmath}[x \sin \beta(\tan \theta_c \sin \theta_c + \cos \theta_c)] \\ & - \hat{k}[x \cos \beta(\tan \theta_c \sin \theta_c + \cos \theta_c)] \end{aligned} \quad (13.41)$$

The first term (i.e., the contribution to the rolling moment) is always zero, since, in the absence of viscous forces, the only forces are the pressure forces that act normal to the surface and therefore act through the axis of symmetry for a body of revolution. The negative sign for the $\hat{k}$ term (i.e., the pitching moment) results because a pressure force acting in the first and fourth quadrants ($\pi/2 \geq \beta \geq -\pi/2$) produces a nose-down (negative) pitching moment. For this example we are interested in the pitching moment about the apex, which is considered positive when nose up,

$$M_0 = \oiint_S \vec{r} \times (p - p_\infty)\hat{n} \, dS \cdot \hat{k} \quad (13.42)$$

Taking the resultant cross-product of the moment arm from the apex $(\vec{r})$ and the inward-directed unit normal to the surface area ($\hat{n}$) with the dot product with $\hat{k}$, we obtain

$$(\vec{r} \times \hat{n}) \cdot \hat{k} = -x \cos \theta_c \cos \beta - x \sin \theta_c \tan \theta_c \cos \beta$$

Thus, the pitching moment coefficient

$$C_{M0} = \frac{-2}{\pi R_b^2 R_b} \int_0^{x_L} \left[\int_0^{\pi} C_p x \tan \theta_c d\beta \frac{dx}{\cos \theta_c} (x \cos \theta_c \cos \beta \right. \\ \left. + x \sin \theta_c \tan \theta_c \cos \beta) \right] \quad (13.43)$$

We can simplify this expression if we focus on the terms containing the cone-half-angle (θ_c), so that

$$\tan \theta_c \frac{1}{\cos \theta_c} (\cos \theta_c + \sin \theta_c \tan \theta_c) = \tan \theta_c (1 + \tan^2 \theta_c) = \frac{\tan \theta_c}{\cos^2 \theta_c}$$

Substituting the modified-Newtonian-flow expression for the pressure coefficient and integrating with respect to β yields

$$C_{M0} = - \frac{2C_{p,t2}}{\pi R_b^2 R_b} \frac{\tan \theta_c}{\cos^2 \theta_c} \int_0^{x_L} x^2 (\sin \alpha \cos \alpha \sin \theta_c \cos \theta_c \pi) \, dx$$

Integrating with respect to x and using the fact that $R_b = x_L \tan \theta_c$, we obtain

$$C_{M0} = - \frac{C_{p,t2} \sin 2\alpha}{3 \tan \theta_c} \tag{13.44}$$

Although the resultant moment is produced by the integration of the distributed aerodynamic forces over the vehicle surface, it can be represented as composed of two components, one effectively due to the normal-force component and the other due to the axial-force component. Thus, we can isolate the two terms of equation (13.43):

$$C_{M0} = - \frac{2}{\pi R_b^2 R_b} \int_0^{x_L} \left(\int_0^\pi C_p x \tan \theta_c \, d\beta \, \frac{dx}{\cos \theta_c} x \cos \theta_c \cos \beta \right)$$

$$- \frac{2}{\pi R_b^2 R_b} \int_0^{x_L} \left(\int_0^\pi C_p x \tan \theta_c \, d\beta \, \frac{dx}{\cos \theta_c} x \sin \theta_c \tan \theta_c \cos \beta \right)$$

Writing the total pitching moment as the sum of the two components, we have

$$C_{M0} = C_{M0,N} + C_{M0,A} \tag{13.45}$$

where

$$C_{M0,N} = - \frac{2 \tan \theta_c}{\pi R_b^2 R_b} \int_0^{x_L} \left(\int_0^\pi C_p \cos \beta \, d\beta \right) x^2 \, dx \tag{13.46}$$

and

$$C_{M0,A} = - \frac{2 \tan^3 \theta_c}{\pi R_b^2 R_b} \int_0^{x_L} \left(\int_0^\pi C_p \cos \beta \, d\beta \right) x^2 \, dx \tag{13.47}$$

Substituting the modified Newtonian-flow expression for the pressure coefficient yields

$$C_{M0,N} = - \frac{2 \tan \theta_c C_{p,t2}}{\pi R_b^2 R_b} \int_0^{x_L} \left[\int_0^\pi (\cos^2\alpha \sin^2\theta_c \cos \beta \right.$$

$$\left. + 2 \sin \alpha \cos \alpha \sin \theta_c \cos \theta_c \cos^2\beta + \sin^2\alpha \cos^2\theta_c \cos^3\beta) \, d\beta \right] x^2 \, dx$$

Integrating gives us

$$C_{M0,N} = - \frac{C_{p,t2} \sin 2\alpha}{R_b^2 R_b} \tan \theta_c \sin \theta_c \cos \theta_c \frac{x_L^3}{3}$$

$$C_{M0,N} = - \frac{C_{p,t2} \sin 2\alpha \cos^2\theta_c}{3} \frac{x_L}{R_b}$$

Using equation (13.36) for the expression for C_N, we can rewrite

$$C_{M0,N} = C_N \left(-\frac{2}{3} \frac{x_L}{R_b} \right)$$ (13.48)

Similarly,

$$C_{M0,A} = \frac{-2 \tan^3\theta_c C_{p,t2}}{\pi R_b^2 R_b} \int_0^{x_L} \left[\int_0^\pi (\cos^2\alpha \sin^2\theta_c \cos \beta \right.$$

$$\left. + 2 \sin \alpha \cos \alpha \sin \theta_c \cos \theta_c \cos^2\beta + \sin^2\alpha \cos^2\theta_c \cos^3\beta) \, d\beta \right] x^2 \, dx$$

Integrating yields

$$C_{M0,A} = -\frac{C_{p,t2} \sin 2\alpha \sin^2\theta_c}{3} \frac{x_L}{R_b}$$

Using equation (13.34) for the expression for C_A, we can rewrite

$$C_{M0,A} = C_A \frac{- \sin 2\alpha \sin^2\theta_c}{3[\sin^2 \theta_c + 0.5 \sin^2 \alpha(1 - 3 \sin^2 \theta_c)]} \frac{x_L}{R_b}$$ (13.49)

Following this division of the pitching moment into components, we can represent it as an effective net force acting at a "center of pressure," as shown in Fig. 13.17. The pitching moment may be represented by

$$M_0 = M_{0,N} + M_{0,A} = -x_{cp}N - y_{cp}A$$ (13.50)

The two terms represent the effects of the normal force and of the axial force, respectively. In terms of the coefficients,

$$C_{M0} = \frac{M_0}{q_\infty SR_b} = -C_N \frac{x_{cp}}{R_b} - C_A \frac{y_{cp}}{R_b}$$ (13.51)

Comparing the formulations presented in equations (13.48), (13.49) and (13.51), we find that

$$x_{cp} = \frac{2}{3} x_L$$ (13.53)

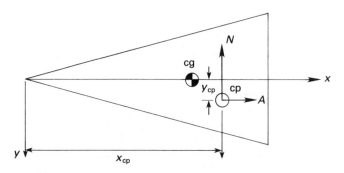

Figure 13.17   Aerodynamic forces acting at the center of pressure as effectively producing the aerodynamic pitching moment.

and

$$y_{cp} = \frac{\sin 2\alpha \sin^2\theta_c x_L}{3[\sin^2\theta_c + 0.5 \sin^2\alpha(1 - 3 \sin^2\theta_c)]} \tag{13.53}$$

If the vehicle is to be statically stable, the center of pressure should be located such that the aerodynamic forces produce a restoring moment when the configuration is perturbed from its "stable orientation." Thus, if the vehicle pitches up (i.e., α increases), the net pitching moment should be negative (causing a nose-down moment), which decreases α.

The parameter

$$\text{S.M.} = \frac{x_{cp} - x_{cg}}{x_L} \tag{13.54}$$

is the static margin. The static margin must be positive for uncontrolled vehicles. For high-performance, hypersonic vehicles, the static margin is usually 3 to 5%. Note, however, as illustrated in Fig. 13.17, both the axial force and the normal force contribute to the pitching moment. Thus, it is possible that the vehicle is statically stable when $x_{cp} = x_{cg}$, if y_{cp} is below y_{cg} (as is the case in the sketch of Fig. 13.17). In this case the axial force will produce the required restoring moment.

Example 13.3

We neglected the viscous forces and the base pressure in estimating the forces acting on the sharp cone of Example 13.2. Consider a sharp cone ($\theta_c = 10°$) exposed to the Mach 8 flow of Tunnel B at the Arnold Engineering Development Center (AEDC), as shown in Fig. 13.18. For this problem, the subscript $t1$ will designate the flow conditions in the nozzle reservoir, the subscript 1 designates those in the test section, the subscript e those at the edge of the boundary layer of the sharp cone (and which are constant along the entire length of the cone) and the subscript w designates conditions at the wall of the cone. If $T_{t1} = 1350°R$, $p_{t1} = 850$ psia, $T_w = 600°R$, and $\alpha = 0°$, develop expressions

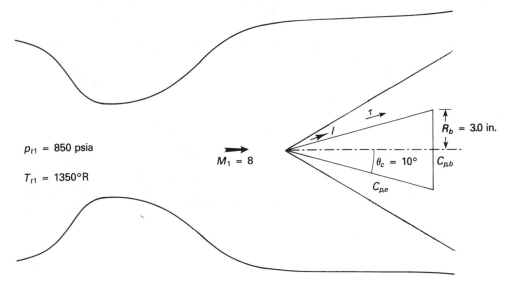

Figure 13.18 Sharp cone mounted in Tunnel B of AEDC.

for the forebody pressure coefficient $(C_{p,e})$, the base pressure coefficient $(C_{p,b})$ and the skin friction. What is the total drag acting on the vehicle?

Solution Let us use Fig. 8.16b and c to obtain values for the pressure coefficient and for the Mach number of the inviscid flow at the edge of the boundary layer for a 10°-half-angle cone in a Mach 8 stream. Inherent in this use of these figures is the assumption that the boundary layer is thin and that there are no significant viscous/invicid interactions which perturb the pressure field. Thus, $M_e = 6$ and $C_{p,e} = 0.07$. Since $q_1 = \frac{1}{2}\rho_1 U_1^2 = (\gamma/2)p_1 M_1^2$,

$$p_e = p_1(1 + \frac{\gamma}{2}M_1^2 C_{p,e})$$

We can use Table 8.1 to calculate the free-stream static pressure (p_1) in the Mach 8 flow of the wind-tunnel test section:

$$p_1 = \frac{p_1}{p_{t1}}p_{t1} = (102 \times 10^{-6})(850) = 0.0867 \text{ psi}$$

Thus,

$$p_e = 0.0867[1.0 + (0.7)(64)(0.07)] = 0.3586 \text{ psi}$$

Since $M_e = 6$, we can use Table 8.1 and the fact that the edge flow is that for an adiabatic flow of a perfect gas, so that $T_{t1} = T_{te}$. Thus,

$$T_e = \frac{T_e}{T_{te}}T_{te} = (0.12195)(1350) = 164.63° R$$

To aid in our decision as to whether the boundary layer is laminar or turbulent, let us calculate the Reynolds number for the flow at the boundary-layer edge.

$$\text{Re}_l = \frac{\rho_e U_e l}{\mu_e} = \left[\frac{p_e}{RT_e}\right][M_e\sqrt{\gamma RT_e}]\left[\frac{T_e + 198.6}{2.27 \times 10^{-8}T_e^{1.5}}\right]l$$

where, as shown in Fig. 13.18, l is the wetted distance along a conical generator. Thus,

$$\text{Re}_l = \left(0.0001828 \frac{\text{lbf} \cdot \text{s}^2}{\text{ft}^4}\right)\left(3773.80 \frac{\text{ft}}{\text{s}}\right)\left(7.575 \times 10^6 \frac{\text{ft}^2}{\text{lbf} \cdot \text{s}}\right)l$$

$$= 5.226 \times 10^6 l$$

Based on the wetted length of a conical ray $(l_c = 17.276 \text{ in})$,

$$\text{Re}_{l_c} = 7.523 \times 10^6$$

It should be noted that, although hypersonic boundary layers are very stable, boundary-layer transition should occur for this flow, probably within the first one-half of the cone length. Thus, we can use the correlation presented in Fig. 13.19 to determine the base pressure:

$$\frac{p_b}{p_e} = 0.02$$

so that $p_b = 0.00717$ psi.

Note that since $p_b = 0.02p_e$,

$$C_{p,b} = \frac{p_b - p_1}{q_1} = \left(\frac{p_b}{p_1} - 1\right)\frac{2}{\gamma M_1^2}$$

$$= \left(\frac{p_b}{p_e}\frac{p_e}{p_1} - 1\right)\frac{2}{\gamma M_1^2} = -0.0205$$

Recall that the assumption for Newtonian flow is $C_{p,b} = 0.0$.

To calculate the skin friction contribution, let us use Reynolds analogy (see Chapter 4). For supersonic flow past a sharp cone, the heat-transfer rate (or equivalently the Stanton number) can be calculated using the relatively simple approximation obtained through the Eckert reference temperature approach (Ref. 13.9). In this approach, the heat-transfer rates are calculated using the incompressible relations [e.g., equation (4.103)] with the temperature-related parameters evaluated at Eckert's reference temperature (T^*). According to Ref. 13.9,

$$T^* = 0.5(T_e + T_w) + 0.22r(T_{te} - T_e)$$

where r is the recovery factor. For turbulent flow, the recovery factor is $\sqrt[3]{Pr}$, which is equal to 0.888. Thus,

$$T^* = 0.5(164.63 + 600) + 0.22(0.888)(1350 - 164.63)$$
$$= 613.89° R$$

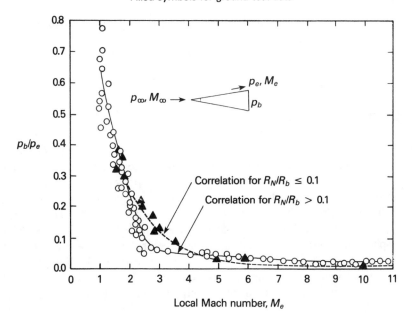

Figure 13.19 Base pressure ratio as a function of the local Mach number for turbulent flow. (From Ref. 13.8.)

Thus,

$$\text{Re}_l^* = \left[\frac{p_e}{RT^*}\right][U_e]\left[\frac{T^* + 198.6}{2.27 \times 10^{-8}T^{*1.5}}\right]l$$

$$= \left(0.00004901 \frac{\text{lbf} \cdot \text{s}^2}{\text{ft}^4}\right)\left(3773.80 \frac{\text{ft}}{\text{s}}\right)\left(2.353 \times 10^6 \frac{\text{ft}^2}{\text{lbf} \cdot \text{s}}\right)l$$

$$= 4.353 \times 10^5 l$$

Thus, using Reynolds' analogy, we have

$$C_f = \frac{0.0583}{(\text{Re}_l^*)^{0.2}} = \frac{0.004344}{l^{0.2}}$$

Thus,

$$\tau = C_f\left(\frac{\gamma}{2}p_eM_e^2\right) = \frac{0.03926}{l^{0.2}}\frac{\text{lbf}}{\text{in}^2}$$

To calculate the total drag,

$$D = \int p_e 2\pi r\, dl \sin\theta_c + \int \tau_w 2\,\pi r\, dl \cos\theta_c - p_b\pi R_b^2$$

Noting that p_e is constant along the length of the conical generator, $dl \sin\theta_c = dr$, and $r = l \sin\theta_c$,

$$D = p_e\pi R_b^2 + 0.03926(2\pi)(\cos\theta_c \sin\theta_c)\int_0^{l_c} l^{0.8}\, dl - p_b\pi R_b^2$$

$$= (0.3586)(9\pi) + (0.04218)\frac{l_c^{1.8}}{1.8} - (0.00717)(9\pi)$$

$$= 10.139 + 3.956 - 0.203 = 13.892\ \text{lbf}$$

AERODYNAMIC HEATING

As noted in the introduction to this chapter, the kinetic energy associated with hypersonic flight is converted to high temperatures within the shock layer. As a result, heat transfer is a very important factor in the design of hypersonic vehicles. Sharp leading edges would experience such extremely large heating rates in hypersonic flight they would quickly melt or ablate (depending on the material used). Therefore, if a low-drag configuration is desired, the design would probably involve a slender cone with a spherically blunted nose. For a manned reentry craft, where the time of flight can be long and the dissipation of kinetic energy at relatively high altitude is desirable, the resultant high-drag configurations may be an airplane-like Space Shuttle that flies at high angles of attack or a blunt, spherical segment such as the Apollo Command Module.

The expression for the modified-Newtonian-flow pressure coefficient [equation (13.17)] can be rearranged to give

$$\frac{p_s}{p_{t2}} = \sin^2\theta_b + \frac{p_\infty}{p_{t2}}\cos^2\theta_b = \cos^2\phi + \frac{p_\infty}{p_{t2}}\sin^2\phi \qquad (13.55)$$

Let us apply equation (13.55) to the relatively simple flow depicted in Fig. 13.20. As the boundary layer grows in the streamwise direction, air is entrained from the inviscid portion of the shock layer. Thus, when determining the fluid properties at the edge of the boundary layer, one must determine the entropy of the streamline at the boundary-layer edge. Note that the streamlines near the axis of symmetry of a blunt-body flow, such as those depicted in Fig. 13.20, have passed through that portion of the bow shock wave which is nearly perpendicular to the free-stream flow. As a result, it can be assumed that all the air particles at the edge of the boundary layer have essentially the same entropy. Therefore, the entropy at the edge of the boundary layer and, as a result, p_{t2} are the same at all streamwise stations. The local flow properties are the same as if the air had passed through a normal shock wave and had undergone an isentropic expansion to the local pressure (designated an NS/IE process). For such an isentropic expansion (IE), the ratio of p_s/p_{t2} can then be used to define the remaining flow conditions (for an equilibrium flow). Note that, if the flow expands isentropically to a point where the local static pressure (p_s) is approximately $0.5p_{t2}$, the flow is essentially sonic for all values of γ. Solving equation (13.55), we find that $p_s \approx 0.5p_{t2}$ when $\theta_b = 45°$ (i.e., the sonic points occur when the local body slope is 45°).

The modified Newtonian pressures for the zero-angle-of-attack Apollo Command Module are compared in Fig. 13.21 with data obtained in Tunnel C at AEDC (Ref. 13.10). The experimental pressures measured in Tunnel C at a nominal free-stream Mach number of 10 and at a Reynolds number ($Re_{\infty,D}$) of 1.1×10^6 have been divided by the calculated value of the stagnation pressure behind a normal shock wave (p_{t2}). For reference, a sketch of the Apollo Command Module is presented in Fig. 13.22. Note that an S/R_b ratio of 0.965 defines the tangency point of the spherical heat shield

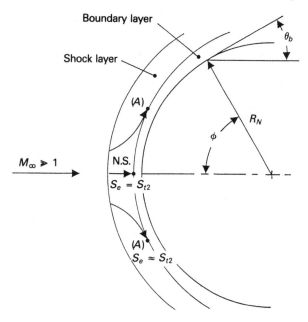

Figure 13.20 Sketch of the normal shock/isentropic expansion flow model for hypersonic flow over a blunt body. The air at the edge of the boundary layer at point A has passed through the (nearly) normal part of the shock wave. Thus, the entropy of the air particles at the edge of the boundary layer (S_e) is essentially equal to S_{t2} at all stations. In essence, the air has passed through the normal part of the shock wave (NS) and has undergone an isentropic expansion (IE) to the local pressures.

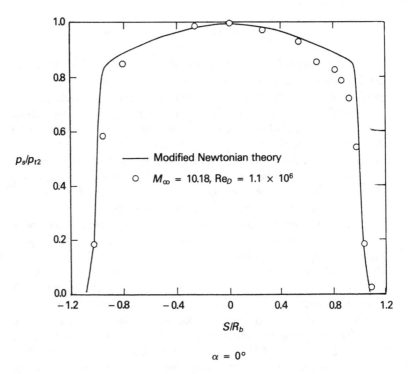

Figure 13.21 Comparison of the modified Newtonian pressures and the experimental pressures for the Apollo Command Module at $\alpha = 0°$. (From Ref. 13.10.)

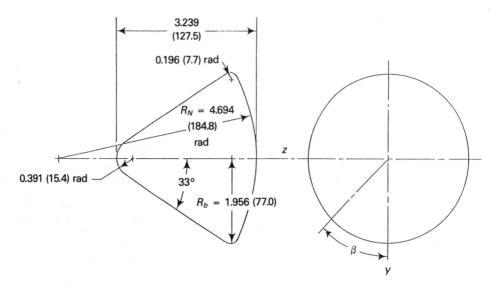

Figure 13.22 Clean (no protuberances) Apollo Command Module. Dimensions in meters (inches).

and the toroidal surface, while an S/R_b ratio of 1.082 corresponds to the maximum body radius. Because the windward heat shield of the Apollo Command Module is a truncated spherical cap, the actual sonic points, which occur near the tangency point of the spherical heat shield and the toroidal surface, are inboard of the locations that they would occupy for a full spherical cap. As a result, the entire flow field in the subsonic portion of the shock layer is modified, and the streamwise velocity gradients are relatively large in order to produce sonic flow at the "corners" of the Command Module. Thus, significant differences exist between the modified Newtonian pressures and the measured values as one approaches the edge of the spherical heat shield. Because the velocity gradient at the stagnation point of a hemispherical segment is increased above the value for a full hemisphere, the stagnation point heating rate will also be increased. Investigations of the stagnation region velocity gradients as a function of R_b/R_N have been reported by Stoney (Ref. 13.11) and by Inouye et al. (Ref. 13.12).

Similarity Solutions for Heat Transfer

The magnitude of the heat transfer from a compressible boundary layer composed of dissociating gases can be approximated by

$$|\dot{q}_w| = \left(k \frac{\partial T}{\partial y} \right)_w + \left(\rho \Sigma D_i h_i \frac{\partial C_i}{\partial y} \right)_w \tag{13.56}$$

where the heat is transported by conduction and by diffusion (Ref. 13.13). Assume that the flow is such that the Lewis number is approximately equal to unity, the hot gas layer is in chemical equilibrium, and the surface temperature is much less than the external stream temperature. For these assumptions, the magnitude of the heat transferred to the wall is

$$|\dot{q}_w| = \left(\frac{k}{c_p} \frac{\partial h}{\partial y} \right)_w \tag{13.57}$$

There are many situations where a coordinate transformation can be used to reduce the governing partial differential equations for a laminar boundary layer to ordinary differential equations (Ref. 13.13). Fay and Riddell (Ref. 13.14) note: "As is usual in boundary-layer problems, one first seeks solutions of restricted form which permit reducing exactly the partial differential equations to ordinary differential form. An easily recognizable case is that of the stagnation point flow, where, because of symmetry, all the dependent variables are chosen to be functions of y alone, except u which must be taken proportional to x times a function of y. This also appears to be the *only case* for which the exact ordinary differential equations may be obtained regardless of the recombination rate."

Using the similarity transformations suggested by Fay and Riddell, we obtain

$$\eta(x_1, y_1) = \frac{\rho_w u_e r^k}{\sqrt{2S}} \int_0^{y_1} \frac{\rho}{\rho_w} \, dy \tag{13.58}$$

and

$$S(x_1) = \int_0^{x_1} \rho_w \mu_w u_e r^{2k} \, dx \tag{13.59}$$

where r is the cross-section radius of the body of revolution and k denotes whether the flow is axisymmetric ($k = 1$), for example, a sphere, or two-dimensional ($k = 0$), for example, a cylinder whose axis is perpendicular to the free stream. Limiting approximations are used to describe the flow in the vicinity of the stagnation point, for example,

$$u_e \approx \left(\frac{du_e}{dx} \right)_{t2} x; \qquad r \approx x; \qquad \text{and } \rho_w = \text{constant}$$

Thus, at the stagnation point,

$$S \approx \rho_w \mu_w \left(\frac{du_e}{dx} \right)_{t2} \int x x^{2k} \, dx = \rho_w \mu_w \left(\frac{du_e}{dx} \right)_{t2} \frac{x^{2(k+1)}}{2(k+1)} \tag{13.60}$$

and

$$\eta \approx \frac{(du_e/dx)_{t2} \, x x^k}{\sqrt{2S}} \int \rho \, dy = \left[\frac{(du_e/dx)_{t2} \, (k+1)}{\rho_w \mu_w} \right]^{0.5} \int \rho \, dy \tag{13.61}$$

Note that using equation (13.61) as the expression for the transformed y- coordinate, together with the definition for the heat transfer as given by equation (13.57), we find that, given the same flow condition,

$$(\dot{q}_{t,\text{ref}})_{\text{axisym}} = \sqrt{2} \, (\dot{q}_{t,\text{ref}})_{2-\text{dim}} \tag{13.62}$$

(i.e., the stagnation-point heat-transfer rate for a sphere is the $\sqrt{2}$ times that for a cylinder).

The stagnation-point heat-transfer rate for a laminar boundary layer of a spherical cap may be written as

$$\dot{q}_{t,\text{ref}} = \frac{\text{Nu}_x}{\sqrt{\text{Re}_x}} \sqrt{\rho_{w,t} \mu_{w,t} \left(\frac{du_e}{dx} \right)_{t2}} \frac{H_{t2} - h_{w,t}}{Pr} \tag{13.63}$$

For a Lewis number of 1, Fay and Riddell (Ref. 13.14) found that

$$\frac{\text{Nu}_x}{\sqrt{\text{Re}_x}} = 0.67 \left(\frac{\rho_{t2} \mu_{t2}}{\rho_{w,t} \mu_{w,t}} \right)^{0.4} \tag{13.64}$$

for velocities between 5800 ft/s (1768 m/s) and 22,800 ft/s (6959 m/s) and at altitudes between 25,000 ft (7620 m) and 120,000 ft (36,576 m).

Let us use Euler's equation to evaluate the velocity gradient at the stagnation point:

$$\left(\frac{dp_e}{dx}\right)_{t2} = -\rho_e u_e \left(\frac{du_e}{dx}\right)_{t2} \approx -\rho_{t2} \left(\frac{du_e}{dx}\right)_{t2}^2 x \quad (13.65)$$

Using the modified Newtonian flow pressure distribution [i.e., equation (13.55)] to evaluate the pressure gradient at the stagnation point,

$$\left(\frac{dp_e}{dx}\right)_{t2} = -2p_{t2} \cos\phi \sin\phi \frac{d\phi}{dx} + 2p_\infty \cos\phi \sin\phi \frac{d\phi}{dx} \quad (13.66)$$

Since $\sin\phi \approx \phi = x/R_N$, $\cos\phi \approx 1$, and $d\phi/dx = 1/R_N$, we find that

$$\left(\frac{du_e}{dx}\right)_{t2} = \frac{1}{R_N} \sqrt{\frac{2(p_{t2} - p_\infty)}{\rho_{t2}}} \quad (13.67)$$

Combining the expressions given by equations (13.63) through (13.67), we obtain the equilibrium correlation of Fay and Riddell (Ref. 13.14):

$$\dot{q}_{t,\text{ref}} = \frac{0.67}{\text{Pr}_{t2}} (\rho_{t2}\,\mu_{t2})^{0.4}\,(\rho_{w,t}\,\mu_{w,t})^{0.1}\,(H_{t2} - h_{w,t}) \left[\frac{1}{R_N}\left(\frac{2(p_{t2} - p_\infty)}{\rho_{t2}}\right)^{0.5}\right]^{0.5} \quad (13.68)$$

This relation can be used either for wind-tunnel flows or for flight conditions. Detra et al. (Ref. 13.15) have developed a correlation:

$$\dot{q}_{t,\text{ref}} = \frac{17,600}{(R_N)^{0.5}} \left(\frac{\rho_\infty}{\rho_{\text{S.L.}}}\right)^{0.5} \left(\frac{U_\infty}{U_{\text{c.o.}}}\right)^{3.15} \quad (13.69)$$

In equation (13.69), R_N is the nose radius in feet, $\rho_{\text{S.L.}}$ is the sea-level density, $U_{\text{c.o.}}$ is the velocity for a circular orbit, and $\dot{q}_{t,\text{ref}}$ is the stagnation point heat-transfer rate in Btu/ft$^2 \cdot$ s.

The heat-transfer rate at the stagnation point of a 1-ft (0.3048-m) sphere flying at 150,000 ft (45,721 m) has been calculated using equation (13.68) for both the perfect-air model and the equilibrium-air model and using equation (13.69). The computed values are presented in Fig. 13.23 as a function of M_∞. Above $M_\infty = 14$, the perfect-gas model no longer provides realistic values of $\dot{q}_{t,\text{ref}}$.

ADVANCED COMPUTATIONAL TECHNIQUES

Solutions of suitable accuracy can be obtained for the windward flow field using two-layer models, for example, dividing the flow field into an inviscid rotational flow between the shock wave and the edge of the boundary layer and a thin viscous, boundary-layer adjacent to the wall. This is especially true if the numerical model for inviscid flow incorporates the Euler equations (e.g., Ref. 13.1). Furthermore, the flow fields can be computed in a relatively short time. However, solutions obtained using the two-layer approach may require considerable interaction by the user, since it may be necessary to process externally the output from the inviscid solution and prepare it as input for the viscous formulation. In addition, in regions where there are strong viscous/

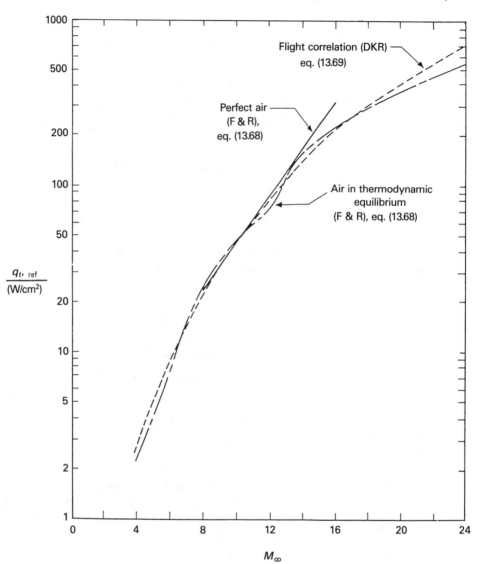

Figure 13.23 Heat-transfer rate at the stagnation point of a sphere ($R_N = 0.3048$ m) at an altitude of 45,721 m.

inviscid interactions, such as occur near the leeward plane of symmetry for vehicles at relatively high angles of attack, the solutions for the two regions should be coupled. Because of such difficulties, it is desirable to develop solution techniques that treat the entire shock layer in a unified fashion.

Davis (Ref. 13.16) was one of the first investigators to use the Navier–Stokes equations to describe the entire viscous shock layer. The compressible Navier–Stokes equations were written in boundary-layer coordinates. Introducing the thin-shock-layer approximation to the normal momentum equation made the equations totally parabolic.

An iterative procedure was then used to reduce possible errors caused by this approximation.

Improved techniques for generating solutions to the viscous shock layer were developed by Lewis and his co-workers. A viscous shock layer code developed by Murray and Lewis (Ref. 13.17) neglects the second-order terms in the normal momentum equation and parabolizes the cross-flow momentum equation. As a result, when cross-flow separation occurred, the solution stopped marching around the body but continued marching downstream. Because of this limitation, solutions of the leeside flow field could not be obtained if cross-flow separation occurred. However, comparisons with experimental data indicated that the viscous shock layer technique could be used to obtain good predictions of the complete flow field for sphere/cones at relatively low angles of attack.

Using the nomenclature established in Appendix A, the general form of the equations of motion in conservation form is

$$\frac{\partial U}{\partial t} + \frac{\partial (E_i - E_v)}{\partial x} + \frac{\partial (F_i - F_v)}{\partial y} + \frac{\partial (G_i - G_v)}{\partial z} = 0 \qquad (13.70)$$

where the subscript i denotes the terms that are included in the equations of motion for an inviscid flow and the subscript v denotes the terms that are unique to the equations of motion when the viscous and heat-transfer effects are included.

Thus, as noted in Appendix A, when the fundamental equations governing the unsteady flow of a gas, without body forces or external heat addition, are written in conservation form, the terms U, E_i, E_v, F_i, F_v, G_i, and G_v can be written by the following vectors:

$$U = \begin{bmatrix} \rho \\ \rho u \\ \rho v \\ \rho w \\ \rho e_t \end{bmatrix}$$

$$E_i = \begin{bmatrix} \rho u \\ p + \rho u^2 \\ \rho u v \\ \rho u w \\ (\rho e_t + p)u \end{bmatrix} \qquad E_v = \begin{bmatrix} 0 \\ \tau'_{xx} \\ \tau_{xy} \\ \tau_{xz} \\ u\tau'_{xx} + v\tau_{xy} + w\tau_{xz} + q_x \end{bmatrix}$$

$$F_i = \begin{bmatrix} \rho v \\ \rho v u \\ p + \rho v^2 \\ \rho v w \\ (\rho e_t + p)v \end{bmatrix} \qquad F_v = \begin{bmatrix} 0 \\ \tau_{yx} \\ \tau'_{yy} \\ \tau_{yz} \\ u\tau_{yx} + v\tau'_{yy} + w\tau_{yz} + q_y \end{bmatrix}$$

$$G_i = \begin{bmatrix} \rho w \\ \rho w u \\ \rho w v \\ p + \rho w^2 \\ (\rho e_t + p)w \end{bmatrix} \qquad G_v = \begin{bmatrix} 0 \\ \tau_{zx} \\ \tau_{zy} \\ \tau'_{zz} \\ u\tau_{zx} + v\tau_{zy} + w\tau'_{zz} + q_z \end{bmatrix}$$

where

$$e_t = u_e + \tfrac{1}{2}(u^2 + v^2 + w^2)$$

The equations are then transformed from physical coordinates (e.g., Cartesian coordinates) into a computational space (e.g., a ξ, η, ζ coordinate system). As noted by Harris (Ref. 13.18): "The purpose of the transformation is to map the physical space into a unit computational plane where the body surface corresponds to $\zeta = 0$, the outer boundary (bow shock) corresponds to $\zeta = 1$, and the leeward and windward symmetry planes map onto $\eta = 0$ and 1, respectively. In addition, it is desirable to incorporate the necessary coordinate stretching within the transformation so that a very fine mesh near the wall in physical space—used to model the large viscous gradients— maps into a uniform grid in the transformed plane, thus simplifying the finite difference expressions."

Tannehill et al. (Ref. 13.19) note that:"At each ξ station, the computational domain [i.e., the (η, ζ) solution surface] is bounded by the bow shock and the body surface. For symmetrical flows, only one half of the computational domain needs to be considered since the other half is known from symmetry. The orientation of each (η, ζ) solution surface is purposely left arbitrary so that the optimum orientation can be selected for a given problem. In general, the optimum orientation occurs when (η, ζ) solution surface is perpendicular to the local flow direction. The location of the grid points in each (η, ζ) solution surface is also left arbitrary so that the optimum distribution can be selected for a given problem." It might be noted that, for Tannehill et al., the wall is defined as the $\eta = 1$ surface. Thus, there are a variety of coordinate transformations that can be used. The specific transformation employed is usually tailored to the application of interest.

The so-called "parabolized" Navier–Stokes (PNS) equations are obtained from the complete Navier–Stokes equations by neglecting the unsteady terms and the streamwise viscous derivative terms. (This assumption is usually valid for large Reynolds number flows over bodies which do not experience severe geometric variations which cause separation in the streamwise direction to occur.) By assuming that the viscous, streamwise derivative terms are small compared with the viscous, normal and circumferential derivatives, a tremendous reduction in computing time and in storage requirements is possible over that required for the time-dependent approaches. Since the equations are parabolic in the streamwise direction, a marching-type numerical-solution technique can be used.

As noted by Prabhu and Tannehill (Ref. 13.20): "The PNS equations are hyperbolic-parabolic in the ξ direction if the inviscid region of the flowfield is completely supersonic, if there is no streamwise flow separation within the domain of interest, and if the streamwise pressure gradient, $dp/d\xi$, in the subsonic portion of the boundary layer is treated in such a manner as to suppress departure or exponentially growing solutions."

As noted by Rakich et al. (Ref. 13.21), the pressure gradient term in the streamwise momentum equation is a main factor that differentiates the parabolic approximation from boundary-layer approximations. If $dp/d\xi$ is retained, an upstream influence is allowed through the subsonic regions of the flow, and a marching method of solution is not well posed. Thus, the numerical treatment of the streamwise pressure gradient becomes an important aspect of the computational flow model.

A PNS code requires a starting solution on an initial data surface where the inviscid flow is supersonic. The presence of an embedded subsonic region over the blunted nose requires the use of an axisymmetric or three-dimensional time-dependent Navier–Stokes code to provide a "starter solution" (i.e., an initial "plane" of data). At the outer boundary, the Rankine–Hugoniot relations are enforced to solve for the shock wave that separates the shock-layer flow from the undisturbed free-stream. Since the flow properties at the node immediately downstream of the shock wave are often computed using an explicit formulation whereas the properties at the other internal points are computed using implicit techniques, there may be significant step-size considerations implicit in the numerical solution procedure. In those cases where the shock-wave boundary conditions for the system of equations are evaluated using an explicit formulation, stability restrictions may be introduced such that the unrestricted step size in the marching direction associated with the "unconditional stability of implicit techniques" is no longer valid. Such is the case in the PNS formulation, when the bow shock wave is fitted using an explicit technique. The stability condition takes the form of a modified CFL number given as

$$\frac{\lambda_{max} \Delta x}{\Delta y} \le 1$$

where Δx is the streamwise marching step and Δy corresponds to the mesh size in the radial direction. The eigenvalue (λ_{max}) is determined from the matrix structure of the numerical algorithm as applied at the shock boundary. If the stability condition given above is violated when marching the PNS equations, oscillations in the shock boundary will occur, leading to inaccurate predictions of the shock wave shape and/or instabilities in the shock-fitting algorithm.

Dwoyer (Ref. 13.22) notes that the hypersonic airplane concept places a somewhat different set of demands on CFD than does a "projectile/glider-like" reentry vehicle. The CFD demands for a hypersonic, air-breathing vehicle stem largely from several dominating issues:

1. The requirement to fully integrate the airframe and propulsion system to achieve high-Mach-number performance
2. The requirement that the vehicle be reuseable with a minimum of refurbishment between flights
3. The requirement to optimize vehicle performance over a wide Mach number range

These considerations require an accurate prediction of the entire shock-layer profile in addition to the prediction of wall-gradient parameters, such as skin friction and heat transfer. It is necessary to predict the shape of the embedded shock waves as well as the bow shock wave. Furthermore, the integration of the propulsion system into the computational scheme requires modeling of a complex inlet/forebody flow-field interaction, the multiple internal/external nozzle system flow and its interaction with the vehicle wing/body flow, and so on. As Dwoyer notes, "Successful solutions of these problems will require an unprecedented level of cooperation between CFD researchers, experimentalists, and theoreticians for many years to come."

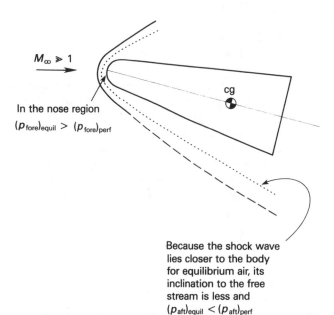

Figure 13.24 Hypersonic flow past an inclined spherically blunted cone comparing perfect-gas and equilibrium-air pressures.

There is a rapidly growing body of literature regarding solutions of the flow fields over hypersonic vehicles. The interested reader should consult the current literature for recent developments in computing hypersonic flow fields.

IMPORTANCE OF INTERRELATING CFD, GROUND-TEST DATA, AND FLIGHT-TEST DATA

Woods et al. (Ref. 13.23) note that preflight predictions based on the aerodynamics in the Aerodynamics Design Data Book (ADDB) indicated that a 7.5° deflection of the body flap would be required to trim the Space Shuttle Orbiter for the center of gravity and for the vehicle configuration of STS-1. In reality, the body flap had to deflect to much larger values ($\delta_{BF} \approx 16°$) to maintain trim at the proper angle of attack ($\alpha = 40°$). Comparisons of equilibrium-air calculations and perfect-gas calculations indicate that at least part of this so-called "hypersonic anomaly" is due to real-gas effects at very high Mach numbers. At Mach 8, the flight data and the ADDB values agreed (Ref. 13.23).

Consider the flow depicted in the sketch of Fig. 13.24. For perfect air ($\gamma = 1.4$), $\rho_2 = 6\rho_1$ across the normal portion of the shock wave, whereas $\rho_2 = 15\rho_1$ for air is thermodynamic equilibrium ($\gamma = 1.14$). Thus, for the equilibrium-air model the shock layer is thinner and the inclination of the bow shock wave relative to the free stream is less than that for the perfect-air model. Tangent-cone theory applied to the

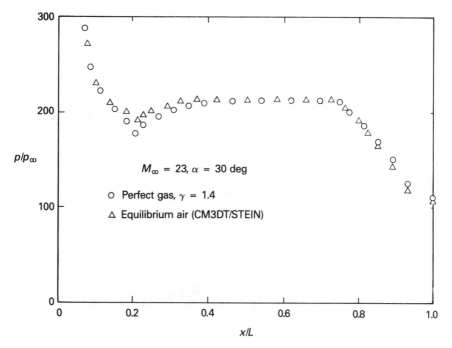

Figure 13.25 Comparison of perfect-gas and equilibrium-air calculations of the windward pitch-plane pressure distribution for the Space Shuttle Orbiter at $\alpha = 30°$. (From Ref. 13.24.)

afterbody region shows a decrease in pressure with decreasing gamma. Using this simplistic flow model, one would expect the equilibrium-air pressures on the aft end of the vehicle to be less than those for perfect air. Computations of the flow field over simplified Orbiter geometries as reported by Woods et al. (Ref. 13.23) and by Maus et al. (Ref. 13.24) indicate that this is the case. The calculations of Maus et al. are reproduced in Fig. 13.25.

Maus et al. (Ref. 13.24) note further that the stagnation pressure increases with decreasing gamma. Thus, as presented in Fig. 13.7, the equilibrium-air value for the stagnation pressure is greater than that for perfect air. This, too, is reflected in the nose region pressures presented in the more rigorous solutions of Woods et al. and Maus et al.

The differences between the equilibrium-air pressure distribution and the perfect-air pressure distribution may appear to be relatively small. Indeed, there is little difference in the normal force coefficients for the equilibrium-air model and for the perfect-air model. However, because the equilibrium-air (real-gas) values are higher at the nose and lower at the tail, the real-gas effects tend to drive C_M more positive. The pitching moments for the Space Shuttle Orbiter at $M_\infty = 23$ presented by Maus et al. (Ref. 13.24) are reproduced in Fig. 13.26. Thus, detailed studies incorporating wind-tunnel data, flight-test data, and CFD solutions, as reported in Refs. 13.23 and 13.24, provide insights into understanding a sophisticated aerodynamic problem.

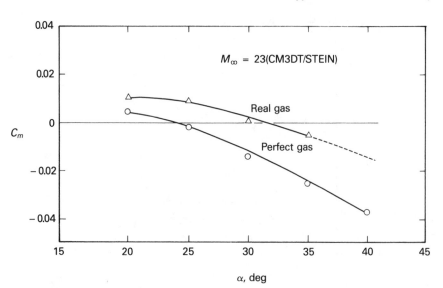

Figure 13.26 Comparison of perfect-gas and equilibrium-air calculations of the pitching moment for Space Shuttle Orbiter at $M_\infty = 23$. (From Ref. 13.24.)

PROBLEMS

13.1. Consider a hypersonic vehicle flying through the earth's atmosphere at 10,000 ft/s at an altitude of 200,000 ft. At 200,000 ft, $p_\infty = 0.4715$ lbf/ft^2, $\rho_\infty = 6.119 \times 10^{-7}$ slug/ft^3, $T_\infty = 449°$R, and $h_\infty = 2.703 \times 10^6$ ft · lbf/slug. Using the thermodynamic properties of equilibrium air presented in Fig. 13.5, what are the static pressure (p_2), the static temperature (T_2), the enthalpy (h_2), and the static density (ρ_2) downstream of a normal shock wave? For comparison, use the normal-shock relations for perfect air, as given in Chapter 8, to calculate p_2, T_2, h_2, and ρ_2.

13.2. Consider a hypersonic vehicle reentering the earth's atmosphere at 20,000 ft/s at an altitude of 200,000 ft. Repeat Problem 13.1., that is, calculate p_2, ρ_2, T_2, and h_2 assuming that the air **(a)** is in thermodynamic equilibrium and **(b)** behaves as a perfect gas.

13.3. Consider a hypersonic vehicle reentering the earth's atmosphere at 26,400 ft/s at an altitude of 200,000 ft. Repeat Problem 13.1., that is, calculate p_2, ρ_2, T_2, and h_2 assuming that the air **(a)** is in thermodynamic equilibrium and **(b)** behaves as a perfect gas.

13.4. If the free-stream Mach number is 6, calculate the pressure coefficient on the surface of a wedge, whose deflection angle is 30°, using the correlations of Fig. 8.13b and using Newtonian theory. How does the altitude influence the resultant values of C_p?

13.5. If the free-stream Mach number is 6, calculate the pressure coefficient on the surface of a sharp cone, whose deflection angle is 30°, using the correlations of Fig. 8.16b and using Newtonian theory.

13.6. In equation (13.22), it was stated without proof that the section drag coefficient (per unit span) for modified Newtonian pressures acting on a right circular cylinder is: $C_{d,p} = \frac{2}{3}C_{p,t2}$. Prove that this equation is correct.

13.7. Consider the sharp cone in a hypersonic stream, which was the subject of Example

13.2. If $M_\infty = 10$ and $\theta_c = 10°$, prepare a graph of C_D as a function of α for $-10° \leq \alpha \leq 10°$.

13.8. Consider the sharp cone in a hypersonic stream, which was the subject of Example 13.2. If $M_\infty = 10$ and $\theta_c = 10°$, prepare a graph of C_L as a function of α for $-10° \leq \alpha \leq 10°$.

13.9. Consider the sharp cone in a hypersonic stream, which was the subject of Example 13.2. If $M_\infty = 10$ and $\theta_c = 10°$, prepare a graph of L/D as a function of α for $-10° \leq \alpha \leq 10°$.

13.10. Consider the sharp cone in a hypersonic stream, which was the subject of Example 13.2. If $M_\infty = 10$ and $\theta_c = 10°$, prepare a graph of C_M as a function of α for $-10° \leq \alpha \leq 10°$. Is the configuration statically stable if $x_{cg} = 0.6x_L$?

13.11. A configuration that would generate lift at low angles of attack is a half-cone model, such as shown in Fig. P13.11. Neglecting the effects of skin friction and assuming the pressure is that of the Newtonian flow model, develop and expression for the lift-to-drag ratio as a function of θ_c for $0 \leq \alpha \leq \theta_c$.

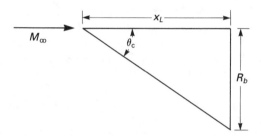

Figure P13.11

13.12. Develop expressions for y_{cp} and x_{cp} for the flow described in Problem 13.11.

13.13. Using the Newtonian flow model to describe the pressure field and neglecting the effects of viscosity, what is the drag coefficient on the $\theta_c = 10°$ sharp cone of Example 13.3? What would the drag coefficient be if you used the modified Newtonian flow model?

13.14. Following the procedure developed in Example 13.3, develop expressions for the forebody pressure coefficient ($C_{p,e}$), the base pressure coefficient ($C_{p,b}$) and the skin friction, if $\theta_c = 20°$ for the sharp cone exposed to the Mach 8 flow of Tunnel B.

13.15 Let us model the flow on the windward surface of a reentry vehicle and its body flap by the double wedge configuration shown in Fig. P13.15. The lower surface of the fuselage is inclined $10°(\alpha = 10°)$ to the free-stream flow producing the supersonic flow in region 2. A body flap is deflected $10°$(so that $\alpha + \delta_{BF} = 20°$).

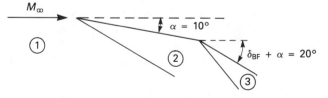

Figure P13.15

(a) Calculate the pressure coefficient for regions 2 and 3 using the Newtonian flow model for $M_1 = 2, 4, 6, 8, 10$, and 20.

(b) Using the correlations of Fig. 8.14, calculate the pressure coefficients for regions 2 and 3 for $M_1 = 2, 4, 6, 8, 10$, and 20. The flow remains supersonic throughout the flow field for these deflection angles. Note that to go from region 2 to 3 using the charts of Fig. 8.14, treat region 2 as an equivalent free-stream flow ("1''); then region 3 corresponds to region 2. Note, however, that $C_{p3} = (p_3 - p_1)/q_1$ that is, the reference conditions are those of the free stream.

REFERENCES

13.1. Marconi, F., M. Salas, and L. Yeager, "Development of a Computer Code for Calculating the Steady Super/Hypersonic Inviscid Flow Around Real Configurations, Vol. I: Computational Technique," *CR-2675*, NASA, Apr. 1976.

13.2. *U.S. Standard Atmosphere, 1962*, Government Printing Office, Washington, D.C., December 1962.

13.3. Moeckel, W. E., and K. C. Weston, "Composition and Thermodynamic Properties of Air in Chemical Equilibrium," *TN 4265*, NACA, Aug. 1958.

13.4. Hansen, C. F., "Approximations for the Thermodynamic and Transport Properties of High-Temperature Air," *Technical Report R-50*, NACA, Nov. 1957.

13.5. Lees, L., "Hypersonic Flow," *Proceedings of the 5th International Aeronautical Conference*, Los Angeles, Institute of the Aeronautical Sciences, 1955, pp. 241–275.

13.6. Isaacson, L. K., and J. W. Jones, "Prediction Techniques for Pressure and Heat-Transfer Distributions over Bodies of Revolution in High Subsonic to Low-Supersonic Flight," *NWC TP 4570*, Nov. 1968, Naval Weapons Center.

13.7. Koppenwallner, G., "Experimentelle Untersuchung der Druckverteilung und des Widerstands von querangestroemten Kreiszylindern bei hypersonischen Machzahlen in Bereich von Kontinuums-bis freier Molekularstroemung," *Zeitschrift fuer Flugwissenschaften*, Oct. 1969, Vol. 17, No. 10, pp. 321–332.

13.8. Cassanto, J. M., "A Base Pressure Experiment for Determining the Atmospheric Profile of Planets," *Journal of Spacecraft and Rockets*, Apr. 1973, Vol. 10, No. 4, pp. 253–261.

13.9. Eckert, E. R. G., "Engineering Relations for Friction and Heat Transfer to Surfaces in High Velocity Flow," *Journal of the Aeronautical Sciences*, Aug. 1955, Vol. 22, No. 8, pp. 585–587.

13.10. Bertin, J. J., "The Effect of Protuberences, Cavities, and Angle of Attack on the Wind-Tunnel Pressure and Heat-Transfer Distribution for the Apollo Command Module," *TMX-1243*, NASA, Sept. 1966.

13.11. Stoney, W. E., Jr., "Aerodynamic Heating of Blunt-Nose Shapes at Mach Numbers up to 14," *RML 58E05a*, NACA, 1958.

13.12. Inouye, M., J. G. Marvin, and A. R. Sinclair, "Comparison of Experimental and Theoretical Shock Shapes and Pressure Distributions on Flat-Faced Cylinders at Mach 10.5," *TN D-4397*, NASA, Feb. 1968.

13.13. Dorrance, W. H., *Viscous Hypersonic Flow*, McGraw-Hill Book Company, New York, 1962.

13.14. Fay, J. A., and F. R. Riddell, "Theory of Stagnation Point Heat Transfer in Dissociated Air," *Journal of the Aeronautical Sciences*, Feb. 1958, Vol. 25, No. 2, pp. 73–85, 121.

13.15. Detra, R. W., N. H. Kemp, and F. R. Riddell, "Addendum to Heat Transfer to Satellite Vehicles Reentering the Atmosphere," *Jet Propulsion*, Dec. 1957, Vol. 27, No. 12, pp. 1256–1257.

13.16. Davis, R. T., "Numerical Solution of the Hypersonic Viscous Shock-Layer Equations," *AIAA Journal*, May 1970, Vol. 8, No. 5, pp. 843–851.

13.17. Murray, A. L., and C. H. Lewis, "Viscous-Shock-Layer Flows over Bodies During Reentry," *AIAA Paper 78-259*, presented at the AIAA 16th Aerospace Sciences Meeting, Huntsville, Ala., Jan. 1978.

13.18. Harris, T. B., "An Efficient Method for Supersonic Viscous Flow Field Calculations," *AIAA Paper 83-0222*, presented at the AIAA 21st Aerospace Sciences Meeting, Reno, Nev., Jan. 1983.

13.19. Tannehill, J. C., E. Venkatapathy, and J. V. Rakich, "Numerical Solution of Supersonic Viscous Flow over Blunt Delta Wings," *AIAA Journal*, Feb. 1982, Volume 20, No. 2, pp. 203–210.

13.20. Prabhu, D. K., and J. C. Tannehill, "Numerical Solutions of Space Shuttle Orbiter Flowfield Including Real-Gas Effects," *Journal of Spacecraft and Rockets*, May–June 1986, Vol. 23, No. 3, pp. 264–272.

13.21. Rakich, J. V., Y. C. Vigneron, and R. Agarwal, "Computation of Supersonic Viscous Flows over Ogive-Cylinders at Angle of Attack," *AIAA Paper 79-0131*, presented at the AIAA 17th Aerospace Sciences Meeting, New Orleans, La., Jan. 1979.

13.22. Dwoyer, D. L., "CFD for Hypersonic, Airbreathing Airplanes," *CP 87-209*, notes from a Short Course in Hypersonics, The University of Texas at Austin, Oct. 1987.

13.23. Woods, W. C., J. P. Arrington, and H. H. Hamilton II, "A Review of Preflight Estimates of Real-Gas Effects on Space Shuttle Aerodynamic Characteristics," *Shuttle Performance*: *Lessons Learned*, CP 2283, Part I, NASA, Mar. 1983.

13.24. Maus, J. R., B. J. Griffith, K. Y. Szema, and J. T. Best, "Hypersonic Mach Number and Real Gas Effects on Space Shuttle Orbiter Aerodynamics," *Journal of Spacecraft and Rockets*, Mar.–Apr. 1984, Vol. 21, No. 2, pp. 136–141.

14

AERODYNAMIC
DESIGN CONSIDERATIONS

In the previous chapters we have discussed techniques for obtaining flow-field solutions when the free-stream Mach number is either low subsonic, high subsonic, transonic, supersonic, or hypersonic. Many airplanes must perform satisfactorily over a wide speed range. Therefore, the thin, low-aspect ratio wings designed to minimize drag during supersonic cruise must deliver sufficient lift at low speeds to avoid unacceptably high landing speeds and/or landing field length. When these moderate aspect ratio, thin, swept wings operate at high angles of attack during high subsonic Mach number maneuvers, their performance is significantly degraded because of shock-induced boundary-layer separation and, at high angles of attack, because of leading edge separation and wing stall. Furthermore, because of possible fuel shortages and sharp fuel price increases, the wings of a high-speed transport may be optimized for minimum fuel consumption instead of for maximum productivity. In this chapter we consider design parameters that improve the aircraft's performance over a wide range of speed.

HIGH-LIFT CONFIGURATIONS

Consider the case where the aerodynamic lifting forces acting on an airplane are equal to its weight:

$$W = L = \tfrac{1}{2}\rho_\infty U_\infty^2 S C_L \tag{14.1}$$

To support the weight of the airplane at relatively low speeds, we could either increase the surface area over which the lift forces act or increase the lift coefficient of the lifting surface.

Increasing the Area

During the early years of aviation, the relatively crude state of the art in structural analysis limited the surface area one could obtain with a single wing. Thus, as discussed in Ref. 14.1: "In the attempt to increase the wing area in order to obtain the greatest lift out of an aerofoil it was found that there was a point beyond which it was not advantageous to proceed. This stage was reached when the extra weight of construction involved in an increase in wing area was just sufficient to counterbalance the increase in lift. The method of using aerofoils in biplanes is desirable in the first palce from the fact that, with a smaller loss in the necessary weight of construction, extra wing area may thus be obtained." Thus, whereas some of the combatants used monoplanes at the start of World War I [e.g., the Morane-Saulnier type N (France) and the Fokker series of E-type fighters (E for Eindecker; Germany)], most of the planes in service at the end of the war were biplanes [e.g., the SE5a (United Kingdom), the Fokker D-VII (Germany) and the SPAD XIII (France)] to carry the increased weight of the engine and of the payload.

Although the serious design of biplanes continued until the late 1930s, with the Fiat C.R. 42 (Italy) making its maiden flight in 1939, the improved performance of monoplane designs brought them to the front. Various methods of changing the wing geometry in flight were proposed in the 1920s and 1930s. Based on a concept proposed by test pilot V. V. Shevchenko, Soviet designer V. V. Nikitin developed a fighter that could translate from a biplane to a monoplane, or vice versa, at the will of the pilot (Ref. 14.2). In the design of Nikitin (known as the *IS-2*), the inboard sections of the lower wing were hinged at their roots, folding upward into recesses in the fuselage sides. The sections outboard of the main undercarriage attachment points also were articulated and, rising vertically and inward, occupied recesses in the upper wing. Thus, the one airplane combined the desirable short-field and low-speed characteristics of a lightly loaded biplane with the higher performance offered by a highly loaded monoplane.

The variable-area concepts included the telescoping wing, an example of which is illustrated in Fig. 14.1. The example is the MAK-10, built in France to the design of an expatriate Russian, Ivan Makhonine. The wing outer panel telescoped into the inner panel to reduce span and wing area for high-speed flight and could be extended for economic cruise and landing.

Trailing-edge flaps, such as the Fowler flap, which extend beyond the normal wing-surface area when deployed are modern examples of design features that increase the wing area for landing. (Aerodynamic data for these flaps will be discussed in the section on multielement airfoils.) The increase in the effective wing area offered by typical multielement, high-lift configurations is illustrated in Fig. 14.2. The area increases available from using a plain (or aileron) type flap, a circular motion flap similar to that used on the Boeing 707, and the extended Fowler flap used on the Boeing 737 are presented in Fig. 14.3. The large increase in area for the 737-type

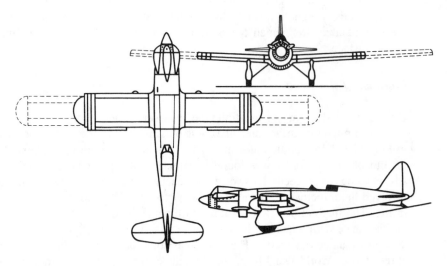

Figure 14.1 Makhonine MAK-10 variable-geometry (telescoping wing) aircraft. (From Ref. 14.2.)

flap is the sum of (1) the aft motion of the entire flap, (2) the aft motion of the main flap from the fore flap, (3) the motion of the auxilliary (aft) flap, and (4) the movement of the leading-edge devices.

Increasing the Lift Coefficient

The progress in developing equivalent straight-wing, nonpropulsive high-lift systems is illustrated in Fig. 14.4. Note the relatively high values obtained by experimental aircraft such as the L-19 of Mississippi State University and the MA4 of Cambridge

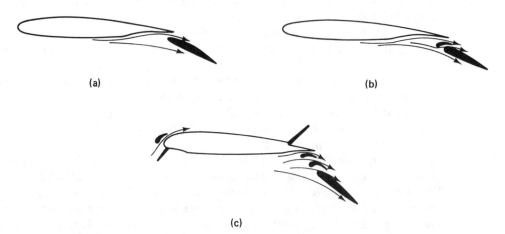

Figure 14.2 Multielement high-lift configurations: (a) Fowler flap; (b) double-slotted flap; (c) leading-edge slat, Krueger leading-edge flap, spoiler, and triple-slotted flaps (representative of Boeing 727 wing section).

Figure 14.2 (*Continued*)   (d) Fowler flaps on the HS 748. (Courtesy of British Aerospace.)

University. Both of these aircraft use distributed suction on the wing so that the flow field approximates that for inviscid flow.

A companion figure from the work of Cleveland (Ref. 14.4) has been included for the interested reader. The parasite drag coefficient, which includes interference drag

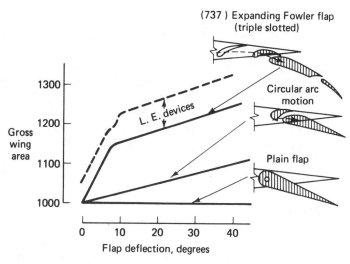

Figure 14.3   Use of flaps to increase the wing area. (From Ref. 14.3.)

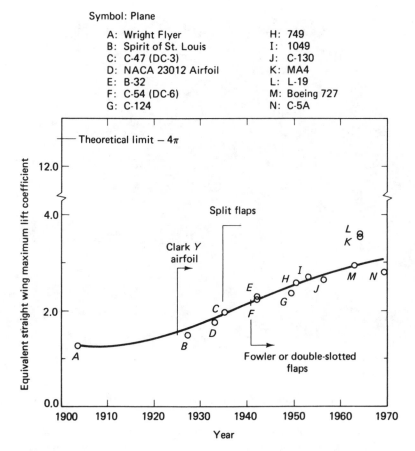

Symbol: Plane

A:	Wright Flyer	H:	749
B:	Spirit of St. Louis	I:	1049
C:	C-47 (DC-3)	J:	C-130
D:	NACA 23012 Airfoil	K:	MA4
E:	B-32	L:	L-19
F:	C-54 (DC-6)	M:	Boeing 727
G:	C-124	N:	C-5A

Figure 14.4   History of nonaugmented maximum lift coefficient. (From Ref. 14.4.)

but does not include induced or compressibility drag, is presented for several airplanes in Fig. 14.5.

The Chance-Vought F8U Crusader offers an interesting design approach for obtaining a sufficiently high lift coefficient for low-speed flight while maintaining good visibility for the pilot during landing on the restricted space of an aircraft carrier deck. As shown in Fig. 14.6, the entire wing could be pivoted about its rear spar to increase its incidence by 7° during takeoff and landing. Thus, while the wing is at a relatively high incidence, the fuselage is nearly horizontal and the pilot has excellent visibility. Furthermore, when the wing is raised, the protruding center section also serves as a large speed brake.

Flap Systems

Olason and Norton (Ref. 14.3) note that "if a clean flaps-up wing did not stall, a flap system would not be needed, except perhaps to reduce nose-up attitude (more correctly, angle of attack) in low-speed flight." Thus, a basic goal of the flap system design is

Symbol: Plane

A: Wright Flyer	H: Bf (Me) 109	O: C-130
B: WWI Bomber	I: B-29	P: P-51
C: WWI Fighter	J: B-17	Q: Comet
D: Spirit of St. Louis	K: Bf (Me) 108	R: Jetstar
E: Lockheed Vega	L: Me 262	S: C-141A
F: Curtiss Navy Fighter	M: XB-19	T: Boeing 747
G: Piper Cub	N: F-80	U: C-5A

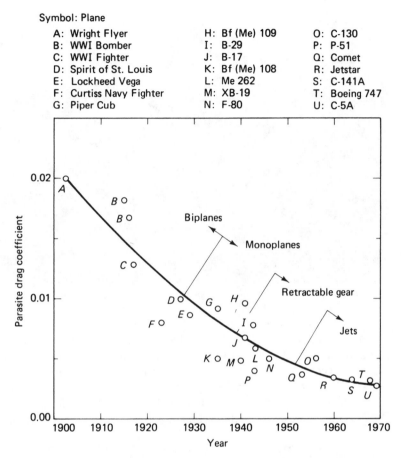

Figure 14.5 History of parasite drag coefficient, where the drag coefficient is based on total surface area. (From Ref. 14.4.)

to attain the highest possible *L/D* ratio at the highest possible lift coefficient, as illustrated in Fig. 14.7. A flap system (1) increases the effective wing area, (2) increases the camber of the airfoil section (thereby increasing the lift produced at a given angle of attack), (3) can provide leading-edge camber to help prevent leading-edge stall, and (4) can include slots which affect the boundary layer and its separation characteristics.

Significant increases in the lift coefficient (and in the drag coefficient) can be obtained by increasing the camber of the airfoil section. The effect of deploying a split flap, which is essentially a plate deflected from the lower surface of the airfoil, is illustrated in the pressure distributions of Fig. 14.8. Deployment of the split flap not only causes an increase in the pressure acting on the lower surface upstream of the flap but also causes a pressure reduction on the upper surface of the airfoil. Thus, the deployment of the split flap produces a marked increase in the circulation around the section and, therefore, increases the lift. The relatively low pressure in the separated region in the wake downstream of the deployed plate causes the drag to be relatively high. The effect is so pronounced that it affects the pressure at the trailing edge of

Figure 14.6 Chance-Vought F8U showing incidence for the wing during landing (or takeoff). (Courtesy of Vought Corp.)

the upper surface (Fig. 14.8). The relatively high drag may not be a disadvantage if the application requires relatively steep landing approaches over obstacles or requires higher power from the engine during approach in order to minimize engine acceleration time in the event of wave-off.

The effect of the flap deflection angle on the lift coefficient is presented in Fig. 14.9. Data are presented both for a plain flap and for a split flap. Both flaps were $0.2c$

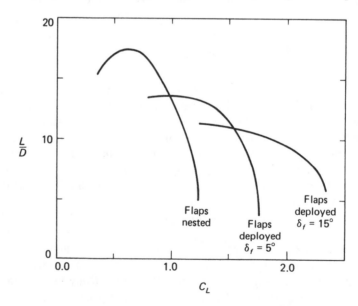

Figure 14.7 Effect of flap deployment on the aerodynamic forces.

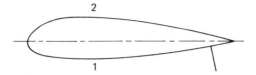

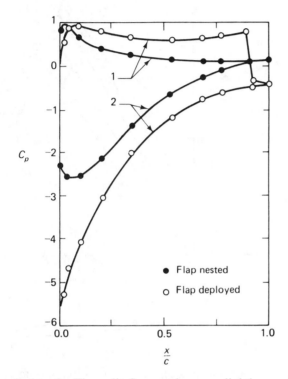

Figure 14.8 Pressure distribution for an airfoil with a split flap. (Data from Ref. 14.5.)

in length. The split flap produces a slightly greater increase in $C_{l,max}$ than does the plain flap.

Simple hinge systems as on a plain flap, even though sealed, can have a significant adverse effect on the separation point and hence on the lift and the drag. The adverse effect of the break at the hinge line is indicated in data presented in Ref. 14.6. Pressure distributions are compared in this reference for a plain flap, which is $0.25c$ in length and is deflected 25°, and for a variable camber flap whose centerline is a circular arc having a final slope of 25°. Although separation occurs for both flaps, it occurs nearer the trailing edge for the variable camber shape, which "turns out to be better only because of its drastic reduction in the suction peak" (The quotes are from Ref. 14.6).

Multielement Airfoils

As noted, the location of separation has a significant effect on the lift, drag, and moment acting on the airfoil section. It has long been recognized that gaps between the main section and the leading edge of the flap can cause a significant increase in

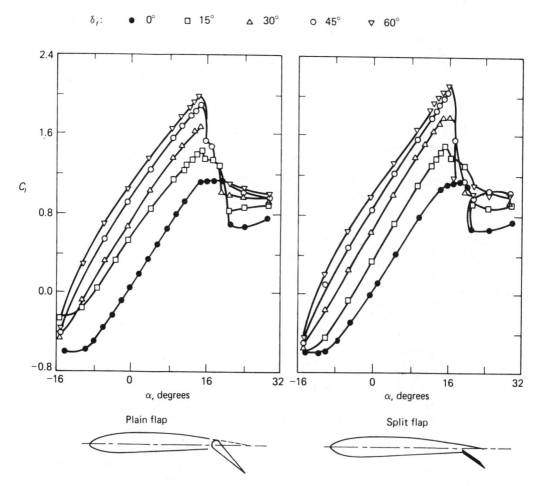

Figure 14.9 Effect of flap angle on the sectional lift coefficient for a NACA 23012 airfoil section, $Re_c = 6 \times 10^5$. (Data from Ref. 14.5.)

$C_{l,max}$ over that for a split flap or a plain flap. Furthermore, the drag for the slotted flap configurations is reduced. Sketches of airfoil sections with leading-edge slats or with slotted flaps are presented in Fig. 14.2.

Smith (Ref. 14.6) notes that the air through the slot cannot really be called high-energy air, since all the air outside the boundary layer has the same total pressure. Smith states that "There appear to be five primary effects of gaps, and here we speak of properly designed aerodynamic slots.

1. *Slat effect.* In the vicinity of the leading edge of a downstream element, the velocities due to circulation on a forward element (e.g., a slat) run counter to the velocities on the downstream element and so reduce pressure peaks on the downstream element.

2. *Circulation effect.* In turn, the downstream element causes the trailing edge of the adjacent upstream element to be in a region of high velocity that is inclined

to the mean line at the rear of the forward element. Such flow inclination induces considerably greater circulation on the forward element.

3. *Dumping effect*. Because the trailing edge of a forward element is in a region of velocity appreciably higher than free stream, the boundary layer "dumps" at a high velocity. The higher discharge velocity relieves the pressure rise impressed on the boundary layer, thus alleviating separation problems or permitting increased lift.

4. *Off-the-surface pressure recovery*. The boundary layer from forward elements is dumped at velocities appreciably higher than free stream. The final deceleration to free-stream velocity is done in an efficient manner. The deceleration of the wake occurs out of contact with a wall. Such a method is more effective than the best possible deceleration in contact with a wall.

5. *Fresh-boundary-layer effect*. Each new element starts out with a fresh boundary layer at its leading edge. Thin boundary layers can withstand stronger adverse gradients than thick ones."

Since the viscous boundary layer is a dominant factor in determining the aerodynamic performance of a high-lift multielement airfoil, inviscid theory is not sufficient. Typical theoretical methods iteratively couple potential-flow solutions with boundary-layer solutions. The potential-flow methods used to determine the velocity at specified locations on the surface of the airfoil usually employ singularity-distribution methods. As discussed in Chapters 3 and 6, singularity-distribution methods, which have been widely used since the advent of the high-speed, high-capacity digital computers needed to solve the large systems of simultaneous equations, can handle arbitrarily shaped airfoils at "any" orientation relative to the free stream. (The word "any" is in quotes since there are limits to the validity of the numerical simulation of the actual flow.) For singularity-distribution methods, either source, sink, or vortex singularities are distributed on the surface of the airfoil and integral equations formulated to determine the resultant velocity induced at a point by the singularities. The airfoil surface is divided into N segments with the boundary condition that the inviscid flow is tangent to the surface at the control point of each and every segment. The integral equations can be approximated by a corresponding system of $N - 1$ simultaneous equations. By satisfying the Kutta condition at the trailing edge of the airfoil, the Nth equation can be formulated and then the singularity strengths determined with a matrix-inversion technique. As noted above, the Kutta condition which is usually employed is that the velocities at the upper and lower surface trailing edge be tangent to the surface and equal in magnitude. The various investigators use diverse combinations of integral and finite-difference techniques to generate solutions to the laminar, the transitional, and the turbulent boundary layers. The interested reader is referred to the brief discussion in Chapters 6 or to one of the numerous analyses of the multielement airfoil problem (e.g., Refs. 14.7 through 14.10).

The lift coefficients for a GA(W)-1 airfoil with a Fowler-type, single-slotted flap which are taken from Ref. 14.11, are presented as a function of the angle of attack in Fig. 14.10. Note the large increases in $C_{l,max}$ which are obtained with the slotted Fowler flap. As shown in the sketch of the airfoil section, the deflected flap segment is moved

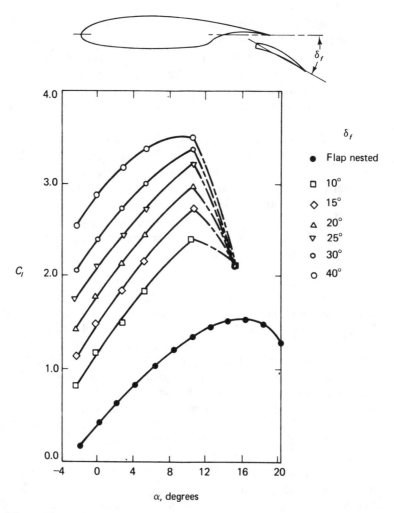

Figure 14.10 Experimental lift coefficient for a GA(W)-1 airfoil with a slotted Fowler flap. (From Ref. 14.11.)

aft along a set of tracks that increases the chord and the effective wing area. Thus, the Fowler flap is characterized by large increases in $C_{l,max}$ with minimum changes in drag. The ability of numerical techniques to predict the pressure distribution is illustrated in Fig. 14.11. Data are presented for the GA(W)-1 airfoil at an angle of attack of 5°, with a flap deflection of 30°.

In this section we have seen how the use of mechanical changes in camber can be used to increase the lift generated by airfoils and wings, independent of the change in angle of attack. Typically, one could expect maximum values of C_L of 2.5 to 3.5 using mechanical systems on wings of commercial and military transports. In addition to the obvious mechanical complexity (e.g., tracks, brackets, and actuators) and weight, mechanical systems are limited aerodynamically in the maximum lift they can generate. One limitation, the Kutta condition, essentially fixes the airfoil circulation (and thus

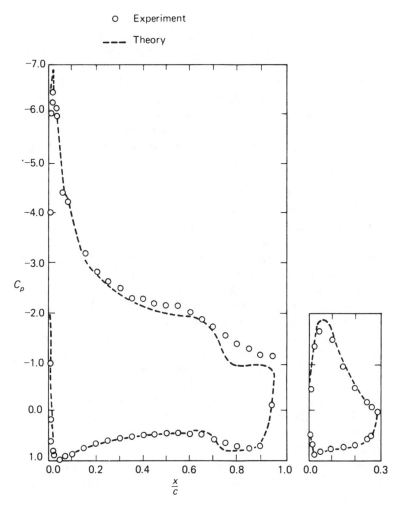

Figure 14.11 Comparison of the theoretical and the experimental pressure distribution for a GA(W)-1 airfoil with a slotted flap, $\alpha = 5°$, $\delta_f = 30°$. (From Ref. 14.11.)

the lift) to the value where the free-system flow leaves the airfoil at the trailing edge. A second limitation, viscous effects, usually reduces the attainable lift to less than that value, because the flow separates from the highly curved flap upper surface before the trailing edge is reached. Slotted flaps, of course, reduce the effects of separation. Nevertheless, concepts employing jet-engine bleed gases are often used to generate increased lift.

Power-Augmented Lift

An additional factor to consider in the comparison of flap types is the aerodynamic moment created by deployment of the flap. Positive camber produces a nose-down pitching moment, which is especially great when applied well aft on the chord, and produces twisting loads on the structure. The pitching moments must be controlled with the horizontal tail. Unfortunately, the flap types which produce the greatest increase in $C_{l,\text{max}}$ usually produce the largest moments. Thus, as shown in the sketches of the MiG 21s presented in Fig. 14.12, the Fowler flap with its extended guides and fairing plates is replaced by blown flaps for some applications. Separation from the surface of the flap is prevented by discharging fluid from the interior of the main airfoil section. The fluid injected tangentially imparts additional energy to the fluid particles in the boundary layer so that the boundary layer remains attached.

The internally blown flap is compared with two other techniques which use engine power to achieve very high lift in Fig. 14.13. The corresponding drag polars (as taken from Ref. 14.13) are included. The externally blown flap (EBF) spreads and turns the jet exhaust directed at the trailing-edge flap. A portion of the flow emerging through the flap slots maintains attachment of the boundary layer over the flap's upper surface. The upper-surface-blowing (USB) concept resembles the externally blown flaps. However, the data indicate "better performance than the externally blown flap if the air-turning process is executed properly. Also, the path of the engine exhaust permits a certain amount of acoustic shielding by the wing, and consequently a significant reduction in noise." The quotes are those of the authors of Ref. 14.13, who work for the Boeing Company, which designed the YC-14 AMST (Advanced Medium STOL Transport), an aircraft that employs USB.

Using suction to remove the decelerated fluid particles from the boundary layer before they separate from the surface in the presence of the adverse pressure gradient is another means of increasing the maximum lift. The "new" boundary layer which is formed downstream of the suction slot can overcome a relatively large adverse pressure gradient without separating. Flight data obtained in the late 1930s at the Aerodynamische Versuchsanstalt at Göttingen are reproduced in Fig. 14.14. The data demonstrate that the application of suction through a slit between the wing and the flap can prevent separation. Since the flaps can operate at relatively large deflection angles and the airfoil at relatively high angles of attack without separation, large increases in lift can be obtained. The maximum lift coefficients for the airplanes

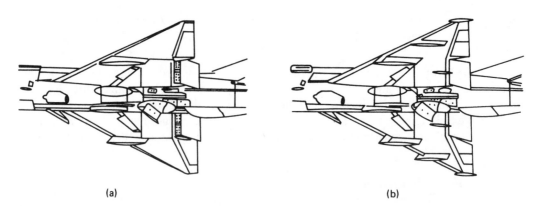

(a)   (b)

Figure 14.12 Fowler-type flaps used on early series MiG 21s and the blown flaps used on later series: (a) Fowler-type flaps employed on the MiG-21 PF; (b) blown flaps employed on the MiG-21 MF. (From Ref. 14.12.)

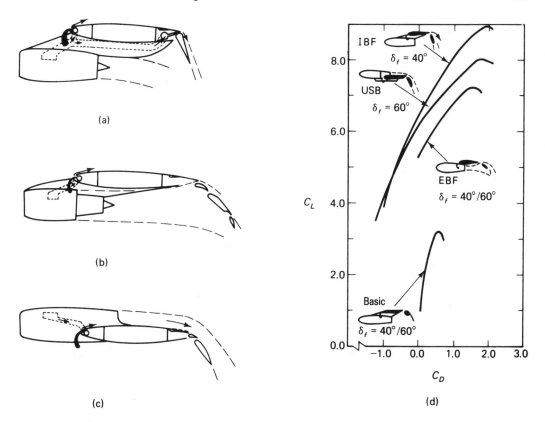

Figure 14.13   Power-augmented high-lift configurations: (a) internally-blown flap (IBF); (b) externally-blown flap (EBF); (c) upper-surface blowing (USB); (d) drag polars for four-engine configuration. $cj = 2.0$ for blown configurations. (From Ref. 14.13.)

equipped with suction are almost twice that for the Fieseler Storch (Fi 156), a famous STOL airplane of the World War II period, which is shown in Fig. 14.15. The entire trailing edge of the Storch wing was hinged; the outer portions acting as statically balanced and slotted ailerons, and the inner portions as slotted camber-changing flaps. A fixed slot occupied the entire leading edge. Initial flight tests showed the speed range of the Fieseler Storch to be 51 to 174 km/h (32 to 108 mi/h) and that the landing run in a 13 km/h (8 mi/h) wind using brakes is 16 m. The interested reader is referred to Ref. 14.15 for more details.

CIRCULATION CONTROL WING

As discussed by Englar (Ref. 14.16), the circulation control wing (CCW) concept avoids the problems of mechanical and blown flaps by replacing the sharp trailing edge with a fixed nondeflecting, round or near-round surface, such as shown in Fig. 14.16. The tangential blowing jet remains attached to the curved surface by creating a balance between the subambient pressure and the centrifugal force. Thus, as shown in Fig.

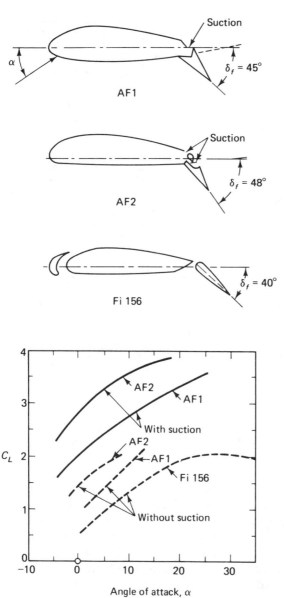

Figure 14.14 Lift of three STOL airplanes for full landing flap deflection (power off). AF1 and AF2 are boundary layer control airplanes of the Aerodynamische Versuchsanstalt, Fi 156 is the Fieseler Storch. (Data from Ref. 14.14.)

14.16, at low values for the blowing momentum coefficient, it serves as a boundary-layer control (BLC) device to entrain the flow field and prevent it from separating. Once the flow is returned to the inviscid condition, the jet continues to turn around the trailing edge. It thus entrains and deflects the flow field, providing pneumatic deflection of the streamlines (equivalent to camber) and resulting in supercirculation lift. In the circulation control region (CC), very high lift can be generated since the jet turning is not limited by a sharp trailing edge. As shown in Fig. 14.17, relatively high lift coefficients can be generated for two-dimensional CCW airfoils with much less blowing than that typical of mechanical blown flaps.

Figure 14.15 Fieseler Storch, Fi 156, showing fixed leading-edge slots and hinged trailing edge. (Courtesy of Jay Miller, ''Aerophile.'')

Blowing momentum coefficient

$$C\mu = \dot{m}V_j/qS$$

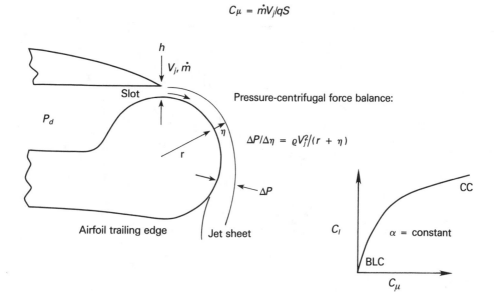

Figure 14.16 Basic principles of circulation control aerodynamics. (From Ref. 14.16.)

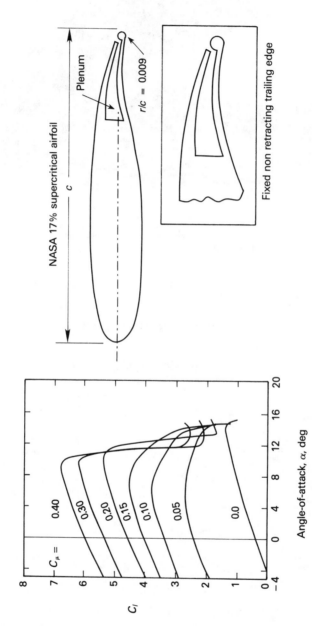

Figure 14.17 Sectional lift characteristics for a 17% thick CCW/supercritical airfoil. (From Ref. 14.16.)

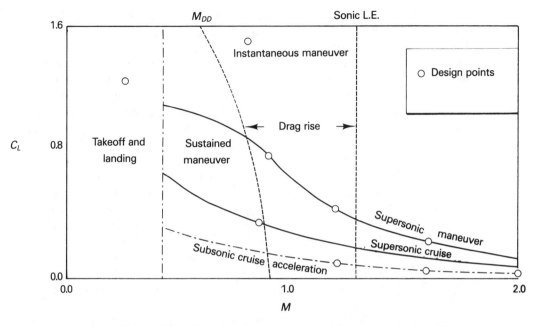

Figure 14.18   Typical performance map for fighters. (From Ref. 14.17.)

DESIGN CONSIDERATIONS FOR TACTICAL MILITARY AIRCRAFT*

As noted by Bradley (Ref. 14.17), the designer of a tactical military aircraft is faced with a multitude of design points throughout the subsonic/supersonic flow regimes plus many off-design constraints. The design goals for a tactical weapons system may include efficient cruise at both subsonic and supersonic Mach number, superior maneuverability at both subsonic and supersonic Mach number, and rapid acceleration.

Bradley states: "The multiple design point requirement turns out to be the major driver for the designer of fighter aircraft. The aerodynamic requirements for each of the design points often present conflicting requirements. For example, the need for rapid acceleration to supersonic flight and efficient supersonic cruise calls for thin wing sections with relatively high sweep and with camber that is designed to trim out the moments resulting from aft ac movement at supersonic flight. However, these requirements are contrary to those requirements for efficient transonic maneuver, where the designer would prefer to have thicker wing sections designed with camber for high C_L operation and a high-aspect-ratio planform to provide a good transonic drag polar. Designers are thus faced with a situation of compromise. These conflicting requirements suggest the obvious solution of variable geometry, that is, variable sweep wing and/ or variable camber. Although this is a satisfactory aerodynamic solution, in many cases the resultant weight increases to a configuration can be prohibitive."

---

* Quotations and data presented in this section are reproduced by permission from General Dynamics Corporation.

"A typical performance spectrum corresponding to mission requirements is illustrated in Fig. 14.18, which presents a map of lift coefficient versus Mach number. The low Mach end of the spectrum throughout the C_L range is typical of takeoff and landing for the configuration. The subsonic cruise and supersonic cruise portions are noted in the moderate lift range. Acceleration to high supersonic speeds occurs at the low lift coefficients. Sustained maneuver takes place in the C_L range of less than one for most fighter configurations. Above this lift coefficient, the aircraft is in the instantaneous maneuver regime. Drag rise occurs, depending on the wing geometry, in the range of 0.8 to 1.2 Mach."

"The particular flow conditions that correspond to the flow map are shown in Fig. 14.19. At the cruise and acceleration points, the aircraft designer is dealing primarily with attached flow, and his design objective is to maintain attached flow for maximum efficiency. At the higher C_L's, corresponding to instantaneous maneuver, separated flow becomes the dominant feature. Current designs take advantage of the separated flow by forming vortex flows in this range. Intermediate C_L's corresponding to sustained maneuver are usually a mixture of separated and attached flow. Consequently, if the aircraft is designed with camber to minimize separation in the maneuver regime, the configuration will have camber drag and may have lower surface separation, which increases drag at the low C_L's needed for acceleration."

"Thus, the aircraft design is a compromise to achieve an optimal flow efficiency considering the numerous design points associated with the mission objectives. It is easily seen that the transonic flow problems that must be addressed in fighter design are driven to a very large extent by the constraints imposed at the other design points—supersonic and subsonic."

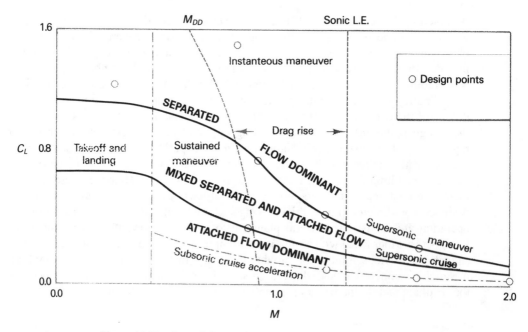

Figure 14.19 General flow regimes encountered for fighters. (From Ref. 14.17.)

Bradley (Ref. 14.17) notes that "Supersonic design enjoys relatively precise computation and optimization thanks to wide applicability of linearized theory." Further, it is noted that "The design of efficient transonic configurations may proceed from two conceptual schools of logic. One acknowledges that the optimum low-drag flow must accelerate rapidly over the airfoil to supercritical flow and decelerate in a nearly isentropic manner, avoiding strong shocks and/or steep gradients that can lead to significant regions of separation. This approach sets attached flow or near fully attached flow as an intuitive design goal and is typically used for aircraft that permit strong emphasis to be placed on transonic cruise or sustained maneuver design points." The design of airfoils for transonic cruise applications was discussed extensively in Chapter 9.

"The second school of thought recognizes the inevitability of significant flow separations at design conditions and adopts a philosophy of controlling certain regions of separation through vortex flows to complement other regions of attached supercritical flow. This approach is appropriate for configurations constrained by multiple design points that emphasize added supersonic requirements. Current tactical fighters that rely on high wing loadings for transonic performance are good examples. The F-16 and

Figure 14.20 Strake vortex system on an F-16. (Photograph provided by General Dynamics Corporation, Fort Worth Division.)

F-18 employ a combination of controlled vortex flow and variable camber to achieve maneuverability.''

Even though a designer has worked carefully to design a wing for attached flow, the flow fields associated with sustained and instantaneous maneuvers involve predominantly separated flows. The manner in which the flow separates and how the separation develops over the configuration will strongly affect the vehicle's drag and its controllability at the higher C_L's. Aircraft, such as the F-16 and the F-18, employ strakes or leading edge extension devices to provide a controlled separated flow. Controlled vortex flow can then be integrated with variable camber devices on the wing surface to provide satisfactory high-lift, stability and control, and buffet characteristics. The resulting flow field is a complex one, combining attached flows over portions of the wing with the vortex flow from the strake.

''Vortex control devices can improve the transonic drag polar, but highly integrated designs are required to avoid other aerodynamic problems and also to minimize the potential weight penalties. The F-16 forebody strake vortex system is clearly visible as it interacts with the wing flow field in Fig. 14.20. Forebody strakes and canards have similar aerodynamic effects—both good and bad. For example, the F-16 forebody strake design required extensive integration with the wing variable leading edge flap system and empennage to achieve significant aerodynamic improvements. The aerodynamic improvements achieved with the forebody strake/variable leading edge flap combination are illustrated in Fig. 14.21 for the YF-16 aircraft. The dashed curve depicts the lift curve and drag polar for the wing without the strake and variable leading edge flaps. The strake with no leading edge flap significantly improves the static

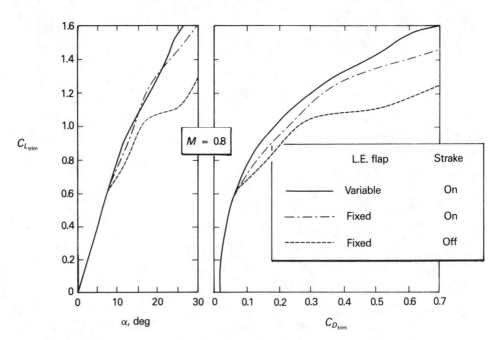

Figure 14.21 Aerodynamic benefits of combined forebody-strake/variable-camber system. (From Ref. 14.17.)

aerodynamic characteristics. Further improvement is seen with the variable leading edge flap in combination with the strake configuration."

Bradley (Ref. 14.17) states: "We should not pass from the separated flow wing design discussion without mentioning a new concept of controlled vortex flow for designing military aircraft having emphasis on supersonic configurations. Wing planforms for supersonic cruise aircraft have higher leading edge sweep and generally lower aspect ratios; these planforms develop vortex flows at relatively low angle of attack. As a result, the transonic drag characteristics are lacking in the maneuver regime since drag polars generally reflect very little leading edge suction recovery. Recently, wings of this type have been designed to take advantage of the separated vortex flows rather than to try to maintain attached flow to higher C_L values."

DRAG REDUCTION

Possible fuel shortages combined with sharp price increases and the requirements of high performance over a wide-speed range emphasize the need for reducing the drag on a vehicle and, therefore, improving the aerodynamic efficiency. Of the various drag reduction concepts, we will discuss:

1. Variable-twist, variable-camber wings
2. Laminar-flow control (LFC)
3. Winglets

Variable-Twist, Variable-Camber Wings

Survivability and mission effectiveness of a supersonic-cruise military aircraft require relatively high lift/drag ratios while retaining adequate maneuverability. The performance of a moderate-aspect-ratio, thin swept-wing is significantly degraded at high lift coefficients at high subsonic Mach numbers because of shock-induced boundary-layer separation and, at higher angles of attack, because of leading-edge separation and wing stall. The resulting degradation in handling qualities significantly reduces the combat effectiveness of such airplanes. There are several techniques to counter leading-edge stall, including leading-edge flaps, slats, and boundary-layer control by suction or by blowing. These techniques, along with trailing-edge flaps, have been used effectively to increase the maximum usable lift coefficient for low-speed landing and for higher subsonic speeds.

Low-thickness-ratio wings incorporating variable camber and twist appear to offer higher performance for fighters with a fixed-wing planform (Ref. 14.18), since the camber can be reduced or reflexed for the supersonic mission and increased to provide the high lift coefficients required for transonic and subsonic maneuverability.

A test program was conducted to determine the effect of variable-twist, variable-camber on the aerodynamic characteristics of a low-thickness-ratio wing (Ref.14.19). The basic wing was planar with a NACA 65A005 airfoil at the root and a NACA 65A004 airfoil at the tip (i.e., there was no camber and no twist). Section camber was varied using four leading-edge segments and four trailing-edge segments, all with spanwise hinge lines. Variable twist was achieved since the leading-edge (or trailing-edge) segments were parallel and were swept more than the leading edge (or trailing edge). Camber and twist could be applied to the wing as shown in Fig. 14.22. Deploying the trailing-edge segments near the root creates a cambered section whose effective chord is at an increased incidence. Similarly, deploying the leading-edge segments near the wing tip creates a cambered section whose local incidence is decreased. Thus, the modified wing could have an effective twist of approximately 8° washout. As noted in Ref. 14.19, use of leading-edge camber lowers the drag substantially for lift coefficients up to 0.4. Furthermore, use of leading-edge camber significantly increased the maximum lift/drag ratio over a Mach number range of 0.6 to 0.9. At the higher lift coefficients (≥ 0.5), the combination of twist and camber achieved using both leading-edge segments and trailing-edge segments was effective in reducing the drag. Trailing-edge camber causes very large increments in C_L with substantial negative shifts in the pitching moment coefficients.

The effectiveness of leading-edge segments and trailing-edge segments in increasing the lift coefficient and in reducing the drag coefficient at these relatively high lift coefficients is illustrated in the data presented in Fig. 14.23. As a result, the maximum lift/drag ratio for this particular configuration at $M_\infty = 0.80$ is 18 and it occurs when $C_L = 0.4$.

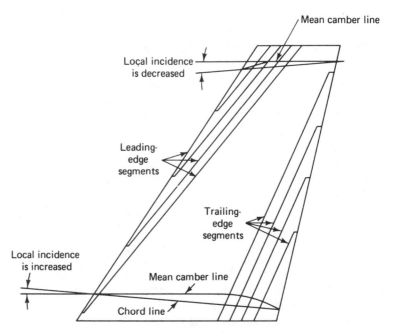

Figure 14.22 Use of leading-edge segments and trailing-edge segments to produce camber and twist on a basic planar wing.

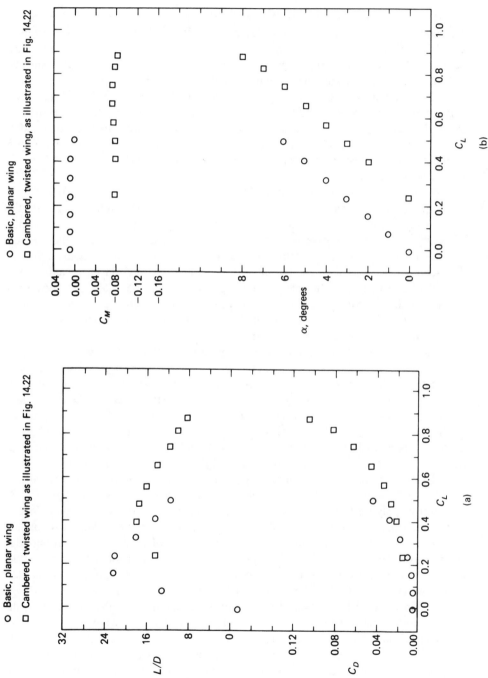

Figure 14.23 Effect of twist and camber on the longitudinal aerodynamic characteristics, $M_\infty = 0.80$, $Re_c = 7.4 \times 10^6$: (a) lift-to-drag ratio and the drag polar; (b) pitching moment and lift coefficients. (From Ref. 14.19.)

Laminar-Flow Control

In previous chapters we have seen that the skin-friction component of the drag is markedly higher when the boundary layer is turbulent. Thus, in an effort to reduce skin friction, which is a major part of the airplane's "parasitic" drag, attempts have been made to maintain laminar flow over substantial portions of the aircraft's surface. Attempts at delaying transition by appropriately shaping the airfoil section geometry were discussed in Chapter 5. However, a natural boundary layer cannot withstand even very small disturbances at the higher Reynolds numbers, making transition difficult to avoid. Theoretical solutions reveal that removing the innermost part of the boundary layer using even very small amounts of suction substantially increases the stability of a laminar boundary layer. Maintaining a laminar profile by suction is termed *laminar-flow control* (LFC).

The aerodynamic analysis of a LFC surface is divided into three parts: (1) the prediction of the inviscid flow field, (2) the calculation of the natural development of the boundary layer, and (3) the suction system analysis. In this approach (which may in reality require an iterative procedure), the first step is to determine the pressure and the velocity distribution of the inviscid flow at the edge of the boundary layer. The second step is to calculate the three-dimensional boundary layer, including both the velocity profiles and the integral thicknesses. It might be noted that because of cross flow, the boundary layer on a swept wing may be more unstable than that on an unswept wing. Finally, the suction required to stabilize the boundary layer must be calculated and the suction system designed.

In 1960, two WB-66 aircraft were adapted to a 30° swept wing with an aspect ratio of 7 and a thickness ratio of approximately 10%. The modified aircraft were designated X-21A. A suction system consisting of turbocompressor units removed boundary-layer air from the wing through many narrowly spaced LFC suction slots. With suction-inflow velocities varying from $0.0001U_\infty$ in regions of negligible pressure gradient to $0.0010U_\infty$ near the wing leading edge, full-chord laminar flows were obtained up to a maximum Reynolds number of 45.7×10^6 (Ref. 14.20). It was concluded that laminar-flow control significantly reduced the wake drag on the wing.

Using the propulsion, structural, flight controls and system technologies predicted for 1985, Jobe, Kulfan, and Vachal (Ref. 14.21) predicted fuel savings from 27 to 30 percent by applying LFC to the design of large subsonic military transports. Jobe et al. assumed that the LFC system used in their design would maintain a laminar boundary layer to $0.70c$, even though full-chord laminarization of a wing with trailing-edge controls is technically feasible. The optimum wing planform for the minimum-fuel airplane has the highest aspect ratio, the lowest thickness-chord ratio, and a quarter-chord sweep of about 12°. The cruise Mach number for this aircraft design is 0.78. As noted in Table 14.1, their sensitivity analysis showed that a high aspect ratio is the most important parameter for minimizing fuel consumption, wing thickness is of secondary importance, and sweep is relatively unimportant. However, since productivity varies linearly with the cruise speed, a maximum productivity airplane requires a relatively high sweep, maximum aspect ratio, and a low thickness ratio for the section. The resultant aircraft cruises at a Mach number of 0.85. The sensitivity analysis of

TABLE 14.1 DESIRABLE LAMINAR-FLOW CONTROL WING PLANFORM
CHARACTERISTICS

Figure of merit	Wing design parameter		
	Aspect ratio	Thickness ratio	Sweep
Performance			
Minimum fuel	High	Low	NMC[a]
Minimum takeoff gross weight	High	NMC	Low
Maximum $\dfrac{\text{maximum payload}}{\text{takeoff gross weight}}$	High	Low	NMC
Ease of laminarization			
Low chord Reynolds number	High	NMC	NMC
Low unit Reynolds number	NMC	NMC	NMC
Minimize cross flow	NMC	Low	Low
Minimize leading-edge contamination	High	Low	Low

[a] NMC, not a major consideration.

Source: Ref. 14.21.

Ref. 14.21 indicates that a low thickness ratio is most important to the design of the wing for a maximum productivity airplane, followed by aspect ratio and sweep.

Winglets

As discussed in Chapter 7, one of the ways of decreasing the induced drag is by increasing the aspect ratio. Although increased wingspan provides improved lift/drag ratios, the higher bending moments at the wing root create the need for a stronger wing structure. Furthermore, there are problems in maneuvering and parking once on the ground. As noted by Thomas (Ref. 14.22), using large-aspect-ratio wings will reduce the induced drag because the tip vortices will be further separated, reducing the strength of the average induced flow between them. However, as Thomas points out, as the aspect ratio is increased for the same chord, there is a weight penalty that may offset the drag reduction. Thus, as noted in Ref. 14.22, ''optimal wing aspect ratio for a transport aircraft varies from 7.5 for minimum acquisition cost, to 9.8 for minimum gross weight, to 12.0 for minimum direct operating cost, and to 15.2 for minimum fuel. At present aspect ratios as large as 15.2 are not structurally feasible but the importance of aspect ratio is clear.''

A possible means of reducing the drag is by the use of fixed winglets. As illustrated in Fig. 14.24, winglets are used on the Gates Learjet Model 28/29, the Longhorn. The drag polars at $M_\infty = 0.7$ and at $M_\infty = 0.8$ for the M28/29 are compared with those for the M25D/F in Fig. 14.25. As can be seen in these data, the greatest improvement is at the lower Mach number. However, this is of no concern for this design application, since the normal cruise speed and long-range cruise speed of this airplane are always less than $M_\infty = 0.8$.

To generate an optimum winglet design for a particular flight condition, we must calculate the flow field for the complex wing. The subsonic aerodynamic load distri-

Figure 14.24   Gates Learjet Model 28/29, the Longhorn, illustrating use of winglets. (Courtesy of Gates Learjet Corp.)

butions for a lifting surface with winglets can be calculated using the vortex-lattice method discussed in Chapter 7. The theoretical lift-curve slopes for a swept wing with end plates which were calculated using a distribution of vortices, such as illustrated in Fig. 14.26, were in good agreement with experimentally determined data (see Ref.

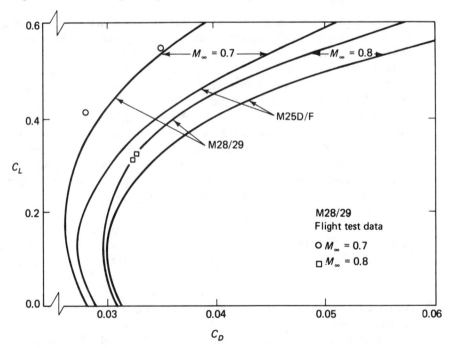

Figure 14.25   Comparison of the drag polars for the Gates Learjet M28/29 with those for the M25D/F. (Unpublished data provided by Gates Learjet Corp.)

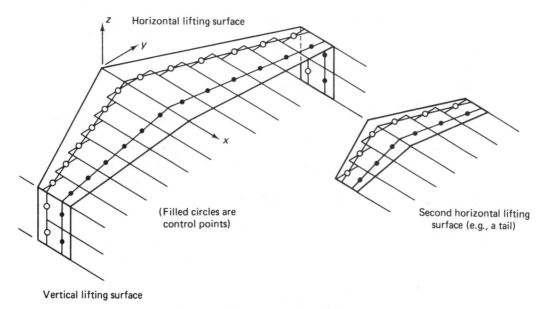

Figure 14.26   Distribution of vortices which can be used to calculate the aerodynamic load distribution for a combination of lifting surfaces. (From Ref. 14.23.)

14.23). The lifting-surface geometry shown in the sketch indicates that the technique can be used to calculate the aerodynamic load distribution for lifting surfaces involving a nonplanar wing, a wing with end plates, and/or a wing and empennage.

For best performance, the proper design of winglets includes the following (Ref. 14.22).

1. For supercritical performance, the winglet should be tapered and swept aft. It should be mounted behind the region of lowest pressure of the main wing to minimize interference effects.
2. Smooth fillets should be used between the wingtip and the winglet or smaller drag reduction benefits might result.
3. Some toe-out of the winglet is needed due to the inflow angles at the wingtip. This is also desirable, since it reduces the likelihood of winglet stall during sideslip.
4. Although the drag reduction increases with winglet span, the decrease is less than linear. Therefore, the optimal winglet height involves a trade-off between improved aerodynamics and the increased moments due to longer moment arms.

DEVELOPMENT OF AN AIRFRAME MODIFICATION TO IMPROVE THE MISSION EFFECTIVENESS OF AN EXISTING AIRPLANE

The EA-6B is a four-place, subsonic, twin-jet, electronic countermeasures airplane designed for land and carrier-based operations. As noted in Ref. 14.24, the EA-6B airplane was difficult to handle when the pilot attempted to maneuver the aircraft at low airspeed (approximately 250 KIAS). These problems relate to the fact that the airplane through its evolution had grown to gross weights above 54,000 lb. To compound the problem further, both the lateral and directional moments become unstable near the stall angle of attack (approximately 16°). These characteristics contribute to the sharp roll-off and directional nose slice that is inherent in the EA-6B at stall.

As noted by Gato and Masiello (Ref. 14.25), the EA-6B Prowler is 10,000 lb heavier than the A-6 Intruder. In addition to additional ECM equipment and structural weight, the EA-6B was provided with more powerful engines. However, in the cruise configuration, there were no aerodynamic changes to account for the extra weight. Since the two aircraft have the same maximum lift capability, when maneuvering at low speeds and/or high altitude, the heavier EA-6B flies much closer to the stall than its predecesor, the A-6. This is illustrated in Fig. 14.27 "where the lift coefficient required to execute a banked 2*g*, 60 degree banked turn at 250 knots and the angle of attack margin to stall are indicated for an A-6, a current production EA-6B, and a future ADVCAP EA-6B at representative landing weights. This is a typical maneuver for these aircraft, always performed while decelerating prior to entering the landing

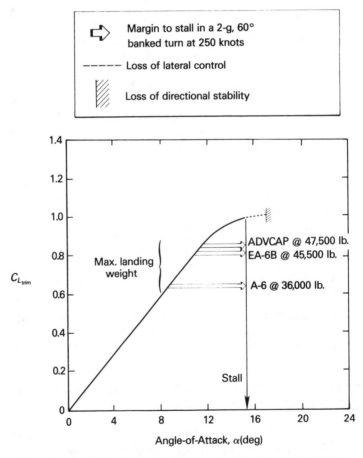

Figure 14.27   EA-6B/A-6 cruise configuration lift. (From Ref. 14.25.)

pattern and initiating landing gear, slat, and flap extension. Incidentally, the stall for these aircraft is defined by loss of lateral control and subsequent roll-off. Further penetration of the stall results in a directional departure which may develop into a spin.''

In order to diagnose the flow mechanism contributing to the directional instability near stall, flow visualization studies were conducted (Ref. 14.26). ''Results from these studies showed that a pair of vortices are generated at the fuselage-wing junctures. These vortices trail behind the wing, close to the fuselage and below the tail at low angles of attack (Fig. 14.28a). As angle of attack is increased, wing downwash maintains the vortex system at the same relative location—that is, low with respect to the vertical tail. At stall angles of attack, flow separation on the wing and consequent downwash breakdown cause the vortices to rise to the level of the vertical tail (Fig. 14.28b). In sideslip, the vortex generated on the windward side of the airplane drifts leeward such that, as angle of attack is increased through stall, the vertical tail becomes immersed in the windward vortex flow field (Fig. 14.28c). Because of the rotational sense of the vortex system, the bottom portion of the vertical tail first becomes immersed in a region of proverse (stabilizing) sidewash—the top portion of the windward vortex. As angle of attack is increased and the vortex system rises further, the vertical tail becomes immersed in the lower portion of the windward vortex where a condition of adverse (destabilizing) sidewash exists. Clearly, the abrupt changes in sidewash that

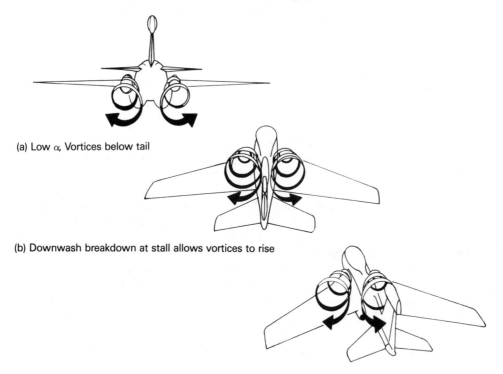

(a) Low α, Vortices below tail

(b) Downwash breakdown at stall allows vortices to rise

(c) In sideslip, windward wing vortex creates adverse/destabilizing flowfield

Figure 14.28   Directional destabilizing vortex system. (From Ref. 14.26.)

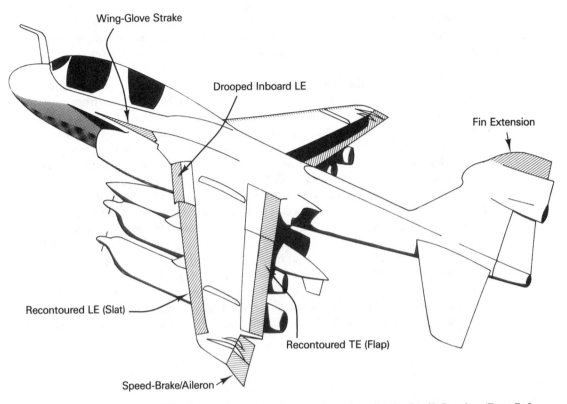

Figure 14.29   Proposed aerodynamics upgrade package for the EA-6B Prowler. (From Ref. 14.25.)

occur as the windward vortex traverses the span of the vertical tail have a direct impact on directional stability. In fact, it is this phenomenon that causes the directional instability the airplane configuration experiences near stall.''

To eliminate these dangers and to provide the EA-6B with electronics growth capability, the Navy undertook a comprehensive Maneuvering Improvement Program. The ground rules imposed by the Navy were that no major changes were allowed to the main wing or airframe structure. The project involved personnel from the Navy, Grumman Aircraft Systems Division, and the Langley Research Center (NASA). The results of the various tasks are described in Refs. 14.24 through 14.28.

As a result of these efforts, an integrated mix of computational and experimental investigations, an aerodynamic upgrade package was proposed for the EA-6B Prowler, as shown in Fig. 14.29. The upgrade package includes the following features:

1. Drooped inboard leading edges (add-on to inboard slat elements)
2. Wing-glove strakes
3. A fin extension
4. Wingtip speed-brake/ailerons

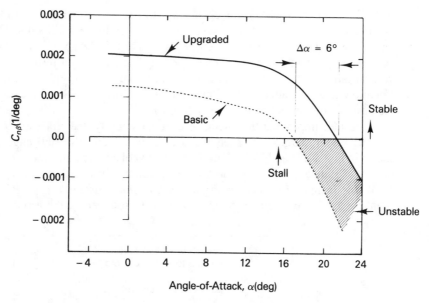

Figure 14.30 Combined effect of fin extension, strakes, and drooped inboard leading edges on lateral-directional stability. (From Ref. 14.25.)

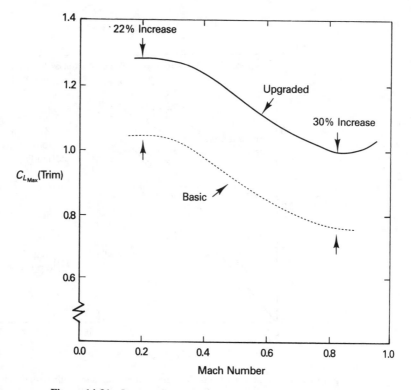

Figure 14.31 Increase in maximum usable lift. (From Ref. 14.25.)

The combined effect of fin extension, strakes, and drooped inboard leading edges on directional stability is illustrated in Fig. 14.30. The unstable directional characteristic was shifted to higher angles of attack, by a total of 6°. The level of directional stability at low angles of attack was also significantly increased as a result of the additional vertical tail area. In fact, the fin extension was sized to provide the EA-6B with the same low angle of attack directional stability as the A-6, thereby compensating for the destabilizing effect of the longer fuselage.

Besides improving the lateral-directional stability, the strakes, and the drooped inboard leading edges increased the maximum lift and the stall angle of attack as shown in Figs 14.31 and 14.32. This combination of aerodynamic modifications/devices provide significant improvements in the maximum lift coefficient. As shown in Fig. 14.31, the increase in cruise configuration maximum usable lift is 22% at low Mach numbers and 30% at higher Mach numbers. Consequently, the aerodynamically up-

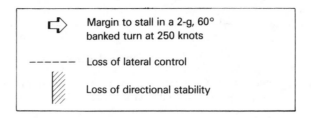

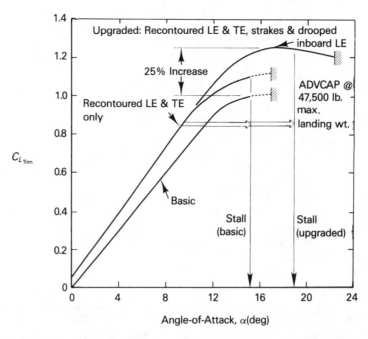

Figure 14.32 Combined effect of recontoured leading and trailing edges, strakes, and drooped inboard leading edges on cruise configuration lift. (From Ref. 14.25.)

graded EA-6B will have essentially the same stall/maneuver margins of the lighter A-6. Improved lateral-directional stability and positive lateral control beyond stall allow full use of this increment in maximum lift.

The ability to modify a design to accommodate the increased weights allows an aircraft to satisfy added missions, handle more ordnance, be fitted with more powerful engines, and so on. The Supermarine Spitfire, a single-engine pursuit aircraft of World War II, grew by the weight of 32 airline passengers and their luggage during its evolution from 1936 through 1946.

PROBLEMS

14.1. Discuss the variation of C_L as a low-speed aircraft consumes fuel during a constant altitude, constant airspeed cruise. What is the variation of α?

14.2. Based on the results of Problem 5.1, discuss ways to increase $(L/D)_{max}$.

14.3. Discuss the desirability of a laminar flow airfoil for:
  (a) A transport aircraft that operates for long periods in cruise flight.
  (b) A stunt or aerobatic-type airplane.

REFERENCES

14.1. Cowley, W. L., and H. Levy, *Aeronautics in Theory and Experiment*, Edward Arnold, Publisher, London, 1918.

14.2. "The Annals of the Polymorph, a Short History of V-G," *Air International*, Mar. 1975, Vol. 8, No. 3, pp. 134–140.

14.3. Olason, M. L., and D. A. Norton, "Aerodynamic Design Philosophy of the Boeing 737," *Journal of Aircraft*, Nov.–Dec. 1966, Vol. 3, No. 6, pp. 524–528.

14.4. Cleveland, F. A., "Size Effects in Conventional Aircraft Design," *Journal of Aircraft*, Nov.–Dec. 1970, Vol. 7, No. 6, pp. 483–512.

14.5. Schlichting, H., and E. Truckenbrodt, *Aerodynamik des Flugzeuges*, Springer-Verlag, Berlin, 1969.

14.6. Smith, A. M. O., "High-Lift Aerodynamics," *Journal of Aircraft*, June 1975, Vol. 12, No. 6, pp. 501–530.

14.7. Stevens, W. A., S. H. Goradia, and J. A. Braden, "Mathematical Model for Two-Dimensional Multi-component Airfoils in Viscous Flow," *CR-1843*, NASA, July 1971.

14.8. Morgan, H. L., Jr., "A Computer Program for the Analysis of Multielement Airfoils in Two-Dimensional Subsonic, Viscous Flow," presented in *Aerodynamic Analysis Requiring Advanced Computers, Part II, SP-347*, NASA, Mar. 1975.

14.9. Olson, L. E., and F. A. Dvorak, "Viscous/Potential Flow About Multi-Element Two-Dimensional and Infinite-Span Swept Wings: Theory and Experiment," *AIAA Paper 76–18*, presented at the AIAA 14th Aerospace Sciences Meeting, Washington, D.C., Jan. 1976.

14.10. Bristow, D. R., "A New Surface Singularity Method for Multi-Element Airfoil Analysis and Design," *AIAA Paper 76–20*, presented at the AIAA 14th Aerospace Sciences Meeting, Washington, D.C., Jan. 1976.

14.11. Wentz, W. H., Jr., and H. C. Seetharam, "Development of a Fowler Flap System for a High Performance General Aviation Airfoil," *CR-2443*, NASA, Dec. 1974.

14.12. "Two Decades of the 'Twenty-One'," *Air Enthusiast International*, May 1974, Vol. 6, No. 5, pp. 226–232.

14.13. Goodmanson, L. T., and L. B. Gratzer, "Recent Advances in Aerodynamics for Transport Aircraft," *Aeronautics and Astronautics*, Dec. 1973, Vol. 11, No. 12, pp. 30–45.

14.14. Schlichting, H., "Some Developments in Boundary Layer Research in the Past Thirty Years," *Journal of the Royal Aeronautical Society*, Feb. 1960, Vol. 64, No. 590, pp. 64–79.

14.15. Green, W., *The Warplanes of the Third Reich*, Doubleday & Company, Garden City, N.Y., 1970.

14.16. Englar, R. J., "Circulation Control Technology for Powered-Lift STOL Aircraft," *Lockheed Horizons*, Sept. 1987, Issue 24, pp. 44–55.

14.17. Bradley, R. G., "Practical Aerodynamic Problems-Military Aircraft," presented at the Transonic Perspective Symposium, NASA/Ames Research Center, Feb. 1981.

14.18. Meyer, R. C., and W. D. Fields, "Configuration Development of a Supersonic Cruise Strike-Fighter," *AIAA Paper 78–148*, presented at the AIAA 16th Aerospace Sciences Meeting, Huntsville, Ala., Jan. 1978.

14.19. Ferris, J. C., "Wind-Tunnel Investigation of a Variable Camber and Twist Wing," *TN D-8475*, NASA, Aug. 1977.

14.20. Kosin, R. E., "Laminar Flow Control by Suction as Applied to the X-21A Airplane," *Journal of Aircraft*, Sept.–Oct. 1965, Vol. 2, No. 5, pp. 384–390.

14.21. Jobe, C. E., R. M. Kulfan, and J. C. Vachal, "Application of Laminar Flow Control to Large Subsonic Military Transport Airplanes," *AIAA Paper 78–95*, presented at the AIAA 16th Aerospace Sciences Meeting, Huntsville, Ala., Jan. 1978.

14.22. Thomas, A. S. W., "Aircraft Drag Reduction Technology," *AGARD Report 723*, 1985 (Quotations from this reference are reproduced by permission of AGARD and A. S. W. Thomas, Lockheed Aeronautical Systems Company, Georgia Division.)

14.23. Blackwell, J. A., Jr., "A Finite-Step Method for Calculation of Theoretical Load Distributions for Arbitrary Lifting-Surface Arrangements at Subsonic Speeds," *TN D-5335*, NASA, July 1969.

14.24. Hanley, R. J., "Development of an Airplane Modification to Improve the Mission Effectiveness of the EA-6B Airplane," *AIAA Paper 87–2358*, presented at the AIAA 5th Applied Aerodynamics Conference, Monterey, Calif., Aug. 1987.

14.25. Gato, W., and M. F. Masiello, "Innovative Aerodynamics: The Sensible Way of Restoring the Growth Capability of the EA-6B Prowler," *AIAA Paper 87–2362*. presented at the AIAA 5th Applied Aerodynamics Conference, Monterey, Calif., Aug. 1987.

14.26. Jordan, F. L., Jr., D. E. Hahne, M. F. Masiello, and W. Gato, "High-Angle-of-Attack Stability and Control Improvements for the EA-6B Prowler," *AIAA Paper 87–2361*, presented at the AIAA 5th Applied Aerodynamics Conference, Monterey, Calif., Aug. 1987.

14.27. Sewall, W. G., R. J. McGhee, and J. C. Ferris, "Wind-Tunnel Test Results of Airfoil Modifications for the EA-6B," *AIAA Paper 87–2359*, presented at the AIAA 5th Applied Aerodynamics Conference, Monterey, Calif., Aug. 1987.

14.28. Waggoner, E. G., and D. O. Allison, "EA-6B High-Lift Wing Modifications," *AIAA Paper 87–2360*, presented at the AIAA 5th Applied Aerodynamics Conference, Monterey, Calif., Aug. 1987.

A

THE EQUATIONS OF MOTION WRITTEN IN CONSERVATION FORM

In the main text the basic equations of motion were developed and were used in the nonconservative form, e.g., the momentum equation [equation (2.12)] and the energy equation [equation (2.32)]. However, the reader who pursues advanced applications of fluid mechanics often encounters the basic equations in conservation form for use in computational fluid dynamics. The conservation form of the basic equations is used for the following reasons (among others). First, it is easier to derive a noniterative, second-order, implicit algorithm if the nonlinear equations are in conservation form. Second, when shock waves and shear layers are expected in the flow field, it is essential that conservative difference approximations be used if one uses a "shock-capturing" method (Ref. A.1). *Shock capturing* is defined as ignoring the presence of embedded discontinuities in the sense that they are not treated as internal boundaries in the difference algorithm.

Let us establish the following nomenclature for the general form of the equations of motion in conservation form:

$$\frac{\partial U}{\partial t} + \frac{\partial (E_i - E_v)}{\partial x} + \frac{\partial (F_i - F_v)}{\partial y} + \frac{\partial (G_i - G_v)}{\partial z} = 0 \qquad (A.1)$$

where the subscript i denotes the terms that are included in the equations of motion for an inviscid flow and the subscript v denotes the terms that are unique to the equations of motion when the viscous and heat-transfer effects are included.

Recall that the continuity equation (2.1) is

$$\frac{\partial \rho}{\partial t} + \frac{\partial (\rho u)}{\partial x} + \frac{\partial (\rho v)}{\partial y} + \frac{\partial (\rho w)}{\partial z} = 0 \qquad (A.2)$$

Note that this equation is already in conservation form. It is valid for an inviscid flow as well as a viscous flow. Therefore, comparing equations (A.1) and (A.2), we obtain

$$U = \rho \qquad E_i = \rho u \qquad F_i = \rho v \qquad G_i = \rho w \quad (A.3a)\text{–}(A.3d)$$

$$E_v = F_v = G_v = 0 \qquad (A.3e\text{–}A.3g)$$

Consider the x-component of the momentum equation, as given by equation (2.11a). Assuming that the body forces are negligible, as is commonly done for gas flows:

$$\rho \frac{\partial u}{\partial t} + \rho u \frac{\partial u}{\partial x} + \rho v \frac{\partial u}{\partial y} + \rho w \frac{\partial u}{\partial z} = -\frac{\partial p}{\partial x} + \frac{\partial (\tau_{xx}')}{\partial x} + \frac{\partial (\tau_{yx})}{\partial y} + \frac{\partial (\tau_{zx})}{\partial z} \quad (A.4)$$

To obtain equation (A.4), we have divided the expression for the normal shear stress τ_{xx}, as defined in Chapter 2, into the sum of the pressure p which is present in the equations for an inviscid flow and a term relating to the viscous flow τ_{xx}'. This is done to satisfy more easily the format of equation (A.1). Thus,

$$\tau_{xx}' = -\frac{2}{3} \mu \nabla \cdot \vec{V} + 2\mu \frac{\partial u}{\partial x}$$

The definitions for τ_{yx} and τ_{zx} remain as defined in Chapter 2. If we multiply the continuity equation (A.2) by u, we obtain

$$u \frac{\partial \rho}{\partial t} + u \frac{\partial (\rho u)}{\partial x} + u \frac{\partial (\rho v)}{\partial y} + u \frac{\partial (\rho w)}{\partial z} = 0 \qquad (A.5)$$

Adding equations (A.4) and (A.5) and rearranging, we obtain

$$\frac{\partial (\rho u)}{\partial t} + \frac{\partial}{\partial x}\left[(p + \rho u^2) - (\tau_{xx}') \right]$$

$$+ \frac{\partial}{\partial y}\left[(\rho u v) - (\tau_{yx}) \right] + \frac{\partial}{\partial z}\left[(\rho u w) - (\tau_{zx}) \right] = 0 \qquad (A.6)$$

Comparing equation (A.6) to the general form of the equations of motion in conservation form, equation (A.1), it is clear that

$$U = \rho u \qquad E_i = p + \rho u^2 \qquad F_i = \rho u v \qquad G_i = \rho u w \quad (A.7a)\text{–}(A.7d)$$

$$E_v = \tau_{xx}' \qquad F_v = \tau_{yx} \qquad G_v = \tau_{zx} \qquad (A.7e)\text{–}(A.7g)$$

Similar manipulations of the y momentum will yield

$$U = \rho v \qquad E_i = \rho v u \qquad F_i = p + \rho v^2 \qquad G_i = \rho v w \quad (A.8a)\text{–}(A.8d)$$

$$E_v = \tau_{xy} \qquad F_v = \tau_{yy}' \qquad G_v = \tau_{zy} \qquad (A.8e)\text{–}(A.8g)$$

where

$$\tau_{yy}' = -\frac{2}{3}\,\mu\nabla\cdot\vec{V} + 2\,\mu\,\frac{\partial v}{\partial y}$$

Similarly, the z-momentum equation can be written in conservation form, if

$$U = \rho w \qquad E_i = \rho wu \qquad F_i = \rho wv \qquad G_i = p + \rho w^2 \quad \text{(A.9a)–(A.9d)}$$

$$E_v = \tau_{xz} \qquad F_v = \tau_{yz} \qquad G_v = \tau_{zz}' \qquad \text{(A.9e)–(A.9g)}$$

where

$$\tau_{zz}' = -\frac{2}{3}\,\mu\nabla\cdot\vec{V} + 2\mu\,\frac{\partial w}{\partial z}$$

The energy equation (2.25) is

$$\rho\dot{q} - \rho\dot{w} = \rho\,\frac{d}{dt}\,(ke) + \rho\,\frac{d}{dt}\,(pe) + \rho\,\frac{d}{dt}\,(u_e) \qquad (2.25)$$

Neglecting the changes in potential energy and using the definitions for $\rho\dot{q}$ and $\rho\dot{w}$, we obtain

$$\frac{\partial}{\partial x}\left(k\frac{\partial T}{\partial x}\right) + \frac{\partial}{\partial y}\left(k\frac{\partial T}{\partial y}\right) + \frac{\partial}{\partial z}\left(k\frac{\partial T}{\partial z}\right)$$

$$-\frac{\partial}{\partial x}\,(up) + \frac{\partial}{\partial x}\,(u\tau_{xx}') + \frac{\partial}{\partial x}\,(v\tau_{xy}) + \frac{\partial}{\partial x}\,(w\tau_{xz})$$

$$+\frac{\partial}{\partial y}\,(u\tau_{yx}) - \frac{\partial}{\partial y}\,(vp) + \frac{\partial}{\partial y}\,(v\tau_{yy}') + \frac{\partial}{\partial y}\,(w\tau_{yz}) \qquad (A.10)$$

$$+\frac{\partial}{\partial z}\,(u\tau_{zx}) + \frac{\partial}{\partial z}\,(v\tau_{zy}) - \frac{\partial}{\partial z}\,(wp) + \frac{\partial}{\partial z}\,(w\tau_{zz}')$$

$$= \rho\,\frac{de_t}{dt} = \rho\,\frac{\partial e_t}{\partial t} + \rho u\,\frac{\partial e_t}{\partial x} + \rho v\,\frac{\partial e_t}{\partial y} + \rho w\,\frac{\partial e_t}{\partial z}$$

where e_t, the specific total energy of the flow (neglecting the potential energy), is given by

$$e_t = u_e + \tfrac{1}{2}\,(u^2 + v^2 + w^2) \qquad (A.11)$$

The continuity equation multiplied by e_t is

$$e_t\,\frac{\partial\rho}{\partial t} + e_t\,\frac{\partial(\rho u)}{\partial x} + e_t\,\frac{\partial(\rho v)}{\partial y} + e_t\,\frac{\partial(\rho w)}{\partial z} = 0 \qquad (A.12)$$

Adding equations (A.10) and (A.12) and rearranging, we obtain

$$\frac{\partial}{\partial t}\,(\rho e_t) + \frac{\partial}{\partial x}\left[(\rho e_t + p)u - (u\tau_{xx}' + v\tau_{xy} + w\tau_{xz} + q_x)\right]$$

$$+ \frac{\partial}{\partial y} \left[(\rho e_t + p)v - (u\tau_{yx} + v\tau'_{yy} + w\tau_{yz} + q_y) \right] \qquad (A.13)$$

$$+ \frac{\partial}{\partial z} \left[(\rho e_t + p)w - (u\tau_{zx} + v\tau_{zy} + w\tau'_{zz} + q_z) \right]$$

Equation (A.13) is the energy equation in the conservation form. Comparing equation (A.13) to the general form, equation (A.1), it is clear that

$$U = \rho e_t \qquad E_i = (\rho e_t + p)u \qquad F_i = (\rho e_t + p)v$$

$$G_i = (\rho e_t + p)w$$

$$(A.14a)–(A.14d)$$

$$E_v = u\tau'_{xx} + v\tau_{xy} + w\tau_{xz} + q_x \qquad (A.14e)$$

$$F_v = u\tau_{yx} + v\tau'_{yy} + w\tau_{yz} + q_y \qquad (A.14f)$$

$$G_v = u\tau_{zx} + v\tau_{zy} + w\tau'_{zz} + q_z \qquad (A.14g)$$

When the fundamental equations governing the unsteady flow of gas, without body forces or external heat addition, are written in conservation form [i.e., equation (A.1)], the terms U, E_i, E_v, F_i, F_v, G_i and G_v can be represented by the following vectors.

$$U = \begin{bmatrix} \rho \\ \rho u \\ \rho v \\ \rho w \\ \rho e_t \end{bmatrix}$$

$$E_i = \begin{bmatrix} \rho u \\ p + \rho u^2 \\ \rho u v \\ \rho u w \\ (\rho e_t + p)u \end{bmatrix} \qquad E_v = \begin{bmatrix} 0 \\ \tau'_{xx} \\ \tau_{xy} \\ \tau_{xz} \\ (u\tau'_{xx} + v\tau_{xy} + w\tau_{xz} + q_x) \end{bmatrix}$$

$$F_i = \begin{bmatrix} \rho v \\ \rho v u \\ p + \rho v^2 \\ \rho v w \\ (\rho e_t + p)v \end{bmatrix} \qquad F_v = \begin{bmatrix} 0 \\ \tau_{yx} \\ \tau'_{yy} \\ \tau_{yz} \\ (u\tau_{yx} + v\tau'_{yy} + w\tau_{yz} + q_y) \end{bmatrix}$$

$$G_i = \begin{bmatrix} \rho w \\ \rho w u \\ \rho w v \\ p + \rho w^2 \\ (\rho e_t + p)w \end{bmatrix} \qquad G_v = \begin{bmatrix} 0 \\ \tau_{zx} \\ \tau_{zy} \\ \tau'_{zz} \\ (u\tau_{zx} + v\tau_{zy} + w\tau'_{zz} + q_z) \end{bmatrix}$$

The successive lines of these vectors are the continuity equation, the x-momentum equation, the y-momentum equation, the z-momentum equation, and the energy equation. Note that considering only U and the inviscid terms (E_i, F_i, and G_i) of lines 2 through 4 yields Euler's equations in conservation form.

Repeating the information of Chapter 2 (with the modifications noted above)

$$\tau_{xx}' = 2\mu\frac{\partial u}{\partial x} - \frac{2}{3}\mu\nabla\cdot\vec{V}$$

$$\tau_{yy}' = 2\mu\frac{\partial v}{\partial y} - \frac{2}{3}\mu\nabla\cdot\vec{V}$$

$$\tau_{zz}' = 2\mu\frac{\partial w}{\partial z} - \frac{2}{3}\mu\nabla\cdot\vec{V}$$

$$\tau_{xy} = \tau_{yx} = \mu\left(\frac{\partial u}{\partial y} + \frac{\partial v}{\partial x}\right)$$

$$\tau_{xz} = \tau_{zx} = \mu\left(\frac{\partial u}{\partial z} + \frac{\partial w}{\partial x}\right)$$

$$\tau_{yz} = \tau_{zy} = \mu\left(\frac{\partial v}{\partial z} + \frac{\partial w}{\partial y}\right)$$

$$q_x = k\frac{\partial T}{\partial x} \qquad q_y = k\frac{\partial T}{\partial y} \qquad q_z = k\frac{\partial T}{\partial z}$$

Similar procedures can be followed to develop the conservative form of the governing equations for flow in cylindrical coordinates.

$$\frac{\partial U}{\partial t} + \frac{\partial(F_i - F_v)}{\partial z} + \frac{\partial(G_i - G_v)}{\partial r} + \frac{\partial(H_i - H_v)}{\partial\theta} + (R_i - R_v) = 0 \qquad (A.15)$$

$$U = \begin{bmatrix} \rho \\ \rho v_z \\ \rho v_r \\ \rho v_\theta \\ \rho e_t \end{bmatrix}$$

$$F_i = \begin{bmatrix} \rho v_z \\ p + \rho v_z^2 \\ \rho v_z v_r \\ \rho v_z v_\theta \\ (\rho e_t + p)v_z \end{bmatrix} \qquad F_v = \begin{bmatrix} 0 \\ \tau_{zz} \\ \tau_{zr} \\ \tau_{z\theta} \\ (v_z\tau_{zz} + v_r\tau_{zr} + v_\theta\tau_{z\theta} + q_z) \end{bmatrix}$$

$$G_i = \begin{bmatrix} \rho v_r \\ \rho v_r v_z \\ p + \rho v_r^2 \\ \rho v_r v_\theta \\ (\rho e_t + p)v_r \end{bmatrix} \qquad G_v = \begin{bmatrix} 0 \\ \tau_{rz} \\ \tau_{rr} \\ \tau_{r\theta} \\ (v_z\tau_{rz} + v_r\tau_{rr} + v_\theta\tau_{r\theta} + q_r) \end{bmatrix}$$

$$H_i = \frac{1}{r} \begin{bmatrix} \rho v_\theta \\ \rho v_\theta \, v_z \\ \rho v_\theta \, v_r \\ p + \rho v_\theta^2 \\ (\rho e_t + p)v_\theta \end{bmatrix} \qquad H_v = \frac{1}{r} \begin{bmatrix} 0 \\ \tau_{\theta z} \\ \tau_{\theta r} \\ \tau_{\theta\theta} \\ (v_z \, \tau_{z\theta} + v_r \, \tau_{r\theta} + v_\theta \, \tau_{\theta\theta} + q_\theta) \end{bmatrix}$$

$$R_i = \frac{1}{r} \begin{bmatrix} \rho v_r \\ \rho v_r \, v_z \\ \rho v_r^2 - \rho v_\theta^2 \\ 2\rho v_r \, v_\theta \\ (\rho e_t + p)v_r \end{bmatrix} \qquad R_v = \frac{1}{r} \begin{bmatrix} 0 \\ \tau_{rz} \\ \tau_{rr} - \tau_{\theta\theta} \\ 2\tau_{r\theta} \\ (v_z \, \tau_{zr} + v_r \, \tau_{rr} + v_\theta \, \tau_{\theta r} + q_r) \end{bmatrix}$$

where

$$\tau_{rr} = 2\mu \frac{\partial v_r}{\partial r} - \frac{2}{3} \mu \nabla \cdot \vec{V}$$

$$\tau_{\theta\theta} = 2\mu \left(\frac{1}{r} \frac{\partial v_\theta}{\partial \theta} + \frac{v_r}{r} \right) - \frac{2}{3} \mu \nabla \cdot \vec{V}$$

$$\tau_{zz} = 2\mu \frac{\partial v_z}{\partial z} - \frac{2}{3} \mu \nabla \cdot \vec{V}$$

$$\tau_{\theta z} = \tau_{z\theta} = \mu \left(\frac{1}{r} \frac{\partial v_z}{\partial \theta} + \frac{\partial v_\theta}{\partial z} \right)$$

$$\tau_{\theta r} = \tau_{r\theta} = \mu \left(\frac{\partial v_\theta}{\partial r} - \frac{v_\theta}{r} + \frac{1}{r} \frac{\partial v_r}{\partial \theta} \right)$$

$$\tau_{zr} = \tau_{rz} = \mu \left(\frac{\partial v_r}{\partial z} + \frac{\partial v_z}{\partial r} \right)$$

$$q_z = k \frac{\partial T}{\partial z} \qquad q_r = k \frac{\partial T}{\partial r} \qquad q_\theta = \frac{k}{r} \frac{\partial T}{\partial \theta}$$

REFERENCE

A.1. Warming, R. F., and R. M. Beam, "On the Construction and Application of Implicit Factored Schemes for Conservation Laws," *SIAM-AMS Proceedings*, Vol. 11, 1978, pp. 85–129.

B

A COLLECTION
OF OFTEN USED TABLES

TABLE 1.2 U.S. STANDARD ATMOSPHERE, 1962
a. Metric Units

Altitude (km)	Pressure (mm Hg)	Temperature (K)	Density (kg/m³)	Viscosity (kg/s·m) × 10⁵	Speed of sound (m/s)
0	760.000	288.150	1.2250	1.7894	340.294
1	674.127	281.651	1.1117	1.7579	336.435
2	596.309	275.154	1.0066	1.7260	332.532
3	525.952	268.659	0.9092	1.6938	328.583
4	462.491	262.166	0.8194	1.6612	324.589
5	405.395	255.676	0.7364	1.6282	320.545
6	354.161	249.187	0.6601	1.5949	316.452
7	308.315	242.700	0.5900	1.5612	312.306
8	267.409	236.215	0.5258	1.5271	308.105
9	231.024	229.733	0.4671	1.4926	303.848
10	198.765	223.252	0.4135	1.4577	299.532
11	170.263	216.774	0.3648	1.4223	295.154
12	145.508	216.650	0.3119	1.4216	295.069
13	124.357	216.650	0.2666	1.4216	295.069
14	106.286	216.650	0.2279	1.4216	295.069

TABLE 1.2 (Continued)

Altitude (km)	Pressure (mm Hg)	Tempera- ture (K)	Density (kg/m³)	Viscosity (kg/s·m) $\times 10^5$	Speed of sound (m/s)
15	90.846	216.650	0.1948	1.4216	295.069
16	77.653	216.650	0.1665	1.4216	295.069
17	66.378	216.650	0.1423	1.4216	295.069
18	56.744	216.650	0.1216	1.4216	295.069
19	48.510	216.650	0.1040	1.4216	295.069
20	41.473	216.650	0.0889	1.4216	295.069
21	35.470	217.581	0.0757	1.4267	295.703
22	30.359	218.574	0.0645	1.4322	296.377
23	26.004	219.567	0.0550	1.4376	297.049
24	22.290	220.560	0.0469	1.4430	297.720
25	19.121	221.552	0.0401	1.4484	298.389
26	16.414	222.544	0.0343	1.4538	299.056
27	14.101	223.536	0.0293	1.4592	299.722
28	12.123	224.527	0.0251	1.4646	300.386
29	10.429	225.518	0.0215	1.4699	301.048
30	8.978	226.509	0.0184	1.4753	301.709

b. English Units[a]

Altitude (10^3 ft)	Pressure, p/p_0	Temperature (°R)	Density, ρ/ρ_0	Viscosity, μ/μ_0	Speed of sound (ft/s)
0	1.00000	518.67	1.0000	1.00000	1116.45
5	0.83209	500.84	0.8617	0.97307	1097.10
10	0.68783	483.03	0.7386	0.94570	1077.40
15	0.56459	465.22	0.6295	0.91786	1057.36
20	0.45991	447.42	0.5332	0.88953	1036.93
25	0.37158	429.62	0.4486	0.86070	1016.10
30	0.29754	411.84	0.3747	0.83135	994.85
35	0.23596	394.06	0.3106	0.80144	973.14
40	0.18577	389.97	0.2471	0.79447	968.08
45	0.14623	389.97	0.1945	0.79447	968.08
50	0.11512	389.97	0.1531	0.79447	968.08
55	0.09063	389.97	0.1206	0.79447	968.08
60	0.07137	389.97	0.0949	0.79447	968.08
65	0.05620	389.97	0.0747	0.79447	968.08
70	0.04429	392.25	0.0586	0.79835	970.90
75	0.03496	394.97	0.0459	0.80298	974.26
80	0.02765	397.69	0.0361	0.80760	977.62
85	0.02190	400.42	0.0284	0.81219	980.95
90	0.01738	403.14	0.0224	0.81678	984.28
95	0.01381	405.85	0.0177	0.82134	987.59
100	0.01100	408.57	0.0140	0.82589	990.90

[a] The corresponding sea-level values are: $p_0 = 2116.22$ lbf/ft², $\rho_0 = 0.076474$ lbm/ft³ $= 0.002376$ slug/ft³, and $\mu_0 = 1.2024 \times 10^{-5}$ lbm/ft·s $= 3.740 \times 10^{-7}$ lbf·s/ft².

TABLE 8.1 CORRELATIONS FOR A ONE-DIMENSIONAL, ISENTROPIC FLOW
OF PERFECT AIR($\gamma = 1.4$)

M	$\dfrac{A}{A^*}$	$\dfrac{p}{p_{t1}}$	$\dfrac{\rho}{\rho_{t1}}$	$\dfrac{T}{T_t}$	$\dfrac{A}{A^*}\dfrac{p}{p_{t1}}$
0	∞	1.00000	1.00000	1.00000	∞
0.05	11.592	0.99825	0.99875	0.99950	11.571
0.10	5.8218	0.99303	0.99502	0.99800	5.7812
0.15	3.9103	0.98441	0.98884	0.99552	3.8493
0.20	2.9635	0.97250	0.98027	0.99206	2.8820
0.25	2.4027	0.95745	0.96942	0.98765	2.3005
0.30	2.0351	0.93947	0.95638	0.98232	1.9119
0.35	1.7780	0.91877	0.94128	0.97608	1.6336
0.40	1.5901	0.89562	0.92428	0.96899	1.4241
0.45	1.4487	0.87027	0.90552	0.96108	1.2607
0.50	1.3398	0.84302	0.88517	0.95238	1.12951
0.55	1.2550	0.81416	0.86342	0.94295	1.02174
0.60	1.1882	0.78400	0.84045	0.93284	0.93155
0.65	1.1356	0.75283	0.81644	0.92208	0.85493
0.70	1.09437	0.72092	0.79158	0.91075	0.78896
0.75	1.06242	0.68857	0.76603	0.89888	0.73155
0.80	1.03823	0.65602	0.74000	0.88652	0.68110
0.85	1.02067	0.62351	0.71361	0.87374	0.63640
0.90	1.00886	0.59126	0.68704	0.86058	0.59650
0.95	1.00214	0.55946	0.66044	0.84710	0.56066
1.00	1.00000	0.52828	0.63394	0.83333	0.52828
1.05	1.00202	0.49787	0.60765	0.81933	0.49888
1.10	1.00793	0.46835	0.58169	0.80515	0.47206
1.15	1.01746	0.43983	0.55616	0.79083	0.44751
1.20	1.03044	0.41238	0.53114	0.77640	0.42493
1.25	1.04676	0.38606	0.50670	0.76190	0.40411
1.30	1.06631	0.36092	0.48291	0.74738	0.38484
1.35	1.08904	0.33697	0.45980	0.73287	0.36697
1.40	1.1149	0.31424	0.43742	0.71839	0.35036
1.45	1.1440	0.29272	0.41581	0.70397	0.33486
1.50	1.1762	0.27240	0.39498	0.68965	0.32039
1.55	1.2115	0.25326	0.37496	0.67545	0.30685
1.60	1.2502	0.23527	0.35573	0.66138	0.29414
1.65	1.2922	0.21839	0.33731	0.64746	0.28221
1.70	1.3376	0.20259	0.31969	0.63372	0.27099
1.75	1.3865	0.18782	0.30287	0.62016	0.26042
1.80	1.4390	0.17404	0.28682	0.60680	0.25044
1.85	1.4952	0.16120	0.27153	0.59365	0.24102
1.90	1.5555	0.14924	0.25699	0.58072	0.23211
1.95	1.6193	0.13813	0.24317	0.56802	0.22367
2.00	1.6875	0.12780	0.23005	0.55556	0.21567
2.05	1.7600	0.11823	0.21760	0.54333	0.20808
2.10	1.8369	0.10935	0.20580	0.53135	0.20087
2.15	1.9185	0.10113	0.19463	0.51962	0.19403
2.20	2.0050	0.09352	0.18405	0.50813	0.18751

TABLE 8.1 *(Continued)*

M	$\dfrac{A}{A^*}$	$\dfrac{p}{p_{t1}}$	$\dfrac{\rho}{\rho_{t1}}$	$\dfrac{T}{T_t}$	$\dfrac{A}{A^*}\dfrac{p}{p_{t1}}$
2.25	2.0964	0.08648	0.17404	0.49689	0.18130
2.30	2.1931	0.07997	0.16458	0.48591	0.17539
2.35	2.2953	0.07396	0.15564	0.47517	0.16975
2.40	2.4031	0.06840	0.14720	0.46468	0.16437
2.45	2.5168	0.06327	0.13922	0.45444	0.15923
2.50	2.6367	0.05853	0.13169	0.44444	0.15432
2.55	2.7630	0.05415	0.12458	0.43469	0.14963
2.60	2.8960	0.05012	0.11787	0.42517	0.14513
2.65	3.0359	0.04639	0.11154	0.41589	0.14083
2.70	3.1830	0.04295	0.10557	0.40684	0.13671
2.75	3.3376	0.03977	0.09994	0.39801	0.13276
2.80	3.5001	0.03685	0.09462	0.38941	0.12897
2.85	3.6707	0.03415	0.08962	0.38102	0.12534
2.90	3.8498	0.03165	0.08489	0.37286	0.12185
2.95	4.0376	0.02935	0.08043	0.36490	0.11850
3.00	4.2346	0.02722	0.07623	0.35714	0.11527
3.50	6.7896	0.01311	0.04523	0.28986	0.08902
4.00	10.719	0.00658	0.02766	0.23810	0.07059
4.50	16.562	0.00346	0.01745	0.19802	0.05723
5.00	25.000	$189(10)^{-5}$	0.01134	0.16667	0.04725
6.00	53.189	$633(10)^{-6}$	0.00519	0.12195	0.03368
7.00	104.143	$242(10)^{-6}$	0.00261	0.09259	0.02516
8.00	190.109	$102(10)^{-6}$	0.00141	0.07246	0.01947
9.00	327.189	$474(10)^{-7}$	0.000815	0.05814	0.01550
10.00	535.938	$236(10)^{-7}$	0.000495	0.04762	0.01263
∞	∞	0	0	0	0

TABLE 8.2 MACH NUMBER AND MACH ANGLE AS A FUNCTION OF PRANDTL-MEYER ANGLE

ν (deg)	M	μ (deg)	ν (deg)	M	μ (deg)
0.0	1.000	90.000	5.0	1.256	52.738
0.5	1.051	72.099	5.5	1.275	51.642
1.0	1.082	67.574	6.0	1.294	50.619
1.5	1.108	64.451	6.5	1.312	49.658
2.0	1.133	61.997	7.0	1.330	48.753
2.5	1.155	59.950	7.5	1.348	47.896
3.0	1.177	58.180	8.0	1.366	47.082
3.5	1.198	56.614	8.5	1.383	46.306
4.0	1.218	55.205	9.0	1.400	45.566
4.5	1.237	53.920	9.5	1.418	44.857

TABLE 8.2   (*Continued*)

ν (deg)	M	μ (deg)	ν (deg)	M	μ (deg)
10.0	1.435	44.177	32.5	2.230	26.646
10.5	1.452	43.523	33.0	2.249	26.397
11.0	1.469	42.894	33.5	2.269	26.151
11.5	1.486	42.287	34.0	2.289	25.908
12.0	1.503	41.701	34.5	2.309	25.668
12.5	1.520	41.134	35.0	2.329	25.430
13.0	1.537	40.585	35.5	2.349	25.196
13.5	1.554	40.053	36.0	2.369	24.965
14.0	1.571	39.537	36.5	2.390	24.736
14.5	1.588	39.035	37.0	2.410	24.510
15.0	1.605	38.547	37.5	2.431	24.287
15.5	1.622	38.073	38.0	2.452	24.066
16.0	1.639	37.611	38.5	2.473	23.847
16.5	1.655	37.160	39.0	2.495	23.631
17.0	1.672	36.721	39.5	2.516	23.418
17.5	1.689	36.293	40.0	2.538	23.206
18.0	1.706	35.874	40.5	2.560	22.997
18.5	1.724	35.465	41.0	2.582	22.790
19.0	1.741	35.065	41.5	2.604	22.585
19.5	1.758	34.673	42.0	2.626	22.382
20.0	1.775	34.290	42.5	2.649	22.182
20.5	1.792	33.915	43.0	2.671	21.983
21.0	1.810	33.548	43.5	2.694	21.786
21.5	1.827	33.188	44.0	2.718	21.591
22.0	1.844	32.834	44.5	2.741	21.398
22.5	1.862	32.488	45.0	2.764	21.207
23.0	1.879	32.148	45.5	2.788	21.017
23.5	1.897	31.814	46.0	2.812	20.830
24.0	1.915	31.486	46.5	2.836	20.644
24.5	1.932	31.164	47.0	2.861	20.459
25.0	1.950	30.847	47.5	2.886	20.277
25.5	1.968	30.536	48.0	2.910	20.096
26.0	1.986	30.229	48.5	2.936	19.916
26.5	2.004	29.928	49.0	2.961	19.738
27.0	2.023	29.632	49.5	2.987	15.561
27.5	2.041	29.340	50.0	3.013	19.386
28.0	2.059	29.052	50.5	3.039	19.213
28.5	2.078	28.769	51.0	3.065	19.041
29.0	2.096	28.491	51.5	3.092	18.870
29.5	2.115	28.216	52.0	3.119	18.701
30.0	2.134	27.945	52.5	3.146	18.532
30.5	2.153	27.678	53.0	3.174	18.366
31.0	2.172	27.415	53.5	3.202	18.200
31.5	2.191	27.155	54.0	3.230	18.036
32.0	2.210	26.899	54.5	3.258	17.873

TABLE 8.2 (*Continued*)

ν (deg)	M	μ (deg)	ν (deg)	M	μ (deg)
55.0	3.287	17.711	76.5	4.955	11.642
55.5	3.316	17.551	77.0	5.009	11.517
56.0	3.346	17.391			
56.5	3.375	17.233	77.5	5.063	11.392
57.0	3.406	17.076	78.0	5.118	11.268
			78.5	5.174	11.145
57.5	3.436	16.920	79.0	5.231	11.022
58.0	3.467	16.765	79.5	5.289	10.899
58.5	3.498	16.611			
59.0	3.530	16.458	80.0	5.348	10.777
59.5	3.562	16.306	80.5	5.408	10.656
			81.0	5.470	10.535
60.0	3.594	16.155	81.5	5.532	10.414
60.5	3.627	16.006	82.0	5.596	10.294
61.0	3.660	15.856			
61.5	3.694	15.708	82.5	5.661	10.175
62.0	3.728	15.561	83.0	5.727	10.056
			83.5	5.795	9.937
62.5	3.762	15.415	84.0	5.864	9.819
63.0	3.797	15.270	84.5	5.935	9.701
63.5	3.832	15.126			
64.0	3.868	14.983	85.0	6.006	9.584
64.5	3.904	14.840	85.5	6.080	9.467
			86.0	6.155	9.350
65.0	3.941	14.698	86.5	6.232	9.234
65.5	3.979	14.557	87.0	6.310	9.119
66.0	4.016	14.417			
66.5	4.055	14.278	87.5	6.390	9.003
67.0	4.094	14.140	88.0	6.472	8.888
			88.5	6.556	8.774
67.5	4.133	14.002	89.0	6.642	8.660
68.0	4.173	13.865	89.5	6.729	8.546
68.5	4.214	13.729			
69.0	4.255	13.593	90.0	6.819	8.433
69.5	4.297	13.459	90.5	6.911	8.320
			91.0	7.005	8.207
70.0	4.339	13.325	91.5	7.102	8.095
70.5	4.382	13.191	92.0	7.201	7.983
71.0	4.426	13.059			
71.5	4.470	12.927	92.5	7.302	7.871
72.0	4.515	12.795	93.0	7.406	7.760
			93.5	7.513	7.649
72.5	4.561	12.665	94.0	7.623	7.538
73.0	4.608	12.535	94.5	7.735	7.428
73.5	4.655	12.406			
74.0	4.703	12.277	95.0	7.851	7.318
74.5	4.752	12.149	95.5	7.970	7.208
			96.0	8.092	7.099
75.0	4.801	12.021	96.5	8.218	6.989
75.5	4.852	11.894	97.0	8.347	6.881
76.0	4.903	11.768			

TABLE 8.2 (*Continued*)

ν (deg)	M	μ (deg)	ν (deg)	M	μ (deg)
97.5	8.480	6.772	100.0	9.210	6.233
98.0	8.618	6.664	100.5	9.371	6.126
98.5	8.759	6.556	101.0	9.536	6.019
99.0	8.905	6.448	101.5	9.708	5.913
99.5	9.055	6.340	102.0	9.885	5.806

TABLE 8.3 CORRELATION OF FLOW PROPERTIES ACROSS A NORMAL SHOCK WAVE AS A FUNCTION OF THE UPSTREAM MACH NUMBER FOR AIR, $\gamma = 1.4$

M_1	M_2	$\dfrac{p_2}{p_1}$	$\dfrac{\rho_2}{\rho_1}$	$\dfrac{T_2}{T_1}$	$\dfrac{p_{t2}}{p_{t1}}$
1.00	1.00000	1.00000	1.00000	1.00000	1.00000
1.05	0.95312	1.1196	1.08398	1.03284	0.99987
1.10	0.91177	1.2450	1.1691	1.06494	0.99892
1.15	0.87502	1.3762	1.2550	1.09657	0.99669
1.20	0.84217	1.5133	1.3416	1.1280	0.99280
1.25	0.81264	1.6562	1.4286	1.1594	0.98706
1.30	0.78596	1.8050	1.5157	1.1909	0.97935
1.35	0.76175	1.9596	1.6027	1.2226	0.96972
1.40	0.73971	2.1200	1.6896	1.2547	0.95819
1.45	0.71956	2.2862	1.7761	1.2872	0.94483
1.50	0.70109	2.4583	1.8621	1.3202	0.92978
1.55	0.68410	2.6363	1.9473	1.3538	0.91319
1.60	0.66844	2.8201	2.0317	1.3880	0.89520
1.65	0.65396	3.0096	2.1152	1.4228	0.87598
1.70	0.64055	3.2050	2.1977	1.4583	0.85573
1.75	0.62809	3.4062	2.2781	1.4946	0.83456
1.80	0.61650	3.6133	2.3592	1.5316	0.81268
1.85	0.60570	3.8262	2.4381	1.5694	0.79021
1.90	0.59562	4.0450	2.5157	1.6079	0.76735
1.95	0.58618	4.2696	2.5919	1.6473	0.74418
2.00	0.57735	4.5000	2.6666	1.6875	0.72088
2.05	0.56907	4.7363	2.7400	1.7286	0.69752
2.10	0.56128	4.9784	2.8119	1.7704	0.67422
2.15	0.55395	5.2262	2.8823	1.8132	0.65105
2.20	0.54706	5.4800	2.9512	1.8569	0.62812
2.25	0.54055	5.7396	3.0186	1.9014	0.60554
2.30	0.53441	6.0050	3.0846	1.9468	0.58331
2.35	0.52861	6.2762	3.1490	1.9931	0.56148
2.40	0.52312	6.5533	3.2119	2.0403	0.54015
2.45	0.51792	6.8362	3.2733	2.0885	0.51932

TABLE 8.3 (*Continued*)

M_1	M_2	$\dfrac{p_2}{p_1}$	$\dfrac{\rho_2}{\rho_1}$	$\dfrac{T_2}{T_1}$	$\dfrac{p_{t2}}{p_{t1}}$
2.50	0.51299	7.1250	3.3333	2.1375	0.49902
2.55	0.50831	7.4196	3.3918	2.1875	0.47927
2.60	0.50387	7.7200	3.4489	2.2383	0.46012
2.65	0.49965	8.0262	3.5047	2.2901	0.44155
2.70	0.49563	8.3383	3.5590	2.3429	0.42359
2.75	0.49181	8.6562	3.6119	2.3966	0.40622
2.80	0.48817	8.9800	3.6635	2.4512	0.38946
2.85	0.48470	9.3096	3.7139	2.5067	0.37330
2.90	0.48138	9.6450	3.7629	2.5632	0.35773
2.95	0.47821	9.986	3.8106	2.6206	0.34275
3.00	0.47519	10.333	3.8571	2.6790	0.32834
3.50	0.45115	14.125	4.2608	3.3150	0.21295
4.00	0.43496	18.500	4.5714	4.0469	0.13876
4.50	0.42355	23.458	4.8119	4.8761	0.09170
5.00	0.41523	29.000	5.0000	5.8000	0.06172
6.00	0.40416	41.833	5.2683	7.941	0.02965
7.00	0.39736	57.000	5.4444	10.469	0.01535
8.00	0.39289	74.500	5.5652	13.387	0.00849
9.00	0.38980	94.333	5.6512	16.693	0.00496
10.00	0.38757	116.50	5.1743	20.388	0.00304
∞	0.37796	∞	6.000	∞	0

INDEX